Solutions Manual for

Algebra 2

An Incremental Development

Second Edition

JOHN H. SAXON, JR.

SAXON PUBLISHERS, INC.

Algebra 2: An Incremental Development
Second Edition
Solutions Manual

Copyright © 1992 by Saxon Publishers, Inc.

Printed in the United States of America

ISBN: 0-939798-99-9

Production supervisor: David Pond
Production coordinator: Joan Coleman

Manufacturing Code: 10S0202

This manual contains solutions to every problem in the *Algebra 2*, Second Edition, textbook by John Saxon. Early solutions of a problem of a particular type contain every step. Later solutions omit steps considered unnecessary.

The following people were instrumental in the development of this solutions manual, and we gratefully acknowledge their contributions: Mikki Hunter and Jonathan Maltz for working the solutions and proofing the various revisions; Julie Webster and Scott Kirby for setting up the manual; Tim Maltz, Letha Steinbron, and Tyra Aldridge for typesetting the manual; and Tim Maltz, John Chitwood, and Tyra Aldridge for creating the artwork.

Reaching us via the Internet

WWW: www.saxonpublishers.com

E-mail: info@saxonpublishers.com

Saxon Publishers, Inc.
2450 John Saxon Blvd.
Norman, OK 73071

PROBLEM SET A

1. $y + 65 = 180$
   $y = 180 - 65 = \mathbf{115}$

2. $x + 40 = 90$
   $x = 90 - 40 = \mathbf{50}$

3. $x + 89 = 180$
   $x = 180 - 89 = \mathbf{91}$
   Since vertical angles are equal:
   $y = \mathbf{89}$
   $p = \mathbf{91}$

4. $z + 100 = 180$
   $z = 180 - 100 = \mathbf{80}$
   Since vertical angles are equal:
   $2x = 80$
   $x = \mathbf{40}$

   $4y = 100$
   $y = \mathbf{25}$

5. Angle + Supplement = $180°$
   Angle + $40° = 180°$
   Angle = $180° - 40°$
   Angle = $\mathbf{140°}$

6. Angle + Complement = $90°$
   Angle + $40° = 90°$
   Angle = $90° - 40°$
   Angle = $\mathbf{50°}$

7. $-2 - (-2) = -2 + 2 = \mathbf{0}$

8. $-3 - [-(-2)] = -3 - 2 = \mathbf{-5}$

9. $-2 - 3(-2 - 2) - 5(-5 + 7)$
   $= -2 - 3(-4) - 5(2)$
   $= -2 - (-12) - (10)$
   $= -2 + 12 - 10 = \mathbf{0}$

10. $-[-2(-5 + 2) - (-2 - 3)]$
    $= -[-2(-3) - (-5)]$
    $= -[6 + 5] = \mathbf{-11}$

11. $-2 + (-2)^3 = -2 + (-8)$
    $= -2 - 8 = \mathbf{-10}$

12. $-3^2 - 3 - (-3)^2$
    $= -9 - 3 - 9 = \mathbf{-21}$

13. $-3(-2 - 3 + 6) - [-5(-2) + 3(-2 - 4)]$
    $= -3(1) - [-5(-2) + 3(-6)]$
    $= -3 - [10 - 18] = -3 - 10 + 18 = \mathbf{5}$

14. $-2 - 2^2 - 2^3 - 2^4$
    $= -2 - 4 - 8 - 16 = \mathbf{-30}$

15. $|-2| - |-4 - 2| + |8| = |-2| - |-6| + |8|$
    $= 2 - 6 + 8 = \mathbf{4}$

16. $-|-3(2) - 3| - 2^2 = -|-6 - 3| - 4$
    $= -|-9| - 4 = -9 - 4 = \mathbf{-13}$

17. $-2^2 - 2^3 - |-2| - 2 = -4 - 8 - 2 - 2$
    $= \mathbf{-16}$

18. $-3[-1 - 2(-1 - 1)][-3(-2) - 1]$
    $= -3[-1 - 2(-2)][6 - 1] = -3[-1 + 4][6 - 1]$
    $= -3[3][5] = \mathbf{-45}$

19. $-3[-3(-4 - 1) - (-3 - 4)]$
    $= -3[-3(-5) - (-7)]$
    $= -3[15 + 7] = -3[22] = \mathbf{-66}$

20. $-2[(-3 + 1) - (-2 - 2)(-1 + 3)]$
    $= -2[(-2) - (-4)(2)] = -2[-2 + 8]$
    $= -2[6] = \mathbf{-12}$

21. $-2\left[-2(-4) - 2^3\right](-|2|) = -2[8 - 8](-2)$
    $= -2[0](-2) = \mathbf{0}$

22. $-8 - 3^2 - (-2)^2 - 3(-2) + 2$
    $= -8 - 9 - 4 + 6 + 2 = \mathbf{-13}$

23. $-\{-[-5(-3 + 2)7]\} = -\{-[-5(-1)7]\}$
    $= -\{-[35]\} = \mathbf{35}$

24. $-5 - |-3 - 4| - (3)^2 - 3$
    $= -5 - |-7| - 9 - 3$
    $= -5 - 7 - 9 - 3 = \mathbf{-24}$

25. $3(-2 + 5) - 2^2(2 - 3) - |-2|$
    $= 3(3) - 4(-1) - 2$
    $= 9 + 4 - 2 = \mathbf{11}$

26. $\dfrac{-5 - (-2) + 8 - 4(5)}{6 - 4(-3)}$
    $= \dfrac{-5 + 2 + 8 - 20}{6 + 12} = \dfrac{-15}{18} = \mathbf{-\dfrac{5}{6}}$

27. $(-2)\left[|-3 - 4 - 5| - 2^3 - (-1)\right]$
    $= (-2)\left[|-12| - 8 + 1\right] = (-2)[12 - 8 + 1]$
    $= -2[5] = \mathbf{-10}$

28. $\dfrac{-3 - (-2) + 9 - (-5)}{7(|-3 + 4|)}$
    $= \dfrac{-3 + 2 + 9 + 5}{7(1)} = \mathbf{\dfrac{13}{7}}$

29. $4(-2)[-(7 - 3)(5 - 2)2] = -8[-(4)(3)(2)]$
    $= -8[-24] = \mathbf{192}$

30. $4 - (-4) - 5(3 - 1) + 3(4)(-2)^3$
    $= 4 + 4 - 5(2) + 3(4)(-8)$
    $= 4 + 4 - 10 - 96 = \mathbf{-98}$

PROBLEM SET B

1. $A_{Shaded} = A_{Square} - A_{Circle}$
$= [(8)(8) - \pi(4)^2]\ m^2$
$= (64 - 16\pi)\ m^2 \approx \mathbf{13.76\ m^2}$

2. $A_{Shaded} = A_{Total} - A_{Not\ Shaded}$
$= \left[\frac{1}{2}(12)(10)\right] m^2 -$
$\left[(3)(4) + \frac{1}{2}(3)(6) + \frac{1}{2}(9)(4)\right] m^2$
$= 60\ m^2 - 39\ m^2 = \mathbf{21\ m^2}$

3. $A_{Shaded} = A_{Circle} - A_{Triangle}$
$= \left[\pi(8)^2 - \frac{1}{2}(16)(8)\right] cm^2$
$= (64\pi - 64)\ cm^2 \approx \mathbf{136.96\ cm^2}$

4. $Perimeter = \left[\frac{1}{2}(2)(\pi)(2) + 14\right] m$
$= (2\pi + 14)\ m \approx \mathbf{20.28\ m}$

5. $Area\ of\ 40°\ sector = \frac{40}{360}(A_{Circle})$
$= \frac{40}{360}[\pi(5)^2]\ m^2$
$= \frac{40}{360}(25\pi)\ m^2 \approx \mathbf{8.72\ m^2}$

6. $V_{Pyramid} = \frac{1}{3}V_{Prism} = \frac{1}{3}\left[A_{Base} \times height\right]$
$= \frac{1}{3}\left[\frac{1}{2}(6)(4)\ cm^2 \times 10\ cm\right]$
$= \frac{1}{3}(120)\ cm^3 = \mathbf{40\ cm^3}$

7. $A_{Base} = \left[(3)(4) + \frac{1}{2}\pi(2)^2\right] m^2$
$= (12 + 2\pi)\ m^2 \approx \mathbf{18.28\ m^2}$

$V_{Cylinder} = A_{Base} \times height$
$= (12 + 2\pi)\ m^2 \times 8\ m$
$= (96 + 16\pi)\ m^3 \approx \mathbf{146.24\ m^3}$

8. $V_{Sphere} = \frac{2}{3}V_{Cylinder} = \frac{2}{3}[A_{Base} \times height]$
$= \frac{2}{3}[\pi(6)^2\ cm^2 \times 12\ cm]$
$= \frac{2}{3}(36\pi)(12)\ cm^3 \approx \mathbf{904.32\ cm^3}$

$S.\,A. = 4\pi r^2 = 4\pi(6)^2\ cm^2 \approx \mathbf{452.16\ cm^2}$

9. $Area\ of\ 72°\ sector = \frac{72}{360}(A_{Circle})$
$= \frac{72}{360}[\pi(10)^2]\ cm^2$
$= \frac{72}{360}(100\pi)\ cm^2 \approx \mathbf{62.8\ cm^2}$

10. $Perimeter = \left[\frac{1}{2}(2)(\pi)(2) + 16 + 4\right] yd$
$+ \left[\frac{1}{2}(2)(\pi)(2)\right] yd$
$= (4\pi + 20)\ yd \approx \mathbf{32.56\ yd}$

11. $z + 70 = 180$
$z = 180 - 70 = \mathbf{110}$
Since vertical angles are equal:
$2x = 70$
$x = \mathbf{35}$

$y = \mathbf{110}$

12. $5A + 40 = 90$
$5A = 90 - 40$
$5A = 50$
$A = \frac{50}{5} = \mathbf{10}$

13. $2B + 140 = 180$
$2B = 180 - 140$
$2B = 40$
$B = \frac{40}{2} = \mathbf{20}$

14. Angle + complement = 90°
Angle + 10° = 90°
Angle = 90° − 10° = **80°**

15. Angle + supplement = 180°
Angle + 60° = 180°
Angle = 180° − 60° = **120°**

16. $-2^2 - 2^3 - (-2)^2 - 2 = -4 - 8 - 4 - 2$
$= \mathbf{-18}$

17. $-2^2 - |-4| + |4| = -4 - 4 + 4 = \mathbf{-4}$

18. $-|-3| - 3 - 3^2 = -3 - 3 - 9 = \mathbf{-15}$

19. $-4 - (-3)^3 - 2^2 + |-4|$
$= -4 + 27 - 4 + 4 = \mathbf{23}$

20. $-3^2 - 2(-4 + 6) = -9 - 2(2) = \mathbf{-13}$

21. $-4(-2^2 - 3) - 5 + |-3| = -4(-4 - 3) - 5 + 3$
$= -4(-7) - 5 + 3 = 28 - 5 + 3 = \mathbf{26}$

22. $-2[-1 - (-5)] - [-6(-2) + 3]$
$= -2[-1 + 5] - [12 + 3]$
$= -2[4] - [15] = -8 - 15 = \mathbf{-23}$

23. $-2^2 - 2^3 - 2 - |-2| = -4 - 8 - 2 - 2 = -16$

24. $-2 - |-3 - 4 + 8| - 2^2 = -2 - |1| - 4$
$= -2 - 1 - 4 = -7$

25. $-|-2 - 3 - 4| - |-2| = -|-9| - |-2|$
$= -9 - 2 = -11$

26. $\dfrac{-5 - (-2) + 8 - 4(5) - 3}{6 - 4(-3)}$

$= \dfrac{-5 + 2 + 8 - 20 - 3}{6 + 12} = \dfrac{-18}{18} = -1$

27. $(-2)\big[|-3 + 4 - 5| - 2^3 - (-1)\big]$
$= (-2)\big[|-4| - 8 + 1\big] = (-2)[4 - 8 + 1]$
$= (-2)(-3) = 6$

28. $\dfrac{-|-5| - (-2) + 6 - 4(3 - |6 - 9|)}{5 - |(4)(-3)|}$

$= \dfrac{-5 + 2 + 6 - 4(3 - 3)}{5 - |-12|}$

$= \dfrac{-5 + 2 + 6 - 0}{5 - 12} = \dfrac{3}{-7} = -\dfrac{3}{7}$

29. $\dfrac{-2 - (-3 - 2) - (-2 + 5)}{-4(2^2 - 3)(-2)}$

$= \dfrac{-2 - (-5) - (3)}{-4(1)(-2)} = \dfrac{-2 + 5 - 3}{8}$

$= \dfrac{0}{8} = 0$

30. $-2(-3 + 4 - 6) - 2^2(-2) - 3(-2) - |-5|$
$= -2(-5) - 4(-2) + 6 - 5$
$= 10 + 8 + 6 - 5 = 19$

PROBLEM SET 1

1. Since angles opposite equal sides are equal angles:
$x = 45$
$y + 45 + 45 = 180$
$\quad\quad\quad y = 180 - 90 = 90$

2. Since angles opposite equal sides are equal angles:
$x = 55$
$y + 55 + 55 = 180$
$\quad\quad\quad y = 180 - 110 = 70$

3. $2C + 70 = 180$
$\quad 2C = 110$
$\quad\quad C = 55$
Since lines are parallel: $B = 110$; $A = 70$

4. $2 \times \overrightarrow{SF} = 3$
$\overrightarrow{SF} = \dfrac{3}{2}$
$3 \times \overrightarrow{SF} = x$
$3\left(\dfrac{3}{2}\right) = x$
$x = \dfrac{9}{2}$

5. $A_{\text{Shaded}} = A_{\text{Top Shaded}} + A_{\text{Bottom Shaded}}$
$A_{\text{Top Shaded}} = \left[\dfrac{1}{2}\pi(4)^2 - \dfrac{1}{2}(8)(4)\right] \text{cm}^2$
$= (8\pi - 16) \text{ cm}^2$

$A_{\text{Bottom Shaded}} = \dfrac{60}{360}[\pi(4)^2] \text{ cm}^2$
$= \dfrac{60}{360}(16\pi) \text{ cm}^2$

$A_{\text{Shaded}} = (8\pi - 16) \text{ cm}^2 + \dfrac{60}{360}(16\pi) \text{ cm}^2$
$= \left(8\pi + \dfrac{8}{3}\pi - 16\right) \text{ cm}^2 \approx 17.49 \text{ cm}^2$

6. $A_{\text{Shaded}} = A_{\text{Total}} - A_{\text{White Triangles}}$

$A_{\text{Total}} = \Big[\dfrac{1}{2}(8)(6) + \dfrac{1}{2}\pi(5)^2$
$\quad\quad + \dfrac{1}{2}\pi(4)^2 + \dfrac{1}{2}\pi(3)^2\Big] \text{cm}^2$
$= (24 + 25\pi) \text{ cm}^2$

$A_{\text{White Triangles}} = \Big[\dfrac{1}{2}(5)(5) + \dfrac{1}{2}(4)(4)$
$\quad\quad + \dfrac{1}{2}(3)(3)\Big] \text{cm}^2$
$= 25 \text{ cm}^2$

$A_{\text{Shaded}} = (24 + 25\pi) \text{ cm}^2 - 25 \text{ cm}^2$
$= (25\pi - 1) \text{ cm}^2 \approx 77.5 \text{ cm}^2$

7. $\text{Perimeter} = \left[\dfrac{1}{2}(2)(\pi)(4) + 48\right] \text{ft}$
$= (4\pi + 48) \text{ ft} \approx 60.56 \text{ ft}$

8. $V_{\text{Cylinder}} = A_{\text{Base}} \times \text{height}$
$= [\pi(4)^2 \text{ ft}^2 \times 8 \text{ ft}]$
$= 128\pi \text{ ft}^3 \approx 401.92 \text{ ft}^3$

$V_{\text{Sphere}} = \dfrac{2}{3}V_{\text{Cylinder}}$

$= \dfrac{2}{3}(128\,\pi) \text{ ft}^3 \approx 267.95 \text{ ft}^3$

9. $p + 30 = 180$
$p = 180 - 30 = \textbf{150}$
$2y + 30 = 90$
$2y = 90 - 30$
$2y = 60$
$y = \textbf{30}$

Since vertical angles are equal angles: $x = \textbf{30}$

10. $p + 60 = 180$
$p = 180 - 60 = \textbf{120}$
Since vertical angles are equal angles:
$10x = 60$
$x = \textbf{6}$

$4y = 120$
$y = \textbf{30}$

11. Angle + complement $= 90°$
Angle $+ 17° = 90°$
Angle $= 90° - 17° = \textbf{73}°$

12. $V_{Cone} = \dfrac{1}{3}(A_{Base} \times height)$

$= \dfrac{1}{3}\left[\dfrac{1}{2}\pi(3)^2 + (5)(6)\right] m^2 \times 7\,m$

$= \dfrac{1}{3}\left(\dfrac{9}{2}\pi + 30\right)(7)\,m^3 \approx \textbf{102.97 m}^3$

13. $r = \dfrac{1}{4}$ side of square

$A_{Square} = s^2$
$s = 4r$
$A_{Square} = (4r)^2 = \textbf{16}r^2$

14. $-[-2(-3-2) - (-2-3)] = -[-2(-5) - (-5)]$
$= -[10 + 5] = -[15] = \textbf{-15}$

15. $-2[-2 - 3(-2-2)][-2(-4) - 3]$
$= -2[-2 - 3(-4)][-2(-4) - 3]$
$= -2[-2 + 12][8 - 3] = -2[10][5] = \textbf{-100}$

16. $-2^2 - 2^3[-2 + 3(-2)] - |-2^3|$
$= -4 - 8[-2 - 6] - |-8| = -4 - 8[-8] - 8$
$= -4 + 64 - 8 = \textbf{52}$

17. $-3 - 2^3 - 4^2 - |-2 - 3(2)|$
$= -3 - 8 - 16 - |-2 - 6|$
$= -3 - 8 - 16 - |-8|$
$= -3 - 8 - 16 - 8 = \textbf{-35}$

18. $-\{-2[(-3+7) - (-2)][-3(-2+1)]\}$
$= -\{-2[4 + 2][-3(-1)]\} = -\{-2[6][3]\}$
$= -\{-36\} = \textbf{36}$

19. $3^2 - 3^3 + 3^4 - (-3)^3 - 3$
$= 9 - 27 + 81 - (-27) - 3$
$= 9 - 27 + 81 + 27 - 3 = \textbf{87}$

20. $-(-4)^2 - 4|-2| - 2^3 + |-11 - 4|$
$= -16 - 4(2) - 8 + |-15|$
$= -16 - 8 - 8 + 15 = \textbf{-17}$

21. $6 - \{[3^2 - 8 + (-2)][-(4 - 6)(-3)^2 + 2]2\}$
$= 6 - \{[9 - 8 - 2][-(-2)(9) + 2]2\}$
$= 6 - \{[9 - 8 - 2][18 + 2]2\}$
$= 6 - \{[-1][20]2\} = 6 - \{-40\}$
$= 6 + 40 = \textbf{46}$

22. $-[-(-2)] - |-4 - 3|2^2 - 4$
$= -[2] - |-7|4 - 4 = -2 - 7(4) - 4$
$= -2 - 28 - 4 = \textbf{-34}$

23. $(-|-3|)[(2 - 7)(-3 - 2) + (-2)^2]$
$= -3[-5(-5) + 4] = -3[25 + 4]$
$= -3[29] = \textbf{-87}$

24. $\dfrac{-|-4| - (-3) + 7 - 6(4 - |7 - 11|)}{7 - |(3)(-2)|}$

$= \dfrac{-4 + 3 + 7 - 6(4 - |-4|)}{7 - |-6|}$

$= \dfrac{-4 + 3 + 7 - 6(4 - 4)}{7 - 6}$

$= \dfrac{-4 + 3 + 7 - 6(0)}{1}$

$= \dfrac{-4 + 3 + 7 - 0}{1} = \dfrac{6}{1} = \textbf{6}$

25. $-(-3 - 2)(-7 - |-3 - 2|) - (-3)^2$
$= -(-5)(-7 - |-5|) - 9 = 5(-7 - 5) - 9$
$= 5(-12) - 9 = -60 - 9 = \textbf{-69}$

26. $(-3)[|-2 - 7 - 2| - (-3)^2 - (-2)]$
$= -3[|-11| - 9 + 2] = -3[11 - 9 + 2]$
$= -3[4] = \textbf{-12}$

27. $\dfrac{-4 - (-3) + 7 - 6(2)}{7 - (3)(-2)}$

$= \dfrac{-4 + 3 + 7 - 12}{7 + 6} = \dfrac{-6}{13} = -\dfrac{\textbf{6}}{\textbf{13}}$

28. $3 - 5 - 2^2 - 4^2(-1)(-3 - |-2 - 5| - 3)$
$= 3 - 5 - 4 - 16(-1)(-3 - |-7| - 3)$
$= 3 - 5 - 4 - 16(-1)(-3 - 7 - 3)$
$= 3 - 5 - 4 - 16(-1)(-13)$
$= 3 - 5 - 4 - 208 = \textbf{-214}$

29. $-8 + (-3)(-2)^2 + (-7) - 2(-4 - 2)$
$= -8 + (-3)(4) - 7 - 2(-6)$
$= -8 - 12 - 7 + 12 = -15$

30. $6(-3)[-(5 - 4)(6 - 2)3] = -18[-1(4)(3)]$
$= -18[-12] = 216$

PROBLEM SET 2

1. $4 \times \overrightarrow{SF} = 6$

$\overrightarrow{SF} = \dfrac{6}{4} = \dfrac{3}{2}$

$3 \times \overrightarrow{SF} = x$

$3 \cdot \dfrac{3}{2} = x$

$x = \dfrac{9}{2}$

2. Since angles opposite equal sides are equal angles:
$x = 38$
$y + 38 + 38 = 180$
$\qquad y = 180 - 38 - 38 = 104$

3. $V_{Cylinder} = A_{Base} \times height$

$= \left[\dfrac{60}{360}(\pi(4)^2) + \dfrac{1}{2}(4)(7) \right] m^2 \times 8\ m$

$= \left[\dfrac{60}{360}(16\pi) + 14 \right](8)\ m^3 \approx 178.99\ m^3$

4. Since vertical angles are equal angles:
$4A = 80$
$\ A = 20$

$2B + 80 = 180$
$\quad 2B = 100$
$\qquad B = 50$
Since lines are parallel:
$2C = 4A$
$2C = 80$
$\ C = 40$

5. $A + 60 = 180$
$\qquad A = 180 - 60 = 120$
Since vertical angles are equal angles:
$2B = 60$
$\ B = 30$

$3C = 120$
$\ C = 40$

6. $A_{Square} = s^2 = 16\ cm^2$

$s = \sqrt{16\ cm^2} = 4\ cm$

$r = \dfrac{1}{4}(s) = \dfrac{1}{4}(4\ cm) = 1\ cm$

$A_{Circle} = \pi r^2 \approx 3.14(1\ cm)^2 = 3.14\ cm^2$

7. $V_{Cylinder} = A_{Base} \times height$
$250\pi\ cm^3 = [\pi(5)^2]\ cm^2 \times height$
$250\pi\ cm^3 = 25\pi\ cm^2 \times height$

$height = \dfrac{250\pi\ cm^3}{25\pi\ cm^2} = 10\ cm$

8. $V_{Cone} = \dfrac{1}{3} \times A_{Base} \times height$

$= \dfrac{1}{3} \left[\dfrac{1}{2}(1)(2) + (2)(2) \right.$

$\left. + \dfrac{1}{2}\pi(1)^2 \right] m^2 \times 4\ m$

$= \dfrac{1}{3}\left(5 + \dfrac{\pi}{2} \right)(4)\ m^3 \approx 8.76\ m^3$

9. $\dfrac{xx^2(x^0y^{-1})^2}{x^2x^{-5}(y^2)^5} = \dfrac{xx^2x^0y^{-2}}{x^2x^{-5}y^{10}} = \dfrac{x^3y^{-2}}{x^{-3}y^{10}}$
$= x^6y^{-12}$

10. $\dfrac{m^2p^0(m^{-2}p)^2}{m^{-2}p^{-1}(m^{-3}p^2)^3} = \dfrac{m^2p^0m^{-4}p^2}{m^{-2}p^{-1}m^{-9}p^6}$
$= \dfrac{m^{-2}p^2}{m^{-11}p^5} = m^9p^{-3}$

11. $\dfrac{(x^2y)^0xy}{x^2(y^{-2})^3} = \dfrac{x^0y^0xy}{x^2y^{-6}} = \dfrac{xy}{x^2y^{-6}} = x^{-1}y^7$

12. $\dfrac{(a^2b^0)^2ab^{-2}}{a^2b^{-2}(ab^{-3})^2} = \dfrac{a^4b^0ab^{-2}}{a^2b^{-2}a^2b^{-6}} = \dfrac{a^5b^{-2}}{a^4b^{-8}}$
$= ab^6$

13. $\dfrac{(xm^{-1})^{-3}x^2m^2}{(x^0y^2)^{-2}xy} = \dfrac{x^{-3}m^3x^2m^2}{x^0y^{-4}xy} = \dfrac{x^{-1}m^5}{xy^{-3}}$
$= \dfrac{m^5y^3}{x^2}$

14. $\dfrac{(c^2d)^{-3}c^{-5}}{(c^2d^0)^{-2}d^3} = \dfrac{c^{-6}d^{-3}c^{-5}}{c^{-4}d^0d^3} = \dfrac{c^{-11}d^{-3}}{c^{-4}d^3}$
$= \dfrac{1}{c^7d^6}$

15. $\dfrac{(m^2n^{-5})^{-2}m(n^0)^2}{(m^2n^{-2})^{-3}m^2} = \dfrac{m^{-4}n^{10}mn^0}{m^{-6}n^6m^2} = \dfrac{m^{-3}n^{10}}{m^{-4}n^6}$
$= mn^4$

16. $\dfrac{(x^{-2}y^5)^3(x^2)^0y}{xy^{-3}x^{-2}} = \dfrac{x^{-6}y^{15}x^0y}{xy^{-3}x^{-2}} = \dfrac{x^{-6}y^{16}}{x^{-1}y^{-3}}$
$= \dfrac{y^{19}}{x^5}$

17. $\dfrac{(b^2c^{-2})^{-3}c^{-3}}{(b^2c^0b^{-2})^4} = \dfrac{b^{-6}c^6c^{-3}}{b^8c^0b^{-8}} = \dfrac{b^{-6}c^3}{1}$

$= \dfrac{c^3}{b^6}$

18. $\dfrac{(abc)^{-3}c^2b}{a^{-4}bc^2a} = \dfrac{a^{-3}b^{-3}c^{-3}c^2b}{a^{-4}bc^2a} = \dfrac{a^{-3}b^{-2}c^{-1}}{a^{-3}bc^2}$

$= b^{-3}c^{-3}$

19. $\dfrac{kL^2k^{-2}}{(k^0L)^2L^{-3}k} = \dfrac{kL^2k^{-2}}{k^0L^2L^{-3}k} = \dfrac{k^{-1}L^2}{kL^{-1}} = \dfrac{k^{-2}}{L^{-3}}$

20. $\dfrac{s^2ym^{-3}}{(s^0t^2)^{-3}m^{-3}st} = \dfrac{s^2ym^{-3}}{s^0t^{-6}m^{-3}st} = \dfrac{s^2ym^{-3}}{sm^{-3}t^{-5}}$

$= \dfrac{1}{s^{-1}t^{-5}y^{-1}}$

21. $\dfrac{(x^{-3}yz^{-3})^2xy^0}{(xy^0z^{-2})^{-3}xy} = \dfrac{x^{-6}y^2z^{-6}xy^0}{x^{-3}y^0z^6xy} = \dfrac{x^{-5}y^2z^{-6}}{x^{-2}yz^6}$

$= \dfrac{x^{-3}z^{-12}}{y^{-1}}$

22. $\dfrac{x^{-3}y^2xy^4}{(x^{-2}y)^3y^{-3}x} = \dfrac{x^{-3}y^2xy^4}{x^{-6}y^3y^{-3}x} = \dfrac{x^{-2}y^6}{x^{-5}}$

$= \dfrac{1}{x^{-3}y^{-6}}$

23. $-3^{-2} = -\dfrac{1}{3^2} = -\dfrac{1}{9}$

24. $\dfrac{1}{-2^{-3}} = -2^3 = -8$

25. $-3^2 - [-2^0 - (3 - 2) - 2]$
$= -9 - [-1 - 1 - 2]$
$= -9 - [-4] = -9 + 4 = -5$

26. $-2\{[-3 - 2(-2)][-2 - 3(-2)]\}$
$= -2\{[-3 + 4][-2 + 6]\} = -2\{[1][4]\} = -8$

27. $2\{-3^0[(-5 - 2)(-3) - 2]\}$
$= 2\{-1[(-7)(-3) - 2]\} = 2\{-1[21 - 2]\}$
$= 2\{-1[19]\} = 2\{-19\} = -38$

28. $-3[4^0 - 7(2 - 3) - 2^2] = -3[1 - 7(-1) - 4]$
$= -3[1 + 7 - 4] = -3[4] = -12$

29. $-|2 - 3| - (-5) - 3^3 = -|-5| + 5 - 27$
$= -5 + 5 - 27 = -27$

30. $-|-3^2 - 2| - 2^0 - (-3) = -|-9 - 2| - 1 + 3$
$= -|-11| - 1 + 3 = -11 - 1 + 3 = -9$

1. $4 \times \overrightarrow{SF} = 6$
$\overrightarrow{SF} = \dfrac{6}{4} = \dfrac{3}{2}$
$5 \times \overrightarrow{SF} = x$
$5 \cdot \dfrac{3}{2} = x$
$x = \dfrac{15}{2}$

Since lines are parallel:
$9z = 81$
$z = 9$
$3y + 9z = 180$
$3y + 81 = 180$
$3y = 99$
$y = 33$

2. $V_{Cone} = \dfrac{1}{3}(A_{Base} \times \text{height})$
$48\pi \text{ m}^3 = \dfrac{1}{3}[\pi r^2 \times 9 \text{ m}]$
$144\pi \text{ m}^3 = \pi r^2 \times 9 \text{ m}$
$r = \sqrt{\dfrac{144\pi \text{ m}^3}{9\pi \text{ m}}} = 4 \text{ m}$
Circumference $= 2\pi r = 2\pi(4 \text{ m}) \approx 25.12 \text{ m}$

3. Since angles opposite equal sides are equal angles:
$A = 40$
$B + 40 + 40 = 180$
$B = 180 - 40 - 40 = 100$

4. $V_{Cone} = \dfrac{1}{3}(A_{Base} \times \text{height})$
$= \dfrac{1}{3}\left[\dfrac{60}{360}(\pi(5)^2) + \dfrac{1}{2}(5)(8)\right] \text{cm}^2 \times 10 \text{ cm}$
$= \dfrac{1}{3}\left[\dfrac{60}{360}(25\pi) + 20\right](10) \text{ cm}^3 \approx 110.28 \text{ cm}^3$

5. Perimeter $= 2r_8 + 2r_8 + 2r_6 + 2r_5$
$= 2(8 \text{ cm}) + 2(8 \text{ cm}) + 2(6 \text{ cm}) + 2(5 \text{ cm})$
$= 16 \text{ cm} + 16 \text{ cm} + 12 \text{ cm} + 10 \text{ cm} = 54 \text{ cm}$

6. $x - |x|y^2 - xy = -2 - |-2|(-3)^2 - (-2)(-3)$
$= -2 - 2(9) - 6 = -2 - 18 - 6 = -26$

7. $(a - b) - a(-b) = (-5 - 3) - (-5)(-3)$
$= -8 - 15 = -23$

8. $-a(a - ax)(x - a)$
$= -(-2)(-2 - (-2)(4))(4 - (-2))$
$= 2(-2 + 8)(4 + 2) = 2(6)(6) = 72$

9. $a^2 - y^3(a - y^2)y$
$= (-2)^2 - (-3)^3(-2 - (-3)^2)(-3)$
$= 4 + 27(-2 - 9)(-3) = 4 + 27(-11)(-3)$
$= 4 + 891 = 895$

10. $-p^2 - p(a - p^2) = -(-3)^2 - (-3)(4 - (-3)^2)$
$= -9 + 3(4 - 9) = -9 + 3(-5) = -9 - 15$
$= \mathbf{-24}$

11. $a^2 - y^3(a - y^2)y^2 = (-2)^2 - 3^3(-2 - 3^2)3^2$
$= 4 - 27(-2 - 9)9 = 4 - 27(-11)9 = \mathbf{2677}$

12. $a^2 - a(x - ax) = 1^2 - 1(2 - 1(2))$
$= 1 - 1(2 - 2) = 1 - 1(0) = 1 - 0 = \mathbf{1}$

13. $\dfrac{p^2x^4}{m^5} - \dfrac{2p^4m^5}{x^{-4}} - \dfrac{3p^2x^2}{x^{-2}m^5} + \dfrac{7p^2m^{-5}m^{10}x^4}{p^{-2}}$
$= \dfrac{p^2x^4}{m^5} - 2p^4m^5x^4 - \dfrac{3p^2x^4}{m^5} + 7p^4x^4m^5$
$= \mathbf{5p^4x^4m^5 - \dfrac{2p^2x^4}{m^5}}$

14. $-\dfrac{m^4x^5}{k^5} + \dfrac{2m^2x^5}{k^5m^{-2}} - \dfrac{3m^3x^2k}{m^{-4}x^3k^4}$
$= -\dfrac{m^4x^5}{k^5} + \dfrac{2m^4x^5}{k^5} - \dfrac{3m^7}{xk^3}$
$= \mathbf{\dfrac{m^4x^5}{k^5} - \dfrac{3m^7}{xk^3}}$

15. $-2x^5y^4 + \dfrac{3xy^3}{x^{-4}y^{-1}} + \dfrac{4x^3y^2}{x^2y}$
$= -2x^5y^4 + 3x^5y^4 + 4xy = \mathbf{x^5y^4 + 4xy}$

16. $2xy^2m - \dfrac{3x^2y^2m^4}{m^3x} + \dfrac{2x^2ym^3}{mx^2}$
$= 2xy^2m - 3xy^2m + 2ym^2 = \mathbf{2ym^2 - xy^2m}$

17. $\dfrac{(x^0y^2)^{-3}y^{-2}p^0}{(x^2)^{-4}(y^2)^0(p^3)^{-2}} = \dfrac{x^0y^{-6}y^{-2}p^0}{x^{-8}y^0p^{-6}}$
$= \dfrac{x^0y^{-8}p^0}{x^{-8}y^0p^{-6}} = \mathbf{x^8y^{-8}p^6}$

18. $\dfrac{(mxy^2p)^2\,p^{-2}x^2}{(p^0xmy^3)^{-2}\,xp^{-2}} = \dfrac{m^2x^2y^4p^2p^{-2}x^2}{p^0x^{-2}m^{-2}y^{-6}xp^{-2}}$
$= \dfrac{m^2x^4y^4}{m^{-2}x^{-1}y^{-6}p^{-2}} = \mathbf{m^4x^5y^{10}p^2}$

19. $\dfrac{xx^2(x^{-2})^{-2}(mpx^2)^{-4}}{(x^0)^2(x^2)^0(x^2mp^{-2})^3} = \dfrac{xx^2x^4m^{-4}p^{-4}x^{-8}}{x^0x^0x^6m^3p^{-6}}$
$= \dfrac{x^{-1}m^{-4}p^{-4}}{x^6m^3p^{-6}} = \mathbf{x^{-7}m^{-7}p^2}$

20. $\dfrac{p^2x^{-4}k^5(p^2k)^{-2}}{(p^2x^{-3})^{-2}} = \dfrac{p^2x^{-4}k^5p^{-4}k^{-2}}{p^{-4}x^6}$
$= \dfrac{p^{-2}x^{-4}k^3}{p^{-4}x^6} = \mathbf{p^2x^{-10}k^3}$

21. $\dfrac{x^2xx^0(x^{-2})^2}{xx^3x^{-14}(x^{-2})^{-3}} = \dfrac{x^2xx^0x^{-4}}{xx^3x^{-14}x^6}$
$= \dfrac{x^{-1}}{x^{-4}} = \mathbf{x^3}$

22. $\dfrac{(x^2y^{-2}p^0)^{-3}p^2}{x^2(x^{-4})^0(p^{-2}y^5)^{-2}} = \dfrac{x^{-6}y^6p^0p^2}{x^2x^0p^4y^{-10}}$
$= \dfrac{x^{-6}y^6p^2}{x^2p^4y^{-10}} = \mathbf{x^{-8}y^{16}p^{-2}}$

23. $\dfrac{(x^{-2}y^2)^5(x^0y^{-2})^{-4}}{(x^{-4}yy^2)^2(p^{-4})^0} = \dfrac{x^{-10}y^{10}x^0y^8}{x^{-8}y^2y^4p^0}$
$= \dfrac{x^{-10}y^{18}}{x^{-8}y^6} = \mathbf{x^{-2}y^{12}}$

24. $-3^{-2} + \dfrac{1}{2^{-3}} = -\dfrac{1}{3^2} + 2^3 = -\dfrac{1}{9} + 8 = \mathbf{7\dfrac{8}{9}}$

25. $-(-2)^3 - \dfrac{1}{(-3)^{-2}} = -(-8) - (-3)^2$
$= 8 - 9 = \mathbf{-1}$

26. $-3[-2 - 2 - (-3)](-2 - 3)$
$= -3[-2 - 2 + 3](-5) = -3[-1](-5) = \mathbf{-15}$

27. $-2(-3 + 7^0) - |(-2 - 3)| = -2(-3 + 1) - |-5|$
$= -2(-2) - 5 = 4 - 5 = \mathbf{-1}$

28. $|-2| - 3^2 - (-3)^3 - 2 = 2 - 9 - (-27) - 2$
$= 2 - 9 + 27 - 2 = \mathbf{18}$

29. $-2\{[(-3 - 2)(-2)](-2 - 3)\}$
$= -2\{[-5(-2)](-5)\} = -2\{[10](-5)\} = \mathbf{100}$

30. $-[(-3)(-2) - (-3)(-2 + 4)] = -[6 + 3(2)]$
$= -[6 + 6] = -[12] = \mathbf{-12}$

PROBLEM SET 4

1. $A_{Circle} = \pi r^2$
   Radius $= r$ cm: Area $= \pi(r \text{ cm})^2 = \mathbf{\pi r^2 \text{ cm}^2}$
   Radius $= 2r$ cm: Area $= \pi(2r \text{ cm})^2 = \mathbf{4\pi r^2 \text{ cm}^2}$

2. Since lines are parallel: $2x = 78 \longrightarrow x = \mathbf{39}$
   $6z + 2x = 180$
   $6z + 78 = 180$
   $6z = 102$
   $z = \mathbf{17}$
   $2 \times \overrightarrow{SF} = 3$
   $\overrightarrow{SF} = \dfrac{3}{2}$
   $y = \dfrac{7}{2} \times \overrightarrow{SF}$
   $y = \dfrac{7}{2} \cdot \dfrac{3}{2} = \mathbf{\dfrac{21}{4}}$

3. $x + y + 100 = 180$
$x + y = 80$
Since angles opposite equal sides are equal angles:
$x = y = \mathbf{40}$
$x + P = 180$
$40 + P = 180$
$P = \mathbf{140}$

$Q + R + P = 180$
$Q + R + 140 = 180$
$Q + R = 40$
Since angles opposite equal sides are equal angles:
$Q = R = \mathbf{20}$

4. $A_{Square} = s^2 = (6 \text{ cm})^2 = \mathbf{36 \text{ cm}^2}$

Radius of circle $= \dfrac{s}{6} = \dfrac{6 \text{ cm}}{6} = \mathbf{1 \text{ cm}}$

$A_{Shaded} = A_{Square} - A_{All\ circles}$
$= 36 \text{ cm}^2 - 9\pi(1)^2 \text{ cm}^2$
$= (36 - 9\pi) \text{ cm}^2 \approx \mathbf{7.74 \text{ cm}^2}$

5. $A_{Circle} = A_{Triangle}$

$\pi r^2 = \dfrac{1}{2} bH$

$\pi(1)^2 \text{ cm}^2 = \dfrac{1}{2}(3 \text{ cm})(H)$

$H = \dfrac{2\pi}{3} \text{ cm} \approx \mathbf{2.09 \text{ cm}}$

6. $15(4 - 5b) = 16(4 - 6b) + 10$
$60 - 75b = 64 - 96b + 10$
$96b - 75b = 64 + 10 - 60$
$21b = 14$
$b = \dfrac{14}{21} = \mathbf{\dfrac{2}{3}}$

7. $3\dfrac{1}{3}x - \dfrac{5}{6} = -\dfrac{2}{3}$

$\dfrac{10}{3}x = -\dfrac{2}{3} + \dfrac{5}{6}$

$\dfrac{10}{3}x = \dfrac{-4}{6} + \dfrac{5}{6}$

$\dfrac{10}{3}x = \dfrac{1}{6}$

$x = \dfrac{1}{6} \cdot \dfrac{3}{10} = \dfrac{3}{60} = \mathbf{\dfrac{1}{20}}$

8. $3(-2x - 3) - 2^2 = -(-3x - 5) - 2$
$-6x - 9 - 4 = 3x + 5 - 2$
$-6x - 13 = 3x + 3$
$-13 - 3 = 3x + 6x$
$-16 = 9x$
$x = \mathbf{-\dfrac{16}{9}}$

9. $-2(2x - 3) - 2^3 - 3 = -x - (-4)$
$-4x + 6 - 8 - 3 = -x + 4$
$-4x - 5 = -x + 4$
$-4 - 5 = -x + 4x$
$-9 = 3x$
$x = \dfrac{-9}{3} = \mathbf{-3}$

10. $4\dfrac{1}{3}x - \dfrac{1}{2} = 3\dfrac{2}{5}$

$\dfrac{13}{3}x = \dfrac{17}{5} + \dfrac{1}{2}$

$\dfrac{13}{3}x = \dfrac{34}{10} + \dfrac{5}{10}$

$\dfrac{13}{3}x = \dfrac{39}{10}$

$x = \dfrac{39}{10} \cdot \dfrac{3}{13} = \dfrac{117}{130} = \mathbf{\dfrac{9}{10}}$

11. $-\dfrac{3}{5}x + \dfrac{2}{7} = 4\dfrac{3}{8}$

$-\dfrac{3}{5}x = \dfrac{35}{8} - \dfrac{2}{7}$

$-\dfrac{3}{5}x = \dfrac{245}{56} - \dfrac{16}{56}$

$-\dfrac{3}{5}x = \dfrac{229}{56}$

$x = \dfrac{229}{56} \cdot -\dfrac{5}{3} = \mathbf{-\dfrac{1145}{168}}$

12. $\dfrac{xy^2}{x^0 x^{-3}}\left(\dfrac{xy^{-2}}{x(y^2)^0} - \dfrac{3y^{-2}}{x^4}\right)$

$= \dfrac{xy^2 xy^{-2}}{x^0 x^{-3} x(y^2)^0} - \dfrac{3xy^2 y^{-2}}{x^0 x^{-3} x^4}$

$= \dfrac{x^2}{x^{-2}} - \dfrac{3x}{x} = \mathbf{x^4 - 3}$

13. $\dfrac{ay^{-4}}{p}\left(\dfrac{p^{-2}}{ay^2} - \dfrac{3a^{-1}y}{p^{-2}}\right)$

$= \dfrac{ay^{-4}p^{-2}}{pay^2} - \dfrac{3ay^{-4}a^{-1}y}{pp^{-2}} = \mathbf{y^{-6}p^{-3} - 3y^{-3}p}$

14. $\dfrac{(3x^2)^{-2}y^0 y^5}{(9y)^{-2}yy^2 x^{-3}} = \dfrac{81x^{-4}y^0 y^5}{9y^{-2}yy^2 x^{-3}} = \dfrac{81x^{-4}y^5}{9yx^{-3}}$
$= \mathbf{9x^{-1}y^4}$

15. $\dfrac{(2yx^{-2})^{-2}yx^2}{(x^2)^0 y^{-3}x^2} = \dfrac{y^{-2}x^4 yx^2}{4x^0 y^{-3}x^2} = \dfrac{y^{-1}x^6}{4y^{-3}x^2}$
$= \mathbf{\dfrac{x^4 y^2}{4}}$

16. $\dfrac{2(x^{-2})^{-2}yx^2y^{-3}}{x^0xx^2x^{-5}(x^2)^3} = \dfrac{2x^4yx^2y^{-3}}{x^0xx^2x^{-5}x^6} = \dfrac{2x^6y^{-2}}{x^4}$

$= \mathbf{2x^2y^{-2}}$

17. $\dfrac{(x^2y2x)^{-2}y}{(x^{-4})^0xxy^2} = \dfrac{x^{-4}y^{-2}x^{-2}y}{4x^0xxy^2} = \dfrac{x^{-6}y^{-1}}{4x^2y^2}$

$= \dfrac{x^{-8}y^{-3}}{4}$

18. $\dfrac{3x^2xy^2x^{-4}}{(x^2y)^{-2}(-2)^{-2}} = \dfrac{12x^2xy^2x^{-4}}{x^{-4}y^{-2}} = \dfrac{12x^{-1}y^2}{x^{-4}y^{-2}}$

$= \mathbf{12x^3y^4}$

19. $\dfrac{2x^2xyx}{x^2y^{-1}} - \dfrac{3x^2y^4}{yy} + \dfrac{7xx^{-3}y^{-2}}{x^{-4}y^{-4}}$

$= 2x^2y^2 - 3x^2y^2 + 7x^2y^2 = \mathbf{6x^2y^2}$

20. $\dfrac{3x}{y} - 7x^2x^{-1}y^{-1} + 2y^2y^{-1}x^{-1}$

$= \dfrac{3x}{y} - \dfrac{7x}{y} + \dfrac{2y}{x} = \dfrac{\mathbf{2y}}{\mathbf{x}} - \dfrac{\mathbf{4x}}{\mathbf{y}}$

21. $\dfrac{2ay^2}{x} + \dfrac{5a^2x^{-1}}{ay^{-2}} + \dfrac{2xy^2}{ay}$

$= \dfrac{2ay^2}{x} + \dfrac{5ay^2}{x} + \dfrac{2xy}{a} = \dfrac{\mathbf{7ay^2}}{\mathbf{x}} + \dfrac{\mathbf{2xy}}{\mathbf{a}}$

22. $a^2(a - ab) = (-2)^2(-2 - (-2)(3)) = 4(-2 + 6)$

$= 4(4) = \mathbf{16}$

23. $x^0yx(xy - x^2) = (-3)^0(-1)(-3)(-3(-1) - (-3)^2)$

$= 3(3 - 9) = 3(-6) = \mathbf{-18}$

24. $a^{-2}b(a - b)(b - a)$

$= (-1)^{-2}(-2)(-1 - (-2))(-2 - (-1))$

$= -2(-1 + 2)(-2 + 1) = -2(1)(-1) = \mathbf{2}$

25. $ab(a^2 - b)a - b$

$= (-3)(-1)((-3)^2 - (-1))(-3) - (-1)$

$= 3(9 + 1)(-3) + 1 = 3(10)(-3) + 1 = \mathbf{-89}$

26. $-3^0 - 2^0 - 2^0(-2 - 3^2) - (-2 + 7)$

$- |-2 - 3|$

$= -1 - 1 - 1(-11) - 5 - |-5|$

$= -1 - 1 + 11 - 5 - 5 = \mathbf{-1}$

27. $-2\{2[(-3 - 2^2) - (-3 + 7)] - 2\}$

$= -2\{2[(-7) - 4] - 2\} = -2\{2[-11] - 2\}$

$= -2\{-22 - 2\} = -2\{-24\} = \mathbf{48}$

28. $-3^0 - (-2 - 3 - 2^0)(-3) - (-2 - 4) + (-6)$

$= -1 - (-6)(-3) - (-6) - 6$

$= -1 - 18 + 6 - 6 = \mathbf{-19}$

29. $-2^{-2}(-16) = -\dfrac{1}{2^2}(-16) = -\dfrac{1}{4}(-16) = \dfrac{16}{4} = \mathbf{4}$

30. $-(-2^{-3}) - \dfrac{1}{(-2)^{-2}} = -\left(\dfrac{1}{-2^3}\right) - (-2)^2$

$= -\left(-\dfrac{1}{8}\right) - 4 = \dfrac{1}{8} - 4 = \mathbf{-3\dfrac{7}{8}}$

PROBLEM SET 5

1. $(2N - 9)4 = 10N - 8$

$8N - 36 = 10N - 8$

$-36 + 8 = 10N - 8N$

$-28 = 2N$

$N = \mathbf{-14}$

2. $3N_D + 11 = 4N_D - 13$

$11 + 13 = 4N_D - 3N_D$

$24 = N_D$

3. $(5N - 8)4 = 6N - 116$

$20N - 32 = 6N - 116$

$20N - 6N = -116 + 32$

$14N = -84$

$N = \mathbf{-6}$

4. $F \times of = is$

$\dfrac{1}{8} \times C = 12$

$C = 12 \cdot \dfrac{8}{1} = \mathbf{96}$

5. $F \times of = is$

$\dfrac{2}{7} \cdot 140{,}000 = N_R$

$\mathbf{40{,}000} = N_R$

6. $A_{PQR} = \dfrac{1}{2}bH$

$27 \text{ in.}^2 = \dfrac{1}{2}(6 \text{ in.})(H)$

$\mathbf{9 \text{ in.}} = H$

$AQ = PQ - PA = 9 \text{ in.} - 6 \text{ in.} = \mathbf{3 \text{ in.}}$

$A_{QAB} = \dfrac{1}{2}bH = \dfrac{1}{2}(2)(3) \text{ in.}^2 = \mathbf{3 \text{ in.}^2}$

7. Since angles opposite equal sides are equal angles:
$y = \mathbf{52}$

$x + y + 52 = 180$

$x + 52 + 52 = 180$

$x = 180 - 52 - 52 = \mathbf{76}$

8. $K + 110 = 180$

$K = 180 - 110 = \mathbf{70}$

Since lines are parallel: $Q = \mathbf{70}$; $P = \mathbf{110}$

$D = x = \mathbf{70}$

Since C is a straight angle: $C = \mathbf{180}$

9. Circumference $= 2\pi r$
    16π in. $= 2\pi r$
    8 in. $= r$

    $A_{Circle} = \pi r^2 = \pi(8 \text{ in.})^2$
    $= 64\pi \text{ in.}^2 \approx \textbf{200.96 in.}^2$

    $V_{Cylinder} = A_{Base} \times \text{height}$
    $= (64\pi) \text{ in.}^2 \times 5 \text{ in.}$
    $= 320\pi \text{ in.}^3 \approx \textbf{1004.8 in.}^3$

10. $-3x^0(2x - 3) - (-2^0) - 2 = 5(x - 3^0)2$
    $-6x + 9 + 1 - 2 = 10x - 10$
    $-6x + 8 = 10x - 10$
    $10 + 8 = 10x + 6x$
    $18 = 16x$
    $x = \dfrac{18}{16} = \dfrac{9}{8}$

11. $-2^2(-2 - x) - x^0(3 - 2) = -2(x + 3)$
    $8 + 4x - 3 + 2 = -2x - 6$
    $4x + 7 = -2x - 6$
    $2x + 4x = -6 - 7$
    $6x = -13$
    $x = -\dfrac{13}{6}$

12. $3\dfrac{1}{2}x + 2\dfrac{1}{4} = -\dfrac{1}{8}$
    $\dfrac{7}{2}x = -\dfrac{1}{8} - \dfrac{9}{4}$
    $\dfrac{7}{2}x = -\dfrac{1}{8} - \dfrac{18}{8}$
    $\dfrac{7}{2}x = \dfrac{-19}{8}$
    $x = \dfrac{-19}{8} \cdot \dfrac{2}{7} = -\dfrac{38}{56} = \boldsymbol{-\dfrac{19}{28}}$

13. $\dfrac{1}{2}(6 - 8x) + \dfrac{3}{4}(8x - 12) = 4x + 6$
    $3 - 4x + 6x - 9 = 4x + 6$
    $2x - 6 = 4x + 6$
    $-6 - 6 = 4x - 2x$
    $-12 = 2x$
    $x = \boldsymbol{-6}$

14. $-3 - 3^0 - 3^2(2x - 5) - (-2x - 3)$
    $= -x^0(x - 3)$
    $-3 - 1 - 18x + 45 + 2x + 3 = -x + 3$
    $-16x + 44 = -x + 3$
    $-16x + x = 3 - 44$
    $-15x = -41$
    $x = \dfrac{41}{15}$

15. $-2^3 - \dfrac{1}{-2^{-2}}(x + 2) - 3x = -2^0(-2x^0 - 4)$
    $-8 + 4x + 8 - 3x = 2 + 4$
    $x = \boldsymbol{6}$

16. $-3[x - 2 - 3(2)] + 2[x - 3(x - 2)]$
    $= 7(x - 5)$
    $-3x + 6 + 18 + 2x - 6x + 12 = 7x - 35$
    $-7x + 36 = 7x - 35$
    $-7x - 7x = -35 - 36$
    $-14x = -71$
    $x = \dfrac{71}{14}$

17. $\dfrac{2ab}{c^2}\left(\dfrac{c^2a^{-1}}{b} - \dfrac{3ac}{b}\right) = \dfrac{2abc^2a^{-1}}{c^2b} - \dfrac{6abac}{c^2b}$
    $= \dfrac{2bc^2}{c^2b} - \dfrac{6a^2bc}{c^2b} = \boldsymbol{2} - \dfrac{\boldsymbol{6a^2}}{\boldsymbol{c}}$

18. $-\dfrac{ax^2}{b}\left(\dfrac{bax^3}{a^2} - 3ax\right) = -\dfrac{ax^2bax^3}{ba^2} + \dfrac{3ax^2ax}{b}$
    $= -\dfrac{a^2bx^5}{ba^2} + \dfrac{3a^2x^3}{b} = \boldsymbol{-x^5} + \dfrac{\boldsymbol{3a^2x^3}}{\boldsymbol{b}}$

19. $\dfrac{(xm^{-2})^0x^0m^0}{xx^2m^0(2x)^{-2}} = \dfrac{4\,x^0m^0\,x^0m^0}{xx^2m^0x^{-2}} = \dfrac{\boldsymbol{4}}{\boldsymbol{x}}$

20. $\dfrac{4c^2dc^{-3}(2cd^{-2})^{-2}}{c^0c^{-3}(c^{-2}d)^2} = \dfrac{4c^2dc^{-3}2^{-2}c^{-2}d^4}{c^0c^{-3}c^{-4}d^2}$
    $= \dfrac{c^{-3}d^5}{c^{-7}d^2} = \boldsymbol{c^4d^3}$

21. $\dfrac{p^2m^5(p^{-3})(2p)^{-3}}{m^6(m^{-2})^2mp^3} = \dfrac{p^2m^5p^{-3}2^{-3}p^{-3}}{m^6m^{-4}mp^3}$
    $= \dfrac{m^5p^{-4}}{8m^3p^3} = \dfrac{\boldsymbol{m^2p^{-7}}}{\boldsymbol{8}}$

22. $\dfrac{x^2xy}{y^{-2}} - \dfrac{3x^5}{xxy^{-3}} + \dfrac{7x^7}{y^3x^4}$
    $= x^3y^3 - 3x^3y^3 + 7x^3y^{-3} = \boldsymbol{-2x^3y^3 + 7x^3y^{-3}}$

23. $-\dfrac{3a^2x^4}{x} + \dfrac{2aax^2}{x} - \dfrac{5x^3}{a^{-2}}$
    $= -3a^2x^3 + 2a^2x - 5a^2x^3 = \boldsymbol{2a^2x - 8a^2x^3}$

24. $mx - m(m - mx^2)$
    $= -2(-1) - (-2)(-2 - (-2)(-1)^2)$
    $= 2 + 2(-2 + 2) = 2 + 2(0) = 2 + 0 = \boldsymbol{2}$

25. $a^2 - b(a - b) = \left(-\dfrac{1}{2}\right)^2 - \dfrac{1}{4}\left(-\dfrac{1}{2} - \dfrac{1}{4}\right)$
    $= \dfrac{1}{4} - \dfrac{1}{4}\left(-\dfrac{2}{4} - \dfrac{1}{4}\right) = \dfrac{1}{4} - \dfrac{1}{4}\left(-\dfrac{3}{4}\right)$
    $= \dfrac{1}{4} + \dfrac{3}{16} = \dfrac{4}{16} + \dfrac{3}{16} = \dfrac{\boldsymbol{7}}{\boldsymbol{16}}$

26. $a - ba(a^2 - b)$

$$= -\frac{1}{2} - \left(-\frac{1}{4}\right)\left(-\frac{1}{2}\right)\left(\left(-\frac{1}{2}\right)^2 - \left(-\frac{1}{4}\right)\right)$$

$$= -\frac{1}{2} - \frac{1}{8}\left(\frac{1}{4} + \frac{1}{4}\right) = -\frac{1}{2} - \frac{1}{8}\left(\frac{2}{4}\right)$$

$$= -\frac{1}{2} - \frac{2}{32} = -\frac{8}{16} - \frac{1}{16} = -\frac{9}{16}$$

27. $-2(-2 - 3^2) - 2[-2(-3)] = -2(-2 - 9) - 2[6]$
$= -2(-11) - 2[6] = 22 - 12 = \mathbf{10}$

28. $-3^2 - (-3)^3 - \dfrac{1}{-2^2} = -9 + 27 + \dfrac{1}{4} = \mathbf{18\dfrac{1}{4}}$

29. $-3^0[-2^0 - 2^2 - 2^3(-2 - 3)]$
$= -1[-1 - 4 - 8(-5)] = -1[-1 - 4 + 40]$
$= -1[35] = \mathbf{-35}$

30. $-3[(-2^0 + 5) - (-3 + 7) - |-2|]$
$= -3[4 - 4 - 2] = -3[-2] = \mathbf{6}$

PROBLEM SET 6

1. $WD \times of = is$
$0.016T = 480$
$T = \mathbf{30{,}000}$

2. If 0.653 were prophetic, then 0.347 were not prophetic.
$1 - 0.653 = 0.347$
$WD \times of = is$
$0.347(3000) = NP$
$\mathbf{1041} = NP$

3. $-3N - 7 = -2N - 4$
$\mathbf{-3} = N$

4. $F \times of = is$
$2\dfrac{1}{2} \times BW = 175$

$BW = 175 \cdot \dfrac{2}{5} = \mathbf{70}$

5. $N \qquad N + 2 \qquad N + 4$
$6(N + N + 4) = 8(N + 2) + 28$
$12N + 24 = 8N + 44$
$4N = 20$
$N = 5$
The desired integers are **5**, **7**, and **9**.

6. $N \qquad N + 1 \qquad N + 2 \qquad N + 3$
$4(N + N + 3) = 6(N + 2) + 24$
$8N + 12 = 6N + 36$
$2N = 24$
$N = 12$
The desired integers are **12**, **13**, **14**, and **15**.

7. Surface Area $= 4\pi r^2 = 46\pi$ cm^2

$$r = \sqrt{\frac{46\pi \text{ cm}^2}{4\pi}}$$

$$r = \frac{\sqrt{46}}{2} \text{ cm}$$

8. Since angles opposite equal sides are equal angles:
$A = B = \mathbf{34}$

$K + A = 180$
$K = 180 - 34 = \mathbf{146}$

Since angles opposite equal sides are equal angles:
$M = \mathbf{17}$

9. $3 \times \overrightarrow{SF} = \dfrac{11}{3}$

$\overrightarrow{SF} = \dfrac{11}{3} \cdot \dfrac{1}{3} = \dfrac{11}{9}$

$\dfrac{15}{2} \times \overrightarrow{SF} = x$

$\dfrac{15}{2} \cdot \dfrac{11}{9} = x$

$\dfrac{55}{6} = x$

Since lines are parallel:
$3B = 130$

$B = \dfrac{130}{3}$

$2A + 3B = 180$
$2A = 180 - 130$
$2A = 50$
$A = \mathbf{25}$

10. $A = 2(180 - A)$
$A = 360 - 2A$
$3A = 360$
$A = \mathbf{120°}$

11. $0.005x + 0.6 = 2.05$
$5x + 600 = 2050$
$5x = 1450$
$x = \mathbf{290}$

12. $3\dfrac{2}{5}x + 1\dfrac{1}{4} = 7\dfrac{1}{3}$

$\dfrac{17}{5}x = \dfrac{22}{3} - \dfrac{5}{4}$

$\dfrac{17}{5}x = \dfrac{88}{12} - \dfrac{15}{12}$

$\dfrac{17}{5}x = \dfrac{73}{12}$

$x = \dfrac{73}{12} \cdot \dfrac{5}{17} = \mathbf{\dfrac{365}{204}}$

13. $-3(x - 2 + 1) - (-2)^2 - 3(x - 2)$

$\quad = 5x^0(2 - x) - 2x$

$\quad -3x + 6 - 3 - 4 - 3x + 6 = 10 - 5x - 2x$

$\quad\quad\quad\quad\quad -6x + 5 = 10 - 7x$

$\quad\quad\quad\quad\quad\quad\quad x = \textbf{5}$

14. $-3 - 2^2 - 2(x - 3) = 2[(x - 5)(2 - 5)]$

$\quad\quad -3 - 4 - 2x + 6 = 2[-3x + 15]$

$\quad\quad\quad\quad -2x - 1 = -6x + 30$

$\quad\quad\quad\quad\quad\quad 4x = 31$

$\quad\quad\quad\quad\quad\quad\quad x = \dfrac{\textbf{31}}{\textbf{4}}$

15. $4(x + 3) - 2^0(-x - 3) = 2x - 4(x^0 - x) - 3^2$

$\quad\quad 4x + 12 + x + 3 = 2x - 4 + 4x - 9$

$\quad\quad\quad\quad 5x + 15 = 6x - 13$

$\quad\quad\quad\quad\quad\quad \textbf{28} = x$

16. $\dfrac{xy}{p}\left(\dfrac{-3p^{-1}}{xy} + \dfrac{2p}{x^{-1}y}\right) = \dfrac{-3xyp^{-1}}{pxy} + \dfrac{2xyp}{px^{-1}y}$

$\quad = \textbf{-3}\boldsymbol{p^{-2}} + \textbf{2}\boldsymbol{x^2}$

17. $-\dfrac{x^0 k}{p}\left(\dfrac{k^0 p}{x} - 2p\right) = -\dfrac{x^0 k k^0 p}{px} + \dfrac{2px^0 k}{p}$

$\quad = -\dfrac{kp}{px} + \dfrac{2pk}{p} = \boldsymbol{-kx^{-1} + 2k}$

18. $\dfrac{(2x^{-2}y^0)^{-2}yx^{-2}}{xxxy^2(y^{-2})^2} = \dfrac{x^4 y^0 yx^{-2}}{4x^3 y^2 y^{-4}} = \dfrac{x^2 y}{4x^3 y^{-2}}$

$\quad = \dfrac{\boldsymbol{y^3}}{\boldsymbol{4x}}$

19. $\dfrac{a^0 bc^0(a^{-1}b^{-1})^2}{ab(ab^0)abc} = \dfrac{ba^{-2}b^{-2}}{abaabc} = \dfrac{a^{-2}b^{-1}}{a^3 b^2 c}$

$\quad = \boldsymbol{a^{-5}b^{-3}c^{-1}}$

20. $\dfrac{(2x^2)^{-3}(xy^0)^{-2}}{2xx^0 x^1 xxy^2} = \dfrac{x^{-6}x^{-2}y^0}{8(2)x^4 y^2} = \dfrac{x^{-8}}{16x^4 y^2}$

$\quad = \dfrac{\boldsymbol{x^{-12}y^{-2}}}{\textbf{16}}$

21. $-2xy + \dfrac{5x^0 xy^{-1}}{y^{-2}} - \dfrac{5xx^{-1}x^2}{(x^{-1})^{-1}}$

$\quad = -2xy + 5xy - 5x = \boldsymbol{3xy - 5x}$

22. $-\dfrac{3x^2 xy^2}{y^4} + \dfrac{2xxx}{y^{-2}} - \dfrac{3xy}{x^{-2}y^{-1}}$

$\quad = -3x^3 y^{-2} + 2x^3 y^2 - 3x^3 y^2 = \boldsymbol{-3x^3 y^{-2} - x^3 y^2}$

23. $xy - x^2 y - y = (-2)(-4) - (-2)^2(-4) - (-4)$

$\quad = 8 + 16 + 4 = \textbf{28}$

24. $a^{-2}b - a(a - b)$

$\quad = \left(-\dfrac{1}{2}\right)^{-2}\left(\dfrac{1}{4}\right) - \left(-\dfrac{1}{2}\right)\left(-\dfrac{1}{2} - \dfrac{1}{4}\right)$

$\quad = 1 + \left(\dfrac{1}{2}\right)\left(-\dfrac{3}{4}\right) = 1 - \dfrac{3}{8} = \dfrac{\textbf{5}}{\textbf{8}}$

25. $m^2 p(mp - p^2) = \left(-\dfrac{1}{4}\right)^2\left(\dfrac{1}{5}\right)\left(-\dfrac{1}{4}\left(\dfrac{1}{5}\right) - \left(\dfrac{1}{5}\right)^2\right)$

$\quad = \dfrac{1}{80}\left(-\dfrac{1}{20} - \dfrac{1}{25}\right) = \dfrac{1}{80}\left(-\dfrac{9}{100}\right) = -\dfrac{\textbf{9}}{\textbf{8000}}$

26. $-3^0[-3^2 - 2(-2 - 3)][-2^0] = -1[-9 + 10][-1]$

$\quad = -1[1][-1] = \textbf{1}$

27. $-3 - (-3)^2 + (-3)(-6) = -3 - 9 + 18 = \textbf{6}$

28. $-3^2 + (-3)^2 - 4^2 - |-2 - 2|$

$\quad = -9 + 9 - 16 - 4 = \textbf{-20}$

29. $-3^{-2} - \dfrac{2}{-2^{-3}} - 2^0 = -\dfrac{1}{9} + 16 - 1$

$\quad = -\dfrac{1}{9} + \dfrac{135}{9} = \dfrac{134}{9} = \textbf{14}\dfrac{\textbf{8}}{\textbf{9}}$

30. $-(-2)^{-3} - 3^{-2} - 3 = \dfrac{1}{8} - \dfrac{1}{9} - 3$

$\quad = \dfrac{9}{72} - \dfrac{8}{72} - \dfrac{216}{72} = -\dfrac{215}{72} = \boldsymbol{-2}\dfrac{\textbf{71}}{\textbf{72}}$

PROBLEM SET 7

1. $\dfrac{P}{100} \times of = is \quad\longrightarrow\quad \dfrac{20}{100} \times WN = 26$

$WN = 26 \cdot \dfrac{100}{20} = \textbf{130}$

Since one part of 130 is 26 for 20%, the other part must be 104 for 80%.

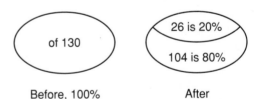

Before, 100%   After

2. $\dfrac{WP}{100} \times of = is \quad\longrightarrow\quad \dfrac{WP}{100} \times 350 = 1400$

$WP = 1400 \cdot \dfrac{100}{350} = \textbf{400\%}$

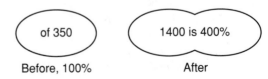

Before, 100%   After

3. $\dfrac{P}{100} \times of = is \quad \longrightarrow \quad \dfrac{20}{100} \times WN = 460$

$WN = 460 \cdot \dfrac{100}{20} = \mathbf{2300}$

Since one part of 2300 is 460 for 20%, the other part must be 1840 for 80%.

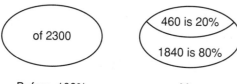

Before, 100%   After

4. $\dfrac{WP}{100} \times of = is \quad \longrightarrow \quad \dfrac{WP}{100} \times 20 = 680$

$WP = 680 \cdot \dfrac{100}{20} = \mathbf{3400\%}$

of 20   680 is 3400%

Before, 100%   After

5. $\dfrac{P}{100} \times of = is \quad \longrightarrow \quad \dfrac{1900}{100} \times WN = 380$

$WN = 380 \cdot \dfrac{100}{1900} = \mathbf{20}$

of 20   380 is 1900%

Before, 100%   After

6. $N \quad N + 2 \quad N + 4$
$7(N + N + 4) = 10(-N - 2) - 120$
$14N + 28 = -10N - 140$
$24N = -168$
$N = -7$
The desired integers are $\mathbf{-7, -5}$, and $\mathbf{-3}$.

7. $N \quad N + 2 \quad N + 4$
$6(N + N + 4) = 14(N + 2) - 8$
$12N + 24 = 14N + 20$
$4 = 2N$
$2 = N$
The desired integers are $\mathbf{2, 4}$, and $\mathbf{6}$.

8. $-2N + 5 = -N$
$\mathbf{5 = N}$

9. $(4x + 15) + (5x + 10) + (x + 5) = 180$
$10x + 30 = 180$
$10x = 150$
$x = \mathbf{15}$

10. Since lines are parallel:
$(3x + 20) + (5x) = 180$
$8x + 20 = 180$
$8x = 160$
$x = \mathbf{20}$

11. Relabel the angles a, b, c, and d as having measures $x°$, $(3x)°$, $(6x)°$, and $(2x)°$, respectively.

$x + 3x + 6x + 2x = 180$
$12x = 180$
$x = 15$

$m\angle a = \mathbf{15°};\ m\angle b = \mathbf{45°};$
$m\angle c = \mathbf{90°};\ m\angle d = \mathbf{30°}$

12. The sum of $x°$, $y°$, $z°$, and the three 60° angles that touch the vertex must equal 360°, a full circle.
$x + y + z + 60 + 60 + 60 = 360$
$x + y + z = 180$
$x° + y° + z° = \mathbf{180°}$

13. $-3p(-2 - 3) + p - 2^2 = -(2p + 4) - p^0$
$6p + 9p + p - 4 = -2p - 4 - 1$
$16p - 4 = -2p - 5$
$18p = -1$
$p = -\dfrac{1}{18}$

14. $0.005x - 0.07 = 0.02x + 0.0032$
$50x - 700 = 200x + 32$
$-732 = 150x$
$x = -\dfrac{732}{150} = -\dfrac{122}{25} = \mathbf{-4.88}$

15. $2\dfrac{1}{5} + 3\dfrac{1}{8} + 2\dfrac{1}{2}x = 4\dfrac{3}{20}$
$\dfrac{5}{2}x = \dfrac{166}{40} - \dfrac{88}{40} - \dfrac{125}{40}$
$\dfrac{5}{2}x = -\dfrac{47}{40}$
$x = -\dfrac{47}{40} \cdot \dfrac{2}{5}$
$x = -\dfrac{94}{200} = -\dfrac{47}{100}$

16. $3x - 2 - 2^0(x - 3) - 2^0 + 2^2 = 5(-x - 2)$
$\quad + 3^0$
$3x - 2 - x + 3 - 1 + 4 = -5x - 10 + 1$
$2x + 4 = -5x - 9$
$7x = -13$
$x = -\dfrac{13}{7}$

17. $\dfrac{2xyp}{y^{-2}}\left(\dfrac{x^{-1}}{y^3 p} - \dfrac{3x}{yyp^2}\right) = \dfrac{2xypx^{-1}}{y^{-2}y^3 p} - \dfrac{6xypx}{y^{-2}yyp^2}$
$= 2 - 6yp^{-1}x^2$

18. $\dfrac{4x^{-2}y}{k}\left(\dfrac{2kx^2}{y} - \dfrac{3xy}{k}\right)$

$= \dfrac{8x^{-2}ykx^2}{ky} - \dfrac{12x^{-2}yxy}{kk} = 8 - 12x^{-1}y^2k^{-2}$

19. $\dfrac{(2x^2y^3)^{-3}y}{(4xy)^{-2}(x^{-2}y)^3y} = \dfrac{16x^{-6}y^{-9}y}{8x^{-2}y^{-2}x^{-6}y^3y}$

$= \dfrac{2x^{-6}y^{-8}}{x^{-8}y^2} = 2x^2y^{-10}$

20. $\dfrac{xx^{-2}y(x^{-3})^2xy^0}{(2xy)^{-2}x^2(y^{-3})^2} = \dfrac{4xx^{-2}yx^{-6}xy^0}{x^{-2}y^{-2}x^2y^{-6}} = \dfrac{4x^{-6}y}{y^{-8}}$

$= 4x^{-6}y^9$

21. $-\dfrac{3x^2xy}{p} + \dfrac{7xyp^{-1}}{x^{-2}} - \dfrac{2xxxp^{-1}}{y^{-1}}$

$= -\dfrac{3x^3y}{p} + \dfrac{7x^3y}{p} - \dfrac{2x^3y}{p} = \dfrac{2x^3y}{p}$

22. $-4xp^2 + \dfrac{3xxp^4}{p^2x^2} - \dfrac{2xp}{p^{-1}} = -4xp^2 + 3p^2 - 2xp^2$

$= -6xp^2 + 3p^2$

23. $-a(a - b) = -\left(-\dfrac{1}{2}\right)\left(-\dfrac{1}{2} - \dfrac{1}{3}\right)$

$= \dfrac{1}{2}\left(-\dfrac{3}{6} - \dfrac{2}{6}\right) = \dfrac{1}{2}\left(-\dfrac{5}{6}\right) = -\dfrac{5}{12}$

24. $-xy(-x^2 - y) = -\left(-\dfrac{1}{2}\right)\left(\dfrac{1}{4}\right)\left(-\left(-\dfrac{1}{2}\right)^2 - \dfrac{1}{4}\right)$

$= \dfrac{1}{8}\left(-\dfrac{1}{4} - \dfrac{1}{4}\right) = \dfrac{1}{8}\left(-\dfrac{2}{4}\right) = -\dfrac{2}{32} = -\dfrac{1}{16}$

25. $x^3 - x(xy - y)$

$= (-2)^3 - (-2)((-2)(-4) - (-4))$

$= -8 + 2(8 + 4) = -8 + 24 = 16$

26. $x - a(a - xa) = 2 - \left(-\dfrac{1}{2}\right)\left(-\dfrac{1}{2} - 2\left(-\dfrac{1}{2}\right)\right)$

$= 2 + \dfrac{1}{2}\left(-\dfrac{1}{2} + 1\right) = 2 + \dfrac{1}{2}\left(\dfrac{1}{2}\right) = 2 + \dfrac{1}{4}$

$= 2\dfrac{1}{4}$

27. $-2\left\{[-2^0 - 3(-2)] - [-2(-3 - 2)(-2)]\right\}$

$= -2\left\{[-1 + 6] - [-2(-5)(-2)]\right\}$

$= -2\left\{[5] - [-20]\right\} = -2\{25\} = -50$

28. $-2^0 - 2 - 2^2 - (-2)^3 - 2(-2 - 2) - 2$

$= -1 - 2 - 4 + 8 + 4 + 4 - 2 = 7$

29. $3^0(-2 - 3)(-2 + 5)(-2) - (-3 + 7)(-4^0 - 3^0)$

$= (1)(-5)(3)(-2) - 4(-2) = 30 + 8 = 38$

30. $2[(-2^0 - 1)(-2^0 - 15^0) - (-2)^2 - 3^0] - 2$

$= 2[-2(-2) - 4 - 1] - 2 = 2[4 - 4 - 1] - 2$

$= 2[-1] - 2 = -2 - 2 = -4$

PROBLEM SET 8

1. $WD \times of = is$

$0.36 \times K = 828$

$K = \dfrac{828}{0.36} = 2300$

2. $N \quad N + 2 \quad N + 4 \quad N + 6$

$10(N + N + 6) = 9(N + 2 + N + 6) + 24$

$20N + 60 = 18N + 96$

$2N = 36$

$N = 18$

The desired integers are **18**, **20**, **22**, and **24**.

3. $3N - 7 = -2N - 72$

$5N = -65$

$N = -13$

4. $F \times of = is$

$\dfrac{7}{16} \times WN = 420$

$WN = 420 \cdot \dfrac{16}{7} = 960$

5. $N \quad N + 2 \quad N + 4$

$5(N + N + 4) = 2(-N - 2) + 108$

$10N + 20 = -2N + 104$

$12N = 84$

$N = 7$

The desired integers are **7**, **9**, and **11**.

6. $\dfrac{P}{100} \times of = is \quad \longrightarrow \quad \dfrac{20}{100} \times WN = 86$

$WN = 86 \cdot \dfrac{100}{20} = 430$

Since one part of 430 is 86 for 20%, the other part must be 344 for 80%.

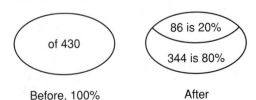

Before, 100%   After

7. $\dfrac{P}{100} \times of = is \longrightarrow \dfrac{340}{100} \times 56 = WN$

$WN = \dfrac{340}{100} \cdot 56 = \mathbf{190.4}$

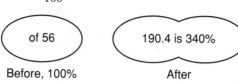

of 56   190.4 is 340%

Before, 100%   After

8. $A = 2(90 - A) + 3$
$A = 180 - 2A + 3$
$3A = 183$
$A = \mathbf{61°}$

9. $3x + 4y = 8$
$4y = -3x + 8$
$y = -\dfrac{3}{4}x + 2$

The y intercept is 2; the slope is $-\dfrac{3}{4}$.

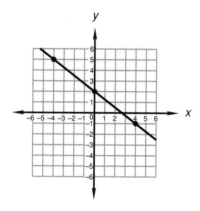

10. $(2x + 2) + (4x + 4) + (3x + 3) = 180$
$9x + 9 = 180$
$9x = 171$
$x = \mathbf{19}$

11. $\text{Area}_{\text{Circle}} = \pi r^2 = 25\pi \text{ m}^2$

$r = \sqrt{\dfrac{25\pi \text{ m}^2}{\pi}} = \mathbf{5\ m}$

$r = BD = AC = \mathbf{5\ m}$

12. Since lines are parallel:
$(5x - 19) + (3x - 1) = 180$
$8x - 20 = 180$
$8x = 200$
$x = \mathbf{25}$

13. $(x + 5) + 155 = 180$
$x = 180 - 155 - 5 = \mathbf{20}$

$(y + 10) + (x + 5) + 40 = 180$
$y + 10 + 20 + 5 + 40 = 180$
$y = 180 - 75 = \mathbf{105}$

14. $0.003x + 0.02x - 0.03 = 0.177$
$3x + 20x - 30 = 177$
$23x = 207$
$x = \mathbf{9}$

15. $2\dfrac{1}{3}x + 1\dfrac{3}{5} = 7\dfrac{2}{5}$

$\dfrac{7}{3}x = \dfrac{37}{5} - \dfrac{8}{5}$

$\dfrac{7}{3}x = \dfrac{29}{5}$

$x = \dfrac{29}{5} \cdot \dfrac{3}{7} = \dfrac{\mathbf{87}}{\mathbf{35}}$

16. $-4^0 - 2^2 - (-2)^3 - 3 - (2 - 2x) - 4$
$\qquad = -3(-2 + 2x)$
$-1 - 4 + 8 - 3 - 2 + 2x - 4 = 6 - 6x$
$\qquad\qquad\qquad 2x - 6 = 6 - 6x$
$\qquad\qquad\qquad\qquad 8x = 12$
$\qquad\qquad\qquad\qquad x = \dfrac{12}{8} = \dfrac{3}{2}$

17. $-\dfrac{1}{2} + 2\dfrac{3}{8} - 7\dfrac{1}{4} + 3\dfrac{1}{2}x = 4\dfrac{1}{16}$

$\dfrac{7}{2}x = \dfrac{65}{16} + \dfrac{8}{16} - \dfrac{38}{16} + \dfrac{116}{16}$

$x = \dfrac{151}{16} \cdot \dfrac{2}{7} = \dfrac{302}{112} = \dfrac{\mathbf{151}}{\mathbf{56}}$

18. $\dfrac{x^{-2}y}{p}\left(\dfrac{-3x^2p}{y} - \dfrac{4xy^2}{p^2}\right)$

$= \dfrac{-3x^{-2}yx^2p}{py} - \dfrac{4x^{-2}yxy^2}{pp^2} = \mathbf{-3 - 4x^{-1}y^3p^{-3}}$

19. $\dfrac{-3^0x^0}{p^0}\left(-3x + \dfrac{5xy}{p^{-2}}\right) = 3x - \dfrac{5xy}{p^{-2}}$

$= \mathbf{3x - 5p^2xy}$

20. $\dfrac{(4x)^{-2}y^0(y^{-2})^2y}{32^{-1}x^2(yx^0)^{-3}} = \dfrac{32x^{-2}y^{-4}y}{16x^2y^{-3}} = \mathbf{2x^{-4}}$

21. $\dfrac{5x^{-2}(y^2x^3)^{-3}}{2^{-2}xyx^2(x^{-2}y)} = \dfrac{20x^{-2}y^{-6}x^{-9}}{xyx^2x^{-2}y} = \dfrac{20x^{-11}y^{-6}}{xy^2}$

$= \mathbf{20x^{-12}y^{-8}}$

22. $3x^{-2}y + 5x^2y^{-1} - \dfrac{3}{xxy^{-1}}$

$= 3x^{-2}y + 5x^2y^{-1} - 3x^{-2}y = \mathbf{5x^2y^{-1}}$

23. $\dfrac{2xxy^{-3}y}{xy} + \dfrac{2y^{-1}y^{-2}}{x^{-1}} - \dfrac{7xy^2}{y}$

$= 2xy^{-3} + 2xy^{-3} - 7xy = \mathbf{4xy^{-3} - 7xy}$

24. $ax - a(x^2) = -\frac{1}{2}\left(\frac{1}{4}\right) - \left(-\frac{1}{2}\right)\left(\frac{1}{4}\right)^2$

$= -\frac{1}{8} + \frac{1}{2}\left(\frac{1}{16}\right) = -\frac{1}{8} + \frac{1}{32} = -\frac{4}{32} + \frac{1}{32}$

$= -\frac{3}{32}$

25. $ab^2(a - ab) = \frac{1}{2}\left(-\frac{1}{3}\right)^2\left(\frac{1}{2} - \frac{1}{2}\left(-\frac{1}{3}\right)\right)$

$= \frac{1}{18}\left(\frac{1}{2} + \frac{1}{6}\right) = \frac{1}{18}\left(\frac{3}{6} + \frac{1}{6}\right) = \frac{1}{18}\left(\frac{2}{3}\right)$

$= \frac{2}{54} = \frac{1}{27}$

26. $mx - (m^2 - x) = -\frac{1}{3}\left(\frac{1}{2}\right) - \left(\left(-\frac{1}{3}\right)^2 - \frac{1}{2}\right)$

$= -\frac{1}{6} - \left(\frac{1}{9} - \frac{1}{2}\right) = -\frac{1}{6} - \left(\frac{2}{18} - \frac{9}{18}\right)$

$= -\frac{1}{6} - \left(-\frac{7}{18}\right) = -\frac{3}{18} + \frac{7}{18} = \frac{4}{18} = \frac{2}{9}$

27. $-3[(-2^0 - 4) - (-2)(-3)]$
$\quad - [(-6^0 - 2) - 2^2(-3)]$
$= -3[-5 - 6] - [-3 + 12] = -3[-11] - [9]$
$= 33 - 9 = 24$

28. $-3 - 3^0 - 3^{-2} + \frac{1}{9} - 3^0(-3 - 3)$

$= -3 - 1 - \frac{1}{9} + \frac{1}{9} + 3 + 3 = 2$

29. $-|-2^0| - 2^{-2} - (-2)^{-2} = -1 - \frac{1}{4} - \frac{1}{4} = -1\frac{1}{2}$

30. $(-1)^{-3} - 1^{-2} - 1^2 - (-1)^3$
$= -1 - 1 - 1 + 1 = -2$

• PROBLEM SET 9

1.

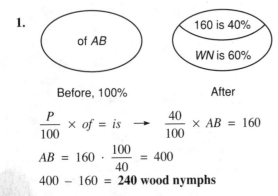

Before, 100%   After

$\frac{P}{100} \times of = is \longrightarrow \frac{40}{100} \times AB = 160$

$AB = 160 \cdot \frac{100}{40} = 400$

$400 - 160 = \textbf{240 wood nymphs}$

2.

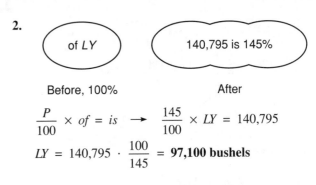

Before, 100%   After

$\frac{P}{100} \times of = is \longrightarrow \frac{145}{100} \times LY = 140{,}795$

$LY = 140{,}795 \cdot \frac{100}{145} = \textbf{97,100 bushels}$

3.

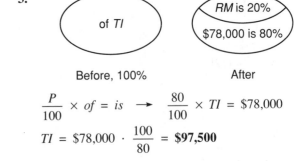

Before, 100%   After

$\frac{P}{100} \times of = is \longrightarrow \frac{80}{100} \times TI = \$78{,}000$

$TI = \$78{,}000 \cdot \frac{100}{80} = \textbf{\$97,500}$

4. $N \quad\quad N + 1 \quad\quad N + 2$
$\quad -5(N + N + 1) = 2(N + 1) - 43$
$\quad\quad -10N - 5 = 2N - 41$
$\quad\quad\quad\quad 36 = 12N$
$\quad\quad\quad\quad 3 = N$
The desired integers are **3**, **4**, and **5**.

5. If 84% staggered to their feet, then 16% did not get up.

$\frac{P}{100} \times of = is$

$\frac{16}{100} \times SP = 40{,}000$

$SP = 40{,}000 \cdot \frac{100}{16} = \textbf{250,000}$

6. If he could not see $\frac{7}{8}$ of the Trojans, then he could see $\frac{1}{8}$ of the Trojans.
$F \times of = is$

$\frac{1}{8} \times T = 1400$

$T = 1400 \cdot \frac{8}{1} = \textbf{11,200}$

7.

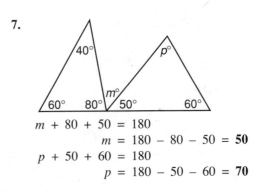

$m + 80 + 50 = 180$
$\quad\quad m = 180 - 80 - 50 = \textbf{50}$
$p + 50 + 60 = 180$
$\quad\quad p = 180 - 50 - 60 = \textbf{70}$

8. $y + 140 = 180 \longrightarrow y = \mathbf{40}$

$(7x + 6) + (6x + 4) + y = 180$

$13x + 10 + 40 = 180$

$13x = 130$

$x = \mathbf{10}$

9. $(5x + 10) + (7x + 50) = 180$

$12x + 60 = 180$

$12x = 120$

$x = \mathbf{10}$

$5x + 10 = 5(10) + 10 = 50 + 10 = \mathbf{60}$

Since vertical angles are equal angles:

$y + 1 = 60$

$y = \mathbf{59}$

$z + 60 = 180$

$z = \mathbf{120}$

10. $\text{Area}_{\text{Circle}} = \pi r^2 = 9\pi\ \text{m}^2$

$$r = \sqrt{\frac{9\pi\ \text{m}^2}{\pi}} = \mathbf{3\ m}$$

$\text{Circumference} = 2\pi r = 2\pi(3\ \text{m}) = \mathbf{6\pi\ m}$

11. $y - 2x + 3 = 0$

$y = 2x - 3$

The y intercept is -3; the slope is $\dfrac{2}{1}$.

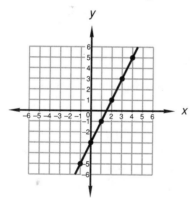

12. $3y + 6 = -x$

$$y = -\frac{1}{3}x - 2$$

The y intercept is -2; the slope is $-\dfrac{1}{3}$.

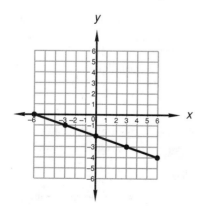

13. $0.02 - 0.003x + x = 5.005$

$20 - 3x + 1000x = 5005$

$997x = 4985$

$x = \mathbf{5}$

14. $-3\dfrac{1}{5}x + 7\dfrac{1}{10} = 4\dfrac{2}{9}$

$$-\frac{16}{5}x = \frac{38}{9} - \frac{71}{10}$$

$$-\frac{16}{5}x = \frac{380}{90} - \frac{639}{90}$$

$$-\frac{16}{5}x = -\frac{259}{90}$$

$$x = -\frac{259}{90}\left(-\frac{5}{16}\right) = \frac{1295}{1440} = \mathbf{\frac{259}{288}}$$

15. $-2[(2 - 3)x + 7(2^0 - 1)] = -3(x - 2)$

$-2[-x + 7(0)] = -3x + 6$

$2x = -3x + 6$

$5x = 6$

$$x = \mathbf{\frac{6}{5}}$$

16. $-2^0(2x - 3) - 4 = 2x - 3^0$

$-2x + 3 - 4 = 2x - 1$

$-2x - 1 = 2x - 1$

$0 = 4x$

$\mathbf{0 = x}$

17. $\dfrac{x^0 y^2}{p^{-2}}\left(\dfrac{p^2}{y^2} - \dfrac{y^{-2}}{p^{-2}}\right) = \dfrac{x^0 y^2 p^2}{p^{-2}y^2} - \dfrac{x^0 y^2 y^{-2}}{p^{-2}p^{-2}}$

$= p^4 - p^4 = \mathbf{0}$

18. $\dfrac{ak^{-2}}{a^{-3}}\left(\dfrac{2k^4}{a^4} - 3k\right) = \dfrac{2ak^{-2}k^4}{a^{-3}a^4} - \dfrac{3ak^{-2}k}{a^{-3}}$

$= \mathbf{2k^2 - 3a^4k^{-1}}$

19. $\dfrac{(2x^2ya)^{-3}ya^3}{x^2y(ay)^{-2}y} = \dfrac{x^{-6}y^{-3}a^{-3}ya^3}{8x^2ya^{-2}y^{-2}y} = \dfrac{x^{-6}y^{-2}}{8x^2a^{-2}}$

$= \mathbf{\dfrac{a^2}{8x^8y^2}}$

20. $\dfrac{(-2xyz)^{-3}}{(x^2z^{-3})^{-3}} = -\dfrac{x^{-3}y^{-3}z^{-3}}{8x^{-6}z^9} = \mathbf{-\dfrac{x^3}{8y^3z^{12}}}$

21. $3x - \dfrac{2xy^2}{y} + \dfrac{4xx^{-2}}{(x^2)^{-1}} = 3x - 2xy + 4x$

$= \mathbf{7x - 2xy}$

22. $\dfrac{2xy}{p} - \dfrac{5xxx}{(x^{-2})^{-1}y^{-1}} + \dfrac{3xp^{-1}}{y^{-1}}$

$= \dfrac{2xy}{p} - 5xy + \dfrac{3xy}{p} = \mathbf{\dfrac{5xy}{p} - 5xy}$

23. $-a^2b - a = -\left(-\dfrac{1}{2}\right)^2\left(\dfrac{1}{4}\right) - \left(-\dfrac{1}{2}\right)$

$= -\dfrac{1}{16} + \dfrac{1}{2} = -\dfrac{1}{16} + \dfrac{8}{16} = \dfrac{7}{16}$

24. $a(a - ab) = -\dfrac{1}{2}\left(-\dfrac{1}{2} - \left(-\dfrac{1}{2}\right)\left(-\dfrac{1}{8}\right)\right)$

$= -\dfrac{1}{2}\left(-\dfrac{1}{2} - \dfrac{1}{16}\right) = -\dfrac{1}{2}\left(-\dfrac{8}{16} - \dfrac{1}{16}\right)$

$= -\dfrac{1}{2}\left(-\dfrac{9}{16}\right) = \dfrac{9}{32}$

25. $a(a - b)(ab - b) = -2(-2 - 3)(-2(3) - 3)$
$= -2(-5)(-6 - 3) = -2(-5)(-9) = \mathbf{-90}$

26. $a^2(x - ax^2) = (-2)^2(-4 - (-2)(-4)^2)$
$= 4(-4 + 32) = 4(28) = \mathbf{112}$

27. $-2(-3 - 2^0) - 2^0(-2^2 - 2) = -2(-4) - 1(-6)$
$= 8 + 6 = \mathbf{14}$

28. $-3[(-5 + 2)(-2) - (3^0 - 2) - 2]$
$= -3[(-3)(-2) - (-1) - 2] = -3[6 + 1 - 2]$
$= -3[5] = \mathbf{-15}$

29. $-2^0(-2 - 3^0) - (-2)^3 - |-3|$
$= -1(-3) + 8 - 3$
$= 3 + 8 - 3 = \mathbf{8}$

30. $-\dfrac{1}{-2^{-3}} + \dfrac{1}{-(-2)^{-3}} - 3^2 = 8 + 8 - 9 = \mathbf{7}$

PROBLEM SET 10

1. If 20% were in a festive mood, then 80% were not in a festive mood.

$\dfrac{P}{100} \times of = is$

$\dfrac{80}{100} \times P = 1400$

$P = 1400 \cdot \dfrac{100}{80} = \mathbf{1750}$

2. $F \times of = is$

$3\dfrac{1}{4} \times S = 26{,}000$

$S = 26{,}000 \cdot \dfrac{4}{13} = \mathbf{8000}$

3. $F \times of = is$

$\dfrac{1}{10} \times M = 590$

$M = 590 \cdot \dfrac{10}{1} = 5900$

$\text{Men}_{\text{After}} = \text{Men}_{\text{Total}} - \text{Men}_{\text{Killed}}$
$\text{Men}_{\text{After}} = 5900 - 590 = \mathbf{5310}$

4. $N \qquad N + 2 \qquad N + 4$
$4N = 3(N + 2 + N + 4) - 8$
$4N = 6N + 10$
$-2N = 10$
$N = -5$
The desired integers are **–5, –3,** and **–1**.

5. $\dfrac{P}{100} \times of = is$

$\dfrac{260}{100} \times OA = 10{,}400$

$OA = 10{,}400 \cdot \dfrac{100}{260} = \mathbf{4000\ minas}$

6. $\dfrac{P}{100} \times of = is$

$\dfrac{14}{100} \times A_{\text{Total}} = 4200$

$A_{\text{Total}} = 4200 \cdot \dfrac{100}{14} = 30{,}000$

$A_{\text{Hidden}} = A_{\text{Total}} - A_{\text{See}}$
$A_{\text{Hidden}} = 30{,}000 - 4200 = \mathbf{25{,}800}$

7. $\text{Circumference} = 2\pi r = 6\pi \text{ cm}$

$r = \dfrac{6\pi}{2\pi} \text{ cm} = 3 \text{ cm}$

$\text{Area} = \pi r^2 = \pi(3 \text{ cm})^2 = \mathbf{9\pi\ cm^2}$

8. Since angles opposite equal sides are equal angles:
$x = \mathbf{60}$
$x + y + 60 = 180$
$y = 180 - 60 - 60 = \mathbf{60}$

$y + \angle BDC = 180$
$\angle BDC = 180 - 60 = 120$
$z + 20 + 120 = 180$
$z = 180 - 20 - 120 = \mathbf{40}$

9. $(17x + 20) + (20x - 25) = 180$
$37x - 5 = 180$
$37x = 185$
$x = \mathbf{5}$
Since vertical angles are equal angles:
$P = 17x + 20$
$P = 17(5) + 20 = \mathbf{105}$
Since lines are parallel:
$Q = P = \mathbf{105}$

10. $\text{Area}_{\text{Square}} = 4 \times \text{Area}_{\text{Shaded}}$
$= 4 \times 9 \text{ m}^2 = \mathbf{36\ m^2}$
$\text{Area}_{\text{Square}} = s^2 = 36 \text{ m}^2$
$s = \sqrt{36 \text{ m}^2} = \mathbf{6\ m}$

11. $2y = 3x + 2$

    $y = \dfrac{3}{2}x + 1$

The y intercept is 1; the slope is $\dfrac{3}{2}$.

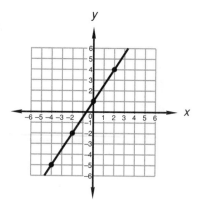

12. $y = -3$

The y intercept is -3; the slope is 0.

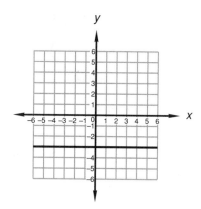

13. $\quad c^2 = a^2 + b^2$

$\quad\ 7^2 = p^2 + 4^2$

$\quad 49 = p^2 + 16$

$\quad 33 = p^2$

$\sqrt{33} = p$

14. $D^2 = 1^2 + 7^2$

$D^2 = 1 + 49$

$D^2 = 50$

$D = \sqrt{50}$

$D = 5\sqrt{2}$

15. $\quad \dfrac{3}{4}x - \dfrac{1}{5}x = 2\dfrac{3}{4}$

$\quad \dfrac{15}{20}x - \dfrac{4}{20}x = \dfrac{11}{4}$

$\quad\quad\ \dfrac{11}{20}x = \dfrac{11}{4}$

$\quad\quad\quad x = \dfrac{11}{4} \cdot \dfrac{20}{11} = \mathbf{5}$

16. $-5.2 + 3y = 0.2(y + 2)$

$\ -52 + 30y = 2y + 4$

$\quad\quad 28y = 56$

$\quad\quad\quad y = \mathbf{2}$

17. $4x(2 - 3^0) + (-2)(x - 5) = -(3x + 2)$

$\quad\ 8x - 4x - 2x + 10 = -3x - 2$

$\quad\quad\quad\quad 2x + 10 = -3x - 2$

$\quad\quad\quad\quad\quad\quad 5x = -12$

$\quad\quad\quad\quad\quad\quad\ x = -\dfrac{\mathbf{12}}{\mathbf{5}}$

18. $\dfrac{4xy}{m^{-2}}\left(\dfrac{3y^{-1}}{m^2 x} - \dfrac{2x}{ym}\right) = \dfrac{12xyy^{-1}}{m^{-2}m^2 x} - \dfrac{8xyx}{m^{-2}ym}$

$= \mathbf{12 - 8x^2 m}$

19. $\dfrac{2x^0 y}{p}\left(\dfrac{2p}{y} - \dfrac{3xy}{p}\right) = \dfrac{4x^0 yp}{py} - \dfrac{6x^0 yxy}{pp}$

$= \mathbf{4 - 6xy^2 p^{-2}}$

20. $\dfrac{(3x^{-2})^{-2}xy}{3^{-3}x^{-2}(yx^0)^{-3}} = \dfrac{27x^4 xy}{9x^{-2}y^{-3}x^0} = \dfrac{3x^5 y}{x^{-2}y^{-3}}$

$= \mathbf{3x^7 y^4}$

21. $\dfrac{(x^2 p^2)^{-3}x^0 p^2}{x^{-2}px^0(xp)^{-3}} = \dfrac{x^{-6}p^{-6}x^0 p^2}{x^{-2}px^0 x^{-3}p^{-3}} = \dfrac{x^{-6}p^{-4}}{x^{-5}p^{-2}}$

$= \mathbf{x^{-1}p^{-2}}$

22. $3x + \dfrac{2x^2 x^{-3}}{x^{-2}y^0} - x^0 = 3x + 2x - 1 = \mathbf{5x - 1}$

23. $\dfrac{5x^2 y}{z} - \dfrac{3z^{-1}y}{x^{-2}} + \dfrac{7xxy^2 z}{yz^2}$

$= \dfrac{5x^2 y}{z} - \dfrac{3x^2 y}{z} + \dfrac{7x^2 y}{z} = \dfrac{\mathbf{9x^2 y}}{\mathbf{z}}$

24. $(A + x + 3) + (B + x + 2)$

$\quad + (C + x + 1) = 30 \text{ cm}$

Since $A = x + 3$; $B = x + 2$; and $C = x + 1$:

$2(x + 3) + 2(x + 2) + 2(x + 1) = 30 \text{ cm}$

$\quad 2x + 6 + 2x + 4 + 2x + 2 = 30 \text{ cm}$

$\quad\quad\quad\quad\quad\quad\quad 6x = 18 \text{ cm}$

$\quad\quad\quad\quad\quad\quad\quad\ x = \mathbf{3 \text{ cm}}$

25. $b(ab - b) = -\dfrac{1}{2}\left(\left(-\dfrac{1}{3}\right)\left(-\dfrac{1}{2}\right) - \left(-\dfrac{1}{2}\right)\right)$

$= -\dfrac{1}{2}\left(\dfrac{1}{6} + \dfrac{1}{2}\right) = -\dfrac{1}{2}\left(\dfrac{1}{6} + \dfrac{3}{6}\right) = -\dfrac{1}{2}\left(\dfrac{2}{3}\right)$

$= -\dfrac{2}{6} = -\dfrac{1}{3}$

26. $ab - a^2b^2 - b$

$= -\dfrac{1}{2}\left(-\dfrac{1}{2}\right) - \left(-\dfrac{1}{2}\right)^2\left(-\dfrac{1}{2}\right)^2 - \left(-\dfrac{1}{2}\right)$

$= \dfrac{1}{4} - \dfrac{1}{16} + \dfrac{1}{2} = \dfrac{4}{16} - \dfrac{1}{16} + \dfrac{8}{16} = \dfrac{11}{16}$

27. $a^2(b - ab)b$

$= \left(-\dfrac{1}{2}\right)^2\left(-\dfrac{1}{2} - \left(-\dfrac{1}{2}\right)\left(-\dfrac{1}{2}\right)\right)\left(-\dfrac{1}{2}\right)$

$= \dfrac{1}{4}\left(-\dfrac{1}{2} - \dfrac{1}{4}\right)\left(-\dfrac{1}{2}\right) = \dfrac{1}{4}\left(-\dfrac{2}{4} - \dfrac{1}{4}\right)\left(-\dfrac{1}{2}\right)$

$= \dfrac{1}{4}\left(-\dfrac{3}{4}\right)\left(-\dfrac{1}{2}\right) = \dfrac{3}{32}$

28. $-\dfrac{1}{2^{-3}} - \dfrac{1}{-2^{-3}} - (-3 - 2^0) - 2$

$= -8 + 8 + 4 - 2 = \mathbf{2}$

29. $-|-2| - |-2^0| - 3^2 - (-3)^3$

$= -2 - 1 - 9 + 27 = \mathbf{15}$

30. $-\dfrac{1}{2} - \left(\dfrac{1}{2}\right)^2 - \left(-\dfrac{1}{2}\right)^3 - \dfrac{1}{2}$

$= -\dfrac{1}{2} - \dfrac{1}{4} + \dfrac{1}{8} - \dfrac{1}{2} = -\dfrac{4}{8} - \dfrac{2}{8} + \dfrac{1}{8} - \dfrac{4}{8}$

$= -\dfrac{9}{8}$

PROBLEM SET 11

1. If 13% believed, 87% did not believe.

$\dfrac{P}{100} \times of = is$

$\dfrac{87}{100} \times L = 5220$

$L = 5220 \cdot \dfrac{100}{87} = 6000$

$L_{\text{Believed}} = L_{\text{Total}} - L_{\text{Not believed}}$

$L_{\text{Believed}} = 6000 - 5220 = \mathbf{780}$

2. $N \quad N + 2 \quad N + 4 \quad N + 6$

$-2(N + N + 6) = -N - 4 - 20$

$-4N - 12 = -N - 24$

$-3N = -12$

$N = 4$

The desired integers are **4, 6, 8**, and **10**.

3. $\dfrac{P}{100} \times of = is$

$\dfrac{220}{100} \times OP = \$5599$

$OP = \$5599 \cdot \dfrac{100}{220} = \$2545$

Increase = $\$5599 - \$2545 = \mathbf{\$3054}$

4. If 30% sat around, 70% worked.

$\dfrac{P}{100} \times of = is$

$\dfrac{70}{100} \times P = 1400$

$P = 1400 \cdot \dfrac{100}{70} = 2000$

$P_{\text{Sat}} = 2000 - 1400 = \mathbf{600}$

5. $\dfrac{P}{100} \times of = is$

$\dfrac{340}{100} \times RP = 6800$

$RP = 6800 \cdot \dfrac{100}{340} = 2000$

Total = $6800 + 2000 = \mathbf{8800}$

6. $N \quad N + 2 \quad N + 4 \quad N + 6$

$-4(N + N + 6) = 10(-N - 4) + 10$

$-8N - 24 = -10N - 30$

$2N = -6$

$N = -3$

The desired integers are **−3, −1, 1**, and **3**.

7. Since angles opposite equal sides are equal angles:

$z = \mathbf{32}$

$y + z + 32 = 180$

$y = 180 - 32 - 32 = \mathbf{116}$

$x + y = 180$

$x + 116 = 180$

$x = \mathbf{64}$

Since the measure of an arc of a circle equals the measure of the central angle: $p = \mathbf{64}$

8. $A_{\text{Shaded}} = A_{\text{Square}} - A_{3 \text{ Triangles}}$

$= (4)(4) - \dfrac{1}{2}(2)(1) - \dfrac{1}{2}(2)(3) - \dfrac{1}{2}(4)(1)$

$= 16 - 1 - 3 - 2$

$= \mathbf{10 \text{ units}^2}$

9. $A_{\text{Circle}} = A_{\text{Triangle}}$

$\pi r^2 = \dfrac{1}{2}bH$

$\pi(\pi \text{ cm})^2 = \dfrac{1}{2}(\pi \text{ cm})(H)$

$H = \dfrac{2\pi^3 \text{ cm}^2}{\pi \text{ cm}} = \mathbf{2\pi^2 \text{ cm}}$

10. $2(3x + 2) + 2(4x - 2) + 2(3x - 2) = 36\,m$

$\qquad 6x + 4 + 8x - 4 + 6x - 4 = 36\,m$

$\qquad\qquad\qquad 20x - 4 = 36\,m$

$\qquad\qquad\qquad 20x = 40\,m$

$\qquad\qquad\qquad x = 2\,m$

$3x + 2 = 3(2\,m) + 2 = \mathbf{8\,m}$

11. $\dfrac{k}{ax} + \dfrac{bc}{x^2} - \dfrac{m}{ax^3} = \dfrac{kx^2}{ax^3} + \dfrac{abcx}{ax^3} - \dfrac{m}{ax^3}$

$\qquad = \dfrac{\mathbf{kx^2 + abcx - m}}{\mathbf{ax^3}}$

12. $\dfrac{p}{ak} - c + \dfrac{3a}{4k} = \dfrac{4p}{4ak} - \dfrac{4ack}{4ak} + \dfrac{3a^2}{4ak}$

$\qquad = \dfrac{\mathbf{4p - 4ack + 3a^2}}{\mathbf{4ak}}$

13. $\dfrac{m^2}{p} - \dfrac{3p}{cx} - \dfrac{5}{4c^2x}$

$\qquad = \dfrac{4c^2m^2x}{4c^2px} - \dfrac{12cp^2}{4c^2px} - \dfrac{5p}{4c^2px}$

$\qquad = \dfrac{\mathbf{4c^2m^2x - 12cp^2 - 5p}}{\mathbf{4c^2px}}$

14. $c^2 = a^2 + b^2$

$\qquad 11^2 = 4^2 + x^2$

$\qquad 121 = 16 + x^2$

$\qquad 105 = x^2$

$\qquad \mathbf{\sqrt{105}} = x$

15. $D^2 = 5^2 + 8^2$

$\qquad D^2 = 25 + 64$

$\qquad D^2 = 89$

$\qquad D = \mathbf{\sqrt{89}}$

16. (a) $4x + 3y - 6 = 0$

$\qquad\qquad 3y = -4x + 6$

$\qquad\qquad y = -\dfrac{4}{3}x + 2$

(b) $x = -3$

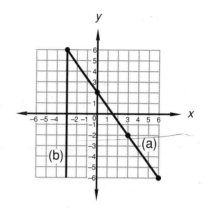

17. $3\dfrac{1}{2} - 2\dfrac{1}{3}x = 3\dfrac{1}{4}$

$\qquad -\dfrac{7}{3}x = \dfrac{13}{4} - \dfrac{14}{4}$

$\qquad -\dfrac{7}{3}x = -\dfrac{1}{4}$

$\qquad x = -\dfrac{1}{4}\left(-\dfrac{3}{7}\right) = \dfrac{\mathbf{3}}{\mathbf{28}}$

18. $0.03x - x + 2 = -0.91$

$\qquad 3x - 100x + 200 = -91$

$\qquad -97x = -291$

$\qquad x = \mathbf{3}$

19. $3x(2 - 3^0) - 7^0 = -2x(3 - 7^0) + 2$

$\qquad 3x - 1 = -4x + 2$

$\qquad 7x = 3$

$\qquad x = \dfrac{\mathbf{3}}{\mathbf{7}}$

20. $\dfrac{a^{-2}}{y}\left(3a^2y - \dfrac{2y}{a^2}\right) = \dfrac{3a^{-2}a^2y}{y} - \dfrac{2a^{-2}y}{ya^2}$

$\qquad = \mathbf{3 - 2a^{-4}}$

21. $\dfrac{2x^0yp}{k}\left(\dfrac{3k}{yp} - \dfrac{2yk}{p}\right) = \dfrac{6x^0ypk}{kyp} - \dfrac{4x^0ypyk}{kp}$

$\qquad = \mathbf{6 - 4y^2}$

22. $\dfrac{2p^2a^{-2}ap^0p^4}{(2pa^{-2})^{-3}ap^0} = \dfrac{16p^6a^{-1}}{p^{-3}a^7} = \mathbf{16p^9a^{-8}}$

23. $\dfrac{xym^2m^{-4}xm}{(2x^2y)^{-3}xy^0x^{-3}y} = \dfrac{8x^2ym^{-1}}{x^{-8}y^{-2}} = \mathbf{8x^{10}y^3m^{-1}}$

24. $\dfrac{xp^{-3}}{y} - \dfrac{3y^{-1}}{x^{-1}p^3} + \dfrac{2x}{pppy}$

$= \dfrac{x}{yp^3} - \dfrac{3x}{yp^3} + \dfrac{2x}{yp^3} = \mathbf{0}$

25. $-3ka + \dfrac{3k^2a^2}{ka} - \dfrac{5a^0k}{a^{-1}} = -3ka + 3ka - 5ka$

$= \mathbf{-5ka}$

26. $-a - ax(a - x)$

$= -\left(-\dfrac{1}{2}\right) - \left(-\dfrac{1}{2}\right)\left(\dfrac{3}{2}\right)\left(-\dfrac{1}{2} - \dfrac{3}{2}\right)$

$= \dfrac{1}{2} + \dfrac{3}{4}(-2) = \dfrac{1}{2} - \dfrac{3}{2} = \mathbf{-1}$

27. $-a^2(b - a) = -\left(-\dfrac{1}{2}\right)^2\left(\dfrac{3}{2} - \left(-\dfrac{1}{2}\right)\right) = -\dfrac{1}{4}(2)$

$= -\dfrac{2}{4} = \mathbf{-\dfrac{1}{2}}$

28. $-2(-3 - 2^0 - 2)(-2 + 5)(-2) = -2(-6)(3)(-2)$

$= \mathbf{-72}$

29. $-\dfrac{1}{-2^0} - \dfrac{1}{-2^2} - \dfrac{1}{-2^{-2}} = 1 + \dfrac{1}{4} + 4 = \mathbf{5\dfrac{1}{4}}$

30. $|-3^0| - |-2 - 3| + (-2^0)(-2 - 5)$

$= 1 - 5 + 7 = \mathbf{3}$

PROBLEM SET 12

1. If 40% were monochromatic, 60% were variegated.

$\dfrac{P}{100} \times of = is$

$\dfrac{60}{100} \times V = 2400$

$V = 2400 \cdot \dfrac{100}{60} = 4000$

Monochromatic = 4000 − 2400 = **1600**

2. $N \quad N + 1 \quad N + 2 \quad N + 3$

$2(N + N + 1 + N + 3) = 3(-N - 2) - 40$

$6N + 8 = -3N - 46$

$9N = -54$

$N = -6$

The desired integers are **−6, −5, −4**, and **−3**.

3. $F \times of = is$

$4\dfrac{1}{4} \times S = 5100$

$S = 5100 \cdot \dfrac{4}{17} = \mathbf{1200}$

4. $F \times of = is$

$2\dfrac{1}{5} \times N = 1$

$N = 1 \cdot \dfrac{5}{11} = \mathbf{\dfrac{5}{11}}$

5. $5(-N) + 25 = 8N + 90$

$-13N = 65$

$N = \mathbf{-5}$

6. If the train completed 30%, then 70% remained.

$\dfrac{P}{100} \times of = is$

$\dfrac{70}{100} \times TL = 6300$

$TL = 6300 \cdot \dfrac{100}{70} = \mathbf{9000\ miles}$

7. $z + (180 - 140) + (180 - 70) = 180$

$z + 40 + 110 = 180$

$z = \mathbf{30}$

8. $(4x + 25) + (7x - 20) = 360 - 40$

$11x + 5 = 320$

$11x = 315$

$x = \mathbf{\dfrac{315}{11}}$

$y = \dfrac{1}{2}(40) = \mathbf{20}$

9. $\text{Area}_{\text{Sector}} = \dfrac{60}{360}(\pi r^2) = 36\pi\ \text{cm}^2$

$r^2 = \dfrac{6(36\pi)\ \text{cm}^2}{\pi}$

$r = \sqrt{216}\ \text{cm} = 6\sqrt{6}\ \text{cm}$

Diameter $= 2r = 2(6\sqrt{6})\ \text{cm} = \mathbf{12\sqrt{6}\ cm}$

10. Relabel angles A, B, and C as having measures of $3x$, $2x$, and x, respectively.

$3x + 2x + x = 180$

$6x = 180$

$x = 30$

$A = \mathbf{90};\ B = \mathbf{60};\ C = \mathbf{30}$

11. $m + \dfrac{x}{c} + \dfrac{c}{x^2b} = \dfrac{mbcx^2}{bcx^2} + \dfrac{bx^3}{bcx^2} + \dfrac{c^2}{bcx^2}$

$= \dfrac{\mathbf{mbcx^2 + bx^3 + c^2}}{\mathbf{bcx^2}}$

12. $\dfrac{a}{b} - \dfrac{3b}{a^2} - \dfrac{2}{abc} = \dfrac{a^3c}{a^2bc} - \dfrac{3b^2c}{a^2bc} - \dfrac{2a}{a^2bc}$

$= \dfrac{\mathbf{a^3c - 3b^2c - 2a}}{\mathbf{a^2bc}}$

13. $1 + \dfrac{a}{b} = \dfrac{b}{b} + \dfrac{a}{b} = \dfrac{\mathbf{b + a}}{\mathbf{b}}$

14. $c^2 = a^2 + b^2$

$13^2 = 5^2 + k^2$

$169 = 25 + k^2$

$144 = k^2$

$12 = k$

15. $D^2 = 9^2 + 6^2$

$D^2 = 81 + 36$

$D^2 = 117$

$D = \sqrt{117}$

$D = 3\sqrt{13}$

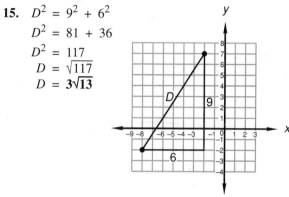

16. (a) $3y + x - 9 = 0$

$3y = -x + 9$

$y = -\dfrac{1}{3}x + 3$

(b) $x = 2$

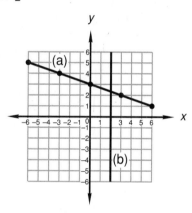

17. (a) The y intercept is +2. The slope is positive, and the rise over the run for any triangle drawn is $\frac{1}{3}$.

$y = \dfrac{1}{3}x + 2$

(b) Every point is 2 units below the x axis.

$y = -2$

18. $8\dfrac{1}{4} + 2\dfrac{1}{2}x = \dfrac{1}{8}$

$\dfrac{5}{2}x = \dfrac{1}{8} - \dfrac{66}{8}$

$\dfrac{5}{2}x = -\dfrac{65}{8}$

$x = -\dfrac{65}{8} \cdot \dfrac{2}{5} = -\dfrac{130}{40} = -\dfrac{13}{4}$

19. $0.001 + 0.02x - 0.1 = 0.002x$

$1 + 20x - 100 = 2x$

$18x = 99$

$x = \dfrac{11}{2} = \mathbf{5.5}$

20. $-3(-2 - 2^0x) - (-2) - 2(-2 - 3x) = -2(x + 4)$

$6 + 3x + 2 + 4 + 6x = -2x - 8$

$9x + 12 = -2x - 8$

$11x = -20$

$x = -\dfrac{20}{11}$

21. $\dfrac{a^{-3}x}{y^{-3}}\left(\dfrac{xxx^{-2}}{y^{-2}yy} - 3\right) = \dfrac{a^{-3}xxxx^{-2}}{y^{-3}y^{-2}yy} - \dfrac{3a^{-3}x}{y^{-3}}$

$= a^{-3}xy^3 - 3a^{-3}xy^3 = \mathbf{-2a^{-3}xy^3}$

22. $\dfrac{x^0x^{-2}}{y}\left(x^2y - \dfrac{2x^2y}{x^4}\right)$

$= \dfrac{x^0x^{-2}x^2y}{y} - \dfrac{2x^0x^{-2}x^2y}{yx^4} = \mathbf{1 - 2x^{-4}}$

23. $\dfrac{x^{-2}(2x^{-3})y^2y^0}{x^{-3}yx^2y^{-7}} = \mathbf{2x^{-4}y^8}$

24. $\dfrac{a^0ba^2a^{-1}a}{(a^2b^{-2})^{-3}} = \dfrac{ba^2}{a^{-6}b^6} = \mathbf{a^8b^{-5}}$

25. $\dfrac{a}{x} - \dfrac{3a^2y^0x^{-1}}{a} + \dfrac{4x^{-1}}{aa^{-2}} = \dfrac{a}{x} - \dfrac{3a}{x} + \dfrac{4a}{x}$

$= \mathbf{2ax^{-1}}$

26. $abc - \dfrac{5a^2c^3}{ab^{-1}c^2} - \dfrac{3}{a^{-1}b^{-1}c^{-1}}$

$= abc - 5abc - 3abc = \mathbf{-7abc}$

27. $a^2(a^0 - ab) = \left(-\dfrac{1}{2}\right)^2\left(\left(-\dfrac{1}{2}\right)^0 - \left(-\dfrac{1}{2}\right)\left(\dfrac{3}{4}\right)\right)$

$= \dfrac{1}{4}\left(1 + \dfrac{3}{8}\right) = \dfrac{1}{4}\left(\dfrac{11}{8}\right) = \mathbf{\dfrac{11}{32}}$

28. $x(x^2 - x^2y) = -\dfrac{1}{2}\left(\left(-\dfrac{1}{2}\right)^2 - \left(-\dfrac{1}{2}\right)^2\left(\dfrac{1}{2}\right)\right)$

$= -\dfrac{1}{2}\left(\dfrac{1}{4} - \dfrac{1}{8}\right) = -\dfrac{1}{2}\left(\dfrac{1}{8}\right) = \mathbf{-\dfrac{1}{16}}$

29. $-ab(a - a^2b - b)$

$= -2(-3)(2 - (2)^2(-3) - (-3))$

$= 6(2 + 12 + 3) = 6(17) = \mathbf{102}$

30. $-2^0(-2^0 - 3^0 - |-2|) - (-2)(-3)$

$- \dfrac{1}{3^{-2}} - 3^2 - 3$

$= -1(-1 - 1 - 2) - 6 - 9 - 9 - 3$

$= -1(-4) - 6 - 9 - 9 - 3$

$= 4 - 6 - 9 - 9 - 3 = \mathbf{-23}$

PROBLEM SET 13

1. If 60% had blue sails, 40% did not have blue sails.

$$\frac{P}{100} \times of = is$$

$$\frac{40}{100} \times B = 300$$

$$B = 300 \cdot \frac{100}{40} = 750$$

Blue sails = $750 - 300 = $ **450**

2. $F \times of = is$

$$\frac{3}{16} \times C = 93{,}750$$

$$C = 93{,}750 \cdot \frac{16}{3} = \textbf{500{,}000}$$

3. $N \quad N + 2 \quad N + 4 \quad N + 6$

$$3(N + N + 6) = 5(N + 4) + 14$$
$$6N + 18 = 5N + 34$$
$$N = 16$$

The desired integers are **16**, **18**, **20** and **22**.

4. $3(N + 14) = 2(-N) + 67$
$$3N + 42 = -2N + 67$$
$$5N = 25$$
$$N = \textbf{5}$$

5. $A_{\text{Shaded}} = A_{\text{Rectangle}} + A_{\text{Triangle}}$

$$= (1)(4) + \frac{1}{2}(4)(3)$$

$$= 4 + 6 = \textbf{10 units}^2$$

6. $(10x + 10) + (6x + 10) = 180$
$$16x + 20 = 180$$
$$16x = 160$$
$$x = \textbf{10}$$

$$4 \times \overrightarrow{SF} = \frac{9}{2}$$

$$\overrightarrow{SF} = \frac{9}{2} \cdot \frac{1}{4} = \frac{9}{8}$$

$$5 \times \overrightarrow{SF} = y$$

$$5 \cdot \frac{9}{8} = y$$

$$\frac{45}{8} = y$$

Since lines are parallel:

$$P = 10x + 10$$
$$P = 10(10) + 10$$
$$P = \textbf{110}$$

7. $5^2 = (AB)^2 + 4^2$
$$25 = (AB)^2 + 16$$
$$9 = (AB)^2$$
$$3 = AB$$

$$AC - AB = BC$$
$$12 - 3 = BC$$
$$9 = BC$$

$$(DC)^2 = 9^2 + 4^2$$
$$(DC)^2 = 81 + 16$$
$$DC = \sqrt{97}$$

8. (a) $x = y + 1$
(b) $3x + 2y = 8$
Substitute (a) into (b) and get:
(b) $3(y + 1) + 2y = 8$
$$3y + 3 + 2y = 8$$
$$5y = 5$$
$$y = 1$$
(a) $x = y + 1$
$$x = (1) + 1 = 2$$
(2, 1)

9. (a) $3x - y = 22$
(a′) $y = 3x - 22$
(b) $2x + 3y = -11$
Substitute (a′) into (b) and get:
(b) $2x + 3(3x - 22) = -11$
$$11x - 66 = -11$$
$$11x = 55$$
$$x = 5$$
(a′) $y = 3(5) - 22 = -7$
(5, −7)

10. (a) $x + y = 20$
(a′) $y = -x + 20$
(b) $5x + 10y = 200$
Substitute (a′) into (b) and get:
(b) $5x + 10(-x + 20) = 200$
$$-5x + 200 = 200$$
$$-5x = 0$$
$$x = 0$$
(a′) $y = -(0) + 20 = 20$
(0, 20)

11. (a) $x + y = 20$
(a′) $y = -x + 20$
(b) $25x + 10y = 395$
Substitute (a′) into (b) and get:
(b) $25x + 10(-x + 20) = 395$
$$15x + 200 = 395$$
$$15x = 195$$
$$x = 13$$
(a′) $y = -(13) + 20 = 7$
(13, 7)

12. $4 + \dfrac{2}{a} = \dfrac{4a}{a} + \dfrac{2}{a} = \dfrac{4a + 2}{a}$

13. $\dfrac{a^2}{k} + k + \dfrac{k}{4} = \dfrac{4a^2}{4k} + \dfrac{4k^2}{4k} + \dfrac{k^2}{4k}$

$\quad = \dfrac{4a^2 + 5k^2}{4k}$

14. $m^2 + \dfrac{m}{p} + \dfrac{m}{ap^2} = \dfrac{am^2 p^2}{ap^2} + \dfrac{amp}{ap^2} + \dfrac{m}{ap^2}$

$\quad = \dfrac{am^2 p^2 + amp + m}{ap^2}$

15. $D^2 = 6^2 + 4^2$

$\quad D^2 = 36 + 16$

$\quad D^2 = 52$

$\quad D = \sqrt{52}$

$\quad D = 2\sqrt{13}$

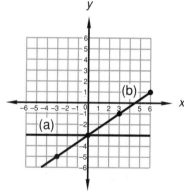

16. (a) $y = -3$

(b) $2x - 3y = 9$

$\quad -3y = -2x + 9$

$\quad y = \dfrac{2}{3}x - 3$

17. (a) Every point is 5 units to the right of the y axis.

$\quad x = 5$

(b) The y intercept is -2. The slope is negative, and the rise over the run for any triangle drawn is $-\dfrac{2}{3}$.

$\quad y = -\dfrac{2}{3}x - 2$

18. $3\dfrac{1}{2}x + 4\dfrac{1}{3} = 7\dfrac{2}{9}$

$\quad \dfrac{7}{2}x = \dfrac{65}{9} - \dfrac{39}{9}$

$\quad x = \dfrac{26}{9} \cdot \dfrac{2}{7} = \dfrac{52}{63}$

19. $0.03 + 0.03x = 0.003$

$\quad 30 + 30x = 3$

$\quad 30x = -27$

$\quad x = -\dfrac{27}{30} = -0.9$

20. $-3^0 - 3^2 - (-2x - |2|) = 7x$

$\quad -1 - 9 + 2x + 2 = 7x$

$\quad -8 = 5x$

$\quad -\dfrac{8}{5} = x$

21. $\dfrac{x^2 a}{3}\left(\dfrac{9a^{-1}}{x^2} - \dfrac{2xa^2}{xa}\right) = \dfrac{3x^2 aa^{-1}}{x^2} - \dfrac{2x^2 axa^2}{3xa}$

$\quad = 3 - \dfrac{2x^2 a^2}{3}$

22. $\dfrac{-2a^0 p}{m}\left(\dfrac{mp}{a^{-2}} - \dfrac{2a^{-4}a}{p^2 m}\right)$

$\quad = \dfrac{-2a^0 pmp}{ma^{-2}} + \dfrac{4a^0 pa^{-4}a}{mp^2 m}$

$\quad = -2a^2 p^2 + 4a^{-3}m^{-2}p^{-1}$

23. $\dfrac{xa^2 (x^0 a^{-2})^4}{(2x^{-2})^{-2}} = \dfrac{4xa^{-6}}{x^4} = 4x^{-3}a^{-6}$

24. $\dfrac{m^2 pxx^{-4}(x^{-2})^2}{(3p^{-2})^{-2} xpx} = \dfrac{9m^2 px^{-7}}{p^5 x^2} = 9m^2 p^{-4}x^{-9}$

25. $xa - \dfrac{3x^2 a^3}{xa^2} + \dfrac{2x}{a^{-1}} = xa - 3xa + 2xa = 0$

26. $\dfrac{amp^{-1}}{m^{-1}} - \dfrac{3a^2 m^2}{pa} + \dfrac{5pa}{m^2}$

$\quad = am^2 p^{-1} - 3am^2 p^{-1} + 5pam^{-2}$

$\quad = -2am^2 p^{-1} + 5pam^{-2}$

27. $-x^3 - x^2 - x(a - x)$

$\quad = -\left(-\dfrac{1}{2}\right)^3 - \left(-\dfrac{1}{2}\right)^2 - \left(-\dfrac{1}{2}\right)\left(2 - \left(-\dfrac{1}{2}\right)\right)$

$\quad = \dfrac{1}{8} - \dfrac{1}{4} + \dfrac{1}{2}\left(2 + \dfrac{1}{2}\right) = \dfrac{1}{8} - \dfrac{1}{4} + \dfrac{1}{2}\left(\dfrac{5}{2}\right)$

$\quad = \dfrac{1}{8} - \dfrac{2}{8} + \dfrac{10}{8} = \dfrac{9}{8}$

28. $a^2 - ax(x - ax) = 4^2 - 4(-3)(-3 - (4)(-3))$

$\quad = 16 + 12(-3 + 12) = 16 + 12(9) = 124$

29. $\dfrac{1}{2^{-3}} - \dfrac{2}{-2^{-2}} - \dfrac{1}{(-2)^{-2}} - 2^0$

$\quad = 8 + 8 - 4 - 1 = 11$

30. $3^0 - 2(-2) - |-2 - 4^0 - 3| = 1 + 4 - 6$

$\quad = -1$

PROBLEM SET 14

1. $\dfrac{P}{100} \times of = is$

$\dfrac{260}{100} \times R = 6578$

$R = 6578 \cdot \dfrac{100}{260} = \mathbf{2530}$

2. $\dfrac{P}{100} \times of = is$

$\dfrac{350}{100} \times 4900 = S$

$S = \dfrac{350}{100} \cdot 4900 = \mathbf{17{,}150}$

3. $-3(2N + 7) = 3(-N) + 9$

$-6N - 21 = -3N + 9$

$-3N = 30$

$N = \mathbf{-10}$

4. $N \qquad N + 2 \qquad N + 4$

$4(N + 2 + N + 4) = 10N + 2$

$8N + 24 = 10N + 2$

$-2N = -22$

$N = 11$

The desired integers are **11**, **13**, and **15**.

5. (a) $3x - 3y = 21$

(b) $2x - y = 12$

(b') $y = 2x - 12$

Substitute (b') into (a) and get:

(a) $3x - 3(2x - 12) = 21$

$3x - 6x + 36 = 21$

$-3x = -15$

$x = 5$

(b') $y = 2(5) - 12 = -2$

(5, –2)

6. (a) $4x - y = 22$

(a') $y = 4x - 22$

(b) $2x + 3y = 4$

Substitute (a') into (b) and get:

(b) $2x + 3(4x - 22) = 4$

$2x + 12x - 66 = 4$

$14x = 70$

$x = 5$

(a') $y = 4(5) - 22 = -2$

(5, –2)

7. (a) $x + y = 28$

(a') $y = -x + 28$

(b) $5x + 10y = 230$

Substitute (a') into (b) and get:

(b) $5x + 10(-x + 28) = 230$

$5x - 10x + 280 = 230$

$-5x = -50$

$x = 10$

(a') $y = -(10) + 28 = 18$

(10, 18)

8. (a) $x + y = 22$

(a') $x = -y + 22$

(b) $100x + 25y = 2050$

Substitute (a') into (b) and get:

(b) $100(-y + 22) + 25y = 2050$

$-100y + 2200 + 25y = 2050$

$-75y = -150$

$y = 2$

(a') $x = -(2) + 22 = 20$

(20, 2)

9. $x + \dfrac{x^2}{y} - \dfrac{3x}{cy^2} = \dfrac{cy^2x}{cy^2} + \dfrac{cyx^2}{cy^2} - \dfrac{3x}{cy^2}$

$= \dfrac{\mathbf{cy^2x + cyx^2 - 3x}}{cy^2}$

10. $\dfrac{m}{x} + 4 = \dfrac{m}{x} + \dfrac{4x}{x} = \dfrac{\mathbf{m + 4x}}{\mathbf{x}}$

11. $4 + \dfrac{c}{x} - cxy = \dfrac{4x}{x} + \dfrac{c}{x} - \dfrac{cx^2y}{x}$

$= \dfrac{\mathbf{4x + c - cx^2y}}{\mathbf{x}}$

12. $9^2 = H^2 + 2^2$

$81 = H^2 + 4$

$77 = H^2$

$\sqrt{77} = H$

$\text{Area} = \dfrac{B \times H}{2} = \dfrac{\left(4 \times \sqrt{77}\right)}{2} = \mathbf{2\sqrt{77}\ cm^2}$

13. $D^2 = 8^2 + 0^2$

$D^2 = 64$

$D = \sqrt{64}$

$D = \mathbf{8}$

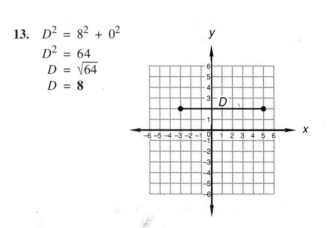

14. (a) $x = -4$

(b) $3y + 2x = 6$

$$3y = -2x + 6$$

$$y = -\frac{2}{3}x + 2$$

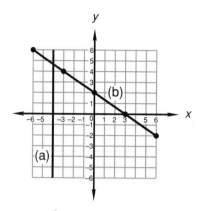

15. (a) The y intercept is $+4$. The slope is positive, and the rise over the run for any triangle drawn is $\frac{2}{3}$.

$$y = \frac{2}{3}x + 4$$

(b) Every point on this line is 4 units below the x axis.

$$y = -4.$$

16. Graph the line to find the slope.

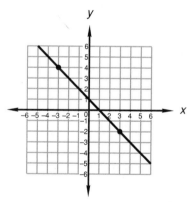

Slope $= \dfrac{-1}{1} = -1$

$y = -1x + b$

$4 = -1(-3) + b$

$1 = b$

Since $m = -1$ and $b = 1$: $y = -x + 1$

17. $y = -\dfrac{3}{4}x + b$

$6 = -\dfrac{3}{4}(-4) + b$

$3 = b$

Since $m = -\dfrac{3}{4}$ and $b = 3$: $y = -\dfrac{3}{4}x + 3$

18. $A + 134 = 180$

$$A = 46$$

$$B + 46 + 90 = 180$$

$$B = 44$$

$$C + 44 = 180$$

$$C = 136$$

19. $A = B = \dfrac{1}{2} \cdot 30 = 15$

$A = B = 15$

$C + 150 = 180$

$$C = 30$$

Since the measure of an arc of a circle is the same as the measure of the central angle: $D = C = 30$

20. $2\dfrac{1}{4}x - \dfrac{3}{5} = -1\dfrac{1}{20}$

$$\dfrac{9}{4}x = -\dfrac{21}{20} + \dfrac{12}{20}$$

$$x = -\dfrac{9}{20} \cdot \dfrac{4}{9} = -\dfrac{36}{180} = -\dfrac{1}{5}$$

21. $0.005x - 0.05 = 0.5$

$$5x - 50 = 500$$

$$5x = 550$$

$$x = 110$$

22. $-2^0 - 3^2 = -2(x - 3^0) - 4(2x - 5)$

$$-1 - 9 = -2x + 2 - 8x + 20$$

$$-10 = -10x + 22$$

$$-32 = -10x$$

$$x = \dfrac{16}{5}$$

23. $\dfrac{-3^{-2}x}{y}\left(\dfrac{9y^0 x}{-x} - \dfrac{3x}{y}\right) = \dfrac{xy^0 x}{yx} + \dfrac{3xx}{9yy}$

$$= \dfrac{x}{y} + \dfrac{x^2}{3y^2}$$

24. $\dfrac{(-2x)^{-2}xy^0y^2(x)^{-2}}{(x^2y)^{-2}xyxy^{-2}} = \dfrac{x^{-2}xy^2x^{-2}}{4x^{-4}y^{-2}xyxy^{-2}}$

$$= \dfrac{x^{-3}y^2}{4x^{-2}y^{-3}} = \dfrac{y^5}{4x}$$

25. $mp^2\left(m^{-1}p^{-2} - \dfrac{4m}{p^2}\right) = mp^2m^{-1}p^{-2} - \dfrac{4mp^2m}{p^2}$

$$= 1 - 4m^2$$

26. $-\dfrac{3xy}{m} + \dfrac{7x^2x^{-1}m^{-1}}{y} - \dfrac{8m^{-1}x^{-1}}{x^{-2}y^{-1}}$

$$= -\dfrac{3xy}{m} + \dfrac{7x}{ym} - \dfrac{8xy}{m} = -\dfrac{11xy}{m} + \dfrac{7x}{ym}$$

27. $x^2 - y - xy^2 = \left(-\dfrac{1}{3}\right)^2 - \dfrac{1}{2} - \left(-\dfrac{1}{3}\right)\left(\dfrac{1}{2}\right)^2$

$= \dfrac{1}{9} - \dfrac{1}{2} + \dfrac{1}{12} = \dfrac{4}{36} - \dfrac{18}{36} + \dfrac{3}{36} = -\dfrac{\mathbf{11}}{\mathbf{36}}$

28. $a^2x - a(xa - a)$

$= (-1)^2(-3) - (-1)(-3(-1) - (-1))$

$= -3 + 1(3 + 1) = -3 + 4 = \mathbf{1}$

29. $-5^0(-2 - 3^0) - 2^2 - \dfrac{1}{(-2)^{-2}}$

$= 3 - 4 - 4 = \mathbf{-5}$

30. $5 - |-2 - 3| - 4^0 - 2|-2^0| - 3$

$= 5 - 5 - 1 - 2 - 3 = \mathbf{-6}$

PROBLEM SET 15

1. $\dfrac{P}{100} \times of = is$

$\dfrac{20}{100} \times P = 1400$

$P = 1400 \cdot \dfrac{100}{20} = \mathbf{7000}$

2. $\dfrac{WP}{100} \times of = is$

$\dfrac{WP}{100} \times 1000 = 200$

$WP = 200 \cdot \dfrac{100}{1000} = \mathbf{20\ percent}$

3. $5(-N) - 7 = 2N - 35$

$-5N - 7 = 2N - 35$

$-7N = -28$

$N = \mathbf{4}$

4. $N \quad N + 1 \quad N + 2$

$-4(N + N + 2) = 7(-N - 1) + 12$

$-8N - 8 = -7N + 5$

$-N = 13$

$N = -13$

The desired integers are **-13**, **-12**, and **-11**.

5. (a) $3x + y = 11$

(b) $3x - 2y = 2$

2(a) $6x + 2y = 22$

(b) $\underline{3x - 2y = 2}$

$9x = 24$

$x = \dfrac{8}{3}$

(a) $3\left(\dfrac{8}{3}\right) + y = 11$

$8 + y = 11$

$y = 3$

$\left(\dfrac{8}{3},\ 3\right)$

6. (a) $3x + 4y = 20$

(b) $-4x + 3y = 15$

4(a) $12x + 16y = 80$

3(b) $\underline{-12x + 9y = 45}$

$25y = 125$

$y = 5$

(a) $3x + 4(5) = 20$

$3x + 20 = 20$

$3x = 0$

$x = 0$

(0, 5)

7. (a) $3x + y = 16$

(a') $y = -3x + 16$

(b) $2x - 3y = -4$

Substitute (a') into (b) and get:

(b) $2x - 3(-3x + 16) = -4$

$2x + 9x - 48 = -4$

$11x = 44$

$x = 4$

(a') $y = -3(4) + 16 = 4$

(4, 4)

8. (a) $x + 3y = -9$

(a') $x = -3y - 9$

(b) $5x - 2y = 23$

Substitute (a') into (b) and get:

(b) $5(-3y - 9) - 2y = 23$

$-15y - 45 - 2y = 23$

$-17y = 68$

$y = -4$

(a') $x = -3(-4) - 9 = 3$

(3, -4)

9. $y + \dfrac{x}{a^2} - \dfrac{mx}{3y^2} = \dfrac{3a^2y^3}{3a^2y^2} + \dfrac{3xy^2}{3a^2y^2} - \dfrac{a^2mx}{3a^2y^2}$

$= \dfrac{\mathbf{3a^2y^3 + 3xy^2 - a^2mx}}{\mathbf{3a^2y^2}}$

10. $4 - \dfrac{3a}{x} = \dfrac{4x}{x} - \dfrac{3a}{x} = \dfrac{\mathbf{4x - 3a}}{\mathbf{x}}$

11. $c + \dfrac{c^2}{x} + ac^2 = \dfrac{cx}{x} + \dfrac{c^2}{x} + \dfrac{ac^2x}{x}$

$= \dfrac{\mathbf{cx + c^2 + ac^2x}}{\mathbf{x}}$

12. $c^2 = a^2 + b^2$

$m^2 = 4^2 + 8^2$

$m^2 = 16 + 64$

$m^2 = 80$

$m = \sqrt{80} = \mathbf{4\sqrt{5}}$

13. $D^2 = 6^2 + 7^2$
$D^2 = 36 + 49$
$D^2 = 85$
$D = \sqrt{85}$

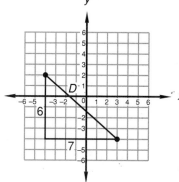

$\text{Slope} = \dfrac{-5}{-6} = \dfrac{5}{6}$

$y = \dfrac{5}{6}x + b$

$-2 = \dfrac{5}{6}(-2) + b$

$-2 = \dfrac{-10}{6} + b$

$\dfrac{5}{3} - \dfrac{6}{3} = b$

$-\dfrac{1}{3} = b$

Since $m = \dfrac{5}{6}$ and $b = -\dfrac{1}{3}$:

$y = \dfrac{5}{6}x - \dfrac{1}{3}$

14. (a) $y = -3$
(b) $3x - 4y = 8$
$-4y = -3x + 8$
$y = \dfrac{3}{4}x - 2$

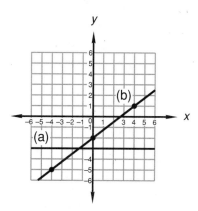

17.
$y = -\dfrac{1}{7}x + b$

$5 = -\dfrac{1}{7}(-2) + b$

$5 = \dfrac{2}{7} + b$

$-\dfrac{2}{7} + \dfrac{35}{7} = b$

$\dfrac{33}{7} = b$

Since $m = -\dfrac{1}{7}$ and $b = \dfrac{33}{7}$:

$y = -\dfrac{1}{7}x + \dfrac{33}{7}$

15. (a) Every point on this line is $\dfrac{9}{2}$ units above the x axis.

$y = \dfrac{9}{2}$

(b) The y intercept is -1. The slope is negative, and the rise over the run for any triangle drawn is $-\dfrac{2}{3}$.

$y = -\dfrac{2}{3}x - 1$

16. Graph the line to find the slope.

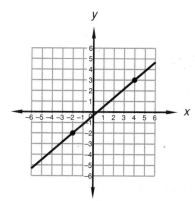

18. $A + 134 = 180$
$A = 46$

$B + 46 + 90 = 180$
$B = 44$

Since vertical angles are equal angles:
$C = B = 44$

$k + 44 + 90 = 180$
$k = 46$

Since vertical angles are equal angles:
$y = k = 46$

19. $A + B + 140 = 180$
$A + B = 40$
Since angles opposite equal sides are equal angles:
$A = B = 20$

$C + 140 = 180$
$C = 40$

Since the measure of an arc of a circle is the same as the measure of the central angle: $D = C = 40$

20. $3x + 2 = \dfrac{9x - 17}{2}$

$6x + 4 = 9x - 17$

$-3x = -21$

$x = 7$

$P = 3(7) + 2 = 23$

21. $\dfrac{-3x^{-1}y^2}{p}\left(\dfrac{-px}{y^2} - \dfrac{3x^2y}{p^{-3}}\right)$

$= \dfrac{3x^{-1}y^2\,px}{py^2} + \dfrac{9\,x^{-1}y^2x^2y}{pp^{-3}} = 3 + 9xy^3p^2$

22. $3\dfrac{2}{5}x - \dfrac{4}{5} = 7\dfrac{3}{10}$

$\dfrac{17}{5}x = \dfrac{73}{10} + \dfrac{8}{10}$

$\dfrac{17}{5}x = \dfrac{81}{10}$

$x = \dfrac{81}{10} \cdot \dfrac{5}{17} = \dfrac{405}{170} = \dfrac{81}{34}$

23. $0.07x - 0.02 = 0.4$

$7x - 2 = 40$

$7x = 42$

$x = 6$

24. $-3x(-2 - 3^0) = -2^0(-x - 3x^0)$

$9x = x + 3$

$8x = 3$

$x = \dfrac{3}{8}$

25. $\dfrac{x^2y^{-2}x^0(y^{-1})}{x^2y^2(x^0)^2} = \dfrac{x^2y^{-3}}{x^2y^2} = y^{-5}$

26. $\dfrac{8x^2yy}{m^{-2}x^2} + \dfrac{3y^2}{m^{-2}} - \dfrac{5m^2x^{-2}}{x^{-2}y^{-2}}$

$= 8m^2y^2 + 3m^2y^2 - 5m^2y^2 = 6m^2y^2$

27. $axy - a^2x = \left(-\dfrac{1}{2}\right)(3)\left(-\dfrac{1}{3}\right) - \left(-\dfrac{1}{2}\right)^2(3)$

$= \dfrac{1}{2} - \dfrac{3}{4} = \dfrac{2}{4} - \dfrac{3}{4} = -\dfrac{1}{4}$

28. $a^0(a^0x - ax) = (-2)^0((-2)^0(-3) - (-2)(-3))$

$= 1(-3 - 6) = -9$

29. $-2^0(2^0 - 3^2) - [-2 - 3 - (-2)]$

$= -1(1 - 9) - [-3] = -1(-8) + 3$

$= 8 + 3 = 11$

30. $3^0 - \dfrac{3}{3^{-2}} + 2 - 5^0 = 1 - 27 + 2 - 1 = -25$

---

1. $F \times of = is$

$2\dfrac{1}{7} \times AN = 900$

$AN = 900 \cdot \dfrac{7}{15} = 420$

2. $WD \times of = is$

$0.016 \times 420{,}000 = D$

$6720 = D$

3. If 14% were uninvited, 86% had been invited.

$\dfrac{P}{100} \times of = is$

$\dfrac{86}{100} \times GA = 903$

$GA = 903 \cdot \dfrac{100}{86} = 1050$

4. $\dfrac{P}{100} \times of = is$

$\dfrac{40}{100} \times C = 8600$

$C = 8600 \cdot \dfrac{100}{40} = 21{,}500$

$N_P = 21{,}500 - N_M = 21{,}500 - 8600 = 12{,}900$

5. (a) $4x - 3y = -1$

(b) $2x + 5y = 19$

$\begin{array}{rl} 1(a) & 4x - 3y = -1 \\ -2(b) & -4x - 10y = -38 \\ \hline & -13y = -39 \\ & y = 3 \end{array}$

(b) $2x + 5(3) = 19$

$2x = 4$

$x = 2$

$(2, 3)$

6. (a) $7x - 2y = 13$

(b) $4x + 7y = 40$

$\begin{array}{rl} 7(a) & 49x - 14y = 91 \\ 2(b) & 8x + 14y = 80 \\ \hline & 57x = 171 \\ & x = 3 \end{array}$

(b) $4(3) + 7y = 40$

$7y = 28$

$y = 4$

$(3, 4)$

7. (a) $3x + y = 2$

(a') $y = -3x + 2$

(b) $2x - 5y = 7$

Substitute (a') into (b) and get:

(b) $2x - 5(-3x + 2) = 7$

$$2x + 15x - 10 = 7$$
$$17x = 17$$
$$x = 1$$

(a') $y = -3(1) + 2 = -1$

(1, –1)

8. (a) $x - 3y = 4$

(a') $x = 3y + 4$

(b) $4x - 7y = 16$

Substitute (a') into (b) and get:

(b) $4(3y + 4) - 7y = 16$
$$12y + 16 - 7y = 16$$
$$5y = 0$$
$$y = 0$$

(a') $x = 3(0) + 4 = 4$

(4, 0)

9. $(2x + 3)(2x^2 - 4x + 3)$
$$= 4x^3 - 8x^2 + 6x + 6x^2 - 12x + 9$$
$$= \mathbf{4x^3 - 2x^2 - 6x + 9}$$

10.

$$\begin{array}{r} 4x^2 + 10x + 34 + \frac{104}{x-3} \\ x - 3 \overline{)\ 4x^3 - 2x^2 + 4x + 2} \\ \underline{4x^3 - 12x^2} \\ 10x^2 + 4x \\ \underline{10x^2 - 30x} \\ 34x + 2 \\ \underline{34x - 102} \\ 104 \end{array}$$

Check: $\dfrac{4x^2(x - 3)}{x - 3} + \dfrac{10x(x - 3)}{x - 3}$

$+ \dfrac{34(x - 3)}{x - 3} + \dfrac{104}{x - 3}$

$= \dfrac{4x^3 - 2x^2 + 4x + 2}{x - 3}$

11.

$$\begin{array}{r} x^2 + 4x + 16 + \frac{56}{x-4} \\ x - 4 \overline{)\ x^3 + 0x^2 + 0x - 8} \\ \underline{x^3 - 4x^2} \\ 4x^2 + 0x \\ \underline{4x^2 - 16x} \\ 16x - 8 \\ \underline{16x - 64} \\ 56 \end{array}$$

Check: $\dfrac{x^2(x - 4)}{x - 4} + \dfrac{4x(x - 4)}{x - 4}$

$+ \dfrac{16(x - 4)}{x - 4} + \dfrac{56}{x - 4}$

$= \dfrac{x^3 - 4x^2 + 4x^2 - 16x + 16x - 8}{x - 4}$

$= \dfrac{x^3 - 8}{x - 4}$

12. $2 + \dfrac{a}{2x^2} = \dfrac{4x^2}{2x^2} + \dfrac{a}{2x^2} = \dfrac{4x^2 + a}{2x^2}$

13. $\dfrac{4}{cx} + c - \dfrac{3}{4c^2x} = \dfrac{16c}{4c^2x} + \dfrac{4c^3x}{4c^2x} - \dfrac{3}{4c^2x}$

$= \dfrac{16c + 4c^3x - 3}{4c^2x}$

14. $2^2 = H^2 + 1^2$

$4 = H^2 + 1$

$3 = H^2$

$\sqrt{3} = H$

Area $= \dfrac{b \times H}{2} = \dfrac{2 \times \sqrt{3}}{2} = \sqrt{3}$ in.^{2}

15. $D^2 = 7^2 + 2^2$

$D^2 = 49 + 4$

$D^2 = 53$

$D = \sqrt{53}$

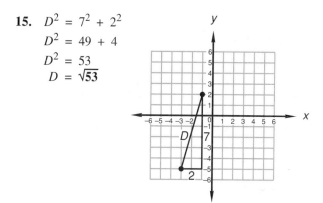

16. (a) $x = -2\dfrac{1}{2}$

(b) $x - 3y = 6$

$-3y = -x + 6$

$y = \dfrac{1}{3}x - 2$

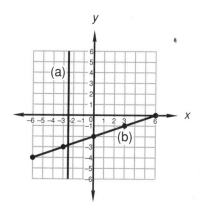

17. (a) Every point on this line is 4 units below the x axis.

$y = -4$

(b) The y intercept is $+3$. The slope is negative, and the rise over the run for any triangle drawn is

$-\dfrac{1}{2}$.

$y = -\dfrac{1}{2}x + 3$

18. Graph the line to find the slope.

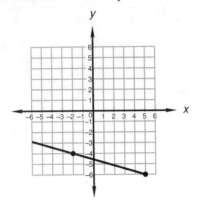

Slope $= \dfrac{+2}{-7} = -\dfrac{2}{7}$

$y = -\dfrac{2}{7}x + b$

$-4 = -\dfrac{2}{7}(-2) + b$

$-4 = \dfrac{4}{7} + b$

$-\dfrac{28}{7} - \dfrac{4}{7} = b$

$-\dfrac{32}{7} = b$

Since $m = -\dfrac{2}{7}$ and $b = -\dfrac{32}{7}$:

$y = -\dfrac{2}{7}x - \dfrac{32}{7}$

19.

$y = \dfrac{3}{5}x + b$

$1 = \dfrac{3}{5}(-4) + b$

$1 = -\dfrac{12}{5} + b$

$\dfrac{5}{5} + \dfrac{12}{5} = b$

$\dfrac{17}{5} = b$

Since $m = \dfrac{3}{5}$ and $b = \dfrac{17}{5}$:

$y = \dfrac{3}{5}x + \dfrac{17}{5}$

20. $\text{Area}_S = \pi r_S{}^2 = \pi \text{ cm}^2$

$r_S = \mathbf{1\ cm}$

$r_L = 2r_S = \mathbf{2\ cm}$

$\text{Area}_L = \pi r_L{}^2 = \pi(2 \text{ cm})^2 = \mathbf{4\pi\ cm^2}$

21. Since angles opposite equal sides are equal angles:
$A = \mathbf{30}$

$B + 30 + 30 = 180$
$B = \mathbf{120}$

$C + 120 = 180$
$\qquad C = \mathbf{60}$
Since the measure of an arc of a circle is the same as the measure of the central angle: $D = C = \mathbf{60}$

22. $3\dfrac{1}{2}x - \dfrac{1}{5} = 3\dfrac{3}{20}$

$\dfrac{7}{2}x = \dfrac{63}{20} + \dfrac{4}{20}$

$\dfrac{7}{2}x = \dfrac{67}{20}$

$x = \dfrac{67}{20} \cdot \dfrac{2}{7} = \dfrac{134}{140} = \dfrac{\mathbf{67}}{\mathbf{70}}$

23. $0.05m - 0.05 = 0.5$
$\qquad 5m - 5 = 50$
$\qquad\quad 5m = 55$
$\qquad\quad\ m = \mathbf{11}$

24. $-4(-x - 2) - 3(-2 - 4) = -4^0(x - 2)$
$\qquad\quad 4x + 8 + 18 = -x + 2$
$\qquad\qquad\qquad\ 5x = -24$

$\qquad\qquad\qquad\ x = -\dfrac{\mathbf{24}}{\mathbf{5}}$

25. $\dfrac{3^{-2}x^{-2}y}{z}\left(\dfrac{18x^2z}{y} - \dfrac{3xyz^2}{x}\right)$

$= \dfrac{2x^{-2}yx^2z}{zy} - \dfrac{x^{-2}yxyz^2}{3xz} = \mathbf{2 - \dfrac{y^2z}{3\,x^2}}$

26. $\dfrac{3^{-2}x^0(x^{-2}y^4)^{-3}}{xxy^2y^0(y^3)^{-4}} = \dfrac{x^6y^{-12}}{9x^2y^{-10}} = \dfrac{\mathbf{x^4}}{\mathbf{9\,y^2}}$

27. $\dfrac{3x^2xxy}{y^{-4}} - \dfrac{2x^2}{(x^{-1})^2y^{-5}} - \dfrac{7xxyx^2}{(y^{-2})^2}$

$= 3x^4y^5 - 2x^4y^5 - 7x^4y^5 = \mathbf{-6x^4y^5}$

28. $p^2 - xp - x(p - x)$

$= \left(\dfrac{1}{2}\right)^2 - (-2)\left(\dfrac{1}{2}\right) - (-2)\left(\dfrac{1}{2} - (-2)\right)$

$= \dfrac{1}{4} + 1 + 2\left(\dfrac{5}{2}\right) = \dfrac{1}{4} + \dfrac{4}{4} + \dfrac{20}{4} = \dfrac{\mathbf{25}}{\mathbf{4}}$

29. $-2^0(-2 - 3) - (-2)^0 - 2[-1 - (-3 - 7)$
$\qquad - 2(-5 + 3)]$
$= 5 - 1 - 2[-1 + 10 + 4]$
$= 5 - 1 - 2[13] = 5 - 1 - 26 = \mathbf{-22}$

30. $2^0 - \dfrac{27}{3^2} - 3^1 - 3^0 - |-3^0 - 2| - (-3)$

$= 1 - 3 - 3 - 1 - 3 + 3 = \mathbf{-6}$

PROBLEM SET 17

1. If $\frac{2}{5}$ doubted, then $\frac{3}{5}$ were nondoubters.

$F \times \text{of} = \text{is}$

$$\frac{3}{5} \times S = 600$$

$$S = 600 \cdot \frac{5}{3} = \mathbf{1000}$$

2. $F \times \text{of} = \text{is}$

$$\frac{7}{13} \times T = 210$$

$$T = 210 \cdot \frac{13}{7} = \mathbf{390}$$

3. $\dfrac{P}{100} \times \text{of} = \text{is}$

$$\frac{360}{100} \cdot 400 = S$$

$$S = \mathbf{1440}$$

4. $N \qquad N + 1 \qquad N + 2$

$7(N + 1 + N + 2) = 10N + 109$

$14N + 21 = 10N + 109$

$4N = 88$

$N = 22$

The desired integers are **22**, **23**, and **24**.

5. (a) $N_N + N_D = 150$

   (b) $5N_N + 10N_D = 450$

$$\begin{array}{rl} -5\text{(a)} & -5N_N - 5N_D = -750 \\ 1\text{(b)} & 5N_N + 10N_D = 450 \\ \hline & 5N_D = -300 \\ & N_D = \mathbf{-60} \end{array}$$

   (a) $N_N + (-60) = 150$

   $N_N = \mathbf{210}$

6. (a) $N_P + N_D = 50$

   (b) $N_P + 10N_D = 140$

   (b') $N_P = -10N_D + 140$

   Substitute (b') into (a) and get:

   (a) $(-10N_D + 140) + N_D = 50$

   $-9N_D = -90$

   $N_D = \mathbf{10}$

   (a) $N_P + (10) = 50$

   $N_P = \mathbf{40}$

7. $(2x + 3)(2x^2 + 2x + 2)$

$= 4x^3 + 4x^2 + 4x + 6x^2 + 6x + 6$

$= \mathbf{4x^3 + 10x^2 + 10x + 6}$

8. $\begin{array}{r} 3x^2 - 3x + 3 - \frac{5}{x+1} \\ x + 1 \overline{\smash{)}\, 3x^3 + 0x^2 + 0x - 2} \\ \underline{3x^3 + 3x^2} \\ -3x^2 + 0x \\ \underline{-3x^2 - 3x} \\ 3x - 2 \\ \underline{3x + 3} \\ -5 \end{array}$

Check: $\dfrac{3x^2(x + 1)}{x + 1} - \dfrac{3x(x + 1)}{x + 1}$

$+ \dfrac{3(x + 1)}{x + 1} - \dfrac{5}{x + 1}$

$= \dfrac{3x^3 + 3x^2 - 3x^2 - 3x + 3x - 2}{x + 1}$

$= \dfrac{3x^3 - 2}{x + 1}$

9. (a) $R_M T_M + R_W T_W = 250$

   (b) $R_M = 50$

   (c) $R_W = 80$

   (d) $T_M + T_W = 5$

   $T_M = 5 - T_W$

   Substitute (b) and (c) into (a) and get:

   (a') $50T_M + 80T_W = 250$

   Substitute $T_M = 5 - T_W$ into (a') and get:

   (a'') $50(5 - T_W) + 80T_W = 250$

   $250 - 50T_W + 80T_W = 250$

   $30T_W = 0$

   $T_W = 0$

   $T_W = \mathbf{0}, T_M = \mathbf{5}, R_W = \mathbf{80}, R_M = \mathbf{50}$

10. (a) $R_E T_E = R_W T_W$

   (b) $R_E = 200$

   (c) $R_W = 250$

   (d) $9 - T_E = T_W$

   Substitute (b) and (c) into (a) and get:

   (a') $200T_E = 250T_W$

   Substitute (d) into (a') and get:

   (a'') $200T_E = 250(9 - T_E)$

   $200T_E = 2250 - 250T_E$

   $450T_E = 2250$

   $T_E = 5$

   $T_E = \mathbf{5}, T_W = \mathbf{4}, R_E = \mathbf{200}, R_W = \mathbf{250}$

11. (a) $R_M T_M = R_R T_R$

(b) $R_M = 8$

(c) $R_R = 2$

(d) $5 - T_M = T_R$

Substitute (b) and (c) into (a) and get:

(a′) $8T_M = 2T_R$

Substitute (d) into (a′) and get:

(a″) $8T_M = 2(5 - T_M)$

$8T_M = 10 - 2T_M$

$10T_M = 10$

$T_M = 1$

$T_M = \mathbf{1}, T_R = \mathbf{4}, R_M = \mathbf{8}, R_R = \mathbf{2}$

12. $4x + \dfrac{3x}{a} = \dfrac{4ax}{a} + \dfrac{3x}{a} = \dfrac{\mathbf{4ax + 3x}}{\mathbf{a}}$

13. $-\dfrac{2x}{y} - cx + \dfrac{7x^2 y}{np^2}$

$= \dfrac{-2np^2 x}{np^2 y} - \dfrac{cnp^2 xy}{np^2 y} + \dfrac{7x^2 y^2}{np^2 y}$

$= \dfrac{\mathbf{-2np^2 x - cnp^2 xy + 7x^2 y^2}}{\mathbf{np^2 y}}$

14. $7^2 = H^2 + 6^2$

$49 = H^2 + 36$

$13 = H^2$

$\sqrt{13} = H$

Area $= \dfrac{b \times H}{2} = \dfrac{12 \text{ m} \times \sqrt{13}\text{ m}}{2}$

$= \mathbf{6\sqrt{13} \text{ m}^2}$

15. $D^2 = 6^2 + 4^2$

$D^2 = 36 + 16$

$D^2 = 52$

$D = \sqrt{52}$

$D = \mathbf{2\sqrt{13}}$

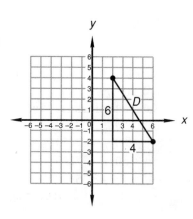

16. (a) $y = -3$

(b) $3x - 5y = 10$

$y = \dfrac{3}{5}x - 2$

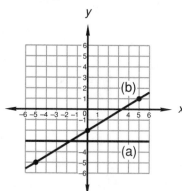

17. (a) The y intercept is -2. The slope is negative, and the rise over the run for any triangle drawn is -2.

$\mathbf{y = -2x - 2}$

(b) Every point on this line is 4 units to the right of the y axis.

$\mathbf{x = 4}$

18. Graph the line to find the slope.

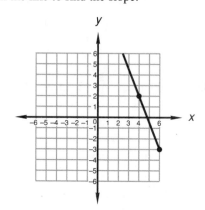

Slope $= \dfrac{-5}{+2} = -\dfrac{5}{2}$

$y = -\dfrac{5}{2}x + b$

$2 = -\dfrac{5}{2}(4) + b$

$12 = b$

Since $m = -\dfrac{5}{2}$ and $b = 12$: $y = \mathbf{-\dfrac{5}{2}x + 12}$

19. $y = -\dfrac{2}{7}x + b$

$5 = -\dfrac{2}{7}(-3) + b$

$b = -\dfrac{6}{7} + \dfrac{35}{7} = \dfrac{29}{7}$

Since $m = -\dfrac{2}{7}$ and $b = \dfrac{29}{7}$: $y = \mathbf{-\dfrac{2}{7}x + \dfrac{29}{7}}$

20. $(2x + 10y) + (-2x + 6y) + 60 = 180$

$$16y = 120$$
$$y = \frac{15}{2}$$

Since vertical angles are equal angles:

$$-2x + 6y = 50$$
$$-2x + 6\left(\frac{15}{2}\right) = 50$$
$$-2x = 5$$
$$x = -\frac{5}{2}$$
$$z = 2x + 10y$$
$$z = 2\left(-\frac{5}{2}\right) + 10\left(\frac{15}{2}\right)$$
$$z = -5 + 75 = \mathbf{70}$$

21. Volume $= \pi r^2 l = 245\pi$ cm^3

$$l = \frac{245\pi \text{ cm}^3}{(5 \text{ cm})^2 \pi}$$
$$= \frac{49}{5} \text{ cm}$$

22. $2\frac{1}{5}x - 3\frac{1}{4} = \frac{7}{20}$

$$\frac{11}{5}x = \frac{7}{20} + \frac{65}{20}$$
$$x = \frac{72}{20} \cdot \frac{5}{11} = \frac{360}{220} = \frac{18}{11}$$

23. $0.003 - 0.03 + 0.3x = 3.3$

$$3 - 30 + 300x = 3300$$
$$300x = 3327$$
$$x = \frac{1109}{100} = \mathbf{11.09}$$

24. $3^0(2x - 5) + (-x - 5) = -3(x^0 - 2)$

$$2x - 5 - x - 5 = -3 + 6$$
$$x = \mathbf{13}$$

25. $\dfrac{x^0 y^{-2} x}{x^3 y}\left(\dfrac{x^2 y}{m} - \dfrac{3x^4 y^2}{m^{-2}}\right)$

$$= \frac{x^3 y^{-1}}{x^3 ym} - \frac{3x^5}{x^3 ym^{-2}} = \mathbf{y^{-2}m^{-1} - 3x^2 y^{-1} m^2}$$

26. $\dfrac{2^{-3} x^0 (x^2)}{x^{-3}xy^{-3}y} = \dfrac{x^2}{8x^{-2}y^{-2}} = \dfrac{x^4 y^2}{8}$

27. $\dfrac{xy}{y^{-2}} - \dfrac{3x^4 y^4}{x^3 y} + \dfrac{7xy^{-2}}{xy^{-3}} = xy^3 - 3xy^3 + 7y$

$$= \mathbf{-2xy^3 + 7y}$$

28. $xy - x(x - y^0) = 2\left(-\frac{1}{2}\right) - 2\left(2 - \left(-\frac{1}{2}\right)^0\right)$

$$= -1 - 2(1) = \mathbf{-3}$$

29. $-(-3 - 2) + 4(-2) + \dfrac{1}{-2^{-3}} - (-2)^{-3}$

$$= 5 - 8 - 8 + \frac{1}{8} = \mathbf{-10\frac{7}{8}}$$

30. $-|-2 - 7| - |-2 - 4| - 3|-2 + 7|$
$$= -9 - 6 - 3(5) = \mathbf{-30}$$

PROBLEM SET 18

1. $\dfrac{2}{21} = \dfrac{A}{420}$

$$2 \cdot 420 = 21A$$
$$840 = 21A$$
$$A = \mathbf{40}$$

2. $\dfrac{2500}{3000} = \dfrac{NS}{6000}$

$$2500 \cdot 6000 = 3000NS$$
$$\mathbf{5000 \text{ kg}} = NS$$

3. $\dfrac{P}{100} \times of = is$

$$\frac{27}{100} \times N = 54,000$$
$$N = 54,000 \cdot \frac{100}{27} = \mathbf{200,000}$$

4. $\dfrac{P}{100} \times of = is$

$$\frac{0.004}{100} \times TP = 40$$
$$TP = 40 \cdot \frac{100}{0.004} = \mathbf{1,000,000}$$

5. (a) $N_D + N_Q = 200$
(b) $10N_D + 25N_Q = 2750$

$$-10(a) \quad -10N_D - 10N_Q = -2000$$
$$1(b) \quad \underline{10N_D + 25N_Q = 2750}$$
$$15N_Q = 750$$
$$N_Q = \mathbf{50}$$

(a) $N_D + (50) = 200$
$$N_D = \mathbf{150}$$

6. (a) $N_P + N_D = 30$

(b) $N_P + 10N_D = 291$

Substitute $N_P = 30 - N_D$ into (b) and get:

(b') $(30 - N_D) + 10N_D = 291$

$$9N_D = 261$$

$$N_D = \mathbf{29}$$

(a) $N_P + (29) = 30$

$$N_P = \mathbf{1}$$

7. $(2x + 4)(3x^2 - 2x - 10)$

$= 6x^3 - 4x^2 - 20x + 12x^2 - 8x - 40$

$= \mathbf{6x^3 + 8x^2 - 28x - 40}$

8.

$$x - 2 \overline{)\,5x^3 + 0x^2 + 0x - 1\,}$$

quotient: $5x^2 + 10x + 20 + \frac{39}{x-2}$

$$\underline{5x^3 - 10x^2}$$

$$10x^2 + 0x$$

$$\underline{10x^2 - 20x}$$

$$20x - 1$$

$$\underline{20x - 40}$$

$$39$$

Check: $\dfrac{5x^2(x - 2)}{x - 2} + \dfrac{10x(x - 2)}{x - 2}$

$+ \dfrac{20(x - 2)}{x - 2} + \dfrac{39}{x - 2}$

$= \dfrac{5x^3 - 10x^2 + 10x^2}{x - 2}$

$+ \dfrac{-20x + 20x - 40 + 39}{x - 2}$

$= \dfrac{5x^3 - 1}{x - 2}$

9. (a) $R_F T_F = R_S T_S$

(b) $T_S = 6$

(c) $T_F = 5$

(d) $R_F - 16 = R_S$

Substitute (b) and (c) into (a) and get:

(a') $5R_F = 6R_S$

Substitute (d) into (a') and get:

(a'') $5R_F = 6(R_F - 16)$

$$5R_F = 6R_F - 96$$

$$R_F = \mathbf{96}$$

(d) $(96) - 16 = R_S = \mathbf{80}$

10. (a) $R_M T_M = R_R T_R$

(b) $R_M = 8$

(c) $R_R = 2$

(d) $T_R = 5 - T_M$

Substitute (b) and (c) into (a) and get:

(a') $8T_M = 2T_R$

Substitute (d) into (a') and get:

(a'') $8T_M = 2(5 - T_M)$

$$8T_M = 10 - 2T_M$$

$$10T_M = 10$$

$$T_M = \mathbf{1}$$

(d) $T_R = 5 - (1) = \mathbf{4}$

11. (a) $R_G T_G + R_B T_B = 100$

(b) $R_G = 4$

(c) $R_B = 10$

(d) $T_B = T_G + 3$

Substitute (b) and (c) into (a) and get:

(a') $4T_G + 10T_B = 100$

Substitute (d) into (a') and get:

(a'') $4T_G + 10(T_G + 3) = 100$

$$4T_G + 10T_G + 30 = 100$$

$$14T_G = 70$$

$$T_G = \mathbf{5}$$

(d) $T_B = (5) + 3 = \mathbf{8}$

12. $7xyz + \dfrac{1}{xyz} = \dfrac{7x^2y^2z^2}{xyz} + \dfrac{1}{xyz}$

$= \dfrac{7x^2y^2z^2 + 1}{xyz}$

13. $-\dfrac{3x}{y} - c + \dfrac{7c}{xy^3} = \dfrac{-3x^2y^2}{xy^3} - \dfrac{cxy^3}{xy^3} + \dfrac{7c}{xy^3}$

$= \dfrac{-3x^2y^2 - cxy^3 + 7c}{xy^3}$

14. $$6^2 = H^2 + 5^2$$

$$36 = H^2 + 25$$

$$11 = H^2$$

$$\sqrt{11}\ \text{cm} = H$$

$\text{Area} = \dfrac{b \times H}{2} = \dfrac{10\ \text{cm} \times \sqrt{11}\ \text{cm}}{2}$

$= \mathbf{5\sqrt{11}\ cm^2}$

15. $$D^2 = 7^2 + 7^2$$

$$D^2 = 49 + 49$$

$$D^2 = 98$$

$$D = \sqrt{98} = \mathbf{7\sqrt{2}}$$

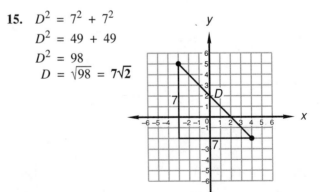

16. (a) $y = 2$
(b) $y = 2x$

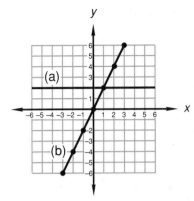

17. (a) Every point on this line is 3 units below the x axis.
$y = -3$
(b) The y intercept is 0. The slope is negative, and the rise over the run for any triangle drawn is -3.
$y = -3x$

18. Graph the line to find the slope.

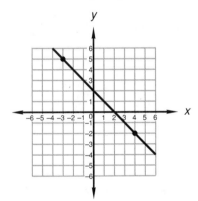

Slope $= \dfrac{-1}{1} = -1$
$y = -x + b$
$5 = -(-3) + b$
$2 = b$
Since $m = -1$ and $b = 2$: $y = -x + 2$

19.
$$y = \frac{5}{3}x + b$$
$$-2 = \frac{5}{3}(4) + b$$
$$-2 = \frac{20}{3} + b$$
$$-\frac{20}{3} - \frac{6}{3} = b$$
$$-\frac{26}{3} = b$$
Since $m = \dfrac{5}{3}$ and $b = -\dfrac{26}{3}$: $y = \dfrac{5}{3}x - \dfrac{26}{3}$

20. $3 \times \overrightarrow{SF} = 9$
$\overrightarrow{SF} = 3$

$a = 4 \times \overrightarrow{SF}$
$a = 4(3) = \mathbf{12}$

$b = 5 \times \overrightarrow{SF}$
$b = 5(3) = \mathbf{15}$

21. $3x + 2 = \dfrac{5x + 10}{2}$
$6x + 4 = 5x + 10$
$x = \mathbf{6}$
$P = 3(6) + 2 = \mathbf{20}$

22. Area $= \dfrac{120}{360}(\pi r^2) = \dfrac{\pi(15 \text{ ft})^2}{3} \approx \mathbf{235.5 \text{ ft}^2}$

23. $4\dfrac{1}{6}x - 2\dfrac{1}{12} = -\dfrac{5}{24}$
$$\frac{25}{6}x = -\frac{5}{24} + \frac{50}{24}$$
$$\frac{25}{6}x = \frac{45}{24}$$
$$x = \frac{15}{8} \cdot \frac{6}{25} = \frac{90}{200} = \mathbf{\frac{9}{20}}$$

24. $-2[(x - 2) - 4x - 3] = -(-4 - 2x)$
$-2[-3x - 5] = 4 + 2x$
$6x + 10 = 4 + 2x$
$4x = -6$
$$x = \mathbf{-\frac{3}{2}}$$

25. $\dfrac{3x^0 y^{-2}}{z^2}\left(\dfrac{2xyz^{-1}}{p} - \dfrac{y^2}{(3x)^{-2}}\right)$
$$= \frac{6xy^{-1}z^{-1}}{z^2 p} - \frac{27}{z^2 x^{-2}} = \frac{6x}{pyz^3} - \frac{27x^2}{z^2}$$

26. $\dfrac{a^0 xy^2 (a^2)^{-2}(xy^{-2})^2}{ax^0(y^{-2})^2} = \dfrac{xy^2 a^{-4}x^2 y^{-4}}{ay^{-4}}$
$$= \frac{x^3 y^2}{a^5}$$

27. $-\dfrac{2x^2}{y^2} - \dfrac{5y^{-2}p^0}{x^{-2}} + \dfrac{7xxy}{y^3}$
$$= -\frac{2x^2}{y^2} - \frac{5x^2}{y^2} + \frac{7x^2}{y^2} = \mathbf{0}$$

28. $ax - a(a - x) = -\dfrac{1}{2}\left(\dfrac{1}{4}\right) - \left(-\dfrac{1}{2}\right)\left(-\dfrac{1}{2} - \dfrac{1}{4}\right)$
$$= -\frac{1}{8} + \frac{1}{2}\left(-\frac{3}{4}\right) = -\frac{1}{8} - \frac{3}{8} = -\frac{4}{8} = \mathbf{-\frac{1}{2}}$$

29. $-2|-2 - 5| + (-3)|-2(-2) - 3| + 7$
$= -2(7) + (-3)(1) + 7 = -14 - 3 + 7 = \mathbf{-10}$

30. $-\dfrac{1}{(-2)^{-3}} + \dfrac{3}{-3^{-2}} = 8 - 27 = \mathbf{-19}$

PROBLEM SET 19

1. (a) $N_N + N_D = 60$
(b) $5N_N + 10N_D = 500$
Substitute $N_D = 60 - N_N$ into (b) and get:
(b') $5N_N + 10(60 - N_N) = 500$
$\qquad 5N_N + 600 - 10N_N = 500$
$\qquad\qquad\qquad -5N_N = -100$
$\qquad\qquad\qquad\quad N_N = \mathbf{20}$

(a) $(20) + N_D = 60$
$\qquad\qquad N_D = \mathbf{40}$

2. (a) $N_C + N_M = 26$
(b) $7N_C + N_M = 86$
Substitute $N_M = 26 - N_C$ into (b) and get:
(b') $7N_C + (26 - N_C) = 86$
$\qquad\qquad\qquad 6N_C = 60$
$\qquad\qquad\qquad\ N_C = \mathbf{10}$

3. $\dfrac{33}{49} = \dfrac{NS}{294}$
$49NS = 294(33)$
$\quad NS = \mathbf{198\ tons}$

4. If 19% was used, 81% remained.

$\dfrac{P}{100} \times of = is$

$\dfrac{81}{100} \times NA = 1134$

$\qquad NA = 1134 \cdot \dfrac{100}{81}$
$\qquad\qquad = \mathbf{1400\ liters\ in\ beginning}$
Amount used $= 1400 - 1134 = \mathbf{266\ liters}$

5. $(-7N - 7)2 = -5N + 4$
$\quad -14N - 14 = -5N + 4$
$\qquad\quad -9N = 18$
$\qquad\qquad N = \mathbf{-2}$

6. (a) $5x + 25y = -160$
(b) $-3x + 2y = -23$

$3(a)\quad 15x + 75y = -480$
$5(b)\ \ \underline{-15x + 10y = -115}$
$\qquad\qquad\quad 85y = -595$
$\qquad\qquad\qquad y = -7$

(b) $-3x + 2(-7) = -23$
$\qquad\qquad -3x = -9$
$\qquad\qquad\quad x = 3$
$\mathbf{(3, -7)}$

7. $(x^2 - 2)(x^3 - 2x^2 - 2x + 4)$
$= x^5 - 2x^4 - 2x^3 + 4x^2 - 2x^3 + 4x^2 + 4x - 8$
$= \mathbf{x^5 - 2x^4 - 4x^3 + 8x^2 + 4x - 8}$

8.
$$x - 2 \overline{)\ -3x^3 + 0x^2 + 0x - 2} \quad\Big|\ -3x^2 - 6x - 12 - \tfrac{26}{x-2}$$

$\qquad\quad \underline{-3x^3 + 6x^2}$
$\qquad\qquad\quad -6x^2 + 0x$
$\qquad\qquad\quad \underline{-6x^2 + 12x}$
$\qquad\qquad\qquad\quad -12x - 2$
$\qquad\qquad\qquad\quad \underline{-12x + 24}$
$\qquad\qquad\qquad\qquad\quad -26$

Check: $\dfrac{-3x^2(x - 2)}{x - 2} - \dfrac{6x(x - 2)}{x - 2}$

$\qquad -\dfrac{12(x - 2)}{x - 2} - \dfrac{26}{x - 2}$

$= \dfrac{-3x^3 + 6x^2 - 6x^2 + 12x}{x - 2}$

$+ \dfrac{-12x + 24 - 26}{x - 2}$

$= \dfrac{-3x^3 - 2}{x - 2}$

9. (a) $R_H T_H + R_S T_S = 180$
(b) $R_H = 70$
(c) $R_S = 20$
(d) $T_H = T_S$

Substitute (b) and (c) into (a) and get:
(a') $70T_H + 20T_S = 180$
Substitute (d) into (a') and get:
(a'') $70T_H + 20(T_H) = 180$
$\qquad\qquad\ 90T_H = 180$
$\qquad\qquad\quad\ T_H = \mathbf{2}$

(d) $T_S = \mathbf{2}$

10. (a) $R_F T_F = R_S T_S$
(b) $T_S = 6$
(c) $T_F = 5$
(d) $R_F - 10 = R_S$

Substitute (b) and (c) into (a) and get:
(a') $5R_F = 6R_S$
Substitute (d) into (a') and get:
(a'') $5R_F = 6(R_F - 10)$
$\qquad\ 5R_F = 6R_F - 60$
$\qquad\quad R_F = \mathbf{60}$

(d) $R_S = (60) - 10 = \mathbf{50}$

11. (a) $R_M T_M = R_R T_R$
(b) $R_M = 8$
(c) $R_R = 2$
(d) $T_R = 5 - T_M$

Substitute (b) and (c) into (a) and get:
(a′) $8T_M = 2T_R$
Substitute (d) into (a′) and get:
(a″) $8T_M = 2(5 - T_M)$
$8T_M = 10 - 2T_M$
$10T_M = 10$
$T_M = \mathbf{1}$
(d) $T_R = 5 - (1) = \mathbf{4}$

12. $4 + \dfrac{3x^2}{7y^2z} = \dfrac{28y^2z}{7y^2z} + \dfrac{3x^2}{7y^2z}$

$= \dfrac{\mathbf{28\,y^2z + 3x^2}}{\mathbf{7\,y^2z}}$

13. $\dfrac{a}{2x^2} - \dfrac{b}{x^2y} - c = \dfrac{ay}{2x^2y} - \dfrac{2b}{2x^2y} - \dfrac{2cx^2y}{2x^2y}$

$= \dfrac{\mathbf{ay - 2b - 2cx^2y}}{\mathbf{2x^2y}}$

14. $10^2 = z^2 + 8^2$
$100 = z^2 + 64$
$36 = z^2$
$\mathbf{6} = z$

$10\overleftrightarrow{SF} = 7$
$\overleftrightarrow{SF} = \dfrac{7}{10}$

$x = 6\left(\dfrac{7}{10}\right) = \dfrac{42}{10} = \dfrac{\mathbf{21}}{\mathbf{5}}$

$y = 8\left(\dfrac{7}{10}\right) = \dfrac{56}{10} = \dfrac{\mathbf{28}}{\mathbf{5}}$

15. $D^2 = 9^2 + 5^2$
$D^2 = 81 + 25$
$D^2 = 106$
$D = \mathbf{\sqrt{106}}$

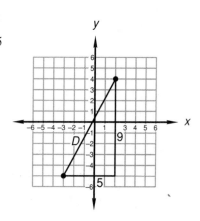

16. (a) $x = -3$
(b) $5x - 3y = 9$
$-3y = -5x + 9$
$y = \dfrac{5}{3}x - 3$

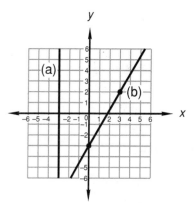

17. (a) Every point on this line is 2 units above the x axis.
$y = \mathbf{2}$
(b) The y intercept is 0. The slope is positive, and the rise over the run for any triangle drawn is 2.
$y = \mathbf{2x}$

18. Graph the line to find the slope.

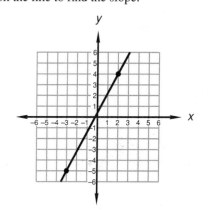

$\text{Slope} = \dfrac{+9}{+5} = \dfrac{9}{5}$

$y = \dfrac{9}{5}x + b$

$-5 = \dfrac{9}{5}(-3) + b$

$-5 = -\dfrac{27}{5} + b$

$\dfrac{27}{5} - \dfrac{25}{5} = b$

$\dfrac{2}{5} = b$

Since $m = \dfrac{9}{5}$ and $b = \dfrac{2}{5}$: $y = \dfrac{\mathbf{9}}{\mathbf{5}}x + \dfrac{\mathbf{2}}{\mathbf{5}}$

19.
$$y = \frac{2}{7}x + b$$
$$-5 = \frac{2}{7}(-3) + b$$
$$-5 = -\frac{6}{7} + b$$
$$-\frac{35}{7} + \frac{6}{7} = b$$
$$-\frac{29}{7} = b$$

Since $m = \frac{2}{7}$ and $b = -\frac{29}{7}$: $y = \frac{2}{7}x - \frac{29}{7}$

20. $5x + 15 = 6x - 5$
$$-x = -20$$
$$x = 20$$
$$3 \times \overrightarrow{SF} = \frac{11}{3}$$
$$\overrightarrow{SF} = \frac{11}{9}$$
$$5 \cdot \frac{11}{9} = y$$
$$\frac{55}{9} = y$$

$$k = 6x - 5$$
$$k = 6(20) - 5 = \mathbf{115}$$

21. Since the measure of an arc of a circle is the same as the measure of the central angle: $A = \mathbf{50}$
$$B + 50 = 180$$
$$B = \mathbf{130}$$
Since angles opposite equal sides are equal angles:
$$C = D$$
$$C + D + 130 = 180$$
$$C + D = 50$$
$$C = D = \mathbf{25}$$

$$\text{Area} = \frac{50}{360} \cdot \pi r^2 = \frac{50}{360} \cdot \pi (3 \text{ cm})^2$$
$$\approx \mathbf{3.93 \text{ cm}^2}$$

22. $3\frac{1}{4}p - \frac{7}{3} = -\frac{5}{12}$
$$\frac{13}{4}p = -\frac{5}{12} + \frac{28}{12}$$
$$p = \frac{23}{12} \cdot \frac{4}{13} = \frac{92}{156} = \mathbf{\frac{23}{39}}$$

23. $0.07x - 0.7 = -7.7$
$$7x - 70 = -770$$
$$7x = -700$$
$$x = \mathbf{-100}$$

24. $-[(-2 - 6)(-2) - 2] = -2[(x - 2)2$
$$- (2x - 3)3]$$
$$-[4 + 12 - 2] = -2[2x - 4 - 6x + 9]$$
$$-14 = -2[-4x + 5]$$
$$-14 = 8x - 10$$
$$-4 = 8x$$
$$-\frac{1}{2} = x$$

25. $-\frac{3x^0 y^2}{p^{-2}}\left(\frac{x}{p^2 y^2} - \frac{3xy}{x^2 p}\right) = -\frac{3xy^2}{y^2} + \frac{9xy^3}{p^{-1}x^2}$
$$= -3x + \frac{9py^3}{x}$$

26. $\frac{x^2 x^0 (xx^{-3})}{x(y^0 y^2)^{-2}} = \frac{1}{xy^{-4}} = \frac{y^4}{x}$

27. $\frac{2x^2 a}{y} - \frac{3xy^{-2}a}{y^{-1}x^{-1}} + \frac{4x^0 y^{-1}}{x^{-2}a^{-1}}$
$$= \frac{2x^2 a}{y} - \frac{3x^2 a}{y} + \frac{4x^2 a}{y} = \frac{3x^2 a}{y}$$

28. $m - m^2 y(m - y)$
$$= -\frac{1}{3} - \left(-\frac{1}{3}\right)^2 \left(\frac{1}{6}\right)\left(-\frac{1}{3} - \frac{1}{6}\right)$$
$$= -\frac{1}{3} - \frac{1}{54}\left(-\frac{1}{2}\right) = -\frac{1}{3} + \frac{1}{108} = -\frac{35}{108}$$

29. $-4^0[(-5 + 2) - |-3 + 7| - 3(-1 - (-2)^0)]$
$$= -1[-3 - 4 - 3(-2)] = -1[-3 - 4 + 6]$$
$$= -1[-1] = \mathbf{1}$$

30. $-\frac{-1}{-(-2)^{-2}} + \frac{3}{-(-2)^{-2}} = -4 - 12 = \mathbf{-16}$

PROBLEM SET 20

1. $\frac{140}{160} = \frac{NC}{640}$
$$160NC = 640(140)$$
$$NC = \mathbf{560}$$

2. If $\frac{3}{10}$ were virtuosos, $\frac{7}{10}$ were not virtuosos.
$$F \times of = is$$
$$\frac{7}{10} \times P = 28$$
$$P = 28 \cdot \frac{10}{7} = \mathbf{40}$$
$$N_V = 40 - 28 = \mathbf{12}$$

3. (a) $N_E = N_W + 3$
(b) $7N_E + 2N_W = 111$
Substitute $N_E = N_W + 3$ into (b) and get:
(b') $7(N_W + 3) + 2N_W = 111$
$$9N_W = 90$$
$$N_W = \mathbf{10}$$
(a) $N_E = 10 + 3 = \mathbf{13}$

4. $\dfrac{P}{100} \times of = is$

$\dfrac{40}{100} \times C = 240$

$C = 240 \cdot \dfrac{100}{40} = 600$

$N_{NB} = 600 - 240 = \mathbf{360}$

5. $N \qquad N + 2 \qquad N + 4 \qquad N + 6$

$-4(N + N + 6) = -(N + 2 + N + 4) + 6$

$-8N - 24 = -2N$

$-6N = 24$

$N = -4$

The desired integers are $-\mathbf{4}, -\mathbf{2}, \mathbf{0}$, and $\mathbf{2}$.

6. (a) $y - 2x = 8$

(b) $2y + 2x = 40$

Substitute $y = 2x + 8$ into (b) and get:

(b′) $2(2x + 8) + 2x = 40$

$6x + 16 = 40$

$6x = 24$

$x = 4$

(a) $y - 2(4) = 8$

$y = 16$

(4, 16)

7.

$$x - 1 \overline{) -2x^3 - 0x^2 - x + 2} \quad \dfrac{-2x^2 - 2x - 3 - \frac{1}{x-1}}{}$$

$$\underline{-2x^3 + 2x^2}$$
$$-2x^2 - x$$
$$\underline{-2x^2 + 2x}$$
$$-3x + 2$$
$$\underline{-3x + 3}$$
$$-1$$

Check: $\dfrac{-2x^2(x - 1)}{x - 1} - \dfrac{2x(x - 1)}{x - 1}$

$- \dfrac{3(x - 1)}{x - 1} - \dfrac{1}{x - 1}$

$= \dfrac{-2x^3 + 2x^2 - 2x^2 + 2x}{x - 1}$

$+ \dfrac{-3x + 3 - 1}{x - 1}$

$= \dfrac{-2x^3 - x + 2}{x - 1}$

8. (a) $R_G T_G + R_B T_B = 100$

(b) $R_G = 4$

(c) $R_B = 10$

(d) $T_B = T_G + 3$

Substitute (b) and (c) into (a) and get:

(a′) $4T_G + 10T_B = 100$

Substitute (d) into (a′) and get:

(a″) $4T_G + 10(T_G + 3) = 100$

$14T_G = 70$

$T_G = \mathbf{5}$

(d) $T_B = (5) + 3 = \mathbf{8}$

9. $3\sqrt{3} \cdot 4\sqrt{12} - 5\sqrt{300} = 3\sqrt{3} \cdot 4\sqrt{2 \cdot 2 \cdot 3}$

$- 5\sqrt{2 \cdot 2 \cdot 3 \cdot 5 \cdot 5}$

$= 3\sqrt{3} \cdot 4\sqrt{2}\sqrt{2}\sqrt{3} - 5\sqrt{2}\sqrt{2}\sqrt{3}\sqrt{5}\sqrt{5}$

$= 3\sqrt{3} \cdot 8\sqrt{3} - 50\sqrt{3} = \mathbf{72 - 50\sqrt{3}}$

10. $4\sqrt{3}(2\sqrt{3} - \sqrt{6}) = 8\sqrt{3}\sqrt{3} - 4\sqrt{3}\sqrt{3}\sqrt{2}$

$= \mathbf{24 - 12\sqrt{2}}$

11. $5\sqrt{5}(2\sqrt{5} - 3\sqrt{10}) = 10\sqrt{5}\sqrt{5} - 15\sqrt{5}\sqrt{5}\sqrt{2}$

$= \mathbf{50 - 75\sqrt{2}}$

12. $\dfrac{m^2}{x^2 a} + \dfrac{5}{ax} - \dfrac{m}{a} = \dfrac{m^2}{ax^2} + \dfrac{5x}{ax^2} - \dfrac{mx^2}{ax^2}$

$= \dfrac{\mathbf{m^2 + 5x - mx^2}}{\mathbf{ax^2}}$

13. $\dfrac{a}{x^2} - a - \dfrac{3x}{2a^4}$

$= \dfrac{2a^5}{2a^4 x^2} - \dfrac{2a^5 x^2}{2a^4 x^2} - \dfrac{3x^3}{2a^4 x^2}$

$= \dfrac{\mathbf{2a^5 - 2a^5 x^2 - 3x^3}}{\mathbf{2a^4 x^2}}$

14. $4^2 = H^2 + 3^2$

$16 = H^2 + 9$

$7 = H^2$

$\sqrt{7} = H$

$\text{Area} = \dfrac{b \times H}{2} = \dfrac{6 \text{ ft} \times \sqrt{7} \text{ ft}}{2} = \mathbf{3\sqrt{7} \text{ ft}^2}$

15. $5 \times \overrightarrow{SF} = 4$

$\overrightarrow{SF} = \dfrac{4}{5}$

$z = 6\left(\dfrac{4}{5}\right) = \dfrac{24}{5}$

$A = 7\left(\dfrac{4}{5}\right) = \dfrac{28}{5}$

16. (a) $y = -2$

(b) $y = -2x - 2$

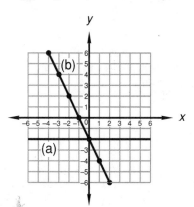

17. (a) Every point on this line is 1 unit above the x axis.

    $y = 1$

    (b) The y intercept is 0. The slope is negative, and the rise over the run for any triangle drawn is -1.

    $y = -x$

18. Write the equation of the given line in slope-intercept form.

    $3y - x = 3$

    $3y = x + 3$

    $y = \dfrac{1}{3}x + 1$

    Since parallel lines have the same slope:

    $y = \dfrac{1}{3}x + b$

    $-1 = \dfrac{1}{3}(2) + b$

    $-\dfrac{3}{3} - \dfrac{2}{3} = b$

    $-\dfrac{5}{3} = b$

    Since $m = \dfrac{1}{3}$ and $b = -\dfrac{5}{3}$: $\mathbf{y = \dfrac{1}{3}x - \dfrac{5}{3}}$

19. $y = -\dfrac{1}{12}x + b$

    $5 = -\dfrac{1}{12}(2) + b$

    $5 = -\dfrac{1}{6} + b$

    $\dfrac{30}{6} + \dfrac{1}{6} = b$

    $\dfrac{31}{6} = b$

    Since $m = -\dfrac{1}{12}$ and $b = \dfrac{31}{6}$:

    $\mathbf{y = -\dfrac{1}{12}x + \dfrac{31}{6}}$

20. $A + 140 = 180$

    $A = \mathbf{40}$

    $B + 40 + 90 = 180$

    $B = \mathbf{50}$

    Since vertical angles are equal angles:

    $C = B = \mathbf{50}$

    $P + 50 + 90 = 180$

    $P = \mathbf{40}$

    Since vertical angles are equal angles:

    $y = P = \mathbf{40}$

21. $3\dfrac{1}{12} - \dfrac{1}{6}x = 2\dfrac{1}{24}$

    $-\dfrac{1}{6}x = \dfrac{49}{24} - \dfrac{74}{24}$

    $x = -\dfrac{25}{24} \cdot -\dfrac{6}{1} = \dfrac{150}{24} = \mathbf{\dfrac{25}{4}}$

22. $-0.04x - x - 0.2x = 6.2$

    $-4x - 100x - 20x = 620$

    $x = \mathbf{-5}$

23. $-2[(-3)(-2 - x) + 2(x - 3)] = -2x$

    $-2[6 + 3x + 2x - 6] = -2x$

    $-10x = -2x$

    $x = \mathbf{0}$

24. $\dfrac{-x^0 y}{y^2 y^{-2}}\left(\dfrac{x}{y} - \dfrac{3xy^2}{y^3}\right) = \dfrac{-xy}{y} + \dfrac{3xy^3}{y^3} = \mathbf{2x}$

25. $\dfrac{(-2x^0)^{-3}x^2 yy^0 y^3}{(xy^{-2})^2(-2x^{-1})^{-2}} = \dfrac{4x^2 y^4}{-8x^4 y^{-4}} = \mathbf{-\dfrac{y^8}{2x^2}}$

26. $-\dfrac{3x}{a} + \dfrac{(a^0)a}{aax^{-1}} - \dfrac{5x^{-1}}{a} = -\dfrac{3x}{a} + \dfrac{x}{a} - \dfrac{5}{xa}$

    $= \mathbf{-\dfrac{2x}{a} - \dfrac{5}{xa}}$

27. $k - kx(k^2 - x)$

    $= -\dfrac{1}{4} - \left(-\dfrac{1}{4}\right)\left(\dfrac{1}{8}\right)\left(\left(-\dfrac{1}{4}\right)^2 - \dfrac{1}{8}\right)$

    $= -\dfrac{1}{4} + \dfrac{1}{32}\left(\dfrac{1}{16} - \dfrac{2}{16}\right) = -\dfrac{1}{4} + \dfrac{1}{32}\left(-\dfrac{1}{16}\right)$

    $= -\dfrac{1}{4} - \dfrac{1}{512} = -\dfrac{128}{512} - \dfrac{1}{512} = \mathbf{-\dfrac{129}{512}}$

28. $-2^0\{(-7 + 3) - |-2 + 9| - 2[-2^0 - (-5)]\}$

    $= -1\{-4 - 7 - 2[4]\} = -1\{-19\} = \mathbf{19}$

29. $-2^0|-3 - 7| + |(-5^0)| - 2(-5) + 2$

    $= -10 + 1 + 10 + 2 = \mathbf{3}$

30. $-\dfrac{1}{-(-3)^{-3}} + \dfrac{2}{-(-3)^{-3}} = -27 + 54 = \mathbf{27}$

PROBLEM SET 21

1. (a) $\dfrac{N}{D} = \dfrac{3}{5} \longrightarrow 5N = 3D$

   (b) $N + D = 96 \longrightarrow D = 96 - N$

   Substitute $D = 96 - N$ into (a) and get:

   (a') $5N = 3(96 - N)$

   $8N = 288$

   $N = \mathbf{36}$

   (b) $(36) + D = 96$

   $D = \mathbf{60}$

2. (a) $L + S = 200$

   (b) $\dfrac{L - S = 66}{2L = 266}$

   $L = \mathbf{133}$

   (a) $(133) + S = 200$

   $S = \mathbf{67}$

3. $\dfrac{1500}{2400} = \dfrac{NA}{3600} \rightarrow 2400NA = 3600(1500)$

$$NA = \textbf{2250 kg}$$

4. If 20% combined, 80% did not combine.

$$\dfrac{80}{100} \times N = 740$$

$$N = 740 \cdot \dfrac{100}{80} = 925$$

$$N_C = 925 - 740 = \textbf{185 kg}$$

5. (a) $5N_N + 10N_D = 575$

(b) $N_N + N_D = 70$

$$\begin{array}{rr} 1(a) & 5N_N + 10N_D = 575 \\ -10(b) & -10N_N - 10N_D = -700 \\ \hline & -5N_N = -125 \\ & N_N = \textbf{25} \end{array}$$

6. $N \quad N + 1 \quad N + 2$

$-5(N + N + 2) = 4(-N - 1) + 24$

$-10N - 10 = -4N + 20$

$-6N = 30$

$N = -5$

The desired integers are $-\textbf{5}, -\textbf{4}$, and $-\textbf{3}$.

7. (a) $8y - 3x = 22$

(b) $2y + 4x = 34$

$$\begin{array}{rr} 1(a) & 8y - 3x = 22 \\ -4(b) & -8y - 16x = -136 \\ \hline & -19x = -114 \\ & x = 6 \end{array}$$

(b) $2y + 4(6) = 34$

$2y = 10$

$y = 5$

$(\textbf{6}, \textbf{5})$

8. $(4x^2 - 2x + 2)(2x - 3)$

$= 8x^3 - 4x^2 + 4x - 12x^2 + 6x - 6$

$= \textbf{8}x^3 - \textbf{16}x^2 + \textbf{10}x - \textbf{6}$

9. (a) $R_K T_K = R_N T_N$

(b) $R_K = 6$

(c) $R_N = 3$

(d) $T_N - T_K = 8$

Substitute (b) and (c) into (a) and get:

(a') $6T_K = 3T_N$

Substitute $T_N = T_K + 8$ into (a') and get:

(a'') $6T_K = 3(T_K + 8)$

$6T_K = 3T_K + 24$

$3T_K = 24$

$T_K = \textbf{8}$

(d) $T_N - (8) = 8$

$T_N = \textbf{16}$

10. $3\sqrt{200} - 5\sqrt{18} + 7\sqrt{50}$

$= 3\sqrt{2}\sqrt{2}\sqrt{2}\sqrt{5}\sqrt{5} - 5\sqrt{2}\sqrt{3}\sqrt{3} + 7\sqrt{2}\sqrt{5}\sqrt{5}$

$= 30\sqrt{2} - 15\sqrt{2} + 35\sqrt{2} = \textbf{50}\sqrt{\textbf{2}}$

11. $2\sqrt{3} \cdot 2\sqrt{2}(6\sqrt{6} - 3\sqrt{2})$

$= 2\sqrt{3} \cdot 2\sqrt{2}(6\sqrt{2}\sqrt{3} - 3\sqrt{2})$

$= 4\sqrt{2}\sqrt{3}(6\sqrt{2}\sqrt{3} - 3\sqrt{2})$

$= 24\sqrt{2}\sqrt{2}\sqrt{3}\sqrt{3} - 12\sqrt{2}\sqrt{2}\sqrt{3} = \textbf{144} - \textbf{24}\sqrt{\textbf{3}}$

12. $4x + \dfrac{1}{p} = \dfrac{4px}{p} + \dfrac{1}{p} = \dfrac{\textbf{4}px + \textbf{1}}{p}$

13. $\dfrac{m^2}{a^2 x^2} - \dfrac{3}{ax} - \dfrac{m}{x} = \dfrac{m^2}{a^2 x^2} - \dfrac{3ax}{a^2 x^2} - \dfrac{a^2 mx}{a^2 x^2}$

$= \dfrac{m^2 - 3ax - a^2 mx}{a^2 x^2}$

14. $\dfrac{(0.0003 \times 10^8)(6000)}{(0.006 \times 10^{15})(2000 \times 10^5)}$

$= \dfrac{(3 \times 10^4)(6 \times 10^3)}{(6 \times 10^{12})(2 \times 10^8)} = \dfrac{18 \times 10^7}{12 \times 10^{20}}$

$= \textbf{1.5} \times \textbf{10}^{-\textbf{13}}$

15. $D^2 = 7^2 + 0^2$

$D^2 = 49$

$D = \textbf{7}$

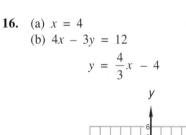

16. (a) $x = 4$

(b) $4x - 3y = 12$

$$y = \dfrac{4}{3}x - 4$$

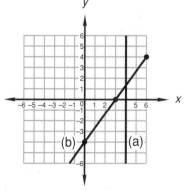

17. (a) Every point on this line is 2 units below the
    x axis.
    $y = -2$
   (b) The y intercept is 0. The slope is negative, and
    the rise over the run for any triangle drawn is
    -2.
    $y = -2x$

18. Graph the line to find the slope.

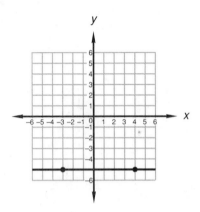

Slope $= \dfrac{0}{7} = 0$

$y = 0x + b$
$-5 = 0(-3) + b$
$-5 = b$
Since $m = 0$ and $b = -5$: $y = -5$

19. Since parallel lines have the same slope:

$$y = -\frac{3}{7}x + b$$

$$2 = -\frac{3}{7}(2) + b$$

$$\frac{14}{7} + \frac{6}{7} = b$$

$$\frac{20}{7} = b$$

Since $m = -\dfrac{3}{7}$ and $b = \dfrac{20}{7}$: $y = -\dfrac{3}{7}x + \dfrac{20}{7}$

20. $x + 4 = 6$
     $x = 2$
   $x + y = 10$
     $y = 10 - (2) = 8$

21. $A + 130 = 180$
       $A = 50$
     $B + 50 + 90 = 180$
       $B = 40$
   Since vertical angles are equal angles:
   $C = B = 40$
   $D + 40 + 90 = 180$
       $D = 50$
   Since vertical angles are equal angles:
   $y = D = 50$

22. $5\dfrac{1}{3}x - \dfrac{3}{4} = \dfrac{7}{8}$

$$\frac{16}{3}x = \frac{7}{8} + \frac{6}{8}$$

$$\frac{16}{3}x = \frac{13}{8}$$

$$x = \frac{13}{8} \cdot \frac{3}{16} = \frac{39}{128}$$

23. $0.03(x - 4) = 0.02(x + 6)$
     $3(x - 4) = 2(x + 6)$
     $3x - 12 = 2x + 12$
         $x = 24$

24. $-[(-4 - 1)(-3) - 6^0] = -2[(y - 4)3 - (2y - 5)]$
       $-[15 - 1] = -2[3y - 12 - 2y + 5]$
         $-[14] = -2[y - 7]$
         $-14 = -2y + 14$
         $-28 = -2y$
           $14 = y$

25. $\dfrac{5y^0 p^{-2}}{x^2}\left(-\dfrac{2p^2}{x^2} - \dfrac{p^2}{x}\right) = -\dfrac{10}{x^4} - \dfrac{5}{x^3}$

26. $\dfrac{x^2(y^{-2})^2(y^0)^2}{(2x^2y^3)^{-2}} = \dfrac{4x^2y^{-4}}{x^{-4}y^{-6}} = 4x^6y^2$

27. $\dfrac{3x^2a}{x} - 5xa + \dfrac{7x^{-2}}{x^{-3}a^{-1}} = 3xa - 5xa + 7xa$
   $= 5xa$

28. $x(x - ax)x = -\dfrac{1}{2}\left(-\dfrac{1}{2} - \dfrac{1}{3}\left(-\dfrac{1}{2}\right)\right)\left(-\dfrac{1}{2}\right)$

   $= \dfrac{1}{4}\left(-\dfrac{1}{2} + \dfrac{1}{6}\right) = \dfrac{1}{4}\left(-\dfrac{3}{6} + \dfrac{1}{6}\right) = \dfrac{1}{4}\left(-\dfrac{1}{3}\right)$

   $= -\dfrac{1}{12}$

29. $-7^0[(-2 + 3) - |-4 + 3|] = -1[1 - 1]$
   $= -1[0] = 0$

30. $-\dfrac{1}{-2^{-2}} - \dfrac{1}{-(-2)^0} = 4 + 1 = 5$

PROBLEM SET 22

1. $D_E = D_C$ so $R_E T_E = R_C T_C$
   $R_E = 15;\ R_C = 30;\ T_C = T_E - 3$

$$15T_E = 30(T_E - 3)$$
$$15T_E = 30T_E - 90$$
$$-15T_E = -90$$
$$T_E = 6$$

$T_C = (6) - 3 = 3$
$D_E = R_E T_E = (15)(6) = $ **90 miles**
$D_C = R_C T_C = (30)(3) = $ **90 miles**

2. $D_A = D_W$ so $R_A T_A = R_W T_W$
$R_A = 9$; $T_A = 4$; $T_W = 2$

$9(4) = R_W(2)$
18 kph $= R_W$

3. (a) $\dfrac{N}{D} = \dfrac{7}{5}$ $\longrightarrow$ $5N = 7D$
(b) $N + D = 960$ $\longrightarrow$ $D = 960 - N$
Substitute $D = 960 - N$ into (a) and get:
(a') $5N = 7(960 - N)$
$12N = 6720$
$N = \mathbf{560}$
(b) $(560) + D = 960$
$D = \mathbf{400}$

4. (a) $T_F = T_S + 500$
(b) $T_F + T_S = 6900$
Substitute (a) into (b) and get:
(b') $(T_S + 500) + T_S = 6900$
$2T_S = 6400$
$T_S = \$3200$
(a) $T_F = (3200) + 500 = \mathbf{\$3700}$

5. (a) $5P_H + 13P_W = 109$
(b) $P_H + P_W = 9$

$\quad$ 1(a) $\quad 5P_H + 13P_W = 109$
-5(b) $\quad -5P_H - 5P_W = -45$
$\rule{4cm}{0.4pt}$
$\quad\quad\quad\quad\quad 8P_W = 64$
$\quad\quad\quad\quad\quad\quad P_W = \mathbf{8}$

6. $\dfrac{570}{600} = \dfrac{NI}{5000}$ $\longrightarrow$ $600NI = 5000(570)$
$NI = \mathbf{4750}$

7. (a) $5x + y = 24$
(b) $7x - 2y = 20$
Substitute $y = 24 - 5x$ into (b) and get:
(b') $7x - 2(24 - 5x) = 20$
$7x - 48 + 10x = 20$
$17x = 68$
$x = 4$
(a) $5(4) + y = 24$
$y = 4$
(4, 4)

8.
$$x - 1 \overline{\smash{\big)}\,x^3 + 0x^2 - 4x + 2} \quad \begin{array}{c} x^2 + x - 3 - \frac{1}{x-1} \end{array}$$
$$\underline{x^3 - x^2}$$
$$x^2 - 4x$$
$$\underline{x^2 - x}$$
$$-3x + 2$$
$$\underline{-3x + 3}$$
$$-1$$

Check: $\dfrac{x^2(x-1)}{x-1} + \dfrac{x(x-1)}{x-1}$
$\quad - \dfrac{3(x-1)}{x-1} - \dfrac{1}{x-1}$

$= \dfrac{x^3 - x^2 + x^2 - x - 3x + 3 - 1}{x-1}$

$= \dfrac{x^3 - 4x + 2}{x-1}$

9. $\dfrac{x}{12} = \dfrac{15}{10}$ $\longrightarrow$ $10x = 180$ $\longrightarrow$ $x = \mathbf{18}$
$\dfrac{y}{11} = \dfrac{15}{10}$ $\longrightarrow$ $10y = 165$ $\longrightarrow$ $y = \dfrac{\mathbf{33}}{\mathbf{2}}$

10. $2\sqrt{27} - 3\sqrt{75} = 2\sqrt{3}\sqrt{3}\sqrt{3} - 3\sqrt{3}\sqrt{5}\sqrt{5}$
$= 6\sqrt{3} - 15\sqrt{3} = \mathbf{-9\sqrt{3}}$

11. $3\sqrt{2}(2\sqrt{2} - \sqrt{6}) \cdot 4\sqrt{3} + 2$
$= 3\sqrt{2}(2\sqrt{2} - \sqrt{2}\sqrt{3}) \cdot 4\sqrt{3} + 2$
$= 12\sqrt{2}\sqrt{3}(2\sqrt{2} - \sqrt{2}\sqrt{3}) + 2$
$= 24\sqrt{2}\sqrt{2}\sqrt{3} - 12\sqrt{2}\sqrt{2}\sqrt{3}\sqrt{3} + 2$
$= 48\sqrt{3} - 72 + 2 = \mathbf{48\sqrt{3} - 70}$

12. $2\sqrt{3}(5\sqrt{3} - 2\sqrt{6}) = 2\sqrt{3}(5\sqrt{3} - 2\sqrt{2}\sqrt{3})$
$= 10\sqrt{3}\sqrt{3} - 4\sqrt{2}\sqrt{3}\sqrt{3} = \mathbf{30 - 12\sqrt{2}}$

13. $2 + \dfrac{1}{x} = \dfrac{2x}{x} + \dfrac{1}{x} = \dfrac{\mathbf{2x + 1}}{\mathbf{x}}$

14. $\dfrac{5x^2}{y} + p^2 - \dfrac{3x}{py} = \dfrac{5x^2 p}{py} + \dfrac{p^3 y}{py} - \dfrac{3x}{py}$

$= \dfrac{\mathbf{5x^2 p + p^3 y - 3x}}{\mathbf{py}}$

15. $\dfrac{(0.0035 \times 10^{-4})(200 \times 10^6)}{(700 \times 10^5)(0.00005)}$

$= \dfrac{(3.5 \times 10^{-7})(2 \times 10^8)}{(7 \times 10^7)(5 \times 10^{-5})} = \dfrac{7 \times 10^1}{35 \times 10^2}$

$= \mathbf{2 \times 10^{-2}}$

16. (a) $x = -5$
(b) $2x - y = 4$
$-y = -2x + 4$
$y = 2x - 4$

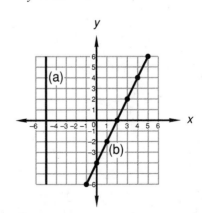

17. (a) Every point on this line is 4 units to the left of the y axis.

$$x = -4$$

(b) The y intercept is +3. The slope is negative, and the rise over the run for any triangle drawn is $-\frac{3}{2}$.

$$y = -\frac{3}{2}x + 3$$

18. Graph the line to find the slope.

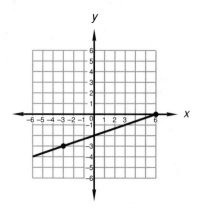

Slope $= \dfrac{+1}{+3} = \dfrac{1}{3}$

$$y = \frac{1}{3}x + b$$

$$0 = \frac{1}{3}(6) + b$$

$$-2 = b$$

Since $m = \dfrac{1}{3}$ and $b = -2$: $\boldsymbol{y = \dfrac{1}{3}x - 2}$

19. $D^2 = 9^2 + 3^2$

$D^2 = 81 + 9$

$D^2 = 90$

$D = \sqrt{90}$

$D = \boldsymbol{3\sqrt{10}}$

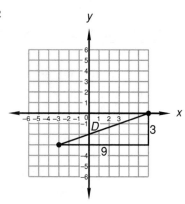

20.

$$y = \frac{2}{5}x + b$$

$$-5 = \frac{2}{5}(3) + b$$

$$-5 = \frac{6}{5} + b$$

$$-\frac{25}{5} - \frac{6}{5} = b$$

$$-\frac{31}{5} = b$$

Since $m = \dfrac{2}{5}$ and $b = -\dfrac{31}{5}$: $\boldsymbol{y = \dfrac{2}{5}x - \dfrac{31}{5}}$

21. $A + 140 = 180$

$A = \boldsymbol{40}$

$B + 40 + 90 = 180$

$B = \boldsymbol{50}$

Since vertical angles are equal angles:

$C = B = \boldsymbol{50}$

$k + 50 + 90 = 180$

$k = \boldsymbol{40}$

$M + 90 + 40 = 180$

$M = \boldsymbol{50}$

22. $3\dfrac{2}{5}x - 4\dfrac{1}{10}x = 2\dfrac{1}{4}$

$$\frac{34}{10}x - \frac{41}{10}x = \frac{9}{4}$$

$$-\frac{7}{10}x = \frac{9}{4}$$

$$x = \frac{9}{4} \cdot -\frac{10}{7} = -\frac{90}{28} = \boldsymbol{-\frac{45}{14}}$$

23. $0.02(p - 2) = 0.03(2p - 6)$

$2(p - 2) = 3(2p - 6)$

$2p - 4 = 6p - 18$

$14 = 4p$

$\boldsymbol{3.5} = p$

24. $-[(-3 - 6)(-1^0) - 6^0] = -4[(x - 3)2]$

$-[9 - 1] = -4[2x - 6]$

$-8 = -8x + 24$

$8x = 32$

$x = \boldsymbol{4}$

25. $\dfrac{xy^{-2}}{z^0 p}\left(\dfrac{py^2}{x} - \dfrac{3xy^{-4}}{py}\right) = \dfrac{xp}{xp} - \dfrac{3x^2y^{-6}}{p^2 y}$

$$= \boldsymbol{1 - 3x^2y^{-7}p^{-2}}$$

26. $\dfrac{(x^{-2}yp)^{-3}(x^0yp)^2}{(2x^2)^{-2}} = \dfrac{4x^6y^{-1}p^{-1}}{x^{-4}} = \boldsymbol{4x^{10}y^{-1}p^{-1}}$

27. $-\dfrac{3x^2y}{xx} + \dfrac{2x^{-2}x^4}{y^{-1}x^2} - \dfrac{5xy^2}{xy} = -3y + 2y - 5y$

$$= \boldsymbol{-6y}$$

28. $ya(y - a)y = -\dfrac{1}{2}\left(\dfrac{1}{5}\right)\left(-\dfrac{1}{2} - \dfrac{1}{5}\right)\left(-\dfrac{1}{2}\right)$

$$= \frac{1}{20}\left(-\frac{5}{10} - \frac{2}{10}\right) = \frac{1}{20}\left(-\frac{7}{10}\right) = \boldsymbol{-\frac{7}{200}}$$

29. $-2^0[(5 - 7 - 2) - |-2 - 7| - 2^0] + 2$
$= -1[-4 - 9 - 1] + 2$
$= -1[-14] + 2 = 14 + 2 = \mathbf{16}$

30. $\dfrac{-3}{-3^{-2}} - \dfrac{2}{-2^{-3}} = 27 + 16 = \mathbf{43}$

PROBLEM SET 23

1. $D_M = D_J$ so $R_M T_M = R_J T_J$
$R_M = 600;\ R_J = 800;\ T_M = T_J + 4$

$600(T_J + 4) = 800T_J$
$600T_J + 2400 = 800T_J$
$2400 = 200T_J$
12 minutes $= T_J$

2. $D_F = D_S$ so $R_F T_F = R_S T_S$
$T_F = 10;\ T_S = 12;\ R_F = R_S + 10$

$(R_S + 10)(10) = R_S(12)$
$10R_S + 100 = 12R_S$
$100 = 2R_S$
$50 = R_S$
$D_F = D_S = (50)(12) = \mathbf{600\ kilometers}$

3. (a) $\dfrac{N}{D} = \dfrac{11}{12} \longrightarrow 12N = 11D$
   (b) $N + D = 230 \longrightarrow D = 230 - N$
$12N = 11(230 - N)$
$12N = 2530 - 11N$
$23N = 2530$
$N = 110$
$D = 230 - 110 = 120$
Fraction $= \dfrac{110}{120}$

4. (a) $N_V + N_S = 300$
   (b) $N_V - N_S = 50$
$\overline{\quad 2N_V \qquad = 350}$
$N_V = \mathbf{175}$

   (b) $(175) - N_S = 50$
$N_S = \mathbf{125}$

5. (a) $5N_N + 25N_Q = 500$
   (b) $N_N - N_Q = 40$

1(a) $\quad 5N_N + 25N_Q = 500$
25(b) $\ 25N_N - 25N_Q = 1000$
$\overline{\quad 30N_N \qquad = 1500}$
$N_N = \mathbf{50}$

(b) $(50) - N_Q = 40$
$N_Q = \mathbf{10}$

6. If 20% was copper sulfate, 80% was not copper sulfate.

$\dfrac{P}{100} \times of = is$

$\dfrac{80}{100} \cdot 400 = NC$

$NC = \mathbf{320\ tons}$

7. (a) $5x + 2y = 70$
   (b) $3x - 2y = 10$
$\overline{\quad 8x \qquad = 80}$
$x = 10$

(b) $3(10) - 2y = 10$
$-2y = -20$
$y = 10$
(10, 10)

8. $(3x^3 - 2x)(2x^2 - x - 4)$
$= 6x^5 - 3x^4 - 12x^3 - 4x^3 + 2x^2 + 8x$
$= \mathbf{6x^5 - 3x^4 - 16x^3 + 2x^2 + 8x}$

9. $4\sqrt{3} \cdot 3\sqrt{12} \cdot 2\sqrt{3} = 4\sqrt{3} \cdot 3\sqrt{2}\sqrt{2}\sqrt{3} \cdot 2\sqrt{3}$
$= 24\sqrt{2}\sqrt{2}\sqrt{3}\sqrt{3}\sqrt{3} = \mathbf{144\sqrt{3}}$

10. $3\sqrt{75} - 4\sqrt{48} = 3\sqrt{3}\sqrt{5}\sqrt{5} - 4\sqrt{2}\sqrt{2}\sqrt{2}\sqrt{2}\sqrt{3}$
$= 15\sqrt{3} - 16\sqrt{3} = \mathbf{-\sqrt{3}}$

11. $2\sqrt{5}(5\sqrt{5} - 3\sqrt{15}) = 2\sqrt{5}(5\sqrt{5} - 3\sqrt{3}\sqrt{5})$
$= 10\sqrt{5}\sqrt{5} - 6\sqrt{3}\sqrt{5}\sqrt{5} = \mathbf{50 - 30\sqrt{3}}$

12. $3xy^2m + \dfrac{4}{x} = \dfrac{3x^2y^2m}{x} + \dfrac{4}{x} = \mathbf{\dfrac{3x^2y^2m + 4}{x}}$

13. $\dfrac{5x^2}{pm} - 4 + \dfrac{c}{p^2m} = \dfrac{5x^2p}{p^2m} - \dfrac{4p^2m}{p^2m} + \dfrac{c}{p^2m}$
$= \mathbf{\dfrac{5x^2p - 4p^2m + c}{p^2m}}$

14. $\dfrac{(0.00003)(0.006 \times 10^{-6})}{(1800 \times 10^{15})(100,000)}$
$= \dfrac{(3 \times 10^{-5})(6 \times 10^{-9})}{(1.8 \times 10^{18})(1 \times 10^5)} = \dfrac{18 \times 10^{-14}}{18 \times 10^{22}}$
$= \mathbf{1 \times 10^{-36}}$

15. (a) $2y - 2x = 8$
$$2y = 2x + 8$$
$$y = x + 4$$
(b) $y + x = -2$
$$y = -x - 2$$

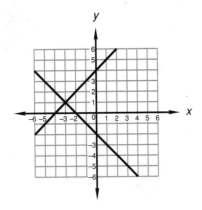

(a) $2y - 2x = 8$
2(b) $\underline{2y + 2x = -4}$
$\ \ 4y = 4$
$\ y = 1$

(b) $(1) + x = -2$
$\ x = -3$

(−3, 1)

16. (a) $y - 2x = 1$
$\ y = 2x + 1$
(b) $y = -2$

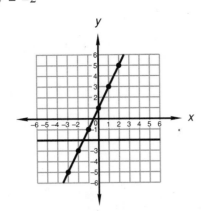

Substitute (b) into (a) and get:
(a′) $(-2) - 2x = 1$
$\ -2x = 3$
$\ x = -\dfrac{3}{2}$

$\left(-\dfrac{3}{2},\ -2\right)$

17. Graph the line to find the slope.

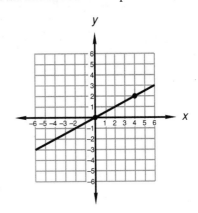

Slope $= \dfrac{+1}{+2} = \dfrac{1}{2}$

$y = \dfrac{1}{2}x + b$

$0 = \dfrac{1}{2}(0) + b$

$0 = b$

Since $m = \dfrac{1}{2}$ and $b = 0$: $\boldsymbol{y = \dfrac{1}{2}x}$

18. $D^2 = 4^2 + 2^2$
$D^2 = 16 + 4$
$D^2 = 20$
$D = \sqrt{20}$
$\boldsymbol{D = 2\sqrt{5}}$

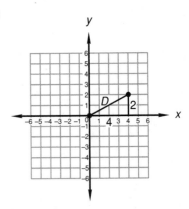

19. $y = -\dfrac{3}{8}x + b$

$4 = -\dfrac{3}{8}(4) + b$

$4 = -\dfrac{12}{8} + b$

$\dfrac{8}{2} + \dfrac{3}{2} = b$

$\dfrac{11}{2} = b$

Since $m = -\dfrac{3}{8}$ and $b = \dfrac{11}{2}$: $\boldsymbol{y = -\dfrac{3}{8}x + \dfrac{11}{2}}$

20. $13^2 = (BC)^2 + 5^2$
$169 = (BC)^2 + 25$
$144 = (BC)^2$
$\textbf{12 m} = BC$

$CM = \dfrac{1}{2}BC = \dfrac{1}{2}(12\text{ m}) = \textbf{6 m}$

$\text{Area}_{\text{ACM}} = \dfrac{6\text{ m} \times 5\text{ m}}{2} = \textbf{15 m}^2$

21. $\dfrac{x}{4} = \dfrac{7}{3} \rightarrow 3x = 28 \rightarrow x = \dfrac{28}{3}$

$\dfrac{y}{6} = \dfrac{3}{7} \rightarrow 7y = 18 \rightarrow y = \dfrac{18}{7}$

22. $3\dfrac{1}{4}x + 5\dfrac{1}{3} = 2\dfrac{1}{6}$

$\dfrac{13}{4}x = \dfrac{13}{6} - \dfrac{32}{6}$

$\dfrac{13}{4}x = -\dfrac{19}{6}$

$x = -\dfrac{19}{6} \cdot \dfrac{4}{13} = -\dfrac{76}{78} = -\dfrac{38}{39}$

23. $0.3(2p - 4) = 0.1(p + 3)$

$3(2p - 4) = 1(p + 3)$

$6p - 12 = p + 3$

$5p = 15$

$p = 3$

24. $-[(-4 - 6)(-4) - 2] = -2^0(x - 4)$

$-[40 - 2] = -x + 4$

$-38 = -x + 4$

$x = 42$

25. $xy^{-2}\left(\dfrac{x^0 y^2}{x}, - \dfrac{3x^0 y^2}{x^2}\right) = \dfrac{x}{x} - \dfrac{3x}{x^2}$

$= 1 - 3x^{-1}$

26. $\dfrac{(-3xy)^{-2} x^2 y^2}{(x^{-3})^2 yx^2} = \dfrac{x^{-2}y^{-2}x^2y^2}{9x^{-6}yx^2} = \dfrac{x^4}{9y}$

27. $\dfrac{3m}{x} - \dfrac{2x^{-1}}{m^0 m^{-1}} + \dfrac{5x^2 m^2}{x^3 m} = \dfrac{3m}{x} - \dfrac{2m}{x} + \dfrac{5m}{x}$

$= \dfrac{6m}{x}$

28. $xy - (x - y) = -\dfrac{1}{2}(2) - \left(-\dfrac{1}{2} - 2\right)$

$= -1 - \left(-\dfrac{1}{2} - \dfrac{4}{2}\right) = -\dfrac{2}{2} + \dfrac{5}{2} = \dfrac{3}{2}$

29. $-3^0[(-2 - 3 + 8) - |-3 - 5| - 5^0]$

$= -1[3 - 8 - 1] = -1[-6] = 6$

30. $\dfrac{-2^0}{-2^2} - \dfrac{(-2)^0}{(-2)^{-3}} = \dfrac{1}{4} + 8 = 8\dfrac{1}{4}$

PROBLEM SET 24

1. $D_F = D_S$ so $R_F T_F = R_S T_S$

$T_F = 6;\ T_S = 8;\ R_F = 60$

$60(6) = R_S(8)$

45 mph $= R_S$

2. $D_1 = D_2$ so $R_1 T_1 = R_2 T_2$

$R_1 = 4;\ R_2 = 5;\ T_1 = T_2 + 1$

$4(T_2 + 1) = 5T_2$

$4T_2 + 4 = 5T_2$

$4 = T_2$

$D_1 = D_2 = 4(5) =$ **20 miles**

3. (a) $\dfrac{N}{D} = \dfrac{5}{7} \rightarrow 7N = 5D$

(b) $N + D = 120$

Substitute $D = 120 - N$ into (a) and get:

(a′) $7N = 5(120 - N)$

$12N = 600$

$N = 50$

(b) $(50) + D = 120$

$D = 70$

The fraction was $\dfrac{50}{70}$.

4. (a) $Q_C + Q_N = 173$

(b) $Q_C - Q_N = 11$

$\overline{2Q_C \qquad\quad = 184}$

$Q_C =$ **92**

(b) $(92) - Q_N = 11$

$Q_N =$ **81**

5. (a) $700M_R + 900M_P = 41{,}000$

(b) $M_R + M_P = 50$

1(a) $\quad 700M_R + 900M_P = \quad 41{,}000$

-900(b) $\underline{-900M_R - 900M_P = -45{,}000}$

$-200M_R \qquad\qquad = \quad -4000$

$M_R =$ **20**

6. $\dfrac{48{,}300}{49{,}000} = \dfrac{NP}{4200}$

$49{,}000NP = 4200(48{,}300)$

$NP =$ **4140 kilograms**

7. (a) $7x + 9y = 119$
(b) $2x + y = 23 \rightarrow y = 23 - 2x$
Substitute $y = 23 - 2x$ into (a) and get:
(a′) $7x + 9(23 - 2x) = 119$
$7x + 207 - 18x = 119$
$-11x = -88$
$x = 8$

(b) $2(8) + y = 23$
$y = 7$

(8, 7)

8.

$$
\begin{array}{r}
-3x^2 + 6x - 12 + \frac{-27}{-x-2} \\
-x - 2 \overline{)\ 3x^3 + 0x^2 + 0x - 3} \\
\underline{3x^3 + 6x^2} \\
-6x^2 + 0x \\
\underline{-6x^2 - 12x} \\
12x - 3 \\
\underline{12x + 24} \\
-27
\end{array}
$$

Check: $-\dfrac{3x^2(-x - 2)}{-x - 2} + \dfrac{6x(-x - 2)}{-x - 2}$

$-\dfrac{12(-x - 2)}{-x - 2} - \dfrac{27}{-x - 2}$

$= \dfrac{3x^3 + 6x^2 - 6x^2 - 12x}{-x - 2}$

$+ \dfrac{12x + 24 - 27}{-x - 2}$

$= \dfrac{3x^3 - 3}{-x - 2}$

9. $4\sqrt{3} \cdot 5\sqrt{2} \cdot 6\sqrt{12} = 4\sqrt{3} \cdot 5\sqrt{2} \cdot 6\sqrt{2}\sqrt{2}\sqrt{3}$
$= 120\sqrt{2}\sqrt{2}\sqrt{2}\sqrt{3}\sqrt{3} = \mathbf{720\sqrt{2}}$

10. $4\sqrt{63} - 3\sqrt{28} = 4\sqrt{3}\sqrt{3}\sqrt{7} - 3\sqrt{2}\sqrt{2}\sqrt{7}$
$= 12\sqrt{7} - 6\sqrt{7} = \mathbf{6\sqrt{7}}$

11. $3\sqrt{2}(5\sqrt{2} - 6\sqrt{12}) = 3\sqrt{2}(5\sqrt{2} - 6\sqrt{2}\sqrt{2}\sqrt{3})$
$= 15\sqrt{2}\sqrt{2} - 18\sqrt{2}\sqrt{2}\sqrt{2}\sqrt{3} = \mathbf{30 - 36\sqrt{6}}$

12. $2\sqrt{2}(5\sqrt{10} - 3\sqrt{2}) = 2\sqrt{2}(5\sqrt{2}\sqrt{5} - 3\sqrt{2})$
$= 10\sqrt{2}\sqrt{2}\sqrt{5} - 6\sqrt{2}\sqrt{2} = \mathbf{20\sqrt{5} - 12}$

13. $4m^2yp + \dfrac{6}{m^2y} = \dfrac{4m^4y^2p}{m^2y} + \dfrac{6}{m^2y}$

$= \dfrac{\mathbf{4m^4y^2p + 6}}{\mathbf{m^2y}}$

14. $\dfrac{k^2}{2p} + c - \dfrac{4}{p^2c} = \dfrac{k^2pc}{2p^2c} + \dfrac{2p^2c^2}{2p^2c} - \dfrac{8}{2p^2c}$

$= \dfrac{k^2pc + 2p^2c^2 - 8}{2p^2c}$

15. $\dfrac{(0.0007 \times 10^{-23})(4000 \times 10^6)}{(0.00004)(7{,}000{,}000)}$

$= \dfrac{(7 \times 10^{-27})(4 \times 10^9)}{(4 \times 10^{-5})(7 \times 10^6)} = \dfrac{28 \times 10^{-18}}{28 \times 10^1}$

$= \mathbf{1 \times 10^{-19}}$

16. (a) $3x + 2y = 12$
$2y = -3x + 12$
$y = -\dfrac{3}{2}x + 6$

(b) $5x - 4y = 8$
$-4y = -5x + 8$
$y = \dfrac{5}{4}x - 2$

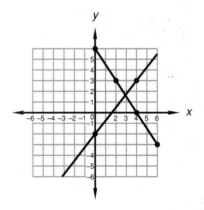

2(a) $6x + 4y = 24$
(b) $\underline{5x - 4y = 8}$
$11x = 32$
$x = \dfrac{32}{11}$

(a) $3\left(\dfrac{32}{11}\right) + 2y = 12$

$2y = \dfrac{132}{11} - \dfrac{96}{11}$

$y = \dfrac{36}{11} \cdot \dfrac{1}{2} = \dfrac{18}{11}$

$\left(\dfrac{\mathbf{32}}{\mathbf{11}}, \dfrac{\mathbf{18}}{\mathbf{11}}\right)$

17. (a) The y intercept is +2. The slope is positive, and the rise over the run for any triangle drawn is $\dfrac{5}{6}$.

$y = \dfrac{\mathbf{5}}{\mathbf{6}}x + \mathbf{2}$

(b) Every point on this line is 4 units below the x axis.

$y = \mathbf{-4}$

18. Since the measure of a central angle is the same as the measure of an arc of a circle:
$A = \mathbf{70}$
$B + 70 = 180$
$B = \mathbf{110}$

Since angles opposite equal sides are equal angles:
$C = D = 35$

Area $= \dfrac{70}{360}\pi r^2 = \dfrac{70}{360}(\pi)(4 \text{ cm})^2 \approx \textbf{9.77 cm}^2$

19. $(\sqrt{21})^2 = H^2 + 2^2$

$21 = H^2 + 4$

$17 = H^2$

$\sqrt{17} = H$

Area $= \dfrac{4 \text{ in.} \times \sqrt{17} \text{ in.}}{2} = \textbf{2}\sqrt{\textbf{17}} \textbf{ in.}^2$

20. Graph the line to find the slope.

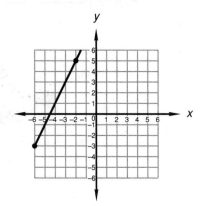

Slope $= \dfrac{+2}{+1} = 2$

$y = 2x + b$

$5 = 2(-2) + b$

$9 = b$

Since $m = 2$ and $b = 9$:

$\textbf{y = 2x + 9}$

21. $y = \dfrac{2}{9}x + b$

$-3 = \dfrac{2}{9}(5) + b$

$-3 = \dfrac{10}{9} + b$

$-\dfrac{37}{9} = b$

Since $m = \dfrac{2}{9}$ and $b = -\dfrac{37}{9}$: $\textbf{y} = \dfrac{\textbf{2}}{\textbf{9}}\textbf{x} - \dfrac{\textbf{37}}{\textbf{9}}$

22. $\dfrac{k + 3}{3} = \dfrac{9}{4} \longrightarrow 4k + 12 = 27$

$4k = 15$

$k = \dfrac{\textbf{15}}{\textbf{4}}$

23. $\dfrac{4x + 2}{3} - \dfrac{3}{4} = \dfrac{1}{2}$

$16x + 8 - 9 = 6$

$16x = 7$

$x = \dfrac{\textbf{7}}{\textbf{16}}$

24. $\dfrac{3x}{2} + \dfrac{8 - 4x}{7} = 3$

$21x + 16 - 8x = 42$

$13x = 26$

$x = \textbf{2}$

25. $0.07 - 0.003x + 0.2 = 1.02$

$70 - 3x + 200 = 1020$

$-3x = 750$

$x = \textbf{-250}$

26. $-2[x - (-2 - 4^0) - 3] + [(x - 2)(-3)] = -x$

$-2[x + 3 - 3] + [-3x + 6] = -x$

$-2x - 3x + 6 = -x$

$-4x = -6$

$x = \dfrac{\textbf{3}}{\textbf{2}}$

27. $\dfrac{xy^{-2}}{p^2}\left(\dfrac{p^2 y^2}{x} + \dfrac{5x^2 y^3}{p^{-2}}\right) = \dfrac{xp^2}{p^2 x} + 5x^3 y$

$= \textbf{1 + 5x}^3\textbf{y}$

28. $\dfrac{4x^2 y^{-2}x}{(-2x^0)^{-3}} = -\dfrac{32x^3 y^{-2}}{x^0} = \textbf{-32x}^3\textbf{y}^{-2}$

29. $\dfrac{3p^2 x^2}{xy} - \dfrac{5ppy^{-1}}{x^{-1}} - \dfrac{5p^2 x^2 x^2}{x^{-3}y}$

$= \dfrac{3p^2 x}{y} - \dfrac{5p^2 x}{y} - \dfrac{5p^2 x^7}{y}$

$= \dfrac{\textbf{-2p}^2\textbf{x}}{\textbf{y}} - \dfrac{\textbf{5p}^2\textbf{x}^7}{\textbf{y}}$

30. $xa(ax - a) = \dfrac{1}{2}\left(-\dfrac{1}{3}\right)\left(-\dfrac{1}{3}\left(\dfrac{1}{2}\right) - \left(-\dfrac{1}{3}\right)\right)$

$= -\dfrac{1}{6}\left(-\dfrac{1}{6} + \dfrac{2}{6}\right) = -\dfrac{1}{6}\left(\dfrac{1}{6}\right) = -\dfrac{\textbf{1}}{\textbf{36}}$

PROBLEM SET 25

1. $\dfrac{38}{14,440} = \dfrac{D}{36,100} \longrightarrow 14,440D = 38(36,100)$

$D = \textbf{95 days}$

2. $D_B = D_F$ so $R_B T_B = R_F T_F$

$T_B = 40; \ T_F = 30; \ R_F = R_B + 6$

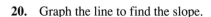

$40R_B = (R_B + 6)30$

$40R_B = 30R_B + 180$

$10R_B = 180$

$R_B = 18 \text{ kph}$

$D_F = D_B = 18(40) = \textbf{720 kilometers}$

3. (a) $N_B = 5N_G$
(b) $N_B = 15N_G - 100$

Substitute (a) into (b) and get:
(b′) $(5N_G) = 15N_G - 100$
$-10N_G = -100$
$N_G = \mathbf{10}$

(a) $N_B = 5(10) = \mathbf{50}$

4. (a) $\dfrac{D_I}{D_A} = \dfrac{5}{2} \rightarrow 2D_I = 5D_A$
(b) $D_I + D_A = 980 \rightarrow D_A = 980 - D_I$

Substitute (b) into (a) and get:
(a′) $2D_I = 5(980 - D_I)$
$7D_I = 4900$
$D_I = \mathbf{700}$

5. (a) $N_Q + N_H = 200$
(b) $25N_Q + 50N_H = 7500$

$-50\text{(a)} \quad -50N_Q - 50N_H = -10{,}000$
$1\text{(b)} \quad \underline{25N_Q + 50N_H = 7500}$
$-25N_Q = -2500$
$N_Q = \mathbf{100}$

(a) $(100) + N_H = 200$
$N_H = \mathbf{100}$

6. If 70% was sodium chloride, 30% was other chemicals.

$\dfrac{P}{100} \times of = is$

$\dfrac{30}{100} \times TW = 660$

$TW = 660 \cdot \dfrac{100}{30} = \mathbf{2200\ grams}$

7.
$$x - 5 \overline{) \begin{array}{l} x^2 + 5x + 25 + \frac{123}{x-5} \\ x^3 + 0x^2 + 0x - 2 \\ \underline{x^3 - 5x^2} \\ 5x^2 + 0x \\ \underline{5x^2 - 25x} \\ 25x - 2 \\ \underline{25x - 125} \\ 123 \end{array}}$$

Check: $\dfrac{x^2(x-5)}{x-5} + \dfrac{5x(x-5)}{x-5}$
$+ \dfrac{25(x-5)}{x-5} + \dfrac{123}{x-5}$
$= \dfrac{x^3 - 5x^2 + 5x^2 - 25x + 25x - 2}{x-5}$
$= \dfrac{x^3 - 2}{x-5}$

8. $5x^2y^2 - 2xy + 10xy^2 = \mathbf{xy(5xy - 2 + 10y)}$

9. $x^2y^3m^5 + 12x^3ym^4 - 3x^2y^2m^2$
$= \mathbf{x^2ym^2(y^2m^3 + 12xm^2 - 3y)}$

10. $16m^2p^3y - 8y^4mp^3 + 4m^2p^2y^2$
$= \mathbf{4mp^2y(4mp - 2py^3 + my)}$

11. $x^3y^2z^3 + x^2yz^2 - 3x^3yz = \mathbf{x^2yz(xyz^2 + z - 3x)}$

12. $p^5x^3 + p^4x^2 - p^3x = \mathbf{p^3x(p^2x^2 + px - 1)}$

13. $2\sqrt{3} \cdot 3\sqrt{6} \cdot 5\sqrt{12} = 2\sqrt{3} \cdot 3\sqrt{2}\sqrt{3} \cdot 5\sqrt{2}\sqrt{2}\sqrt{3}$
$= 30\sqrt{2}\sqrt{2}\sqrt{2}\sqrt{3}\sqrt{3}\sqrt{3} = \mathbf{180\sqrt{6}}$

14. $6\sqrt{18} + 5\sqrt{8} - 3\sqrt{50}$
$= 6\sqrt{2}\sqrt{3}\sqrt{3} + 5\sqrt{2}\sqrt{2}\sqrt{2} - 3\sqrt{2}\sqrt{5}\sqrt{5}$
$= 18\sqrt{2} + 10\sqrt{2} - 15\sqrt{2} = \mathbf{13\sqrt{2}}$

15. $2\sqrt{5}(3\sqrt{15} - 2\sqrt{5}) = 2\sqrt{5}(3\sqrt{3}\sqrt{5} - 2\sqrt{5})$
$= 6\sqrt{3}\sqrt{5}\sqrt{5} - 4\sqrt{5}\sqrt{5} = \mathbf{30\sqrt{3} - 20}$

16. $a + \dfrac{a}{b} = \dfrac{ab}{b} + \dfrac{a}{b} = \mathbf{\dfrac{ab + a}{b}}$

17. $\dfrac{ax^2}{m^2p} - c + \dfrac{2}{m} = \dfrac{ax^2}{m^2p} - \dfrac{cm^2p}{m^2p} + \dfrac{2mp}{m^2p}$
$= \mathbf{\dfrac{ax^2 - cm^2p + 2mp}{m^2p}}$

18. $\dfrac{(38{,}000 \times 10^3)(300 \times 10^{-4})}{0.00019 \times 10^{-5}}$
$= \dfrac{(38 \times 10^6)(3 \times 10^{-2})}{19 \times 10^{-10}} = \dfrac{114 \times 10^4}{19 \times 10^{-10}}$
$= \mathbf{6 \times 10^{14}}$

19. (a) $3x - 2y = 10$
$$-2y = -3x + 10$$
$$y = \frac{3}{2}x - 5$$

(b) $y = -\frac{1}{2}$

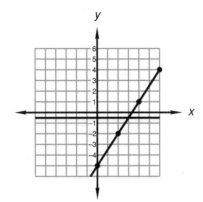

Substitute (b) into (a) and get:

(a′) $3x - 2\left(-\frac{1}{2}\right) = 10$
$$3x + 1 = 10$$
$$3x = 9$$
$$x = 3$$

$$\left(3, -\frac{1}{2}\right)$$

20. Since parallel lines have the same slope:

$$y = \frac{1}{6}x + b$$

$$5 = \frac{1}{6}(3) + b$$

$$\frac{9}{2} = b$$

Since $m = \frac{1}{6}$ and $b = \frac{9}{2}$: $\mathbf{y = \frac{1}{6}x + \frac{9}{2}}$

21. (a) $6x - 2y = 100$
(b) $\underline{3x + 2y = \ \ 80}$
$$9x \qquad = 180$$
$$x = \mathbf{20}$$

(b) $3(20) + 2y = 80$
$$2y = 20$$
$$y = \mathbf{10}$$

$$7 \times \overrightarrow{SF} = 11$$
$$\overrightarrow{SF} = \frac{11}{7}$$
$$9 \times \frac{11}{7} = P$$
$$\frac{\mathbf{99}}{\mathbf{7}} = P$$

22. Arc length $= \dfrac{280}{360} \cdot 2\pi r = \dfrac{280}{360} \cdot 2\pi(12 \text{ cm})$
$$\approx \mathbf{58.61 \text{ cm}}$$

23. $\dfrac{x + 1}{4} - \dfrac{3}{2} = \dfrac{2x - 9}{10}$
$$5x + 5 - 30 = 4x - 18$$
$$x = \mathbf{7}$$

24. $\dfrac{4x - 8}{5} + \dfrac{2x - 4}{2} = 9$
$$8x - 16 + 10x - 20 = 90$$
$$18x = 126$$
$$x = \mathbf{7}$$

25. $\dfrac{n + 3}{6} - \dfrac{1}{3} = \dfrac{2n - 2}{5}$
$$5n + 15 - 10 = 12n - 12$$
$$-7n = -17$$
$$n = \frac{\mathbf{17}}{\mathbf{7}}$$

26. $\dfrac{5x + 3}{2} - \dfrac{3}{4} = \dfrac{5}{2}$
$$10x + 6 - 3 = 10$$
$$10x = 7$$
$$x = \frac{\mathbf{7}}{\mathbf{10}}$$

27. $\dfrac{8x^5 + 12x^3}{4x^3} = \dfrac{4x^3(2x^2 + 3)}{4x^3} = \mathbf{2x^2 + 3}$

28. $\dfrac{x^2 y^{-2}}{z^2} \cdot \left(\dfrac{z^2}{y^2(2x^{-2})^{-1}} - \dfrac{4x^2 y^0}{z^{-2}}\right)$

$$= \dfrac{2x^2 y^{-2} z^2}{z^2 y^2 x^2} - 4x^4 y^{-2} = \mathbf{2y^{-4} - 4x^4 y^{-2}}$$

29. $-3^0[(-2 - 4 - 2^2 - 2^0) - |-3 - 2|]$
$$= -1[-11 - 5] = -1[-16] = \mathbf{16}$$

30. $km(m^2 k - k) = \dfrac{1}{3}\left(-\dfrac{1}{4}\right)\left(\left(-\dfrac{1}{4}\right)^2\left(\dfrac{1}{3}\right) - \dfrac{1}{3}\right)$

$$= -\dfrac{1}{12}\left(\dfrac{1}{48} - \dfrac{16}{48}\right) = -\dfrac{1}{12}\left(-\dfrac{5}{16}\right) = \frac{\mathbf{5}}{\mathbf{192}}$$

PROBLEM SET 26

1. $D_S = D_O$ so $R_S T_S = R_O T_O$

$T_S = 20$; $T_O = 8$; $R_O = R_S + 60$

$20R_S = (R_S + 60)8$

$20R_S = 8R_S + 480$

$12R_S = 480$

$R_S = 40$

$D_S = D_O = (40)(20) =$ **800 miles**

2. (a) $N_B = 5N_R + 50$

(b) $N_R = N_B - 210$

Substitute (a) into (b) and get:

(b') $N_R = (5N_R + 50) - 210$

$-4N_R = -160$

$N_R =$ **40**

(a) $N_B = 5(40) + 50 =$ **250**

3. (a) $5N_F + 10N_T = 7000$

(b) $N_F + N_T = 1200$

$\begin{array}{ll} 1(a) & 5N_F + 10N_T = 7000 \\ -10(b) & -10N_F - 10N_T = -12{,}000 \\ \hline & -5N_F = -5000 \\ & N_F = \textbf{1000} \end{array}$

(b) $(1000) + N_T = 1200$

$N_T =$ **200**

4. If 14% were belligerent, 86% were eristic.

$\dfrac{P}{100} \times of = is$

$\dfrac{86}{100} \times N_T = 4300$

$N_T = 4300 \cdot \dfrac{100}{86} = 5000$

$N_B = 5000 - 4300 =$ **700**

5. $\dfrac{1820}{1960} = \dfrac{NC}{2240}$

$1960NC = 1820(2240)$

$NC =$ **2080**

6. If 10% were nonabsorbent, 90% were absorbent.

$\dfrac{P}{100} \times of = is$

$\dfrac{90}{100} \times C = 7290$

$C = 7290 \cdot \dfrac{100}{90} =$ **8100**

7. Divide $x^4 - 2$ by $x + 1$:

$$x + 1 \overline{)\, x^4 + 0x^3 + 0x^2 + 0x - 2} \quad x^3 - x^2 + x - 1 - \tfrac{1}{x+1}$$

$\dfrac{x^4 + x^3}{}$

$-x^3 + 0x^2$

$\dfrac{-x^3 - x^2}{}$

$x^2 + 0x$

$\dfrac{x^2 + x}{}$

$-x - 2$

$\dfrac{-x - 1}{}$

-1

Check: $\dfrac{x^3(x + 1)}{x + 1} - \dfrac{x^2(x + 1)}{x + 1}$

$+ \dfrac{x(x + 1)}{x + 1} - \dfrac{1(x + 1)}{x + 1} - \dfrac{1}{x + 1}$

$= \dfrac{x^4 + x^3 - x^3 - x^2 + x^2}{x + 1}$

$+ \dfrac{x - x - 1 - 1}{x + 1}$

$= \dfrac{x^4 - 2}{x + 1}$

8. $35x^7y^5m - 7x^5m^2y^2 + 14y^7x^4m^2$

$= 7x^4y^2m(5x^3y^3 - xm + 2y^5m)$

9. $6x^2ym^5 - 2x^2ym + 4xym$

$= 2xym(3xm^4 - x + 2)$

10. $4x^2y^4p^6 - 2xp^5y^7 + 8x^4p^5y^5$

$= 2xy^4p^5(2xp - y^3 + 4x^3y)$

11. $x^2 + x - 6 = (x + 3)(x - 2)$

12. $x^2 - 6x + 8 = (x - 4)(x - 2)$

13. $-2ab + abx + abx^2 = ab(x + 2)(x - 1)$

14. $\dfrac{6x^2y - xy}{xy} = \dfrac{xy(6x - 1)}{xy} = 6x - 1$

15. $3\sqrt{2} \cdot 2\sqrt{6} \cdot 3\sqrt{6} = 3\sqrt{2} \cdot 2\sqrt{2}\sqrt{3} \cdot 3\sqrt{2}\sqrt{3}$

$= 18\sqrt{2}\sqrt{2}\sqrt{2}\sqrt{3}\sqrt{3} = \textbf{108}\sqrt{2}$

16. $-3\sqrt{12} + 5\sqrt{27} - 8\sqrt{25}$

$= -3\sqrt{2}\sqrt{2}\sqrt{3} + 5\sqrt{3}\sqrt{3}\sqrt{3} - 8\sqrt{5}\sqrt{5}$

$= -6\sqrt{3} + 15\sqrt{3} - 40 = \textbf{9}\sqrt{3} - \textbf{40}$

17. $3\sqrt{2}(5\sqrt{3} - 2\sqrt{2}) = 15\sqrt{2}\sqrt{3} - 6\sqrt{2}\sqrt{2}$

$= \textbf{15}\sqrt{6} - \textbf{12}$

18. $\dfrac{a^2m}{x^2} - x^2 - \dfrac{x^2}{c} = \dfrac{a^2cm}{cx^2} - \dfrac{cx^4}{cx^2} - \dfrac{x^4}{cx^2}$

$= \dfrac{a^2cm - cx^4 - x^4}{cx^2}$

19. $\dfrac{(3000 \times 10^{-14})(0.00008)}{(0.0002 \times 10^5)(200,000)}$

$= \dfrac{(3 \times 10^{-11})(8 \times 10^{-5})}{(2 \times 10^1)(2 \times 10^5)} = \dfrac{24 \times 10^{-16}}{4 \times 10^6}$

$= \mathbf{6 \times 10^{-22}}$

20. (a) $2y - x = 6$

$\quad\quad 2y = x + 6$

$\quad\quad y = \dfrac{1}{2}x + 3$

(b) $y - 2x = -3$

$\quad\quad y = 2x - 3$

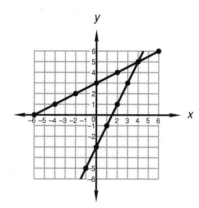

Substitute $y = 2x - 3$ into (a) and get:

(a′) $2(2x - 3) - x = 6$

$\quad\quad 4x - 6 - x = 6$

$\quad\quad\quad 3x = 12$

$\quad\quad\quad x = 4$

(b) $y - 2(4) = -3$

$\quad\quad y = 5$

$\mathbf{(4, 5)}$

21. $D^2 = 7^2 + 3^2$

$D^2 = 49 + 9$

$D^2 = 58$

$D = \mathbf{\sqrt{58}}$

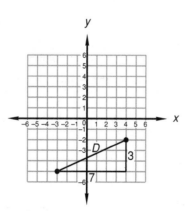

22. $A^2 = 5^2 + 4^2$

$A^2 = 25 + 16$

$A^2 = 41$

$A = \mathbf{\sqrt{41}}$

$4 \times \overrightarrow{SF} = 7$

$\overrightarrow{SF} = \dfrac{7}{4}$

$\sqrt{41} \cdot \dfrac{7}{4} = B + \sqrt{41}$

$7\sqrt{41} = 4B + 4\sqrt{41}$

$3\sqrt{41} = 4B$

$\dfrac{\mathbf{3\sqrt{41}}}{\mathbf{4}} = B$

$5 \cdot \dfrac{7}{4} = C$

$\dfrac{\mathbf{35}}{\mathbf{4}} = C$

23. $x + 10 = 90$

$\quad\quad x = \mathbf{80}$

$y + 80 + 70 = 180$

$\quad\quad\quad y = \mathbf{30}$

$m + 30 = 180$

$\quad\quad m = \mathbf{150}$

$z + 10 + 150 = 180$

$\quad\quad\quad z = \mathbf{20}$

24. (a) $4x + y = 70$

(b) $2x + 6y - 1 = 115$

(b′) $2x + 6y = 116$

$\begin{array}{r} -6\text{(a)} \quad -24x - 6y = -420 \\ \text{(b′)} \quad 2x + 6y = 116 \\ \hline -22x \quad\quad = -304 \end{array}$

$x = \dfrac{\mathbf{152}}{\mathbf{11}}$

(a) $4\left(\dfrac{152}{11}\right) + y = 70$

$y = \dfrac{770}{11} - \dfrac{608}{11} = \dfrac{\mathbf{162}}{\mathbf{11}}$

25. $\text{Area}_{\text{Triangle}} = \dfrac{b \times H}{2} = 54 \text{ m}^2$

$b = \dfrac{54(2) \text{ m}^2}{9 \text{ m}}$

$b = \mathbf{12 \text{ m}} = BC$

$\text{Area}_{\text{Circle}} = \pi r^2 = \pi(12 \text{ m})^2 \approx \mathbf{452.16 \text{ m}^2}$

26. $\dfrac{2x}{3} + \dfrac{6 - 3x}{4} = 7$

$8x + 18 - 9x = 84$

$-x = 66$

$x = \mathbf{-66}$

27. $\dfrac{4 - x}{3} + \dfrac{1}{6} = 5$

$8 - 2x + 1 = 30$

$-2x = 21$

$x = -\dfrac{\mathbf{21}}{\mathbf{2}}$

28. $-2x - \dfrac{3^0 - x}{2} + \dfrac{x - 5^0}{7} = 2$

$-28x - 7 + 7x + 2x - 2 = 28$

$-19x = 37$

$x = -\dfrac{37}{19}$

29. $\dfrac{x^2(yz^0)^{-2}}{(-2x^2y)^{-3}} = \dfrac{-8x^2y^{-2}}{x^{-6}y^{-3}} = -8x^8y$

30. $-\dfrac{1}{-(-3)^{-2}} + \dfrac{1}{-2^{-3}}(a - ba)$

$= 9 - 8\left(-\dfrac{1}{2} - 3\left(-\dfrac{1}{2}\right)\right) = 9 - 8\left(-\dfrac{1}{2} + \dfrac{3}{2}\right)$

$= 9 - 8(1) = 9 - 8 = \mathbf{1}$

PROBLEM SET 27

1. $D_H = D_T$ so $R_H T_H = R_T T_T$

$R_H = 4;\ R_T = 20;\ T_H + T_T = 18$

$4T_H = 20(18 - T_H)$

$4T_H = 360 - 20T_H$

$24T_H = 360$

$T_H = 15$

$D_T = D_H = 15(4) = \mathbf{60\ miles}$

2. $D_H = D_R$ so $R_H T_H = R_R T_R$

$R_H = 5;\ R_R = 20;\ T_H + T_R = 10$

$5T_H = 20(10 - T_H)$

$5T_H = 200 - 20T_H$

$25T_H = 200$

$T_H = 8$

$D_H = D_R = 5(8) = \mathbf{40\ kilometers}$

3. (a) $N_G = 2N_B + 6$

(b) $N_G + N_B = 36$

Substitute (a) into (b) and get:

(b') $(2N_B + 6) + N_B = 36$

$3N_B = 30$

$N_B = \mathbf{10}$

(b) $N_G + (10) = 36$

$N_G = \mathbf{26}$

4. $7N \qquad 7N + 7 \qquad 7N + 14$

$-3(7N + 7N + 14) = 5(-7N - 7) - 21$

$-42N - 42 = -35N - 56$

$-7N = -14$

$N = 2$

The desired integers are **14**, **21**, and **28**.

5. $\dfrac{600}{3000} = \dfrac{B}{9000}$

$3000B = 600(9000)$

$B = \mathbf{1800\ grams}$

6. If 70% was silver iodide, 30% was not silver iodide.

$\dfrac{P}{100} \times of = is$

$\dfrac{30}{100} \cdot 2000 = NS$

$NS = \mathbf{600\ grams}$

7.

$$x - 2 \overline{\smash{)}\,x^3 + 0x^2 + 0x - 6} \quad \mathbf{x^2 + 2x + 4} + \tfrac{\mathbf{2}}{\mathbf{x-2}}$$

$\underline{x^3 - 2x^2}$

$\qquad 2x^2 + 0x$

$\qquad \underline{2x^2 - 4x}$

$\qquad\qquad 4x - 6$

$\qquad\qquad \underline{4x - 8}$

$\qquad\qquad\qquad 2$

Check: $\dfrac{x^2(x - 2)}{x - 2} + \dfrac{2x(x - 2)}{x - 2}$

$\qquad + \dfrac{4(x - 2)}{x - 2} + \dfrac{2}{x - 2}$

$= \dfrac{x^3 - 2x^2 + 2x^2}{x - 2}$

$\qquad + \dfrac{-4x + 4x - 8 + 2}{x - 2}$

$= \dfrac{x^3 - 6}{x - 2}$

8. $9m^2x^4p^2 + 3x^2p^6m^4 - 6x^4m^3p^2$

$= \mathbf{3m^2x^2p^2(3x^2 + p^4m^2 - 2x^2m)}$

9. $mx^4y - mx^2y^3 - 4mx^2y = \mathbf{mx^2y(x^2 - y^2 - 4)}$

10. $a^2x^3p - 4a^3x^3p - a^2x^4p = \mathbf{a^2x^3p(1 - 4a - x)}$

11. $4ax + ax^2 - 5a = \mathbf{a(x + 5)(x - 1)}$

12. $8x^2 - x^3 - 15x = \mathbf{-x(x - 5)(x - 3)}$

13. $24ax - 5ax^2 - ax^3 = \mathbf{-ax(x + 8)(x - 3)}$

14. $-ax^4 + 4ax^3 + 5ax^2 = \mathbf{-ax^2(x - 5)(x + 1)}$

15. $56p - 15px + px^2 = \mathbf{p(x - 8)(x - 7)}$

16. $\dfrac{4xa + 4x}{4x} = \dfrac{4x(a + 1)}{4x} = \mathbf{a + 1}$

17. $3\sqrt{2} - 2\sqrt{3} \cdot 3\sqrt{12} = 3\sqrt{2} - 2\sqrt{3} \cdot 3\sqrt{2}\sqrt{2}\sqrt{3}$

$= 3\sqrt{2} - 6\sqrt{2}\sqrt{2}\sqrt{3}\sqrt{3} = \mathbf{3\sqrt{2} - 36}$

18. $-3\sqrt{20} + 2\sqrt{125} + 5\sqrt{45}$

$= -3\sqrt{2}\sqrt{2}\sqrt{5} + 2\sqrt{5}\sqrt{5}\sqrt{5} + 5\sqrt{3}\sqrt{3}\sqrt{5}$

$= -6\sqrt{5} + 10\sqrt{5} + 15\sqrt{5} = \mathbf{19\sqrt{5}}$

19. $2\sqrt{3}(3\sqrt{2} - 3\sqrt{3}) = 6\sqrt{2}\sqrt{3} - 6\sqrt{3}\sqrt{3} = \mathbf{6\sqrt{6} - 18}$

20. $\dfrac{2}{x} + \dfrac{3}{x + p} = \dfrac{2(x + p)}{x(x + p)} + \dfrac{3(x)}{x(x + p)}$

$= \dfrac{2x + 2p + 3x}{x(x + p)} = \dfrac{\mathbf{5x + 2p}}{\mathbf{x(x + p)}}$

21. $\dfrac{x + 3}{x + 6} + 5 + \dfrac{3}{x^2}$

$= \dfrac{x^2(x + 3)}{x^2(x + 6)} + \dfrac{5x^2(x + 6)}{x^2(x + 6)} + \dfrac{3(x + 6)}{x^2(x + 6)}$

$= \dfrac{x^3 + 3x^2 + 5x^3 + 30x^2 + 3x + 18}{x^2(x + 6)}$

$= \dfrac{\mathbf{6x^3 + 33x^2 + 3x + 18}}{\mathbf{x^2(x + 6)}}$

22. $\dfrac{(0.00056 \times 10^4)(7 \times 10^3)}{(0.00049 \times 10^{16})(0.00002 \times 10^{-5})}$

$= \dfrac{(56 \times 10^{-1})(7 \times 10^3)}{(49 \times 10^{11})(2 \times 10^{-10})} = \dfrac{392 \times 10^2}{98 \times 10^1}$

$= \mathbf{4 \times 10^1}$

23. (a) $3x + 2y = 12$

$2y = -3x + 12$

$y = -\dfrac{3}{2}x + 6$

(b) $8x - 2y = 10$

$-2y = -8x + 10$

$y = 4x - 5$

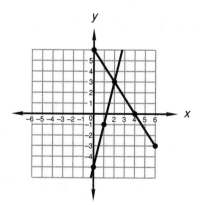

(a) $3x + 2y = 12$
(b) $\underline{8x - 2y = 10}$
$\ 11x\ = 22$
$x = 2$

(a) $3(2) + 2y = 12$

$2y = 6$

$y = 3$

(2, 3)

24. Since parallel lines have the same slope:

$y = -\dfrac{3}{8}x + b$

$-3 = -\dfrac{3}{8}(-3) + b$

$-3 = \dfrac{9}{8} + b$

$-\dfrac{33}{8} = b$

Since $m = -\dfrac{3}{8}$ and $b = -\dfrac{33}{8}$:

$\mathbf{y = -\dfrac{3}{8}x - \dfrac{33}{8}}$

25. $A^2 = 7^2 + 4^2$

$A^2 = 49 + 16$

$A^2 = 65$

$A = \sqrt{65}$

$7 \times \overrightarrow{SF} = 12$

$\overrightarrow{SF} = \dfrac{12}{7}$

$\sqrt{65} \cdot \dfrac{12}{7} = B + \sqrt{65}$

$12\sqrt{65} = 7B + 7\sqrt{65}$

$5\sqrt{65} = 7B$

$\dfrac{\mathbf{5\sqrt{65}}}{\mathbf{7}} = B$

$4 \cdot \dfrac{12}{7} = C$

$\dfrac{\mathbf{48}}{\mathbf{7}} = C$

26. $\dfrac{x}{6} = \dfrac{3}{5} \;\longrightarrow\; 5x = 18 \;\longrightarrow\; x = \dfrac{\mathbf{18}}{\mathbf{5}}$

$\dfrac{5}{8} = \dfrac{7}{y} \;\longrightarrow\; 5y = 56 \;\longrightarrow\; y = \dfrac{\mathbf{56}}{\mathbf{5}}$

27. $\dfrac{6 + x}{5} + \dfrac{1}{3} = 8$

$18 + 3x + 5 = 120$

$3x = 97$

$x = \dfrac{\mathbf{97}}{\mathbf{3}}$

28. $\dfrac{3x - 2}{3} + 4 = \dfrac{1}{6}$

$6x - 4 + 24 = 1$

$6x = -19$

$x = -\dfrac{\mathbf{19}}{\mathbf{6}}$

29. $\dfrac{3a^{-2}y}{b} + 7b^{-1}ya^{-2} - \dfrac{5b^{-1}}{a^2y^{-1}}$

$= \dfrac{3y}{a^2b} + \dfrac{7y}{a^2b} - \dfrac{5y}{a^2b} = \dfrac{5y}{a^2b}$

30. $-3^0(-3 - 2^2) - 4(-2)ax - a$
$= -(-7) + 8(-2)(4) - (-2)$
$= 7 - 64 + 2 = \mathbf{-55}$

PROBLEM SET 28

1. $D_R = D_P$ so $R_RT_R = R_PT_P$
$R_R = 60;\ R_P = 3;\ T_R + T_P = 21$

$\begin{array}{c} \xrightarrow{D_R} \\ \xleftarrow{D_P} \end{array}$

$\quad 60(21 - T_P) = 3T_P$
$1260 - 60T_P = 3T_P$
$\quad\quad 1260 = 63T_P$
$\quad\quad\ \ 20 = T_P$
$D_P = 3(20) = \mathbf{60\ miles}$

2. $3N \quad\quad 3N + 3 \quad\quad 3N + 6 \quad\quad 3N + 9$
$5(3N + 3N + 9) = 13(3N + 6) - 6$
$\quad\quad 30N + 45 = 39N + 72$
$\quad\quad\quad\ -9N = 27$
$\quad\quad\quad\quad\ N = -3$
The desired integers are $\mathbf{-9, -6, -3,}$ and $\mathbf{0}$.

3. (a) $N_G = 3N_B - 1$
(b) $N_G + N_B = 15$
Substitute (a) into (b) and get:
(b′) $(3N_B - 1) + N_B = 15$
$\quad\quad\quad\quad\quad 4N_B = 16$
$\quad\quad\quad\quad\quad\ N_B = \mathbf{4}$
(a) $N_G = 3(4) - 1 = \mathbf{11}$

4. (a) $5N_N + 10N_D = 3000$
(b) $N_N + N_D = 500$

$\quad\ \ 1(a) \quad\ 5N_N + 10N_D = \ \ 3000$
$-10(b) \ -10N_N - 10N_D = -5000$
$\quad\quad\quad\quad\ \overline{\ -5N_N \quad\quad\quad = -2000}$
$\quad\quad\quad\quad\quad\quad\quad\ N_N = \mathbf{400}$

(b) $(400) + N_D = 500$
$\quad\quad\quad\quad\ N_D = \mathbf{100}$

5. If 10 had seen a Dane, 190 had never seen a Dane.
$\dfrac{190}{200} = \dfrac{NS}{150{,}000}$
$200NS = 190(150{,}000)$
$\quad\ NS = \mathbf{142{,}500}$

6. If 16% was arsenic, 84% was silicon.

$\dfrac{P}{100} \times of = is$

$\dfrac{84}{100} \times EM = 7350$

$\quad\quad EM = 7350 \cdot \dfrac{100}{84}$
$\quad\quad\quad\quad = \mathbf{8750\ kg} = $ Entire mixture
$M_A = 8750 - 7350 = \mathbf{1400\ kg}$

7.
$$x - 5\ \overline{)\,x^3 + 0x^2 + 0x - 7}\ \ \ \dfrac{x^2 + 5x + 25 + \frac{118}{x-5}}{}$$

$\quad\quad\ \underline{x^3 - 5x^2}$
$\quad\quad\quad\ 5x^2 + 0x$
$\quad\quad\quad\ \underline{5x^2 - 25x}$
$\quad\quad\quad\quad\quad 25x - \ \ 7$
$\quad\quad\quad\quad\quad \underline{25x - 125}$
$\quad\quad\quad\quad\quad\quad\quad\quad 118$

Check: $\dfrac{x^2(x - 5)}{x - 5} + \dfrac{5x(x - 5)}{x - 5}$

$\quad\quad + \dfrac{25(x - 5)}{x - 5} + \dfrac{118}{x - 5}$

$= \dfrac{x^3 - 5x^2 + 5x^2 - 25x}{x - 5}$

$\quad\quad + \dfrac{25x - 125 + 118}{x - 5}$

$= \dfrac{x^3 - 7}{x - 5}$

8. $2x^2y - 8x^4y^4 = \mathbf{2x^2y(1 - 4x^2y^3)}$

9. $4x^2y^3p^3 - 16x^2y^3p - x^4y^3p^4$
$= \mathbf{x^2y^3p(4p^2 - 16 - x^2p^3)}$

10. $-35xy + 2x^2y + x^3y = \mathbf{xy(x + 7)(x - 5)}$

11. $-8a - 7ax + ax^2 = \mathbf{a(x - 8)(x + 1)}$

12. $2m^2 + 3xm^2 + m^2x^2 = \mathbf{m^2(x + 2)(x + 1)}$

13. $-a^2 - a^2x^2 - 2xa^2 = \mathbf{-a^2(x + 1)(x + 1)}$

14. $\dfrac{4x^2 + x}{x} = \dfrac{x(4x + 1)}{x} = \mathbf{4x + 1}$

15. $4\sqrt{27} - 3\sqrt{48} + 2\sqrt{75}$
$= 4\sqrt{3}\sqrt{3}\sqrt{3} - 3\sqrt{2}\sqrt{2}\sqrt{2}\sqrt{2}\sqrt{3} + 2\sqrt{3}\sqrt{5}\sqrt{5}$
$= 12\sqrt{3} - 12\sqrt{3} + 10\sqrt{3} = \mathbf{10\sqrt{3}}$

16. $3\sqrt{5}(\sqrt{15} - 2\sqrt{5}) = 3\sqrt{5}(\sqrt{3}\sqrt{5} - 2\sqrt{5})$
$= 3\sqrt{3}\sqrt{5}\sqrt{5} - 6\sqrt{5}\sqrt{5} = \mathbf{15\sqrt{3} - 30}$

17. $\dfrac{(0.00077 \times 10^{-3})(40 \times 10^{6})}{(0.00011 \times 10^{5})(140,000)}$

$= \dfrac{(77 \times 10^{-8})(4 \times 10^{7})}{(11)(14 \times 10^{4})} = \dfrac{308 \times 10^{-1}}{154 \times 10^{4}}$

$= \mathbf{2 \times 10^{-5}}$

18. $\dfrac{\dfrac{x}{m + p}}{\dfrac{y}{m + p}} \cdot \dfrac{\dfrac{m + p}{y}}{\dfrac{m + p}{y}} = \dfrac{x}{y}$

19. $\dfrac{3}{4\sqrt{15}} \cdot \dfrac{\sqrt{15}}{\sqrt{15}} = \dfrac{3\sqrt{15}}{60} = \dfrac{\sqrt{15}}{\mathbf{20}}$

20. $\dfrac{4a}{a + x} + \dfrac{6}{a} = \dfrac{4a(a)}{a(a + x)} + \dfrac{6(a + x)}{a(a + x)}$

$= \dfrac{\mathbf{4a^{2} + 6a + 6x}}{\mathbf{a(a + x)}}$

21. $\dfrac{2x}{x^{2} + 2x + 1} + \dfrac{3}{x + 1}$

$= \dfrac{2x}{(x + 1)(x + 1)} + \dfrac{3(x + 1)}{(x + 1)(x + 1)}$

$= \dfrac{2x + 3x + 3}{(x + 1)(x + 1)} = \dfrac{\mathbf{5x + 3}}{\mathbf{(x + 1)(x + 1)}}$

22. (a) $5x + 2y = 6$

$2y = -5x + 6$

$y = -\dfrac{5}{2}x + 3$

(b) $y = \dfrac{1}{2}x$

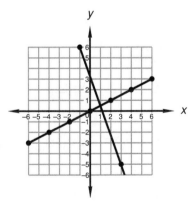

Substitute (b) into (a) and get:

(a) $5x + 2\left(\dfrac{1}{2}x\right) = 6$

$6x = 6$

$x = 1$

(b) $y = \dfrac{1}{2}(1) = \dfrac{1}{2}$

$\left(\mathbf{1, \dfrac{1}{2}}\right)$

23. (a) Every point on this line is 4 units below the x axis.

$\mathbf{y = -4}$

(b) The y intercept is +2. The slope is negative, and the rise over the run for any triangle drawn is $-\dfrac{1}{3}$.

$\mathbf{y = -\dfrac{1}{3}x + 2}$

24. $A + 150 = 180$

$A = \mathbf{30}$

$B + 30 + 90 = 180$

$B = \mathbf{60}$

$C = B = \mathbf{60}$

$D + 60 + 90 = 180$

$D = \mathbf{30}$

$E + 30 + 90 = 180$

$E = \mathbf{60}$

$F + 60 + 40 = 180$

$F = \mathbf{80}$

25. $3x + 90 = 180$

$3x = 90$

$x = \mathbf{30}$

$2x = 2(30) = \mathbf{60}$

$\text{Area}_{\text{Shaded}} = \dfrac{60}{360}(\pi(2\text{ m})^{2}) \approx \mathbf{2.09\ m^{2}}$

$\text{Length of } \widehat{BC} = \dfrac{60}{360}(2\pi(2\text{ m})) \approx \mathbf{2.09\ m}$

26. $\dfrac{2x}{3} + \dfrac{x - 3}{5} = 1$

$10x + 3x - 9 = 15$

$13x = 24$

$x = \dfrac{\mathbf{24}}{\mathbf{13}}$

27. $4\dfrac{1}{3} - 2\dfrac{1}{5}x = -\dfrac{3}{10}$

$-\dfrac{11}{5}x = -\dfrac{9}{30} - \dfrac{130}{30}$

$x = -\dfrac{139}{30} \cdot -\dfrac{5}{11} = \dfrac{\mathbf{139}}{\mathbf{66}}$

28. $D^2 = 8^2 + 4^2$
$D^2 = 64 + 16$
$D^2 = 80$
$D = \sqrt{80}$
$D = 4\sqrt{5}$

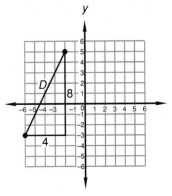

29. $\dfrac{3x^{-2}y^2}{z}\left(\dfrac{x^2 z}{3y^2} - \dfrac{4x^{-2}y}{z^{-3}}\right) = \dfrac{3y^2 z}{3y^2 z} - \dfrac{12\,x^{-4}y^3}{z^{-2}}$

$= 1 - 12x^{-4}y^3z^2$

30. $x^0 - xy^0(x - y)$

$= \left(\dfrac{1}{3}\right)^0 - \dfrac{1}{3}\left(-\dfrac{1}{4}\right)^0\left(\dfrac{1}{3} - \left(-\dfrac{1}{4}\right)\right)$

$= 1 - \dfrac{1}{3}\left(\dfrac{4}{12} + \dfrac{3}{12}\right) = 1 - \dfrac{1}{3}\left(\dfrac{7}{12}\right)$

$= \dfrac{36}{36} - \dfrac{7}{36} = \dfrac{29}{36}$

PROBLEM SET 29

1. $R_W T_W + R_R T_R = 76;$
$R_W = 4;\ R_R = 15;$
$T_W + T_R = 8$

$D_W \qquad D_R$
$\overset{\longleftarrow\qquad\longrightarrow}{\underset{76}{}}$

$4T_W + 15(8 - T_W) = 76$
$4T_W + 120 - 15T_W = 76$
$-11T_W = -44$
$T_W = \textbf{4 hr}$

2. $R_C T_C + R_D T_D = 63;$
$R_C = 3;\ T_C = 9;\ T_D = 6$

$D_C \qquad D_D$
$\overset{\longrightarrow\;\longleftarrow}{\underset{63}{}}$

$3(9) + 6R_D = 63$
$6R_D = 36$
$R_D = \textbf{6 mph}$

3. $R_R T_R + R_M T_M = 7900;$
$R_M = R_R - 400;\ T_R = 6;$
$T_M = 5$

$D_R \qquad D_M$
$\overset{\longleftarrow\qquad\longrightarrow}{\underset{7900}{}}$

$6R_R + (R_R - 400)5 = 7900$
$6R_R + 5R_R - 2000 = 7900$
$11R_R = 9900$
$R_R = \textbf{900 kph}$
$R_M = (900) - 400 = \textbf{500 kph}$

4. (a) $N_R = 2N_P + 15$
(b) $N_R + N_P = 255$
Substitute (a) into (b) and get:
(b′) $(2N_P + 15) + N_P = 255$
$3N_P = 240$
$N_P = 80$
(a) $N_R = 2(80) + 15 = \textbf{175}$

5. $\dfrac{130}{156{,}000} = \dfrac{3250}{H}$
$130H = 3250(156{,}000)$
$H = \textbf{3,900,000}$

6. If 17% contained xenophobes, 83% did not.

$\dfrac{P}{100} \times of = is$

$\dfrac{83}{100} \times HP = 116{,}200$

$HP = 116{,}200 \cdot \dfrac{100}{83} = \textbf{140,000}$

7.
$$\require{enclose}$$
$$\begin{array}{r}
x^2 + 2x + 5 \\
x + 1 \enclose{longdiv}{x^3 + 3x^2 + 7x + 5} \\
\underline{x^3 + x^2} \\
2x^2 + 7x \\
\underline{2x^2 + 2x} \\
5x + 5 \\
\underline{5x + 5}
\end{array}$$

Check: $\dfrac{x^2(x + 1)}{x + 1} + \dfrac{2x(x + 1)}{x + 1}$

$+ \dfrac{5(x + 1)}{x + 1}$

$= \dfrac{x^3 + x^2 + 2x^2 + 2x + 5x + 5}{x + 1}$

$= \dfrac{x^3 + 3x^2 + 7x + 5}{x + 1}$

8. $16x^3y^2z^3 - 8x^2y^2z^2 = \textbf{8}\boldsymbol{x^2y^2z^2}\textbf{(2}\boldsymbol{xz}\textbf{ − 1)}$

9. $2x^2yp^4 - 6x^3yp^3 - 2x^2yp^2$
$= \textbf{2}\boldsymbol{x^2yp^2}\textbf{(}\boldsymbol{p^2}\textbf{ − 3}\boldsymbol{xp}\textbf{ − 1)}$

10. $12a^2x + a^2x^2 + 35a^2 = \boldsymbol{a^2}\textbf{(}\boldsymbol{x}\textbf{ + 7)(}\boldsymbol{x}\textbf{ + 5)}$

11. $-2m^2x - m^2 - m^2x^2 = \boldsymbol{-m^2}\textbf{(}\boldsymbol{x}\textbf{ + 1)(}\boldsymbol{x}\textbf{ + 1)}$

12. $x^2k + 3kx - 40k = \boldsymbol{k}\textbf{(}\boldsymbol{x}\textbf{ + 8)(}\boldsymbol{x}\textbf{ − 5)}$

13. $\dfrac{x^2 + ax^2}{x^2} = \dfrac{x^2(1 + a)}{x^2} = \mathbf{1 + a}$

14. $2\sqrt{75} - 5\sqrt{48} + 2\sqrt{12}$
$= 2\sqrt{3}\sqrt{5}\sqrt{5} - 5\sqrt{2}\sqrt{2}\sqrt{2}\sqrt{2}\sqrt{3} + 2\sqrt{2}\sqrt{2}\sqrt{3}$
$= 10\sqrt{3} - 20\sqrt{3} + 4\sqrt{3} = \mathbf{-6\sqrt{3}}$

15. $2\sqrt{3}(3\sqrt{6} - 4\sqrt{3}) = 2\sqrt{3}(3\sqrt{2}\sqrt{3} - 4\sqrt{3})$
$= 6\sqrt{2}\sqrt{3}\sqrt{3} - 8\sqrt{3}\sqrt{3} = \mathbf{18\sqrt{2} - 24}$

16. $\dfrac{(0.00052 \times 10^{-4})(5000 \times 10^{7})}{(0.0026 \times 10^{21})(10{,}000 \times 10^{-42})}$

$= \dfrac{(52 \times 10^{-9})(5 \times 10^{10})}{(26 \times 10^{17})(1 \times 10^{-38})} = \dfrac{260 \times 10^{1}}{26 \times 10^{-21}}$

$= \mathbf{1 \times 10^{23}}$

17. $\dfrac{\dfrac{\dfrac{m}{x}}{\dfrac{m + x}{x}}} \cdot \dfrac{\dfrac{x}{m + x}}{\dfrac{x}{m + x}} = \dfrac{\mathbf{m}}{\mathbf{m + x}}$

18. $\dfrac{\dfrac{\dfrac{a}{m + x}}{\dfrac{b}{m + x}}} \cdot \dfrac{\dfrac{m + x}{b}}{\dfrac{m + x}{b}} = \dfrac{\mathbf{a}}{\mathbf{b}}$

19. $\dfrac{3}{5\sqrt{12}} \cdot \dfrac{\sqrt{12}}{\sqrt{12}} = \dfrac{3\sqrt{12}}{60} = \dfrac{\mathbf{\sqrt{3}}}{\mathbf{10}}$

20. $\dfrac{14}{3\sqrt{75}} \cdot \dfrac{\sqrt{75}}{\sqrt{75}} = \dfrac{14\sqrt{75}}{225} = \dfrac{70\sqrt{3}}{225} = \dfrac{\mathbf{14\sqrt{3}}}{\mathbf{45}}$

21. $\dfrac{4x}{x + 4} + \dfrac{6}{x + 2}$

$= \dfrac{4x(x + 2)}{(x + 4)(x + 2)} + \dfrac{6(x + 4)}{(x + 4)(x + 2)}$

$= \dfrac{4x^2 + 8x + 6x + 24}{(x + 4)(x + 2)}$

$= \dfrac{\mathbf{4x^2 + 14x + 24}}{\mathbf{(x + 4)(x + 2)}}$

22. $\dfrac{3m}{m^2 + 3m + 2} - \dfrac{5m}{m + 1}$

$= \dfrac{3m}{(m + 2)(m + 1)} - \dfrac{5m(m + 2)}{(m + 2)(m + 1)}$

$= \dfrac{3m - 5m^2 - 10m}{(m + 2)(m + 1)} = \dfrac{\mathbf{-5m^2 - 7m}}{\mathbf{(m + 2)(m + 1)}}$

23. (a) $y - x = 3$
$\qquad y = x + 3$

(b) $y + 2x = 6$
$\qquad y = -2x + 6$

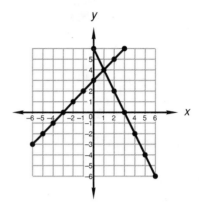

Substitute $y = x + 3$ into (b) and get:
(b) $(x + 3) + 2x = 6$
$\qquad 3x = 3$
$\qquad\quad x = 1$

(a) $y - (1) = 3$
$\qquad y = 4$

$\mathbf{(1, 4)}$

24. Since parallel lines have the same slope:

$y = -\dfrac{3}{8}x + b$

$-3 = -\dfrac{3}{8}(2) + b$

$-3 = -\dfrac{3}{4} + b$

$-\dfrac{9}{4} = b$

Since $m = -\dfrac{3}{8}$ and $b = -\dfrac{9}{4}$:

$y = -\dfrac{3}{8}x - \dfrac{9}{4}$

25. Area $= \pi r^2 = 2500\pi \text{ m}^2$

$r = \sqrt{\dfrac{2500\pi \text{ m}^2}{\pi}} = \mathbf{50 \text{ m}}$

Arc length $= \dfrac{36}{360}(2\pi(50 \text{ m})) \approx \mathbf{31.4 \text{ m}}$

26. $A^2 = 9^2 + 6^2$
$A^2 = 81 + 36$
$A^2 = 117$
$A = \sqrt{117} = \mathbf{3\sqrt{13}}$

$9 \times \overrightarrow{SF} = 13$
$\overrightarrow{SF} = \dfrac{13}{9}$

$3\sqrt{13} \cdot \dfrac{13}{9} = B + 3\sqrt{13}$

$\dfrac{4\sqrt{13}}{3} = B$

$C = 6 \cdot \dfrac{13}{9} = \dfrac{\mathbf{26}}{\mathbf{3}}$

27. $\dfrac{3x}{2} - \dfrac{5}{7} = \dfrac{x+2}{3}$
$63x - 30 = 14x + 28$
$49x = 58$
$x = \dfrac{\mathbf{58}}{\mathbf{49}}$

28. $2\dfrac{1}{5} - \dfrac{1}{10}x = \dfrac{3}{15}$
$-\dfrac{1}{10}x = \dfrac{3}{15} - \dfrac{33}{15}$
$x = -2 \cdot -\dfrac{10}{1} = \mathbf{20}$

29. $\dfrac{(x^2 y^{-2} z)^{-3} x^0}{(x^0 y^{-3} z^2)^3} = \dfrac{x^{-6} y^6 z^{-3}}{y^{-9} z^6} = \mathbf{x^{-6} y^{15} z^{-9}}$

30. $x^2 - y^2(x - y) = \left(\dfrac{1}{2}\right)^2 - \left(\dfrac{1}{3}\right)^2\left(\dfrac{1}{2} - \dfrac{1}{3}\right)$

$= \dfrac{1}{4} - \dfrac{1}{9}\left(\dfrac{1}{6}\right) = \dfrac{1}{4} - \dfrac{1}{54} = \dfrac{27}{108} - \dfrac{2}{108}$

$= \dfrac{\mathbf{25}}{\mathbf{108}}$

PROBLEM SET 30

1. $\dfrac{40}{280} = \dfrac{S}{3360}$
$280S = 40(3360)$
$S = \mathbf{480\ grams}$

2. $\dfrac{P}{100} \times of = is$

$\dfrac{86}{100} \times M = 430$

$M = 430 \cdot \dfrac{100}{86} = \mathbf{500\ grams}$

3. $R_F T_F + R_B T_B = 68;$
$R_F = 6;\ T_F = 6;\ T_B = 4$

$6(6) + 4R_B = 68$
$4R_B = 32$
$R_B = \mathbf{8\ mph}$

4. (a) $5N_D = 2N_P$
(b) $N_D + N_P = 35$
Substitute $N_P = 35 - N_D$ into (a) and get:
(a') $5N_D = 2(35 - N_D)$
$5N_D = 70 - 2N_D$
$7N_D = 70$
$N_D = \mathbf{10}$
(b) $(10) + N_P = 35$
$N_P = \mathbf{25}$

5. (a) $N_D + N_Q = 15$
(b) $10N_D + 25N_Q = 225$

$-10(a)\ -10N_D - 10N_Q = -150$
$(b)\ \underline{\ \ 10N_D + 25N_Q = \ \ 225\ }$
$15N_Q = \ \ \ 75$
$N_Q = \mathbf{5}$

(a) $N_D + (5) = 15$
$N_D = \mathbf{10}$

6. 120 in. = 10 ft
(a) Area $= l \times w = 40\ ft \times 10\ ft = \mathbf{400\ ft^2}$
(b) Perimeter $= 40\ ft + 40\ ft + 10\ ft + 10\ ft$
$= \mathbf{100\ ft}$

7.

$$
\begin{array}{r}
x^2 - x + 1 + \dfrac{1}{x+1} \\[2pt]
x+1\,\overline{\smash{)}\,x^3 + 0x^2 + 0x + 2} \\
\underline{x^3 + x^2} \\
- x^2 + 0x \\
\underline{- x^2 - x} \\
x + 2 \\
\underline{x + 1} \\
1
\end{array}
$$

Check: $\dfrac{x^2(x+1)}{x+1} - \dfrac{x(x+1)}{x+1}$

$+ \dfrac{1(x+1)}{x+1} + \dfrac{1}{x+1}$

$= \dfrac{x^3 + x^2 - x^2 - x + x + 1 + 1}{x+1}$

$= \dfrac{x^3 + 2}{x+1}$

8. $35a^2 + 2a^2 x - a^2 x^2 = \mathbf{-a^2(x-7)(x+5)}$

9. $30 + 3x^2 - 21x = \mathbf{3(x-5)(x-2)}$

10. $-x^2ab - 25ab + 10axb = -ab(x - 5)(x - 5)$

11. $14a^2b^2 + 9a^2xb^2 + x^2a^2b^2$
$= a^2b^2(x + 7)(x + 2)$

12. $\dfrac{p - 4px}{p} = \dfrac{p(1 - 4x)}{p} = 1 - 4x$

13. $5\sqrt{18} - 10\sqrt{50} + 3\sqrt{72}$
$= 5\sqrt{2}\sqrt{3}\sqrt{3} - 10\sqrt{2}\sqrt{5}\sqrt{5} + 3\sqrt{2}\sqrt{6}\sqrt{6}$
$= 15\sqrt{2} - 50\sqrt{2} + 18\sqrt{2} = -17\sqrt{2}$

14. $3\sqrt{12}(4\sqrt{2} - 2\sqrt{3}) = 3\sqrt{2}\sqrt{2}\sqrt{3}(4\sqrt{2} - 2\sqrt{3})$
$= 12\sqrt{2}\sqrt{2}\sqrt{2}\sqrt{3} - 6\sqrt{2}\sqrt{2}\sqrt{3}\sqrt{3} = 24\sqrt{6} - 36$

15. $\dfrac{(0.00035)(5000 \times 10^{42})}{0.00025 \times 10^{-4}}$

$= \dfrac{(35 \times 10^{-5})(5 \times 10^{45})}{25 \times 10^{-9}} = \dfrac{175 \times 10^{40}}{25 \times 10^{-9}}$

$= 7 \times 10^{49}$

16. $\dfrac{\frac{x}{y}}{\frac{x + y}{y}} \cdot \dfrac{\frac{y}{x + y}}{\frac{y}{x + y}} = \dfrac{x}{x + y}$

17. $\dfrac{\frac{a}{a + b}}{\frac{p}{a + b}} \cdot \dfrac{\frac{a + b}{p}}{\frac{a + b}{p}} = \dfrac{a}{p}$

18. $\dfrac{2}{3\sqrt{6}} \cdot \dfrac{\sqrt{6}}{\sqrt{6}} = \dfrac{2\sqrt{6}}{18} = \dfrac{\sqrt{6}}{9}$

19. $\dfrac{2}{5\sqrt{18}} \cdot \dfrac{\sqrt{18}}{\sqrt{18}} = \dfrac{2\sqrt{18}}{90} = \dfrac{6\sqrt{2}}{90} = \dfrac{\sqrt{2}}{15}$

20. $\dfrac{4a}{a + 4} + \dfrac{a + 2}{2a}$

$= \dfrac{4a(2a)}{2a(a + 4)} + \dfrac{(a + 2)(a + 4)}{2a(a + 4)}$

$= \dfrac{8a^2 + a^2 + 6a + 8}{2a(a + 4)}$

$= \dfrac{9a^2 + 6a + 8}{2a(a + 4)}$

21. $\dfrac{4x}{x^2 + 5x + 6} + \dfrac{2}{x + 2}$

$= \dfrac{4x}{(x + 2)(x + 3)} + \dfrac{2(x + 3)}{(x + 2)(x + 3)}$

$= \dfrac{4x + 2x + 6}{(x + 2)(x + 3)} = \dfrac{6x + 6}{(x + 2)(x + 3)}$

22. (a) $3x + 2y = 8$

$\quad y = -\dfrac{3}{2}x + 4$

(b) $2x + 3y = 6$

$\quad y = -\dfrac{2}{3}x + 2$

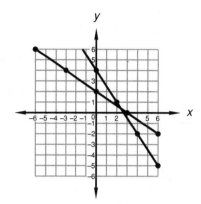

$3(a)\quad 9x + 6y = 24$
$-2(b)\quad \underline{-4x - 6y = -12}$
$5x = 12$

$x = \dfrac{12}{5}$

(a) $y = -\dfrac{3}{2}\left(\dfrac{12}{5}\right) + 4 = -\dfrac{36}{10} + \dfrac{40}{10} = \dfrac{2}{5}$

$\left(\dfrac{12}{5}, \dfrac{2}{5}\right)$

23. Graph the line to find the slope.

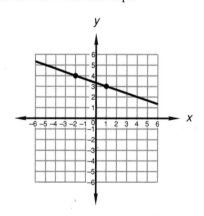

Slope $= \dfrac{-1}{+3} = -\dfrac{1}{3}$

$y = -\dfrac{1}{3}x + b$

$4 = -\dfrac{1}{3}(-2) + b$

$\dfrac{10}{3} = b$

Since $m = -\dfrac{1}{3}$ and $b = \dfrac{10}{3}$: $y = -\dfrac{1}{3}x + \dfrac{10}{3}$

24. Lateral Surface Area = Perimeter × Height

$$= \left[3\left(\frac{2\pi(1 \text{ m})}{2}\right) + 2 \text{ m} \right] \times 10 \text{ m}$$

$$= (3\pi + 2)(10) \text{ m}^2 \approx \mathbf{114.2 \text{ m}^2}$$

25. (A) $\angle 1 + \angle 2 = 180°$; TRUE. Because $\overline{AB}$ is a straight angle.
    (B) $\angle 4 = \angle 7$; TRUE. Because alternate interior angles are equal.
    (C) $\angle 2 + \angle 3 = 180°$; **FALSE**. Because it is not shown that $\angle 6 = \angle 7$.
    (D) $\angle 2 = \angle 6$; TRUE. Because corresponding angles are equal.

26. $\dfrac{2x}{3} - \dfrac{2}{5} = \dfrac{x-5}{2}$

    $20x - 12 = 15x - 75$

    $5x = -63$

    $x = -\dfrac{63}{5}$

27. $3\dfrac{2}{3} + \dfrac{1}{5}x = \dfrac{1}{15}$

    $\dfrac{1}{5}x = \dfrac{1}{15} - \dfrac{55}{15}$

    $x = -\dfrac{54}{15} \cdot \dfrac{5}{1} = \mathbf{-18}$

28. $\dfrac{2x-3}{2} = -\dfrac{2}{5}$

    $10x - 15 = -4$

    $10x = 11$

    $x = \dfrac{11}{10}$

29. $\dfrac{4a^2 x^2 y}{p} - \dfrac{3p^{-1} x^4 y}{a^{-2} x^2} + \dfrac{2xa}{a^{-1} y^{-1} p} - \dfrac{2a^2 yx^2}{p}$

    $= \dfrac{4a^2 x^2 y}{p} - \dfrac{3a^2 x^2 y}{p} + \dfrac{2a^2 xy}{p} - \dfrac{2a^2 x^2 y}{p}$

    $= \dfrac{2a^2 xy}{p} - \dfrac{a^2 x^2 y}{p}$

30. $x^2 - y(x-y) = \left(-\dfrac{1}{2}\right)^2 - \dfrac{1}{4}\left(-\dfrac{1}{2} - \dfrac{1}{4}\right)$

    $= \dfrac{1}{4} - \dfrac{1}{4}\left(-\dfrac{3}{4}\right) = \dfrac{4}{16} + \dfrac{3}{16} = \dfrac{7}{16}$

PROBLEM SET 31

1. $5N \qquad 5N + 5 \qquad 5N + 10 \qquad 5N + 15$

   $6(5N) = 2(5N + 5 + 5N + 15) + 40$

   $30N = 20N + 80$

   $10N = 80$

   $N = 8$

   The desired integers are **40, 45, 50,** and **55**.

2. $R_B T_B + R_O T_O = 960$;
   $R_B = 40$; $R_O = 70$;
   $T_B = T_O + 2$

   $40(T_O + 2) + 70T_O = 960$

   $110T_O = 880$

   $T_O = 8 \text{ hr}$

   Since the *Orange Blossom Special* left at noon, it was **8 p.m.** when it and the bus were 960 miles apart.

3. $R_J T_J = R_R T_R$; $R_J = 6$;
   $R_R = 30$; $T_J + T_R = 12$

   $6T_J = 30(12 - T_J)$

   $36T_J = 360$

   $T_J = 10$

   Distance $= R_J T_J = 6(10) = \mathbf{60 \text{ miles}}$

4. (a) $N_P + N_T = 70$
   (b) $4N_P + 6N_T = 360$
   Substitute $N_T = 70 - N_P$ into (b) and get:
   (b′) $4N_P + 6(70 - N_P) = 360$

   $-2N_P = -60$

   $N_P = \mathbf{30}$

   (a) $N_T = \mathbf{40}$

5. $\dfrac{P}{100} \times of = is$

   $\dfrac{10}{100} \times N = 1200$

   $N = 1200 \cdot \dfrac{100}{10} = \mathbf{12,000 \text{ kilograms}}$

6. $\dfrac{40}{1440} = \dfrac{P}{4320}$

   $1440P = 40(4320)$

   $P = \mathbf{120}$

7. Since the slopes of perpendicular lines are negative reciprocals of each other:

   $y = -3x + b$

   $-3 = -3(2) + b$

   $3 = b$

   Since $m = -3$ and $b = 3$: $y = \mathbf{-3x + 3}$

8. $\dfrac{(0.0006 \times 10^{-42})(2000 \times 10^{-4})}{(0.004 \times 10^{-13})}$

   $= \dfrac{(6 \times 10^{-46})(2 \times 10^{-1})}{4 \times 10^{-16}} = \mathbf{3 \times 10^{-31}}$

9. $\dfrac{\frac{a}{b}}{a+b} \cdot \dfrac{\frac{b}{a+b}}{\frac{b}{a+b}} = \dfrac{a}{a+b}$

10. $\dfrac{\dfrac{4}{\dfrac{x+y}{m}}}{x+y} \cdot \dfrac{\dfrac{x+y}{m}}{x+y} = \dfrac{4}{m}$

11. $\dfrac{3}{2\sqrt{5}} \cdot \dfrac{\sqrt{5}}{\sqrt{5}} = \dfrac{3\sqrt{5}}{10}$

12. $\dfrac{7}{3\sqrt{2}} \cdot \dfrac{\sqrt{2}}{\sqrt{2}} = \dfrac{7\sqrt{2}}{6}$

13. $5\sqrt{75} - 3\sqrt{300} + 2\sqrt{27}$
$= 5\sqrt{3}\sqrt{5}\sqrt{5} - 3\sqrt{2}\sqrt{2}\sqrt{3}\sqrt{5}\sqrt{5} + 2\sqrt{3}\sqrt{3}\sqrt{3}$
$= 25\sqrt{3} - 30\sqrt{3} + 6\sqrt{3} = \mathbf{\sqrt{3}}$

14. $3\sqrt{2}(5\sqrt{2} - 4\sqrt{6}) = 3\sqrt{2}(5\sqrt{2} - 4\sqrt{2}\sqrt{3})$
$= \mathbf{30 - 24\sqrt{3}}$

15. $\dfrac{x + 4x^2}{x} = \dfrac{x(1 + 4x)}{x} = \mathbf{1 + 4x}$

16. $\dfrac{x}{x+2} + \dfrac{3+x}{x^2 + 4x + 4}$
$= \dfrac{x(x+2)}{(x+2)(x+2)} + \dfrac{3+x}{(x+2)(x+2)}$
$= \dfrac{\mathbf{x^2 + 3x + 3}}{\mathbf{(x+2)(x+2)}}$

17. $\dfrac{x}{x-3} + \dfrac{2x}{x^2 - 3x}$
$= \dfrac{x^2}{x(x-3)} + \dfrac{2x}{x(x-3)} = \dfrac{x^2 + 2x}{x(x-3)}$
$= \dfrac{x(x+2)}{x(x-3)} = \dfrac{\mathbf{x+2}}{\mathbf{x-3}}$

18. $-x^3 + 5x^2 - 6x = \mathbf{-x(x-3)(x-2)}$

19. $2ax^3 - 18ax^2 + 40ax = \mathbf{2ax(x-5)(x-4)}$

20. $-3pax + pax^2 + 2pa = \mathbf{ap(x-2)(x-1)}$

21. $-10mc + 3mxc + mx^2c = \mathbf{cm(x+5)(x-2)}$

22. $\dfrac{2x+3}{6} - \dfrac{x}{2} = 1$
$2x + 3 - 3x = 6$
$-x = 3$
$x = \mathbf{-3}$

23. $\dfrac{3x+2}{3} - \dfrac{2}{5} = \dfrac{x+2}{6}$
$30x + 20 - 12 = 5x + 10$
$25x = 2$
$x = \dfrac{\mathbf{2}}{\mathbf{25}}$

24. (a) $2x + 3y = 18$
$y = -\dfrac{2}{3}x + 6$
(b) $-12x + 6y = -18$
$y = 2x - 3$

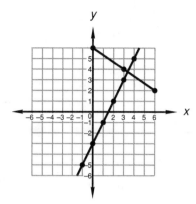

Substitute $y = 2x - 3$ into (a) and get:
(a') $2x + 3(2x - 3) = 18$
$8x = 27$
$x = \dfrac{27}{8}$

(b) $y = 2\left(\dfrac{27}{8}\right) - \dfrac{24}{8} = \dfrac{15}{4}$

$\left(\dfrac{\mathbf{27}}{\mathbf{8}}, \dfrac{\mathbf{15}}{\mathbf{4}}\right)$

25. $\quad x + 1 \overline{\smash{\big)}\, x^4 + 0x^3 + 0x^2 + 0x - 2}$ with quotient $x^3 - x^2 + x - 1 - \dfrac{1}{x+1}$
$\qquad \underline{x^4 + x^3}$
$\qquad -x^3 + 0x^2$
$\qquad \underline{-x^3 - x^2}$
$\qquad\qquad x^2 + 0x$
$\qquad\qquad \underline{x^2 + x}$
$\qquad\qquad\quad -x - 2$
$\qquad\qquad\quad \underline{-x - 1}$
$\qquad\qquad\qquad -1$

26. (a) $6x + 4y = 11$
(b) $2x - 3y = -5$

$\quad 1(a) \quad 6x + 4y = 11$
$-3(b) \quad \underline{-6x + 9y = 15}$
$\qquad\qquad 13y = 26$
$\qquad\qquad y = 2$

(b) $2x - 3(2) = -5$
$2x = 1$
$x = \dfrac{1}{2}$

$\left(\dfrac{\mathbf{1}}{\mathbf{2}}, \mathbf{2}\right)$

27. $a^2 = 24^2 + 7^2$

$a^2 = 625$

$a = 25$

$24 \times \overrightarrow{SF} = 30$

$\overrightarrow{SF} = \dfrac{5}{4}$

$25 \cdot \dfrac{5}{4} = 25 + c$

$125 = 100 + 4c$

$\dfrac{25}{4} = c$

28. $-2(-x^0 - 3) + 4(-x - 5^0) = -3^0(2x - 5)$

$2 + 6 - 4x - 4 = -2x + 5$

$-2x = 1$

$x = -\dfrac{1}{2}$

29. $B = 180 - 30 = 150$

$F + 150 = 180$

$F = 30$

$P + 30 + 90 = 180$

$P = 60$

$A + 60 = 180$

$A = 120$

$C + 60 + 90 = 180$

$C = 30$

$D = C = 30$

$E + 30 + 90 = 180$

$E = 60$

$A = \mathbf{120}$; $B = \mathbf{150}$; $C = \mathbf{30}$; $D = \mathbf{30}$; $E = \mathbf{60}$;
$F = \mathbf{30}$; $P = \mathbf{60}$

30. $\text{Area}_{\text{Circle}} = \dfrac{36\pi \text{ m}^2}{4} = \mathbf{9\pi \text{ m}^2}$

$d = 2r = 2\left(\sqrt{\dfrac{9\pi}{\pi} \text{ m}^2}\right) = \mathbf{6 \text{ m}}$

$\text{Area}_{\text{Square}} = (2d)^2 = (12 \text{ m})^2 = \mathbf{144 \text{ m}^2}$

PROBLEM SET 32

1. $R_W T_W + R_J T_J = 56$;

$R_W = 4$; $R_J = 8$;

$T_W + T_J = 10$

$4T_W + 8(10 - T_W) = 56$

$-4T_W = -24$

$T_W = 6$

$D_W = 4(6) = \mathbf{24 \text{ miles}}$

$D_J = 8(4) = \mathbf{32 \text{ miles}}$

2. (a) $N_L = 3N_S + 30$

(b) $\dfrac{N_S}{N_L} = \dfrac{1}{6} \quad \rightarrow \quad 6N_S = N_L$

Substitute $6N_S = N_L$ into (a) and get:

(a′) $(6N_S) = 3N_S + 30$

$3N_S = 30$

$N_S = \mathbf{10}$

(a) $N_L = 3(10) + 30 = \mathbf{60}$

3. (a) $5N_N + 10N_D = 900$

(b) $N_D - N_N = 30$

Substitute $N_D = N_N + 30$ into (a) and get:

(a′) $5N_N + 10(N_N + 30) = 900$

$15N_N = 600$

$N_N = \mathbf{40}$

(b) $N_D - 40 = 30$

$N_D = \mathbf{70}$

4. $\dfrac{P}{100} \times of = is$

$\dfrac{60}{100} \times M = 300$

$M = 300 \cdot \dfrac{100}{60} = \mathbf{500 \text{ kilograms}}$

5. $\dfrac{20}{24} = \dfrac{NM}{1440}$

$24NM = 20(1440)$

$NM = \mathbf{1200 \text{ grams}}$

6. $WD \times of = is$

$0.87 \times S = 1914$

$S = \dfrac{1914}{0.87} = \mathbf{2200}$

7. $\sqrt{\dfrac{2}{3}} + \sqrt{\dfrac{3}{2}} = \dfrac{\sqrt{2}}{\sqrt{3}}\dfrac{\sqrt{3}}{\sqrt{3}} + \dfrac{\sqrt{3}}{\sqrt{2}}\dfrac{\sqrt{2}}{\sqrt{2}}$

$= \dfrac{\sqrt{6}}{3} + \dfrac{\sqrt{6}}{2} = \dfrac{2\sqrt{6}}{6} + \dfrac{3\sqrt{6}}{6} = \mathbf{\dfrac{5\sqrt{6}}{6}}$

8. $2\sqrt{\dfrac{5}{7}} - 3\sqrt{\dfrac{7}{5}} = \dfrac{2\sqrt{5}}{\sqrt{7}}\dfrac{\sqrt{7}}{\sqrt{7}} - \dfrac{3\sqrt{7}}{\sqrt{5}}\dfrac{\sqrt{5}}{\sqrt{5}}$

$= \dfrac{2\sqrt{35}}{7} - \dfrac{3\sqrt{35}}{5} = \dfrac{10\sqrt{35}}{35} - \dfrac{21\sqrt{35}}{35}$

$= \mathbf{\dfrac{-11\sqrt{35}}{35}}$

9. $2\sqrt{\dfrac{3}{5}} - 5\sqrt{\dfrac{5}{3}} = \dfrac{2\sqrt{3}}{\sqrt{5}}\dfrac{\sqrt{5}}{\sqrt{5}} - \dfrac{5\sqrt{5}}{\sqrt{3}}\dfrac{\sqrt{3}}{\sqrt{3}}$

$= \dfrac{2\sqrt{15}}{5} - \dfrac{5\sqrt{15}}{3} = \dfrac{6\sqrt{15}}{15} - \dfrac{25\sqrt{15}}{15}$

$= \mathbf{\dfrac{-19\sqrt{15}}{15}}$

10. $\text{Area}_{\text{Square}} = s^2 = 64 \text{ m}^2$

$s = \textbf{8 m}$

$r = \dfrac{1}{4}(8 \text{ m}) = \textbf{2 m}$

$\text{Area}_{\text{Circle}} = \pi(2 \text{ m})^2 \approx \textbf{12.56 m}^2$

11. $(180 - A) = 3(90 - A) - 30$

$180 - A = 270 - 3A - 30$

$2A = 60$

$A = \textbf{30}°$

12. Since the slopes of perpendicular lines are negative reciprocals of each other:

$y = -\dfrac{1}{3}x + b$

$2 = -\dfrac{1}{3}(2) + b$

$\dfrac{6}{3} = -\dfrac{2}{3} + b$

$\dfrac{8}{3} = b$

Since $m = -\dfrac{1}{3}$ and $b = \dfrac{8}{3}$: $\boldsymbol{y = -\dfrac{1}{3}x + \dfrac{8}{3}}$

13. $\dfrac{(0.0035 \times 10^{15})(0.002 \times 10^{17})}{7000 \times 10^{33}}$

$= \dfrac{(3.5 \times 10^{12})(2 \times 10^{14})}{7 \times 10^{36}} = \dfrac{7 \times 10^{26}}{7 \times 10^{36}}$

$= \textbf{1} \times \textbf{10}^{-10}$

14. $\dfrac{\dfrac{x + 4y}{y}}{\dfrac{x + y}{y}} \cdot \dfrac{\dfrac{y}{x + y}}{\dfrac{y}{x + y}} = \boldsymbol{\dfrac{x + 4y}{x + y}}$

15. $\dfrac{xy + 4x^2y^2}{xy} = \dfrac{xy(1 + 4xy)}{xy} = \textbf{1} + \textbf{4}\boldsymbol{xy}$

16. $3\sqrt{125} - 2\sqrt{45} + 3\sqrt{20}$

$= 3\sqrt{5}\sqrt{5}\sqrt{5} - 2\sqrt{3}\sqrt{3}\sqrt{5} + 3\sqrt{2}\sqrt{2}\sqrt{5}$

$= 15\sqrt{5} - 6\sqrt{5} + 6\sqrt{5} = \textbf{15}\sqrt{\textbf{5}}$

17. $4\sqrt{5}(2\sqrt{10} - 3\sqrt{5}) = 4\sqrt{5}(2\sqrt{2}\sqrt{5} - 3\sqrt{5})$

$= \textbf{40}\sqrt{\textbf{2}} - \textbf{60}$

18. $\dfrac{x}{x + 3} - \dfrac{2x - 2}{x^2 + 5x + 6}$

$= \dfrac{x(x + 2)}{(x + 3)(x + 2)} - \dfrac{2x - 2}{(x + 3)(x + 2)}$

$= \dfrac{x^2 + 2x - 2x + 2}{(x + 3)(x + 2)}$

$= \boldsymbol{\dfrac{x^2 + 2}{(x + 3)(x + 2)}}$

19. $\dfrac{m}{m - 5} - \dfrac{2}{m^2 - 5m}$

$= \dfrac{m^2}{m(m - 5)} - \dfrac{2}{m(m - 5)} = \boldsymbol{\dfrac{m^2 - 2}{m(m - 5)}}$

20. $-2x^3 + 8x^2 - 6x = \boldsymbol{-2x(x - 3)(x - 1)}$

21. $-14x^3 + 5x^4 + x^5 = \boldsymbol{x^3(x + 7)(x - 2)}$

22. $7ax + ax^3 - 8ax^2 = \boldsymbol{ax(x - 7)(x - 1)}$

23. $-12py + px^2y + 4xpy = \boldsymbol{py(x + 6)(x - 2)}$

24. $\dfrac{x + 2}{5} - \dfrac{3x - 3}{2} = 4$

$2x + 4 - 15x + 15 = 40$

$-13x = 21$

$x = \boldsymbol{-\dfrac{21}{13}}$

25. $\dfrac{x}{2} - \dfrac{3x + 2}{4} = 7$

$2x - 3x - 2 = 28$

$-x = 30$

$x = \textbf{-30}$

26. (a) $-x + 2y = 4$

$y = \dfrac{1}{2}x + 2$

(b) $x + y = -2$

$y = -x - 2$

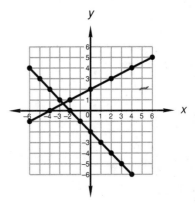

Substitute $y = (-x - 2)$ into (a) and get:

(a′) $-x + 2(-x - 2) = 4$

$-3x = 8$

$x = -\dfrac{8}{3}$

(b) $y = -\left(-\dfrac{8}{3}\right) - \dfrac{6}{3} = \dfrac{2}{3}$

$\left(-\dfrac{8}{3}, \dfrac{2}{3}\right)$

27. $(4x + 2)(x^3 - 2x + 4)$

$= 4x^4 - 8x^2 + 16x + 2x^3 - 4x + 8$

$= \boldsymbol{4x^4 + 2x^3 - 8x^2 + 12x + 8}$

28. (a) $x + 2y = 5$

(b) $3x - y = 7$

(b') $y = 3x - 7$

Substitute $y = 3x - 7$ into (a) and get:

(a') $x + 2(3x - 7) = 5$

$$7x = 19$$

$$x = \frac{19}{7}$$

(b') $y = 3\left(\frac{19}{7}\right) - \frac{49}{7} = \frac{8}{7}$

$$\left(\frac{19}{7}, \frac{8}{7}\right)$$

29. $D^2 = 7^2 + 7^2$

$D^2 = 49 + 49$

$D = \sqrt{98}$

$D = \mathbf{7\sqrt{2}}$

30. $-3\left[(-2^0 - 3^0 - 2)(-3^0) - 2^2\right] - |3 - 2^0|$

$= -3\left[(-4)(-1) - 4\right] - 2 = -3[0] - 2 = \mathbf{-2}$

PROBLEM SET 33

1. $\dfrac{4}{17} = \dfrac{104}{H}$

$4H = 17(104)$

$H = \mathbf{442}$

2. $\dfrac{7}{9} = \dfrac{M}{3600}$

$9M = 7(3600)$

$M = \mathbf{2800 \text{ grams}}$

3. $\dfrac{P}{100} \times of = is$

$\dfrac{80}{100} \times L = 1620$

$L = 1620 \cdot \dfrac{100}{80} = \mathbf{2025 \text{ grams}}$

4. (a) $N_R = 8N_B + 10$

(b) $N_R = 11N_B - 5$

Substitute (b) into (a) and get:

(a') $11N_B - 5 = 8N_B + 10$

$$3N_B = 15$$

$$N_B = \mathbf{5}$$

(a) $N_R = 8(5) + 10 = \mathbf{50}$

5. $R_M T_M + R_C T_C = 540;$

$R_M = 40; \ R_C = 60;$

$T_M + T_C = 11$

$40T_M + 60(11 - T_M) = 540$

$$-20T_M = -120$$

$$T_M = 6$$

$D_M = 40(6) = \mathbf{240 \text{ miles}}$

6. $\dfrac{\dfrac{m}{p} + \dfrac{3}{xp}}{\dfrac{y}{p}} = \dfrac{\dfrac{m}{p}\left(\dfrac{x}{x}\right) + \dfrac{3}{xp}}{\dfrac{y}{p}}$

$= \dfrac{\dfrac{mx + 3}{xp}}{\dfrac{y}{p}} \cdot \dfrac{p}{y} = \dfrac{mx + 3}{xy}$

7. $\dfrac{\dfrac{s}{a + b} + \dfrac{x}{b}}{\dfrac{a}{s + x}}$

$= \dfrac{\dfrac{s}{a + b}\left(\dfrac{b}{b}\right) + \dfrac{x}{b}\left(\dfrac{a + b}{a + b}\right)}{\dfrac{a}{s + x}}$

$= \dfrac{\dfrac{bs + ax + bx}{ab + b^2}}{\dfrac{a}{s + x}} \cdot \dfrac{s + x}{\dfrac{a}{s + x}}$

$= \dfrac{(s + x)(bs + ax + bx)}{a(ab + b^2)}$

8. $x + 40 = 180$

$x = \mathbf{140}$

$2y + 140 = 180$

$2y = 40$

$y = \mathbf{20}$

$z = \mathbf{40}$

Arc length $= \dfrac{40}{360} \cdot 2(\pi)(6 \text{ cm}) \approx \mathbf{4.19 \text{ cm}}$

9. $z = 360 - 150 - 70$

$z = \mathbf{140}$

$x = \dfrac{150}{2} = \mathbf{75}$

$y = \dfrac{140}{2} = \mathbf{70}$

10. Since the slopes of perpendicular lines are negative reciprocals of each other:

$2 = \dfrac{5}{3}(-4) + b$

$\dfrac{26}{3} = b$

Since $m = \dfrac{5}{3}$ and $b = \dfrac{26}{3}$: $y = \dfrac{5}{3}x + \dfrac{26}{3}$

11. $3\sqrt{\dfrac{5}{2}} - 2\sqrt{\dfrac{2}{5}} = \dfrac{3\sqrt{5}}{\sqrt{2}} - \dfrac{2\sqrt{2}}{\sqrt{5}}$

$= \dfrac{3\sqrt{5}}{\sqrt{2}}\dfrac{\sqrt{2}}{\sqrt{2}} - \dfrac{2\sqrt{2}}{\sqrt{5}}\dfrac{\sqrt{5}}{\sqrt{5}}$

$= \dfrac{3\sqrt{10}}{2} - \dfrac{2\sqrt{10}}{5} = \dfrac{15\sqrt{10}}{10} - \dfrac{4\sqrt{10}}{10}$

$= \dfrac{\mathbf{11\sqrt{10}}}{\mathbf{10}}$

12. $4\sqrt{\dfrac{5}{6}} - 2\sqrt{\dfrac{6}{5}} = \dfrac{4\sqrt{5}}{\sqrt{6}} - \dfrac{2\sqrt{6}}{\sqrt{5}}$

$= \dfrac{4\sqrt{5}}{\sqrt{6}}\dfrac{\sqrt{6}}{\sqrt{6}} - \dfrac{2\sqrt{6}}{\sqrt{5}}\dfrac{\sqrt{5}}{\sqrt{5}} = \dfrac{4\sqrt{30}}{6} - \dfrac{2\sqrt{30}}{5}$

$= \dfrac{20\sqrt{30}}{30} - \dfrac{12\sqrt{30}}{30} = \dfrac{8\sqrt{30}}{30} = \dfrac{\mathbf{4\sqrt{30}}}{\mathbf{15}}$

13. Perimeter $= 11.28 \text{ cm} + 6 \text{ cm} + 6 \text{ cm}$
$\qquad\qquad = \mathbf{23.28 \text{ cm}}$

14. $3(90 - A) = (180 - A) + 50$
$270 - 3A = 230 - A$
$\qquad -2A = -40$
$\qquad\quad A = \mathbf{20°}$

15. $\dfrac{(0.0027 \times 10^{15})(500 \times 10^{-20})}{900 \times 10^{14}}$

$= \dfrac{(2.7 \times 10^{12})(5 \times 10^{-18})}{9 \times 10^{16}}$

$= \dfrac{13.5 \times 10^{-6}}{9 \times 10^{16}} = \mathbf{1.5 \times 10^{-22}}$

16. $\dfrac{x + \dfrac{4xy}{x}}{\dfrac{1}{x} - y} = \dfrac{\dfrac{x^2 + 4xy}{x}}{\dfrac{1 - xy}{x}} \cdot \dfrac{\dfrac{x}{1 - xy}}{\dfrac{x}{1 - xy}}$

$= \dfrac{\mathbf{x^2 + 4xy}}{\mathbf{1 - xy}}$

17. $3\sqrt{18} + 2\sqrt{50} - \sqrt{98}$
$= 3\sqrt{2}\sqrt{3}\sqrt{3} + 2\sqrt{2}\sqrt{5}\sqrt{5} - \sqrt{2}\sqrt{7}\sqrt{7}$
$= 9\sqrt{2} + 10\sqrt{2} - 7\sqrt{2} = \mathbf{12\sqrt{2}}$

18. $\dfrac{4x + 4xy}{4x} = \dfrac{4x(1 + y)}{4x} = \mathbf{1 + y}$

19. $\dfrac{a}{x(x + y)} + \dfrac{b}{x^2} + \dfrac{cx + 4}{x + y}$

$= \dfrac{ax}{x^2(x + y)} + \dfrac{b(x + y)}{x^2(x + y)}$

$\quad + \dfrac{x^2(cx + 4)}{x^2(x + y)}$

$= \dfrac{\mathbf{ax + b(x + y) + x^2(cx + 4)}}{\mathbf{x^2(x + y)}}$

20. $\dfrac{4}{x + 4} - \dfrac{6x - 2}{x^2 + 2x - 8}$

$= \dfrac{4(x - 2)}{(x + 4)(x - 2)} - \dfrac{6x - 2}{(x + 4)(x - 2)}$

$= \dfrac{4x - 8 - 6x + 2}{(x + 4)(x - 2)} = \dfrac{\mathbf{-2x - 6}}{\mathbf{(x + 4)(x - 2)}}$

21. $5x^2 + 4x^3 - x^4 = \mathbf{-x^2(x - 5)(x + 1)}$

22. $10k^2 - 7k^2x + k^2x^2 = \mathbf{k^2(x - 5)(x - 2)}$

23. $apx^2 - 20ap - apx = \mathbf{ap(x - 5)(x + 4)}$

24. $\dfrac{x + 2}{3} - \dfrac{2x - 2}{4} = 5$
$\quad 4x + 8 - 6x + 6 = 60$
$\qquad\qquad\quad -2x = 46$
$\qquad\qquad\quad\ \ x = \mathbf{-23}$

25. $\dfrac{3x - 2}{2} - \dfrac{2x + 3}{3} = 4$
$\quad 9x - 6 - 4x - 6 = 24$
$\qquad\qquad\qquad 5x = 36$
$\qquad\qquad\qquad\ x = \dfrac{\mathbf{36}}{\mathbf{5}}$

26. (a) $2x - y = -5$
$\qquad\qquad y = 2x + 5$
$\quad$ (b) $x + y = 1$
$\qquad\qquad y = -x + 1$

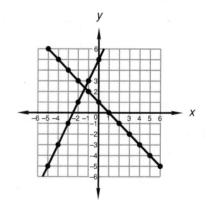

Substitute $y = -x + 1$ into (a) and get:
(a′) $2x - (-x + 1) = -5$
$\qquad\qquad\quad 3x = -4$
$\qquad\qquad\quad\ x = -\dfrac{4}{3}$

(b) $y = -\left(-\dfrac{4}{3}\right) + \dfrac{3}{3} = \dfrac{7}{3}$

$\left(-\dfrac{4}{3}, \dfrac{7}{3}\right)$

27. Divide $2x^4 - x$ by $x - 2$:

$$x - 2 \overline{)\,2x^4 + 0x^3 + 0x^2 - x + 0\,}^{\;2x^3 + 4x^2 + 8x + 15 + \frac{30}{x-2}}$$

$$\underline{2x^4 - 4x^3}$$
$$4x^3 + 0x^2$$
$$\underline{4x^3 - 8x^2}$$
$$8x^2 - x$$
$$\underline{8x^2 - 16x}$$
$$15x + 0$$
$$\underline{15x - 30}$$
$$30$$

Check: $\dfrac{2x^3(x-2)}{x-2} + \dfrac{4x^2(x-2)}{x-2}$

$+ \dfrac{8x(x-2)}{x-2} + \dfrac{15(x-2)}{x-2} + \dfrac{30}{x-2}$

$= \dfrac{2x^4 - x}{x-2}$

28. $-3^0 - x - y^0 - y^2(-2^0 - 3) - |-2 - x|$

$= -3^0 - (-2) - (-3)^0 - (-3)^2(-4)$
$\quad - |-2 - (-2)|$
$= -1 + 2 - 1 - 9(-4) - 0$
$= -1 + 2 - 1 + 36 = \mathbf{36}$

29. Perimeter $= 2(2\text{ ft}) + 2(4\text{ ft}) = = \mathbf{12\ ft}$

30. Area $= l \times w = 2\text{ ft} \times 4\text{ ft} = \mathbf{8\ ft^2}$

PROBLEM SET 34

1. $R_B T_B + 60 = R_E T_E$;
$R_E = 40;\ R_B = 50;$
$T_E = T_B + 3$
$50T_B + 60 = 40(T_B + 3)$
$\quad 10T_B = 60$
$\quad\quad T_B = 6\text{ hr}$

Since Benita left at 11 a.m. and her time was 6 hours, she got within 60 kilometers of Elliot at **5 p.m.**

2. $R_R T_R + 14 = R_C T_C$;
$R_R = 7.5;\ R_C = 11;$
$T_R = T_C$
$7.5(T_C) + 14 = 11T_C$
$\quad -3.5T_C = -14$
$\quad\quad T_C = \mathbf{4\ hr}$

3. $R_K T_K = R_Y T_Y$; $R_K = 10$;
$R_Y = 3$; $T_K + T_Y = 13$

$10T_K = 3(13 - T_K)$
$13T_K = 39$
$\quad T_K = 3$
$D_K = 10(3) = \mathbf{30\ miles}$

4. $\dfrac{P}{100} \times of = is$

$\dfrac{70}{100} \times S = 42$

$S = 42 \cdot \dfrac{100}{70} = \mathbf{60\ tons}$

5. $\dfrac{50}{100} = \dfrac{S}{300}$
$100S = 50(300)$
$\quad S = \mathbf{150\ grams}$

6. $N \quad\quad N + 1 \quad\quad N + 2$
$5(N + N + 2) = 8(N + 1) + 14$
$\quad 10N + 10 = 8N + 22$
$\quad\quad 2N = 12$
$\quad\quad\quad N = 6$
The desired integers are **6, 7,** and **8.**

7. $\dfrac{\dfrac{y}{ab} - ab}{\dfrac{1}{a} - \dfrac{a}{b}} = \dfrac{\dfrac{y - a^2 b^2}{ab}}{\dfrac{b - a^2}{ab}} \cdot \dfrac{\dfrac{ab}{b - a^2}}{\dfrac{ab}{b - a^2}}$

$= \dfrac{y - a^2 b^2}{b - a^2}$

8. $\dfrac{\dfrac{1}{x} - b}{x} = \dfrac{\dfrac{1 - bx}{x}}{x} \cdot \dfrac{\dfrac{1}{x}}{\dfrac{1}{x}} = \dfrac{1 - bx}{x^2}$

9. $V_{Cone} = \dfrac{1}{3}(A_{Base} \times Height)$

$= \dfrac{1}{3}\Big[(2)(4) + (4)(4) + (12)(4)$

$\quad + \dfrac{1}{2}(\pi)(4)^2\Big]\,m^2 \times 8\ m$

$= \dfrac{1}{3}(72 + 8\pi)(8)\ m^3 \approx \mathbf{258.99\ m^3}$

10. $(4x + 20) = 2(x + 20)$
$4x + 20 = 2x + 40$
$\quad 2x = 20$
$\quad\quad x = \mathbf{10}$

11. Since the slopes of perpendicular lines are negative reciprocals of each other:

$y = \dfrac{1}{2}x + b$

$-1 = \dfrac{1}{2}(-2) + b$

$0 = b$

Since $m = \dfrac{1}{2}$ and $b = 0$: $y = \dfrac{1}{2}x$

12. $2\sqrt{\dfrac{2}{9}} - 3\sqrt{\dfrac{9}{2}} = \dfrac{2\sqrt{2}}{\sqrt{9}}\dfrac{\sqrt{9}}{\sqrt{9}} - \dfrac{3\sqrt{9}}{\sqrt{2}}\dfrac{\sqrt{2}}{\sqrt{2}}$

$= \dfrac{2\sqrt{18}}{9} - \dfrac{3\sqrt{18}}{2} = \dfrac{4\sqrt{18}}{18} - \dfrac{27\sqrt{18}}{18}$

$= -\dfrac{23\sqrt{18}}{18} = -\dfrac{23\sqrt{2}}{6}$

13. $-3\sqrt{\dfrac{2}{3}} + 2\sqrt{\dfrac{3}{2}} = -\dfrac{3\sqrt{2}}{\sqrt{3}}\dfrac{\sqrt{3}}{\sqrt{3}} + \dfrac{2\sqrt{3}}{\sqrt{2}}\dfrac{\sqrt{2}}{\sqrt{2}}$

$= -\dfrac{3\sqrt{6}}{3} + \dfrac{2\sqrt{6}}{2} = -\sqrt{6} + \sqrt{6} = \mathbf{0}$

14. $\text{Area}_{\text{Shaded}} = \text{Area}_{\text{Circle, 22}} - \text{Area}_{\text{Circle, 16}}$
$\quad - \text{Area}_{\text{Circle, 6}}$

$= \pi(22\text{ cm})^2 - \pi(16\text{ cm})^2 - \pi(6\text{ cm})^2$

$= (484\pi - 256\pi - 36\pi)\text{ cm}^2$

$\approx \mathbf{602.88\ 1\text{–cm–square floor tiles}}$

15. $\dfrac{(0.000032 \times 10^4)(700 \times 10^{-14})}{16,000}$

$= \dfrac{(3.2 \times 10^{-1})(7 \times 10^{-12})}{1.6 \times 10^4}$

$= \dfrac{2.24 \times 10^{-12}}{1.6 \times 10^4} = \mathbf{1.4 \times 10^{-16}}$

16. $\dfrac{4 + \dfrac{x}{y^2}}{3 - \dfrac{1}{y^2}} = \dfrac{\dfrac{4y^2 + x}{y^2}}{\dfrac{3y^2 - 1}{y^2}} \cdot \dfrac{\dfrac{y^2}{3y^2 - 1}}{\dfrac{y^2}{3y^2 - 1}}$

$= \dfrac{\mathbf{4y^2 + x}}{\mathbf{3y^2 - 1}}$

17. $8\sqrt{27} - 2\sqrt{75} + 2\sqrt{147}$
$= 8\sqrt{3}\sqrt{3}\sqrt{3} - 2\sqrt{3}\sqrt{5}\sqrt{5} + 2\sqrt{3}\sqrt{7}\sqrt{7}$
$= 24\sqrt{3} - 10\sqrt{3} + 14\sqrt{3} = \mathbf{28\sqrt{3}}$

18. $\dfrac{x^2y - 5x^2y^2}{x^2y} = \dfrac{x^2y(1 - 5y)}{x^2y} = \mathbf{1 - 5y}$

19. $\dfrac{a}{x(x + y)} + \dfrac{bx}{x^2(x + y)} + \dfrac{cx}{x^3}$

$= \dfrac{ax^2}{x^3(x + y)} + \dfrac{bx^2}{x^3(x + y)} + \dfrac{cx(x + y)}{x^3(x + y)}$

$= \dfrac{\mathbf{ax^2 + bx^2 + cx(x + y)}}{\mathbf{x^3(x + y)}}$

20. $\dfrac{x - 4}{x - 3} - \dfrac{2x - 1}{x^2 - 6x + 9}$

$= \dfrac{(x - 4)(x - 3)}{(x - 3)(x - 3)} - \dfrac{2x - 1}{(x - 3)(x - 3)}$

$= \dfrac{x^2 - 7x + 12 - 2x + 1}{(x - 3)^2}$

$= \dfrac{\mathbf{x^2 - 9x + 13}}{(x - 3)^2}$

21. $-4x^2 + 2x^3 + 2x^4 = \mathbf{2x^2(x + 2)(x - 1)}$

22. $ax^2p - 8pa - 2axp = \mathbf{ap(x - 4)(x + 2)}$

23. $yx^2 - 4xy + 4y = \mathbf{y(x - 2)(x - 2)}$

24. $\dfrac{x - 3}{2} - \dfrac{3x + 4}{2} = 3$
$\quad x - 3 - 3x - 4 = 6$
$\quad\quad\quad\quad -2x = 13$
$\quad\quad\quad\quad\quad\quad x = -\dfrac{\mathbf{13}}{\mathbf{2}}$

25. $\dfrac{x}{3} - \dfrac{2x - 4}{2} = 5$
$\quad 2x - 6x + 12 = 30$
$\quad\quad\quad\quad -4x = 18$
$\quad\quad\quad\quad\quad x = -\dfrac{\mathbf{9}}{\mathbf{2}}$

26. (a) $x - 3y = 6$
$\quad\quad y = \dfrac{1}{3}x - 2$
$\quad$ (b) $2x + y = 2$
$\quad\quad y = -2x + 2$

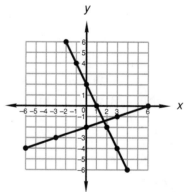

Substitute $y = -2x + 2$ into (a) and get:
(a′) $x - 3(-2x + 2) = 6$
$\quad\quad\quad\quad 7x = 12$
$\quad\quad\quad\quad\quad x = \dfrac{12}{7}$

(b) $y = -2\left(\dfrac{12}{7}\right) + \dfrac{14}{7} = -\dfrac{10}{7}$

$\left(\dfrac{\mathbf{12}}{\mathbf{7}}, -\dfrac{\mathbf{10}}{\mathbf{7}}\right)$

27. $(x^2 + x)(x^2 + 2x + 3)$
$= x^4 + 2x^3 + 3x^2 + x^3 + 2x^2 + 3x$
$= \mathbf{x^4 + 3x^3 + 5x^2 + 3x}$

28. (a) $3x - 2y = 2$
(b) $x - 3y = 4$
(b') $x = 3y + 4$
Substitute $x = 3y + 4$ into (a) and get:
(a') $3(3y + 4) - 2y = 2$
$$7y = -10$$
$$y = -\frac{10}{7}$$
(b') $x = 3\left(-\frac{10}{7}\right) + \frac{28}{7} = -\frac{2}{7}$
$$\left(-\frac{2}{7}, -\frac{10}{7}\right)$$

29. $(\sqrt{17})^2 = H^2 + 3^2$
$$17 - 9 = H^2$$
$$8 = H^2$$
$$2\sqrt{2} = H$$
$$\text{Area} = \frac{6 \text{ cm} \times 2\sqrt{2} \text{ cm}}{2} = \mathbf{6\sqrt{2} \text{ cm}^2}$$

30. $-2^0[-2^0 - 2^2 - (-2)^3 - 2](-2^0 - 2) + x - xy$
$= -1[-1 - 4 + 8 - 2](-3) + (-3) - (-3)(-4)$
$= -1[1](-3) - 3 - 12 = 3 - 3 - 12 = \mathbf{-12}$

PROBLEM SET 35

1. $R_M T_M + 1200 = R_L T_L;$
$R_L = 3R_M;$
$T_M = T_L = 30$

$$30R_M + 1200 = 3R_M(30)$$
$$-60R_M = -1200$$
$$R_M = \mathbf{20 \text{ yards per minute}}$$
$$R_L = 3(20) = \mathbf{60 \text{ yards per minute}}$$

2. $R_T T_T + R_W T_W = 56;$
$R_T = 4; R_W = 6;$
$T_T + T_W = 12$

$$4T_T + 6(12 - T_T) = 56$$
$$-2T_T = -16$$
$$T_T = \mathbf{8 \text{ hr}}$$
$$T_W = 12 - (8) = \mathbf{4 \text{ hr}}$$

3. $R_D T_D = R_B T_B; T_D = 12;$
$T_B = 4; R_B = R_D + 6$

$$12R_D = 4(R_D + 6)$$
$$8R_D = 24$$
$$R_D = 3$$
$$D_D = D_B = 3(12) = \mathbf{36 \text{ miles}}$$

4. (a) $5N_B + 8N_Y = 82$
(b) $N_B = 2N_Y + 2$
Substitute (b) into (a) and get:
(a') $5(2N_Y + 2) + 8N_Y = 82$
$$10N_Y + 10 + 8N_Y = 82$$
$$18N_Y = 72$$
$$N_Y = \mathbf{4}$$
(b) $N_B = 2(4) + 2 = \mathbf{10}$

5. $\frac{P}{100} \times of = is$
$$\frac{40}{100} \times A = 40$$
$$A = 40 \cdot \frac{100}{40} = \mathbf{100 \text{ tons}}$$

6. $\frac{20}{20} = \frac{V}{400}$
$$20V = 20(400)$$
$$V = \mathbf{400 \text{ grams}}$$

7. Lateral S.A. = Perimeter × Height
$$= \left[\left(\frac{1}{2}(2)(\pi)(2)\right) + \left(\frac{1}{2}(2)(\pi)(2)\right) + 20\right] m$$
$$\times 10 \text{ m}$$
$$= (4\pi + 20)(10) \text{ m}^2 \approx \mathbf{325.6 \text{ m}^2}$$

8. $16^{-1/2} = \frac{1}{16^{1/2}} = \frac{1}{4}$

9. $27^{-1/3} = \frac{1}{27^{1/3}} = \frac{1}{3}$

10. $9^{3/2} = (9^{1/2})^3 = 3^3 = \mathbf{27}$

11. $-64^{-2/3} = -\frac{1}{64^{2/3}} = -\frac{1}{(64^{1/3})^2} = -\frac{1}{16}$

12. (a) $3x - 4y = 84$
(b) $x + 10y = 96$

$$\begin{array}{r} 1(a) \quad 3x - 4y = 84 \\ -3(b) \quad -3x - 30y = -288 \\ \hline -34y = -204 \\ y = 6 \end{array}$$

(b) $x + 10(6) = 96$
$$x = \mathbf{36}$$

$$5 \times \overrightarrow{SF} = 9$$
$$\overrightarrow{SF} = \frac{9}{5}$$
$$7 \cdot \frac{9}{5} = A$$
$$\frac{63}{5} = A$$

13. $3\sqrt{\dfrac{7}{5}} + 2\sqrt{\dfrac{5}{7}} = \dfrac{3\sqrt{7}}{\sqrt{5}}\dfrac{\sqrt{5}}{\sqrt{5}} + \dfrac{2\sqrt{5}}{\sqrt{7}}\dfrac{\sqrt{7}}{\sqrt{7}}$

$= \dfrac{3\sqrt{35}}{5} + \dfrac{2\sqrt{35}}{7} = \dfrac{21\sqrt{35}}{35} + \dfrac{10\sqrt{35}}{35}$

$= \dfrac{31\sqrt{35}}{35}$

14. $2\sqrt{\dfrac{2}{5}} - 9\sqrt{\dfrac{5}{2}} = \dfrac{2\sqrt{2}}{\sqrt{5}}\dfrac{\sqrt{5}}{\sqrt{5}} - \dfrac{9\sqrt{5}}{\sqrt{2}}\dfrac{\sqrt{2}}{\sqrt{2}}$

$= \dfrac{2\sqrt{10}}{5} - \dfrac{9\sqrt{10}}{2} = \dfrac{4\sqrt{10}}{10} - \dfrac{45\sqrt{10}}{10}$

$= -\dfrac{41\sqrt{10}}{10}$

15. $\dfrac{\dfrac{a}{b} - 4}{\dfrac{xy}{b}} = \dfrac{\dfrac{a - 4b}{b}}{\dfrac{xy}{b}} \cdot \dfrac{\dfrac{b}{xy}}{\dfrac{b}{xy}} = \dfrac{a - 4b}{xy}$

16. $\dfrac{\dfrac{x}{x+y} + 6}{\dfrac{4}{x+y}} = \dfrac{\dfrac{x + 6(x+y)}{x+y}}{\dfrac{4}{x+y}} \cdot \dfrac{\dfrac{x+y}{4}}{\dfrac{x+y}{4}}$

$= \dfrac{x + 6(x+y)}{4} = \dfrac{7x + 6y}{4}$

17. $2x + 10 = 80$
$2x = 70$
$x = \mathbf{35}$
$3y + 40 = 130$
$3y = 90$
$y = \mathbf{30}$
$k = 360 - 130 - 80 = \mathbf{150}$

18. $2x + 332 = 360$
$2x = 28$
$x = \mathbf{14}$
$y + 40 + 14 = 180$
$y = \mathbf{126}$

19. $x = 2(105) - 112 = \mathbf{98}$
$y = \dfrac{x + 88}{2} = \dfrac{98 + 88}{2} = \mathbf{93}$
$p = \dfrac{88 + 62}{2} = \mathbf{75}$

20. Graph the line to find the slope.

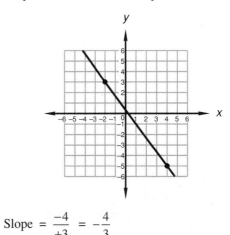

Slope $= \dfrac{-4}{+3} = -\dfrac{4}{3}$

$y = -\dfrac{4}{3}x + b$

$\dfrac{9}{3} = -\dfrac{4}{3}(-2) + b$

$\dfrac{1}{3} = b$

Since $m = -\dfrac{4}{3}$ and $b = \dfrac{1}{3}$: $y = -\dfrac{4}{3}x + \dfrac{1}{3}$

21. $\dfrac{4}{x^2(x+y)} + \dfrac{2x - 2}{x(x+y)}$

$= \dfrac{4}{x^2(x+y)} + \dfrac{2x^2 - 2x}{x^2(x+y)}$

$= \dfrac{2x^2 - 2x + 4}{x^2(x+y)}$

22. $\dfrac{3x}{x-2} - \dfrac{2x}{x^2 + x - 6}$

$= \dfrac{3x(x+3)}{(x-2)(x+3)} - \dfrac{2x}{(x-2)(x+3)}$

$= \dfrac{3x^2 + 7x}{(x-2)(x+3)}$

23. $35a - ax^2 - 2xa = -a(x+7)(x-5)$

24. $8x^2 - 2x^3 - x^4 = -x^2(x+4)(x-2)$

25. $2\sqrt{3} \cdot \sqrt{12} - 3\sqrt{2} \cdot \sqrt{6} + 4\sqrt{2}(3\sqrt{2} - \sqrt{6})$
$= 2\sqrt{3}\sqrt{2}\sqrt{2}\sqrt{3} - 3\sqrt{2}\sqrt{2}\sqrt{3} + 12\sqrt{2}\sqrt{2} - 4\sqrt{2}\sqrt{2}\sqrt{3}$
$= 12 - 6\sqrt{3} + 24 - 8\sqrt{3} = \mathbf{36 - 14\sqrt{3}}$

26. $\dfrac{(7000 \times 10^{14})(0.0002 \times 10^{-11})}{1400 \times 10^{-10}}$

$= \dfrac{(7 \times 10^{17})(2 \times 10^{-15})}{14 \times 10^{-8}} = \dfrac{14 \times 10^2}{14 \times 10^{-8}}$

$= \mathbf{1 \times 10^{10}}$

27. $\dfrac{x-3}{7} - \dfrac{2x}{4} = 5$

$4x - 12 - 14x = 140$

$-10x = 152$

$x = -\dfrac{76}{5}$

28. $0.002x = 0.02 + 0.04$

$2x = 20 + 40$

$x = \mathbf{30}$

29. $2\dfrac{1}{3}x - 2x^0 = 3\dfrac{1}{4}$

$\dfrac{7}{3}x = \dfrac{13}{4} + \dfrac{8}{4}$

$x = \dfrac{21}{4} \cdot \dfrac{3}{7} = \dfrac{9}{4}$

30. $D^2 = 11^2 + 1^2$

$D^2 = 121 + 1$

$D^2 = 122$

$D = \sqrt{122}$

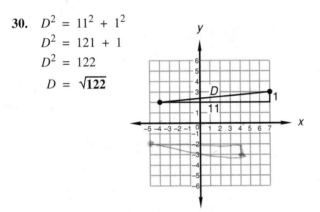

PROBLEM SET 36

1. $\dfrac{P}{100} \times \text{of} = \text{is}$

$\dfrac{340}{100} \cdot 400 = W$

$W = \mathbf{1360}$

2. (a) $4N_F + 21N_L = 290$

(b) $N_F + N_L = 30$

Substitute $N_L = 30 - N_F$ into (a) and get:

(a′) $4N_F + 21(30 - N_F) = 290$

$-17N_F = -340$

$N_F = \mathbf{20}$

(b) $(20) + N_L = 30$

$N_L = \mathbf{10}$

3. (a) $\dfrac{N}{D} = \dfrac{3}{5} \longrightarrow 5N = 3D$

(b) $N + D = 40$

Substitute $D = 40 - N$ into (a) and get:

(a′) $5N = 3(40 - N)$

$8N = 120$

$N = 15$

(b) $(15) + D = 40$

$D = 25$

The fraction was $\dfrac{15}{25}$.

4. $R_D T_D = R_H T_H; \ T_D = 10;$

$T_H = 8; \ R_H = R_D + 10$

$10R_D = 8(R_D + 10)$

$2R_D = 80$

$R_D = 40$

$D_D = D_H = 40(10) = \mathbf{400 \ miles}$

5. $R_J T_J + 80 = R_B T_B;$

$T_J = T_B = 4; \ R_J = 30$

$30(4) + 80 = 4R_B$

$200 = 4R_B$

$\mathbf{50 \ mph} = R_B$

6. $7N \qquad 7N + 7 \qquad 7N + 14 \qquad 7N + 21$

$4(7N + 2) = 3(7N + 14) + 15$

$28N + 8 = 21N + 57$

$7N = 49$

$N = 7$

The desired integers are **49, 56, 63,** and **70.**

7. $2(x + 4) + 2\left(\dfrac{x}{2}\right) + 2(x + 1) + 2x = 31$ in.

$2x + 8 + x + 2x + 2 + 2x = 31$ in.

$7x = 21$ in.

$x = \mathbf{3 \ in.}$

8. $\dfrac{(x + 5)(x - 4)}{(x - 2)(x + 1)} \cdot \dfrac{(x + 8)(x + 2)}{(x + 8)(x - 5)}$

$= \dfrac{(x - 4)(x + 2)(x + 5)}{(x - 2)(x + 1)(x - 5)}$

9. $\dfrac{(x - 6)(x + 3)}{(x - 8)(x + 4)} \cdot \dfrac{(x - 5)(x + 4)}{(x - 6)(x + 4)}$

$= \dfrac{(x + 3)(x - 5)}{(x - 8)(x + 4)}$

10. $\dfrac{1}{-3^{-2}} = -3^2 = \mathbf{-9}$

11. $-27^{-2/3} = \dfrac{1}{-27^{2/3}} = -\dfrac{1}{(27^{1/3})^2} = -\dfrac{1}{9}$

12. $\dfrac{1}{81^{-3/4}} = 81^{3/4} = \left(81^{1/4}\right)^3 = \mathbf{27}$

13. $(-27)^{-2/3} = \dfrac{1}{(-27)^{2/3}} = \dfrac{1}{\left((-27)^{1/3}\right)^2} = \dfrac{1}{9}$

14. $\dfrac{\dfrac{1}{x} + \dfrac{4}{y}}{3 + \dfrac{1}{xy}} = \dfrac{\dfrac{y + 4x}{xy}}{\dfrac{3xy + 1}{xy}} \cdot \dfrac{\dfrac{xy}{3xy + 1}}{\dfrac{xy}{3xy + 1}}$

$= \dfrac{y + 4x}{3xy + 1}$

15. $\dfrac{\dfrac{4}{x} - 3}{\dfrac{7}{x} + 2} = \dfrac{\dfrac{4 - 3x}{x}}{\dfrac{7 + 2x}{x}} \cdot \dfrac{\dfrac{x}{7 + 2x}}{\dfrac{x}{7 + 2x}} = \dfrac{4 - 3x}{7 + 2x}$

16. $3\sqrt{\dfrac{5}{3}} - 2\sqrt{\dfrac{3}{5}} = \dfrac{3\sqrt{5}}{\sqrt{3}}\dfrac{\sqrt{3}}{\sqrt{3}} - \dfrac{2\sqrt{3}}{\sqrt{5}}\dfrac{\sqrt{5}}{\sqrt{5}}$

$= \dfrac{3\sqrt{15}}{3} - \dfrac{2\sqrt{15}}{5} = \dfrac{15\sqrt{15}}{15} - \dfrac{6\sqrt{15}}{15} = \dfrac{3\sqrt{15}}{5}$

17. $\dfrac{(6 \times 10^{17})(3 \times 10^{-20})}{18 \times 10^{-10}} = \dfrac{18 \times 10^{-3}}{18 \times 10^{-10}}$

$= 1 \times 10^{7}$

18. $2\sqrt{\dfrac{7}{3}} - 3\sqrt{\dfrac{3}{7}} = \dfrac{2\sqrt{7}}{\sqrt{3}}\dfrac{\sqrt{3}}{\sqrt{3}} - \dfrac{3\sqrt{3}}{\sqrt{7}}\dfrac{\sqrt{7}}{\sqrt{7}}$

$= \dfrac{2\sqrt{21}}{3} - \dfrac{3\sqrt{21}}{7} = \dfrac{14\sqrt{21}}{21} - \dfrac{9\sqrt{21}}{21} = \dfrac{5\sqrt{21}}{21}$

19. $4\sqrt{12}(3\sqrt{2} - 4\sqrt{3}) = 4\sqrt{2}\sqrt{2}\sqrt{3}(3\sqrt{2} - 4\sqrt{3})$

$= 24\sqrt{6} - 96$

20. $2\sqrt{28} - 3\sqrt{63} + 2\sqrt{175}$
$= 2\sqrt{2}\sqrt{2}\sqrt{7} - 3\sqrt{3}\sqrt{3}\sqrt{7} + 2\sqrt{5}\sqrt{5}\sqrt{7}$
$= 4\sqrt{7} - 9\sqrt{7} + 10\sqrt{7} = 5\sqrt{7}$

21. $\dfrac{6}{x^2(x + 2)} - \dfrac{3}{x^2 + 3x + 2}$

$= \dfrac{6(x + 1)}{x^2(x + 2)(x + 1)} - \dfrac{3x^2}{x^2(x + 2)(x + 1)}$

$= \dfrac{-3x^2 + 6x + 6}{x^2(x + 2)(x + 1)}$

22. $\dfrac{p}{ax^2} + \dfrac{cx + a}{ax^3} + \dfrac{mx + b}{a^2 x^4}$

$= \dfrac{pax^2}{a^2 x^4} + \dfrac{ax(cx + a)}{a^2 x^4} + \dfrac{mx + b}{a^2 x^4}$

$= \dfrac{pax^2 + ax(cx + a) + mx + b}{a^2 x^4}$

23. $\dfrac{3x - 2}{7} - \dfrac{x}{4} - \dfrac{x - 3}{2} = 1$
$12x - 8 - 7x - 14x + 42 = 28$
$-9x = -6$
$x = \dfrac{2}{3}$

24. $3x^0 - 2(x - 3^0) - |-11 - 2| = 4x(2 - 5^0) - 7x$
$3 - 2x + 2 - 13 = 8x - 4x - 7x$
$-2x - 8 = -3x$
$x = 8$

25. (a) Every point on this line is 2 units above the x axis.
$y = 2$

(b) The y intercept is -2. The slope is positive, and the rise over the run for any triangle drawn is $\frac{1}{3}$.

$y = \dfrac{1}{3}x - 2$

26. $A = 180 - 140 = 40$
$B = 180 - 40 - 100 = 40$
$C = 180 - 100 = 80$
$D = B = 40$
$E = 180 - 90 - 40 = 50$
$M = 180 - 80 - 50 = 50$

27.

$$x + 1 \overline{)\, x^4 + 0x^3 + 0x^2 + 0x - 2 \quad}$$

quotient: $x^3 - x^2 + x - 1 - \dfrac{1}{x+1}$

$x^4 + x^3$
$- x^3 + 0x^2$
$- x^3 - x^2$
$x^2 + 0x$
$x^2 + x$
$- x - 2$
$- x - 1$
$- 1$

Check: $\dfrac{x^3(x + 1)}{x + 1} - \dfrac{x^2(x + 1)}{x + 1}$

$+ \dfrac{x(x + 1)}{x + 1} - \dfrac{1(x + 1)}{x + 1} - \dfrac{1}{x + 1}$

$= \dfrac{x^4 - 2}{x + 1}$

28. $D^2 = 9^2 + 7^2$
$D^2 = 81 + 49$
$D^2 = 130$
$D = \sqrt{130}$

29. $\dfrac{x^{-2} y}{zx^4}\left(3x^6 y^{-1} z - \dfrac{3x^{-3} y^2}{z^{-1} x^{-4}}\right) = \dfrac{3x^4 z}{x^4 z} - 3x^{-5} y^3$

$= 3 - 3x^{-5} y^3$

30. $-3^0(2 - 4^0) - |-3| - 2 - 4^2 - (-2)^3 - 2$
$= -1 - 3 - 2 - 16 + 8 - 2 = -16$

PROBLEM SET 37

1. Hydrogen: $2 \times 1 = 2$
Oxygen: $1 \times 16 = 16$
Total: $= 18$

$$\frac{16}{18} = \frac{Ox}{5400}$$
$18Ox = 16(5400)$
$Ox = \textbf{4800 grams}$

2. Nitrogen: $1 \times 14 = 14$
Hydrogen: $3 \times 1 = 3$
Total: $= 17$

$$\frac{14}{17} = \frac{N}{850}$$
$17N = 14(850)$
$N = \textbf{700 grams}$

3. Nitrogen: $1 \times 14 = 14$
Hydrogen: $4 \times 1 = 4$
Chlorine: $1 \times 35 = 35$
Total: $= 53$

$$\frac{35}{53} = \frac{Cl}{795}$$
$53Cl = 35(795)$
$Cl = \textbf{525 grams}$

4. $R_B T_B + 100 = R_T T_T;$
$T_B = T_T = 4;\ R_T = 2R_B$

$4R_B + 100 = 4(2R_B)$
$-4R_B = -100$
$R_B = 25$
$R_T = 2(25) = 50$
$D_B = 25(4) = \textbf{100 miles}$
$D_T = 50(4) = \textbf{200 miles}$

5. $\quad F \times of = is$

$2\frac{3}{4} \times GR = 550$

$GR = 550 \cdot \dfrac{4}{11} = \textbf{200}$

6. $\dfrac{(x + 3)(x - 2)}{x(x + 4)(x + 3)} \cdot \dfrac{x(x + 4)(x + 1)}{(x + 4)(x - 2)}$

$= \dfrac{x + 1}{x + 4}$

7. $\dfrac{ax(x - 4)(x + 3)}{(x + 4)(x + 3)} \cdot \dfrac{(x + 4)(x - 2)}{ax(x - 4)}$

$= x - 2$

8. $-8^{-4/3} = -\dfrac{1}{8^{4/3}} = -\dfrac{1}{(8^{1/3})^4} = -\dfrac{1}{16}$

9. $(-27)^{-4/3} = \dfrac{1}{(-27)^{4/3}} = \dfrac{1}{\left((-27)^{1/3}\right)^4} = \dfrac{1}{81}$

10. $-27^{-4/3} = -\dfrac{1}{27^{4/3}} = -\dfrac{1}{(27^{1/3})^4} = -\dfrac{1}{81}$

11. $\dfrac{-3}{-9^{-3/2}} = (-3)(-9^{3/2}) = (-3)(-9^{1/2})^3 = \textbf{81}$

12. $m\angle PON = 180 - 85 = \textbf{95°}$
$x - 3 = 11$
$\quad x = \textbf{14}$
$y + 4 = 9$
$\quad y = \textbf{5}$

13. $x = 2(90) - 80 = \textbf{100}$

$y = \dfrac{360 - 180}{2} = \dfrac{180}{2} = \textbf{90}$

$z = \dfrac{100 + 60}{2} = \dfrac{160}{2} = \textbf{80}$

14. $\dfrac{x + \dfrac{1}{x^2}}{x^2 - \dfrac{2}{x^2}} = \dfrac{\dfrac{x^3 + 1}{x^2}}{\dfrac{x^4 - 2}{x^2}} \cdot \dfrac{\dfrac{x^2}{x^4 - 2}}{\dfrac{x^2}{x^4 - 2}}$

$= \dfrac{x^3 + 1}{x^4 - 2}$

15. $\dfrac{a + \dfrac{y}{x}}{a - \dfrac{my}{x}} = \dfrac{\dfrac{ax + y}{x}}{\dfrac{ax - my}{x}} \cdot \dfrac{\dfrac{x}{ax - my}}{\dfrac{x}{ax - my}}$

$= \dfrac{ax + y}{ax - my}$

16. $\dfrac{3\sqrt{2}}{\sqrt{7}} \dfrac{\sqrt{7}}{\sqrt{7}} - \dfrac{5\sqrt{7}}{\sqrt{2}} \dfrac{\sqrt{2}}{\sqrt{2}} = \dfrac{3\sqrt{14}}{7} - \dfrac{5\sqrt{14}}{2}$

$= \dfrac{6\sqrt{14}}{14} - \dfrac{35\sqrt{14}}{14} = -\dfrac{29\sqrt{14}}{14}$

17. $\dfrac{2\sqrt{11}}{\sqrt{3}} \dfrac{\sqrt{3}}{\sqrt{3}} - \dfrac{5\sqrt{3}}{\sqrt{11}} \dfrac{\sqrt{11}}{\sqrt{11}} = \dfrac{2\sqrt{33}}{3} - \dfrac{5\sqrt{33}}{11}$

$= \dfrac{22\sqrt{33}}{33} - \dfrac{15\sqrt{33}}{33} = \dfrac{7\sqrt{33}}{33}$

18. Graph the line to find the slope.

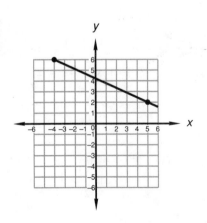

$m = \dfrac{-4}{+9} = -\dfrac{4}{9}$

Since the slopes of perpendicular lines are negative reciprocals of each other:

$m_\perp = \dfrac{9}{4}$

$y = \dfrac{9}{4}x + b$

$2 = \dfrac{9}{4}(-4) + b$

$11 = b$

Since $m_\perp = \dfrac{9}{4}$ and $b = 11$: $\mathbf{y = \dfrac{9}{4}x + 11}$

19. $2x + 280 = 360$
$2x = 80$
$x = \mathbf{40}$
$y + (40 + 30) = 180$
$y = \mathbf{110}$

20. (a) $3x - 3y = -6$
$y = x + 2$
(b) $3x + y = 6$
$y = -3x + 6$

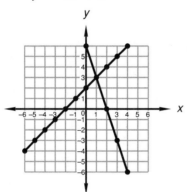

Substitute $y = -3x + 6$ into (a) and get:
(a′) $3x - 3(-3x + 6) = -6$
$12x = 12$
$x = 1$

(a) $y = (1) + 2 = 3$

(1, 3)

21. $\dfrac{x}{x + 5} - \dfrac{3x}{x^2 + 4x - 5}$

$= \dfrac{x(x - 1)}{(x + 5)(x - 1)} - \dfrac{3x}{(x + 5)(x - 1)}$

$= \dfrac{x^2 - 4x}{(x + 5)(x - 1)}$

22. $\dfrac{4}{x(x + 2)} + \dfrac{6}{x} = \dfrac{4}{x(x + 2)} + \dfrac{6(x + 2)}{x(x + 2)}$

$= \dfrac{6x + 16}{x(x + 2)}$

23. $\dfrac{3x + 2}{5} - \dfrac{x - 3}{7} = 2$
$21x + 14 - 5x + 15 = 70$
$16x = 41$
$x = \dfrac{41}{16}$

24. $\dfrac{4x}{3} - \dfrac{2x}{4} + x = 5$
$16x - 6x + 12x = 60$
$22x = 60$
$x = \dfrac{30}{11}$

25. $3\sqrt{2}(5\sqrt{12} - 2\sqrt{2}) = 3\sqrt{2}(5\sqrt{2}\sqrt{2}\sqrt{3} - 2\sqrt{2})$
$= \mathbf{30\sqrt{6} - 12}$

26. $4\sqrt{20}(3\sqrt{2} - 2\sqrt{5}) = 4\sqrt{2}\sqrt{2}\sqrt{5}(3\sqrt{2} - 2\sqrt{5})$
$= \mathbf{24\sqrt{10} - 80}$

27. $\dfrac{(x^{-2})^{-1}(y^{-2}x)^{-3}x^0 y}{(x^{-2})^2 x^4 xx^0 y^0 x^2} = \dfrac{x^2 y^6 x^{-3} y}{x^3} = \mathbf{x^{-4}y^7}$

28. $-2^2 - (-2)^3 - (-2) - 2^0 - 2$
$= -4 + 8 + 2 - 1 - 2 = \mathbf{3}$

29. $\dfrac{(3.5 \times 10^{-18})(3 \times 10^2)}{2.1 \times 10^{-36}} = \dfrac{10.5 \times 10^{-16}}{2.1 \times 10^{-36}}$
$= \mathbf{5 \times 10^{20}}$

30. $\dfrac{4x^{-2}y^{-2}}{z^2}\left(\dfrac{x^2 y}{z^{-2}} - \dfrac{3x^2 y^2 z^2}{p}\right) = \mathbf{4y^{-1} - 12p^{-1}}$

Problem Set 38

1. $3N_P = -4(-N_P) - 15$
$N_P = \mathbf{15}$

2. $\dfrac{2}{9} = \dfrac{I}{1440}$
$9I = 2(1440)$
$I = \mathbf{320 \text{ kilograms}}$

3. Hydrogen: $2 \times 1 = 2$
Sulfur: $1 \times 32 = 32$
Oxygen: $4 \times 16 = 64$
Total: $= 98$

$\dfrac{32}{98} = \dfrac{S}{196}$
$98S = 32(196)$
$S = \mathbf{64 \text{ grams}}$

4. (a) $\dfrac{N}{D} = \dfrac{7}{2} \rightarrow 2N = 7D$

(b) $10D = 2N + 84 \rightarrow N = 5D - 42$

Substitute (b) into (a) and get:

(a′) $2(5D - 42) = 7D$

$$3D = 84$$
$$D = \mathbf{28}$$

(b) $N = 5(28) - 42 = \mathbf{98}$

5. $R_C T_C = R_T T_T;\ R_C = 30;$
$R_T = 20;$
$T_C + T_T = 10$

$$30T_C = 20(10 - T_C)$$
$$50T_C = 200$$
$$T_C = 4$$

$$D_C = 30(4) = \mathbf{120\ miles}$$

6. $(x + 5)^3 = (x + 5)(x + 5)(x + 5)$
$= (x^2 + 10x + 25)(x + 5)$
$= x^3 + 10x^2 + 25x + 5x^2 + 50x + 125$
$= \mathbf{x^3 + 15x^2 + 75x + 125}$

7. $(x + 4)^3 = (x + 4)(x + 4)(x + 4)$
$= (x^2 + 8x + 16)(x + 4)$
$= x^3 + 8x^2 + 16x + 4x^2 + 32x + 64$
$= \mathbf{x^3 + 12x^2 + 48x + 64}$

8. $-x + x^2 = 12$
$x^2 - x - 12 = 0$
$(x - 4)(x + 3) = 0$
$x = \mathbf{4, -3}$

9. $-48x = -2x^2 - x^3$
$x^3 + 2x^2 - 48x = 0$
$x(x + 8)(x - 6) = 0$
$x = \mathbf{0, -8, 6}$

10. $2x^2 + 2x - 112 = 0$
$2(x + 8)(x - 7) = 0$
$x = \mathbf{-8, 7}$

11. $\dfrac{(x + 5)(x + 2)}{x(x + 7)(x + 2)} \cdot \dfrac{x(x + 7)(x + 4)}{(x + 5)(x - 3)}$

$= \dfrac{x + 4}{x - 3}$

12. $-16^{-1/4} = -\dfrac{1}{16^{1/4}} = -\dfrac{1}{2}$

13. $-16^{-3/4} = -\dfrac{1}{16^{3/4}} = -\dfrac{1}{(16^{1/4})^3} = -\dfrac{1}{8}$

14. $8^{2/3} = (8^{1/3})^2 = \mathbf{4}$

15. $(-8)^{1/3} = \mathbf{-2}$

16. $(\sqrt{11})^2 = H^2 + \left(\dfrac{3}{2}\right)^2$

$11 = H^2 + \dfrac{9}{4}$

$\dfrac{35}{4} = H^2$

$\dfrac{\sqrt{35}}{2} = H$

$\text{Area} = \dfrac{3\ \text{in.} \times \dfrac{\sqrt{35}}{2}\ \text{in.}}{2} = \dfrac{3\sqrt{35}}{4}\ \text{in.}^2$

17. Length $\overset{\frown}{DEF} = \dfrac{320}{360} \cdot 2(\pi)(20\ \text{m}) \approx \mathbf{111.64\ m}$

18. $\dfrac{x^2 - \dfrac{a}{x}}{a^2 - \dfrac{a}{x}} = \dfrac{\dfrac{x^3 - a}{x}}{\dfrac{a^2 x - a}{x}} \cdot \dfrac{\dfrac{x}{a^2 x - a}}{\dfrac{x}{a^2 x - a}}$

$= \dfrac{x^3 - a}{a^2 x - a}$

19. $\dfrac{\dfrac{mp^2}{4} - 5}{4p^2 - \dfrac{p^2}{4}} = \dfrac{\dfrac{mp^2 - 20}{4}}{\dfrac{16p^2 - p^2}{4}} \cdot \dfrac{\dfrac{4}{16p^2 - p^2}}{\dfrac{4}{16p^2 - p^2}}$

$= \dfrac{mp^2 - 20}{15p^2}$

20. $\dfrac{(7 \times 10^{-28})(3 \times 10^{-7})}{7 \times 10^{-25}} = \mathbf{3 \times 10^{-10}}$

21. $\dfrac{3\sqrt{2}}{\sqrt{11}} \dfrac{\sqrt{11}}{\sqrt{11}} + \dfrac{5\sqrt{11}}{\sqrt{2}} \dfrac{\sqrt{2}}{\sqrt{2}} = \dfrac{3\sqrt{22}}{11} + \dfrac{5\sqrt{22}}{2}$

$= \dfrac{6\sqrt{22}}{22} + \dfrac{55\sqrt{22}}{22} = \dfrac{\mathbf{61\sqrt{22}}}{\mathbf{22}}$

22. $-\dfrac{2\sqrt{11}}{\sqrt{3}} \dfrac{\sqrt{3}}{\sqrt{3}} + \dfrac{7\sqrt{3}}{\sqrt{11}} \dfrac{\sqrt{11}}{\sqrt{11}} = -\dfrac{2\sqrt{33}}{3} + \dfrac{7\sqrt{33}}{11}$

$= -\dfrac{22\sqrt{33}}{33} + \dfrac{21\sqrt{33}}{33} = -\dfrac{\sqrt{33}}{\mathbf{33}}$

23. $3\sqrt{24}(2\sqrt{6} - 3\sqrt{12})$
$= 3\sqrt{2}\sqrt{2}\sqrt{2}\sqrt{3}(2\sqrt{2}\sqrt{3} - 3\sqrt{2}\sqrt{2}\sqrt{3})$
$= \mathbf{72 - 108\sqrt{2}}$

24. Graph the line to find the slope.

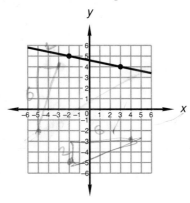

Slope $= \dfrac{-1}{+5} = -\dfrac{1}{5}$

$y = -\dfrac{1}{5}x + b$

$\dfrac{25}{5} = -\dfrac{1}{5}(-2) + b$

$\dfrac{23}{5} = b$

Since $m = -\dfrac{1}{5}$ and $b = \dfrac{23}{5}$: $\boldsymbol{y = -\dfrac{1}{5}x + \dfrac{23}{5}}$

25. $\dfrac{5x - 2}{3} - \dfrac{x}{4} = 7$

$20x - 8 - 3x = 84$

$17x = 92$

$x = \boldsymbol{\dfrac{92}{17}}$

26.

$$
\begin{array}{r}
x^3 - x^2 + x - 1 \\
x + 1 \overline{\smash{)}\, x^4 + 0x^3 + 0x^2 + 0x - 1} \\
\underline{x^4 + x^3} \\
-x^3 + 0x^2 \\
\underline{-x^3 - x^2} \\
x^2 + 0x \\
\underline{x^2 + x} \\
-x - 1 \\
\underline{-x - 1}
\end{array}
$$

27. $D^2 = 5^2 + 1^2$

$D^2 = 25 + 1$

$D^2 = 26$

$D = \boldsymbol{\sqrt{26}}$

28. $\dfrac{1}{x + 3} + \dfrac{3x}{x + 2} + \dfrac{2x + 1}{x^2 + 5x + 6}$

$= \dfrac{x + 2}{(x + 3)(x + 2)} + \dfrac{3x(x + 3)}{(x + 3)(x + 2)}$

$\quad + \dfrac{2x + 1}{(x + 3)(x + 2)}$

$= \dfrac{x + 2 + 3x^2 + 9x + 2x + 1}{(x + 3)(x + 2)}$

$= \boldsymbol{\dfrac{3x^2 + 12x + 3}{(x + 3)(x + 2)}}$

29. $\dfrac{(p^2 y^{-2})^{-3}\, p^{-2}\, (y^0)^{-2}}{(p^{-2} p^0 py)^{-3}\, (yp^{-2})^{-4}\, p} = \dfrac{p^{-6} y^6 p^{-2}}{p^6 p^{-3} y^{-3} y^{-4} p^8 p}$

$= \boldsymbol{p^{-20} y^{13}}$

30. $|3^0 - 3^2| - \dfrac{1}{2^{-2}} - 3^0(-3^3 - 3^2)$

$= 8 - 4 + 36 = \boldsymbol{40}$

PROBLEM SET 39

1. $F \times of = is$

$\dfrac{3}{16} \times S = 420$

$S = 420 \cdot \dfrac{16}{3} = 2240$

Ulterior motives $= 2240 - 420 = \boldsymbol{1820}$

2. $\dfrac{P}{100} \times of = is$

$\dfrac{28}{100} \times P = 9576$

$P = 9576 \cdot \dfrac{100}{28} = 34{,}200$

Ukases $= 34{,}200 - 9576 = \boldsymbol{24{,}624}$

3. Carbon: $1 \times 12 = 12$
Oxygen: $2 \times 16 = 32$
Total: $= 44$

$\dfrac{12}{44} = \dfrac{C}{528}$

$44C = 12(528)$

$C = \boldsymbol{144 \text{ grams}}$

4. $\dfrac{P}{100} \times of = is$

$\dfrac{86}{100} \times M = 688$

$M = 688 \cdot \dfrac{100}{86} = 800$

Sr $= 800 - 688 = \boldsymbol{112 \text{ grams}}$

5. $R_R T_R = R_W T_W; \; R_R = 5;$
    $R_W = 3; \; T_R + T_W = 8$

    $5T_R = 3(8 - T_R)$
    $8T_R = 24$
    $T_R = 3$
    $D_R = 5(3) = \textbf{15 miles}$

6. $7(90 - A) = 2(180 - A) + 110$
    $630 - 7A = 360 - 2A + 110$
    $-5A = -160$
    $A = \textbf{32}°$

7. $\text{Area} = \dfrac{bh}{2} = 52 \text{ ft}^2$

    $h = MP = \dfrac{2(52 \text{ ft}^2)}{13 \text{ ft}} = \textbf{8 ft}$

    $\text{Area}_{\text{Circle}} = \pi r^2 = (\pi)(8 \text{ ft})^2 = \textbf{200.96 ft}^2$

8. $x^2 - 9 = 0$
    $(x - 3)(x + 3) = 0$
    $x = \textbf{3, -3}$

9. $36x^2 - 36 = 0$
    $36(x^2 - 1) = 0$
    $36(x - 1)(x + 1) = 0$
    $x = \textbf{1, -1}$

10. $24x = -11x^2 - x^3$
    $x^3 + 11x^2 + 24x = 0$
    $x(x + 8)(x + 3) = 0$
    $x = \textbf{0, -3, -8}$

11. $(x - 1)^3 = (x - 1)(x - 1)(x - 1)$
    $= (x^2 - 2x + 1)(x - 1)$
    $= x^3 - 2x^2 + x - x^2 + 2x - 1$
    $= \textbf{x}^3 - \textbf{3x}^2 + \textbf{3x} - \textbf{1}$

12. $\dfrac{x(x + 5)(x + 1)}{(x + 5)(x - 3)} \cdot \dfrac{(x + 7)(x - 2)}{x(x + 7)(x + 1)}$

    $= \dfrac{\textbf{x} - \textbf{2}}{\textbf{x} - \textbf{3}}$

13. $\dfrac{2^0}{-4^{-3/2}} = -2^0 4^{3/2} = -(4^{1/2})^3 = \textbf{-8}$

14. $\dfrac{-3^0}{-27^{-2/3}} = -1[-27^{2/3}] = (27^{1/3})^2 = \textbf{9}$

15. $\dfrac{ax^2 - \dfrac{4}{a}}{\dfrac{x^2}{a} + 6} = \dfrac{\dfrac{a^2 x^2 - 4}{a}}{\dfrac{x^2 + 6a}{a}} \cdot \dfrac{\dfrac{a}{x^2 + 6a}}{\dfrac{a}{x^2 + 6a}}$

    $= \dfrac{a^2 x^2 - 4}{x^2 + 6a}$

16. $\dfrac{\dfrac{m^2 p}{x} - 6}{m^2 p - \dfrac{4}{x}} = \dfrac{\dfrac{m^2 p - 6x}{x}}{\dfrac{m^2 px - 4}{x}} \cdot \dfrac{\dfrac{x}{m^2 px - 4}}{\dfrac{x}{m^2 px - 4}}$

    $= \dfrac{m^2 p - 6x}{m^2 px - 4}$

17. $\dfrac{(3 \times 10^{-38})(8 \times 10^6)}{2.4 \times 10^{14}} = \dfrac{2.4 \times 10^{-31}}{2.4 \times 10^{14}}$
    $= \textbf{1} \times \textbf{10}^{-45}$

18. $\dfrac{2\sqrt{3}}{\sqrt{13}} \dfrac{\sqrt{13}}{\sqrt{13}} - \dfrac{5\sqrt{13}}{\sqrt{3}} \dfrac{\sqrt{3}}{\sqrt{3}} = \dfrac{2\sqrt{39}}{13} - \dfrac{5\sqrt{39}}{3}$

    $= \dfrac{6\sqrt{39}}{39} - \dfrac{65\sqrt{39}}{39} = -\dfrac{\textbf{59}\sqrt{\textbf{39}}}{\textbf{39}}$

19. $\dfrac{5\sqrt{3}}{\sqrt{2}} \dfrac{\sqrt{2}}{\sqrt{2}} - \dfrac{2\sqrt{2}}{\sqrt{3}} \dfrac{\sqrt{3}}{\sqrt{3}} = \dfrac{5\sqrt{6}}{2} - \dfrac{2\sqrt{6}}{3}$

    $= \dfrac{15\sqrt{6}}{6} - \dfrac{4\sqrt{6}}{6} = \dfrac{\textbf{11}\sqrt{\textbf{6}}}{\textbf{6}}$

20. $5\sqrt{45} - 2\sqrt{75} + 2\sqrt{108}$
    $= 5\sqrt{3}\sqrt{3}\sqrt{5} - 2\sqrt{3}\sqrt{5}\sqrt{5} + 2\sqrt{2}\sqrt{2}\sqrt{3}\sqrt{3}\sqrt{3}$
    $= 15\sqrt{5} - 10\sqrt{3} + 12\sqrt{3} = \textbf{15}\sqrt{\textbf{5}} + \textbf{2}\sqrt{\textbf{3}}$

21. (a) $2x + 4y = 120$
    (b) $3x - 2y = 60$

    1(a) $2x + 4y = 120$
    2(b) $\dfrac{6x - 4y = 120}{}$
    $8x = 240$
    $x = \textbf{30}$

    (a) $2(30) + 4y = 120$
    $4y = 60$
    $y = \textbf{15}$

22. $m\angle WXY = m\angle YZW = \textbf{68}°$
    $2x + 3 = 13$
    $2x = 10$
    $x = \textbf{5}$
    $3y + 4 = 11$
    $3y = 7$
    $y = \dfrac{\textbf{7}}{\textbf{3}}$

23. $3\sqrt{12}(4\sqrt{3} - 3\sqrt{3}) = 3\sqrt{2}\sqrt{2}\sqrt{3}(4\sqrt{3} - 3\sqrt{3})$
    $= 72 - 54 = \textbf{18}$

24. Graph the line to find the slope.

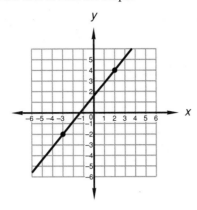

$$m = \frac{+6}{+5} = \frac{6}{5}$$

Since the slopes of perpendicular lines are negative reciprocals of each other:

$$m_\perp = -\frac{5}{6}$$

$$y = -\frac{5}{6}x + b$$

$$\frac{24}{6} = -\frac{5}{6}(2) + b$$

$$\frac{17}{3} = b$$

Since $m_\perp = -\frac{5}{6}$ and $b = \frac{17}{3}$:

$$y = -\frac{5}{6}x + \frac{17}{3}$$

25. $2\frac{1}{4}x - 3\frac{1}{2} = -\frac{1}{16}$

$$\frac{9}{4}x = -\frac{1}{16} + \frac{56}{16}$$

$$x = \frac{55}{16} \cdot \frac{4}{9} = \frac{55}{36}$$

26. $0.002x - 0.02 = 6.6$

$$2x - 20 = 6600$$
$$2x = 6620$$
$$x = 3310$$

27. $\dfrac{3}{x+1} + \dfrac{2x}{y(x+1)} + \dfrac{3x+2}{x^2+2x+1}$

$$= \frac{3y(x+1)}{y(x+1)^2} + \frac{2x(x+1)}{y(x+1)^2} + \frac{y(3x+2)}{y(x+1)^2}$$

$$= \frac{3xy + 3y + 2x^2 + 2x + 3xy + 2y}{y(x+1)^2}$$

$$= \frac{2x^2 + 2x + 5y + 6xy}{y(x+1)^2}$$

28. $\dfrac{4xy + 4x^2y^2}{4xy} = \dfrac{4xy(1+xy)}{4xy} = \mathbf{1 + xy}$

29. $\dfrac{4x^{-2}y^{-2}}{z^2}\left(\dfrac{3x^2y^2z^2}{4} + \dfrac{2x^0y^{-2}}{z^2y^2}\right)$

$$= 3 + 8x^{-2}y^{-6}z^{-4}$$

30. $-2^0 - 3^0(-2 - 5^0) - \dfrac{1}{-2^{-2}} + x^2y - xy$

$$= -1 + 3 + 4 + (-2)^2(3) - (-2)(3)$$
$$= -1 + 3 + 4 + 12 + 6 = \mathbf{24}$$

PROBLEM SET 40

1. $\dfrac{P}{100} \times \text{of} = \text{is}$

$$\frac{164}{100} \cdot 1{,}200{,}000 = ME$$

$$ME = \mathbf{\$1{,}968{,}000}$$

2. $\dfrac{P}{100} \times \text{of} = \text{is}$

$$\frac{380}{100} \cdot 4800 = MS$$

$$MS = \mathbf{18{,}240}$$

3. Iron: $1 \times 56 = 56$
Sulfur: $1 \times 32 = 32$
Total: $= 88$

$$\frac{56}{88} = \frac{448}{\text{FeS}}$$
$$56\text{FeS} = 88(448)$$
$$\text{FeS} = \mathbf{704 \text{ grams}}$$

4. (a) $20N_S + 60N_B = 8000$
(b) $N_B = 3N_S$
Substitute (b) into (a) and get:
(a') $20N_S + 60(3N_S) = 8000$
$$200N_S = 8000$$
$$N_S = \mathbf{40}$$
(b) $N_B = 3(40) = \mathbf{120}$

5. $R_AT_A + 120 = R_KT_K;$
$R_K = 15; R_A = 3;$
$T_A = T_K$

$$3T_K + 120 = 15T_K$$
$$120 = 12T_K$$
$$10 = T_K$$
$$D_K = 15(10) = \mathbf{150 \text{ miles}}$$

6. $\dfrac{c}{x} + \dfrac{m}{a} = c$
$$ac + mx = acx$$
$$mx = acx - ac$$
$$mx = a(cx - c)$$
$$\frac{mx}{cx - c} = a$$

7. $\dfrac{x}{m} + c = \dfrac{y}{a}$

$ax + acm = my$

$a(x + cm) = my$

$$a = \dfrac{my}{x + cm}$$

8. $\dfrac{6}{p} - ax = \dfrac{m}{y} + k$

$6y - apxy = mp + kpy$

$6y - apxy - kpy = mp$

$y(6 - apx - kp) = mp$

$$y = \dfrac{mp}{6 - apx - kp}$$

9. $\dfrac{a}{c} - b = \dfrac{m}{k}$

$ak - bck = cm$

$k(a - bc) = cm$

$$k = \dfrac{cm}{a - bc}$$

10. $(x - 3)^3 = (x - 3)(x - 3)(x - 3)$

$= (x^2 - 6x + 9)(x - 3)$

$= x^3 - 6x^2 + 9x - 3x^2 + 18x - 27$

$= \mathbf{x^3 - 9x^2 + 27x - 27}$

11. $X = \dfrac{360 - 318}{2} = \mathbf{21}$

Since diagonal bisectors of a rhombus are perpendicular: $Z = \mathbf{90}$

$Y + 21 + 90 = 180$

$Y = \mathbf{69}$

12. $4x^2 - 49 = 0$

$(2x - 7)(2x + 7) = 0$

$$x = \mathbf{\dfrac{7}{2}, -\dfrac{7}{2}}$$

13. $x^3 + 3x^2 = 18x$

$x^3 + 3x^2 - 18x = 0$

$x(x + 6)(x - 3) = 0$

$x = \mathbf{0, 3, -6}$

14. $\dfrac{x(x - 5)(x - 2)}{(x + 6)(x - 2)} \cdot \dfrac{(x + 6)(x - 3)}{x(x - 5)(x - 3)} = \mathbf{1}$

15. $\dfrac{(-4^0)^2}{4^{-3/2}} = (1)\left(4^{3/2}\right) = \left(4^{1/2}\right)^3 = \mathbf{8}$

16. $\dfrac{1}{16^{-1/4}} = 16^{1/4} = \mathbf{2}$

17. $\dfrac{\dfrac{4x^2 a}{y^2} + \dfrac{1}{a^2}}{2 - \dfrac{2}{y^2 a^2}}$

$= \dfrac{\dfrac{4x^2 a^3 + y^2}{y^2 a^2}}{\dfrac{2y^2 a^2 - 2}{y^2 a^2}} \cdot \dfrac{\dfrac{y^2 a^2}{2y^2 a^2 - 2}}{\dfrac{y^2 a^2}{2y^2 a^2 - 2}}$

$= \dfrac{\mathbf{4x^2 a^3 + y^2}}{\mathbf{2y^2 a^2 - 2}}$

18. $\dfrac{\dfrac{xy^2}{p} - 4}{a^2 y - \dfrac{1}{p}} = \dfrac{\dfrac{xy^2 - 4p}{p}}{\dfrac{a^2 py - 1}{p}} \cdot \dfrac{\dfrac{p}{a^2 py - 1}}{\dfrac{p}{a^2 py - 1}}$

$= \dfrac{\mathbf{xy^2 - 4p}}{\mathbf{a^2 py - 1}}$

19. $\dfrac{(4 \times 10^{17})(7 \times 10^{-26})}{1.4 \times 10^{-16}} = \dfrac{2.8 \times 10^{-8}}{1.4 \times 10^{-16}}$

$= \mathbf{2 \times 10^8}$

20. $3\sqrt{50} - 2\sqrt{72} + 3\sqrt{162}$

$= 3\sqrt{2}\sqrt{5}\sqrt{5} - 2\sqrt{2}\sqrt{6}\sqrt{6} + 3\sqrt{2}\sqrt{3}\sqrt{3}\sqrt{3}\sqrt{3}$

$= 15\sqrt{2} - 12\sqrt{2} + 27\sqrt{2} = \mathbf{30\sqrt{2}}$

21. $\dfrac{3\sqrt{3}}{\sqrt{7}}\dfrac{\sqrt{7}}{\sqrt{7}} + \dfrac{2\sqrt{7}}{\sqrt{3}}\dfrac{\sqrt{3}}{\sqrt{3}} = \dfrac{3\sqrt{21}}{7} + \dfrac{2\sqrt{21}}{3}$

$= \dfrac{9\sqrt{21}}{21} + \dfrac{14\sqrt{21}}{21} = \dfrac{\mathbf{23\sqrt{21}}}{\mathbf{21}}$

22. $A = \mathbf{30}$

$B + 30 = 180$

$B = \mathbf{150}$

$2C + 150 = 180$

$2C = 30$

$C = \mathbf{15}$

23. $\text{Volume}_{\text{Cylinder}} = \pi r^2 \times h = 32\pi \text{ cm}^3$

$$r^2 = \dfrac{32\pi \text{ cm}^3}{8\pi \text{ cm}}$$

$$r = \sqrt{4 \text{ cm}^2} = \mathbf{2 \text{ cm}}$$

24. (a) Every point on this line is 1 unit to the left of the y axis.

$\mathbf{x = -1}$

(b) The y intercept is -2. The slope is positive, and the rise over the run for any triangle drawn is $+1$.

$\mathbf{y = x - 2}$

25. $\dfrac{3 - 2x}{4} + \dfrac{x}{3} = 5$

$9 - 6x + 4x = 60$

$-2x = 51$

$x = -\dfrac{51}{2}$

26. $0.004x - 0.02 = 2.02$

$4x - 20 = 2020$

$4x = 2040$

$x = \mathbf{510}$

27. $\dfrac{3}{x} + \dfrac{2}{x + 2} + \dfrac{3x}{x^2 + 3x + 2}$

$= \dfrac{3(x + 1)(x + 2)}{x(x + 1)(x + 2)} + \dfrac{2x(x + 1)}{x(x + 1)(x + 2)}$

$+ \dfrac{3x^2}{x(x + 1)(x + 2)}$

$= \dfrac{3x^2 + 9x + 6 + 2x^2 + 2x + 3x^2}{x(x + 1)(x + 2)}$

$= \dfrac{\mathbf{8x^2 + 11x + 6}}{\mathbf{x(x + 1)(x + 2)}}$

28. $\dfrac{x + 4x}{x} = \dfrac{x(1 + 4)}{x} = \mathbf{5}$

29. $\dfrac{x^{-2}y}{p}\left(\dfrac{x^2 p}{y} - \dfrac{3x^2 y}{p}\right) = \mathbf{1 - 3y^2p^{-2}}$

30. $x^2 - xy - x^3 = \left(\dfrac{1}{2}\right)^2 - \dfrac{1}{2}\left(\dfrac{1}{3}\right) - \left(\dfrac{1}{2}\right)^3$

$= \dfrac{1}{4} - \dfrac{1}{6} - \dfrac{1}{8} = \dfrac{6}{24} - \dfrac{4}{24} - \dfrac{3}{24} = \mathbf{-\dfrac{1}{24}}$

PROBLEM SET 41

1. $N \qquad N + 2 \qquad N + 4 \qquad N + 6$

$4(N + N + 6) = 3(N + 2 + N + 4) + 12$

$8N + 24 = 6N + 30$

$2N = 6$

$N = 3$

The desired integers are **3, 5, 7,** and **9.**

2. Chromium: $1 \times 52 = 52$

Chlorine: $3 \times 35 = 105$

Total: $= 157$

$\dfrac{105}{157} = \dfrac{\text{Cl}}{1256}$

$157\text{Cl} = 105(1256)$

$\text{Cl} = \mathbf{840\ grams}$

3. $\dfrac{284}{100} \times TR = 1136$

$TR = 1136 \cdot \dfrac{100}{284} = \mathbf{400\ tons}$

4. (a) $5N_N + 10N_D = 700$

(b) $N_N + N_D = 100$

Substitute $N_D = 100 - N_N$ into (a) and get:

(a') $5N_N + 10(100 - N_N) = 700$

$-5N_N = -300$

$N_N = \mathbf{60}$

(b) $N_D = 100 - (60) = \mathbf{40}$

5. $R_R T_R + R_W T_W = 76;$

$R_R = 16;\ R_W = 4;$

$T_R + T_W = 7$

$16(7 - T_W) + 4T_W = 76$

$-12T_W = -36$

$T_W = 3$

$T_R = 4$

$D_W = 4(3) = \mathbf{12\ miles}$

$D_R = 16(4) = \mathbf{64\ miles}$

6. $87\text{ ft}^2 \times \dfrac{12\text{ in.}}{1\text{ ft}} \times \dfrac{12\text{ in.}}{1\text{ ft}} = \mathbf{87(12)(12)\ in.^2}$

7. $61\text{ yd}^2 \times \dfrac{3\text{ ft}}{1\text{ yd}} \times \dfrac{3\text{ ft}}{1\text{ yd}} \times \dfrac{12\text{ in.}}{1\text{ ft}} \times \dfrac{12\text{ in.}}{1\text{ ft}}$

$= \mathbf{61(3)(3)(12)(12)\ in.^2}$

8. $32\text{ mi}^3 \times \dfrac{5280\text{ ft}}{1\text{ mi}} \times \dfrac{5280\text{ ft}}{1\text{ mi}} \times \dfrac{5280\text{ ft}}{1\text{ mi}}$

$\times \dfrac{12\text{ in.}}{1\text{ ft}} \times \dfrac{12\text{ in.}}{1\text{ ft}} \times \dfrac{12\text{ in.}}{1\text{ ft}}$

$= \mathbf{32(5280)(5280)(5280)(12)(12)(12)\ in.^3}$

9. $\dfrac{x}{p} - \dfrac{k}{m} = c$

$mx - kp = cmp$

$mx = cmp + kp$

$\dfrac{\mathbf{mx}}{\mathbf{cm + k}} = p$

10. $\dfrac{xy}{p} - \dfrac{k}{c} = m$

$cxy - kp = cmp$

$cxy = cmp + kp$

$\dfrac{\mathbf{cxy}}{\mathbf{cm + k}} = p$

11. $\dfrac{4p}{x} - \dfrac{xk}{c} = \dfrac{y}{m}$

$4cmp - x^2 km = cxy$

$4cmp - cxy = x^2 km$

$c = \dfrac{\mathbf{x^2\ km}}{\mathbf{4mp - xy}}$

12. $8 \times \overrightarrow{SF} = 12$

$\overrightarrow{SF} = \dfrac{3}{2}$

$17 \cdot \dfrac{3}{2} = m$

$\dfrac{51}{2} = m$

$8 \times \overrightarrow{SF} = 20$

$\overrightarrow{SF} = \dfrac{5}{2}$

$15 \times \dfrac{5}{2} = p$

$\dfrac{75}{2} = p$

13. $2X + 40 = 180$

$2X = 140$

$X = \mathbf{70}$

Since $\triangle DEF$ is equilateral: $Y = \mathbf{60}$

$K + 70 + 50 = 180$

$K = \mathbf{60}$

14. $16x = -x^3 + 10x^2$

$x^3 - 10x^2 + 16x = 0$

$x(x - 8)(x - 2) = 0$

$x = \mathbf{0, 2, 8}$

15. $4x^2 - 9x = 0$

$x(4x - 9) = 0$

$x = \mathbf{0, \dfrac{9}{4}}$

16.
$$x - 2 \overline{)\, 2x^3 + 0x^2 + 0x - 1} \quad \mathbf{2x^2 + 4x + 8} + \frac{\mathbf{15}}{x-2}$$

$\underline{2x^3 - 4x^2}$

$4x^2 + 0x$

$\underline{4x^2 - 8x}$

$8x - 1$

$\underline{8x - 16}$

15

17. $4^2 = H^2 + \left(\dfrac{3}{2}\right)^2$

$16 = H^2 + \dfrac{9}{4}$

$\dfrac{55}{4} = H^2$

$\dfrac{\sqrt{55}}{2} = H$

Area $= \dfrac{3 \times \dfrac{\sqrt{55}}{2}}{2} + \dfrac{\pi(2)^2}{2} \approx \mathbf{11.84 \ units^2}$

18. $\dfrac{x(x + 5)(x + 3)}{(x - 2)(x - 2)} \cdot \dfrac{(x + 4)(x - 2)}{x(x + 4)(x + 3)}$

$= \dfrac{x + 5}{x - 2}$

19. $32^{-2/5} = \dfrac{1}{32^{2/5}} = \dfrac{1}{(32^{1/5})^2} = \dfrac{1}{4}$

20. $\dfrac{a^2 x - \dfrac{a}{x}}{ax - \dfrac{4}{x}} = \dfrac{\dfrac{a^2 x^2 - a}{x}}{\dfrac{ax^2 - 4}{x}} \cdot \dfrac{\dfrac{x}{ax^2 - 4}}{\dfrac{x}{ax^2 - 4}}$

$= \dfrac{a^2 x^2 - a}{ax^2 - 4}$

21. $\dfrac{(2.1 \times 10^{-38})(5 \times 10^5)}{1.5 \times 10^{-11}} = \dfrac{1.05 \times 10^{-32}}{1.5 \times 10^{-11}}$

$= \mathbf{7 \times 10^{-22}}$

22. $\dfrac{3\sqrt{5}}{\sqrt{7}}\dfrac{\sqrt{7}}{\sqrt{7}} - \dfrac{6\sqrt{7}}{\sqrt{5}}\dfrac{\sqrt{5}}{\sqrt{5}} = \dfrac{3\sqrt{35}}{7} - \dfrac{6\sqrt{35}}{5}$

$= \dfrac{15\sqrt{35}}{35} - \dfrac{42\sqrt{35}}{35} = -\dfrac{\mathbf{27\sqrt{35}}}{\mathbf{35}}$

23. $4\sqrt{24}(2\sqrt{6} - 3\sqrt{2}) = 4\sqrt{2}\sqrt{2}\sqrt{2}\sqrt{3}(2\sqrt{2}\sqrt{3} - 3\sqrt{2})$

$= \mathbf{96 - 48\sqrt{3}}$

24. Write the equation of the given line in slope-intercept form.

$y = -\dfrac{1}{3}x - \dfrac{2}{3}$

Since the slopes of perpendicular lines are negative reciprocals of each other:

$y = 3x + b$

$-5 = 3(-2) + b$

$1 = b$

$\mathbf{y = 3x + 1}$

25. $\dfrac{-3 - x}{2} - \dfrac{x}{2} = 7$

$-3 - x - x = 14$

$-2x = 17$

$x = -\dfrac{\mathbf{17}}{\mathbf{2}}$

26. $2\dfrac{1}{3}x - \dfrac{1}{9} = -\dfrac{1}{18}$

$\dfrac{7}{3}x = -\dfrac{1}{18} + \dfrac{2}{18}$

$x = \dfrac{1}{18} \cdot \dfrac{3}{7}$

$x = \dfrac{\mathbf{1}}{\mathbf{42}}$

27. $\dfrac{2}{x^2(x-2)} - \dfrac{2x+2}{(x^2-4)}$

$= \dfrac{2(x+2)}{x^2(x-2)(x+2)}$

$\quad - \dfrac{x^2(2x+2)}{x^2(x-2)(x+2)}$

$= \dfrac{2x+4-2x^3-2x^2}{x^2(x-2)(x+2)}$

$= \dfrac{-2x^3-2x^2+2x+4}{x^2(x-2)(x+2)}$

28. $\dfrac{4x+8x^2}{4x} = \dfrac{4x(1+2x)}{4x} = 1+2x$

29. $\dfrac{x^2 x^0 x^{-1}(x^{-2})^2 yx^{-3}}{(x^2 y)^{-3} xyx^{-2} x^2} = \dfrac{x^{-6} y}{x^{-5} y^{-2}} = x^{-1}y^3$

30. $x^3 - xy + x^2 = \left(\dfrac{1}{2}\right)^3 - \dfrac{1}{2}\left(\dfrac{1}{3}\right) + \left(\dfrac{1}{2}\right)^2$

$= \dfrac{3}{24} - \dfrac{4}{24} + \dfrac{6}{24} = \dfrac{5}{24}$

PROBLEM SET 42

1. $\dfrac{3}{17} \cdot 2244 = P$

$\qquad 396 = P$

2. $\dfrac{80}{100} \times S = 4800$

$\qquad S = 4800 \cdot \dfrac{100}{80} = 6000$

3. $\dfrac{600}{3600} = \dfrac{P}{43,200}$

$3600P = 600(43,200)$

$\qquad P = 7200 \text{ grams}$

4. Potassium: $\quad 1 \times 39 = 39$
Chlorine: $\qquad 1 \times 35 = 35$
Oxygen: $\qquad 3 \times 16 = 48$
Total: $\qquad\qquad\quad\; = 122$

$\dfrac{39}{122} = \dfrac{K}{488}$

$122K = 39(488)$

$\quad K = 156 \text{ grams}$

5. (a) $N_L = 4N_S + 2$
(b) $N_L = 8N_S - 6$
Substitute (b) into (a) and get:
(a') $(8N_S - 6) = 4N_S + 2$
$\qquad\qquad 4N_S = 8$
$\qquad\qquad\; N_S = 2$
(a) $N_L = 4(2) + 2 = 10$

6. $(24)(5280)(5280)(5280)(12)(12)(12)$

$\approx (2 \times 10^1)(5 \times 10^3)(5 \times 10^3)(5 \times 10^3)$
$\quad (1 \times 10^1)(1 \times 10^1)(1 \times 10^1)$

$= 250 \times 10^{13} \approx 3 \times 10^{15}$

7. $\dfrac{(2472)(570,185 \times 10^{-12})}{(243,195)(0.0003128 \times 10^{-6})}$

$\approx \dfrac{(2 \times 10^3)(6 \times 10^{-7})}{(2 \times 10^5)(3 \times 10^{-10})} = 2 \times 10^1$

8. $\dfrac{(0.0319743 \times 10^{-15})(61,853 \times 10^{37})}{6934 \times 10^{-29}}$

$\approx \dfrac{(3 \times 10^{-17})(6 \times 10^{41})}{7 \times 10^{-26}} \approx 3 \times 10^{50}$

9. $40 \text{ yd}^3 \times \dfrac{3 \text{ ft}}{1 \text{ yd}} \times \dfrac{3 \text{ ft}}{1 \text{ yd}} \times \dfrac{3 \text{ ft}}{1 \text{ yd}}$

$\quad \times \dfrac{12 \text{ in.}}{1 \text{ ft}} \times \dfrac{12 \text{ in.}}{1 \text{ ft}} \times \dfrac{12 \text{ in.}}{1 \text{ ft}}$

$= 40(3)(3)(3)(12)(12)(12) \text{ in.}^3$

10. $\dfrac{m}{c} - x = \dfrac{p}{m}$

$m^2 - cmx = cp$

$\qquad m^2 = cp + cmx$

$\dfrac{m^2}{p+mx} = c$

11. $\dfrac{ax}{y} + m = \dfrac{pc}{d}$

$adx + dmy = pcy$

$\qquad adx = pcy - dmy$

$\dfrac{adx}{pc-dm} = y$

12.

$$x + 2 \;\overline{\big)\; 4x^3 + 0x^2 + 0x - 1} \quad \Big| \; 4x^2 - 8x + 16 - \tfrac{33}{x+2}$$

$\qquad\quad \underline{4x^3 + 8x^2}$

$\qquad\qquad\;\; -8x^2 + 0x$

$\qquad\qquad\;\; \underline{-8x^2 - 16x}$

$\qquad\qquad\qquad\quad 16x - 1$

$\qquad\qquad\qquad\quad \underline{16x + 32}$

$\qquad\qquad\qquad\qquad\quad -33$

13. Area $= \dfrac{240}{360} \cdot \pi(2 \text{ cm})^2 \approx 8.37 \text{ cm}^2$

14. $\qquad\qquad -20x = x^3 - 9x^2$

$x^3 - 9x^2 + 20x = 0$

$x(x-5)(x-4) = 0$

$\qquad\qquad\qquad x = 0, 4, 5$

15.
$$4x^2 - 25 = 0$$
$$(2x - 5)(2x + 5) = 0$$
$$x = \frac{5}{2}, -\frac{5}{2}$$

16.
$$\frac{x(x - 2)(x - 1)}{(x + 4)(x - 1)} \cdot \frac{(x + 5)(x + 3)}{x(x + 3)(x - 2)}$$
$$= \frac{x + 5}{x + 4}$$

17. $-81^{-3/4} = -\dfrac{1}{81^{3/4}} = -\dfrac{1}{(81^{1/4})^3} = -\dfrac{1}{27}$

18.
$$\frac{\dfrac{m}{x} - 4}{6 - \dfrac{1}{x}} = \frac{\dfrac{m - 4x}{x}}{\dfrac{6x - 1}{x}} \cdot \frac{\dfrac{x}{6x - 1}}{\dfrac{x}{6x - 1}}$$
$$= \frac{m - 4x}{6x - 1}$$

19.
$$\frac{2\sqrt{3}}{\sqrt{11}} \frac{\sqrt{11}}{\sqrt{11}} - \frac{5\sqrt{11}}{\sqrt{3}} \frac{\sqrt{3}}{\sqrt{3}} = \frac{2\sqrt{33}}{11} - \frac{5\sqrt{33}}{3}$$
$$= \frac{6\sqrt{33}}{33} - \frac{55\sqrt{33}}{33}$$
$$= -\frac{49\sqrt{33}}{33}$$

20. $3\sqrt{6}(2\sqrt{6} - 4\sqrt{2}) = 3\sqrt{2}\sqrt{3}(2\sqrt{2}\sqrt{3} - 4\sqrt{2})$
$$= 36 - 24\sqrt{3}$$

21. $3\sqrt{20} + 2\sqrt{45} - \sqrt{245}$
$$= 3\sqrt{2}\sqrt{2}\sqrt{5} + 2\sqrt{3}\sqrt{3}\sqrt{5} - \sqrt{5}\sqrt{7}\sqrt{7}$$
$$= 6\sqrt{5} + 6\sqrt{5} - 7\sqrt{5} = 5\sqrt{5}$$

22. $16 \times \overrightarrow{SF} = 24$
$$\overrightarrow{SF} = \frac{3}{2}$$
$$30 \cdot \frac{3}{2} = a$$
$$45 = a$$
$$34 \cdot \frac{3}{2} = b$$
$$51 = b$$

23. (a) $3x + 2y = 80$
(b) $4x - 4y = 100$

2(a) $6x + 4y = 160$
(b) $\underline{4x - 4y = 100}$
$\,10x = 260$
$$x = 26$$

(b) $4(26) - 4y = 100$
$$y = 1$$

24. Graph the line to find the slope.

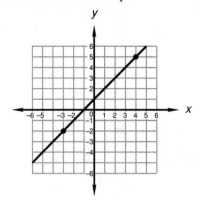

$$m = \frac{+1}{+1} = 1$$

Since the slopes of perpendicular lines are negative reciprocals of each other:
$$m_\perp = -1$$
$$y = -x + b$$
$$-2 = -(-3) + b$$
$$-5 = b$$
$$y = -x - 5$$

25.
$$\frac{-2x - 4}{2} - \frac{x}{3} = 5$$
$$-6x - 12 - 2x = 30$$
$$-8x = 42$$
$$x = -\frac{21}{4}$$

26. $3\dfrac{1}{4}x - \dfrac{1}{8} = \dfrac{3}{16}$
$$\frac{13}{4}x = \frac{3}{16} + \frac{2}{16}$$
$$x = \frac{5}{16} \cdot \frac{4}{13} = \frac{5}{52}$$

27.
$$\frac{3}{x + 2} - \frac{4}{x^2 - 4} - \frac{3}{x - 2}$$
$$= \frac{3(x - 2)}{(x - 2)(x + 2)} - \frac{4}{(x - 2)(x + 2)}$$
$$-\frac{3(x + 2)}{(x - 2)(x + 2)}$$
$$= \frac{3x - 6 - 4 - 3x - 6}{(x - 2)(x + 2)}$$
$$= \frac{-16}{(x - 2)(x + 2)}$$

28. $-\dfrac{x^{-2}}{y^2}\left(y^2 x^2 - \dfrac{3x^{-2}}{y^{-2}}\right) = -1 + 3x^{-4}$

29. $\dfrac{4x^2 - 4x^4}{4x^2} = \dfrac{4x^2(1 - x^2)}{4x^2} = \mathbf{1 - x^2}$

30. $-3x - x^{-2} - x^{-3} = -3\left(\dfrac{1}{2}\right) - \dfrac{1}{\left(\dfrac{1}{2}\right)^2} - \dfrac{1}{\left(\dfrac{1}{2}\right)^3}$

$\qquad = -\dfrac{3}{2} - 4 - 8 = -\dfrac{3}{2} - \dfrac{8}{2} - \dfrac{16}{2} = \mathbf{-\dfrac{27}{2}}$

PROBLEM SET 43

1. $0.0032 \times TP = 1280$

$\qquad TP = \dfrac{1280}{0.0032} = \mathbf{400{,}000}$

2. $\dfrac{232}{100} \times LM = 9280$

$\qquad LM = 9280 \cdot \dfrac{100}{232} = \mathbf{4000}$

3. Sodium: $\quad 1 \times 23 = 23$
Oxygen: $\quad 1 \times 16 = 16$
Hydrogen: $\quad 1 \times 1 = 1$
Total: $\qquad\qquad = 40$

$\dfrac{23}{40} = \dfrac{Na}{320}$
$40Na = 23(320)$
$\quad Na = \mathbf{184\ grams}$

4. $\dfrac{77}{100} \cdot 3000 = NC$

$\qquad NC = \mathbf{2310\ grams}$

5. $R_C T_C + 20 = R_L T_L;$
$T_C = T_L = 5;\ R_L = 2R_C$

$\dfrac{1}{2}R_L(5) + 20 = 5R_L$

$\qquad 20 = \dfrac{5}{2}R_L$

$\qquad 8 = R_L$

$D_L = 8(5) = \mathbf{40\ miles}$

6. $\text{sine } A = \dfrac{\textbf{opposite}}{\textbf{hypotenuse}};\ \cos A = \dfrac{\textbf{adjacent}}{\textbf{hypotenuse}};$

$\text{tangent } A = \dfrac{\textbf{opposite}}{\textbf{adjacent}}$

7. (a) $417 \cos 51.5° \approx \mathbf{259.59}$
(b) $32.6 \tan 86.3° \approx \mathbf{504.12}$

8. (a) $\sin A = \dfrac{\text{opposite}}{\text{hypotenuse}} = \dfrac{5}{7.6} = \mathbf{0.66}$

(b) $\cos B = \dfrac{\text{adjacent}}{\text{hypotenuse}} = \dfrac{7}{9.2} = \mathbf{0.76}$

(c) $\tan C = \dfrac{\text{opposite}}{\text{adjacent}} = \dfrac{8}{6} = \mathbf{1.33}$

9. $4\ \text{ft}^3 \times \dfrac{12\ \text{in.}}{1\ \text{ft}} \times \dfrac{12\ \text{in.}}{1\ \text{ft}} \times \dfrac{12\ \text{in.}}{1\ \text{ft}}$
$\qquad = \mathbf{4(12)(12)(12)\ in.^3}$

10. $\dfrac{a}{b} - x = \dfrac{p}{z}$
$az - bxz = pb$
$\qquad az = pb + bxz$

$\qquad \dfrac{az}{p + xz} = b$

11. $\dfrac{xz}{p} - k = \dfrac{m}{c}$
$cxz - ckp = mp$
$\qquad cxz = mp + ckp$

$\qquad \dfrac{cxz}{m + ck} = p$

12. $\dfrac{x}{m} - \dfrac{k}{c} = \dfrac{p}{z}$
$cxz - kmz = cmp$
$\qquad cxz = cmp + kmz$

$\qquad \dfrac{cxz}{cp + kz} = m$

13. $a^2 = 4^2 + \left(\dfrac{7}{2}\right)^2$

$a^2 = 16 + \dfrac{49}{4}$

$a^2 = \dfrac{113}{4}$

$a = \dfrac{\sqrt{113}}{2}$

$4 \times \overrightarrow{SF} = 6$

$\qquad \overrightarrow{SF} = \dfrac{3}{2}$

$\dfrac{7}{2} \cdot \dfrac{3}{2} = b$

$\dfrac{21}{4} = b$

$\dfrac{\sqrt{113}}{2} \cdot \dfrac{3}{2} = c + \dfrac{\sqrt{113}}{2}$

$\qquad 3\sqrt{113} = 4c + 2\sqrt{113}$

$\qquad \sqrt{113} = 4c$

$\qquad \dfrac{\sqrt{113}}{4} = c$

14. $\qquad -x^3 - 9x^2 = 20x$
$x^3 + 9x^2 + 20x = 0$
$x(x + 5)(x + 4) = 0$
$\qquad\qquad x = \mathbf{0, -5, -4}$

15.
$$4x^2 - 81 = 0$$
$$(2x - 9)(2x + 9) = 0$$
$$x = \frac{9}{2}, \ -\frac{9}{2}$$

16. $(x - 3)^3 = (x - 3)(x - 3)(x - 3)$
$$= (x^2 - 6x + 9)(x - 3)$$
$$= x^3 - 6x^2 + 9x - 3x^2 + 18x - 27$$
$$= \boldsymbol{x^3 - 9x^2 + 27x - 27}$$

17. $\dfrac{x(x - 5)(x - 1)}{(x + 3)(x - 1)} \cdot \dfrac{(x - 3)(x + 3)}{x(x - 5)(x - 3)} = 1$

18. $16^{-3/4} = \dfrac{1}{16^{3/4}} = \dfrac{1}{(16^{1/4})^3} = \dfrac{\boldsymbol{1}}{\boldsymbol{8}}$

19. $\dfrac{4x + \dfrac{1}{x}}{\dfrac{ay^2}{x} - 4} = \dfrac{\dfrac{4x^2 + 1}{x}}{\dfrac{ay^2 - 4x}{x}} \cdot \dfrac{\dfrac{x}{ay^2 - 4x}}{\dfrac{x}{ay^2 - 4x}}$

$$= \dfrac{\boldsymbol{4x^2 + 1}}{\boldsymbol{ay^2 - 4x}}$$

20. $\sqrt{\dfrac{3}{7}} - 4\sqrt{\dfrac{7}{3}} = \dfrac{\sqrt{3}}{\sqrt{7}}\dfrac{\sqrt{7}}{\sqrt{7}} - \dfrac{4\sqrt{7}}{\sqrt{3}}\dfrac{\sqrt{3}}{\sqrt{3}}$

$$= \dfrac{\sqrt{21}}{7} - \dfrac{4\sqrt{21}}{3} = \dfrac{3\sqrt{21}}{21} - \dfrac{28\sqrt{21}}{21}$$

$$= -\dfrac{\boldsymbol{25\sqrt{21}}}{\boldsymbol{21}}$$

21. $3\sqrt{2}(5\sqrt{12} - \sqrt{2}) = 3\sqrt{2}(5\sqrt{2}\sqrt{2}\sqrt{3} - \sqrt{2})$
$$= \boldsymbol{30\sqrt{6} - 6}$$

22. Since the triangle is isosceles: $A = B$
$$A^2 = 2^2 + 4^2$$
$$A^2 = 4 + 16$$
$$A = \sqrt{20} = 2\sqrt{5}$$
$$A = B = \boldsymbol{2\sqrt{5}}$$

23.
$$y = -\frac{1}{7}x + b$$
$$-5 = -\frac{1}{7}(-2) + b$$
$$-\frac{37}{7} = b$$
$$y = -\frac{1}{7}x - \frac{37}{7}$$

24. $\dfrac{5x - 7}{2} - \dfrac{3x - 2}{5} = 4$
$$25x - 35 - 6x + 4 = 40$$
$$19x = 71$$
$$x = \frac{\boldsymbol{71}}{\boldsymbol{19}}$$

25. $\dfrac{(47,816 \times 10^5)(4923 \times 10^{-14})}{403,000}$

$$\approx \dfrac{(5 \times 10^9)(5 \times 10^{-11})}{4 \times 10^5} = 6.25 \times 10^{-7}$$

$$\approx \boldsymbol{6 \times 10^{-7}}$$

26. (a) $2x - 3y = -9$
$$-3y = -2x - 9$$
$$y = \frac{2}{3}x + 3$$
(b) $x + y = 2$
$$y = -x + 2$$

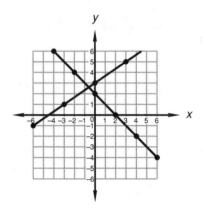

Substitute $y = -x + 2$ into (a) and get:
(a′) $2x - 3(-x + 2) = -9$
$$5x = -3$$
$$x = -\frac{3}{5}$$

(b) $y = -\left(-\dfrac{3}{5}\right) + 2 = \dfrac{13}{5}$

$$\left(-\frac{\boldsymbol{3}}{\boldsymbol{5}}, \frac{\boldsymbol{13}}{\boldsymbol{5}}\right)$$

27. $\dfrac{3}{x + 1} - \dfrac{2}{x^2(x + 1)} + \dfrac{3x + 2}{(x^2 - 1)}$

$$= \dfrac{3x^2(x - 1)}{x^2(x + 1)(x - 1)} - \dfrac{2(x - 1)}{x^2(x + 1)(x - 1)}$$

$$+ \dfrac{x^2(3x + 2)}{x^2(x + 1)(x - 1)}$$

$$= \dfrac{3x^3 - 3x^2 - 2x + 2 + 3x^3 + 2x^2}{x^2(x + 1)(x - 1)}$$

$$= \dfrac{\boldsymbol{6x^3 - x^2 - 2x + 2}}{\boldsymbol{x^2(x + 1)(x - 1)}}$$

28. $\dfrac{4x^{-2}y}{p^{-2}}\left(\dfrac{2p^{-2}x^2}{y} - \dfrac{4x^{-2}y}{p^2}\right) = \boldsymbol{8 - 16x^{-4}y^2}$

29. $D^2 = 6^2 + 1^2$
$D^2 = 36 + 1$
$D = \sqrt{37}$

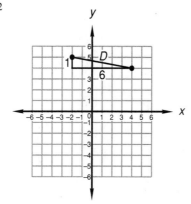

30. $x^2 - yx - (x - y)$

$= \left(\dfrac{1}{2}\right)^2 - \dfrac{1}{4}\left(\dfrac{1}{2}\right) - \left(\dfrac{1}{2} - \dfrac{1}{4}\right)$

$= \dfrac{1}{4} - \dfrac{1}{8} - \dfrac{1}{4} = -\dfrac{1}{8}$

PROBLEM SET 44

1. $-2(N + 5) = 6(-N) + 18$
$4N = 28$
$N = 7$

2. $N \quad N + 2 \quad N + 4 \quad N + 6$
$5(N + N + 4) = 8(N + 2 + N + 6) + 22$
$10N + 20 = 16N + 86$
$-6N = 66$
$N = -11$
The desired integers are **−11, −9, −7**, and **−5**.

3. $\dfrac{270}{700} = \dfrac{T}{2800}$
$700T = 270(2800)$
$T = \textbf{1080 tons}$

4. Carbon: $1 \times 12 = 12$
Chlorine: $2 \times 35 = 70$
Flourine: $2 \times 19 = 38$
Total: $= 120$

$\dfrac{12}{120} = \dfrac{C}{550}$
$120C = 12(550)$
$C = \textbf{55 grams}$

5. $R_R T_R + R_D T_D = 200$;
$R_R = 25$; $R_D = 50$;
$T_R + T_D = 5$

$25T_R + 50(5 - T_R) = 200$
$-25T_R = -50$
$T_R = 2$
$D_R = 25(2) = \textbf{50 miles}$

6. $B + 28 + 90 = 180$
$B = \textbf{62}$

$\cos 28° = \dfrac{9.5}{H}$

$H = \dfrac{9.5}{\cos 28°} \approx \textbf{10.76}$

$\tan 28° = \dfrac{y}{9.5}$
$y = 9.5 \tan 28° \approx \textbf{5.05}$

7. $B + 31 + 90 = 180$
$B = \textbf{59}$

$\cos 31° = \dfrac{m}{14}$
$m = 14 \cos 31° \approx \textbf{12.00}$

$\sin 31° = \dfrac{x}{14}$
$x = 14 \sin 31° \approx \textbf{7.21}$

8. $\cos A = \dfrac{4}{7}$
$A \approx \textbf{55.15}$

$B + 55.15 + 90 \approx 180$
$B \approx \textbf{34.85}$

$\sin 55.15° = \dfrac{x}{7}$
$x = 7 \sin 55.15° \approx \textbf{5.74}$

9. $4 \text{ mi}^2 \times \dfrac{5280 \text{ ft}}{1 \text{ mi}} \times \dfrac{5280 \text{ ft}}{1 \text{ mi}} = \textbf{4(5280)(5280) ft}^2$

10. $\dfrac{k}{m} - c = \dfrac{p}{d}$
$dk - cdm = pm$
$dk = pm + cdm$
$\dfrac{\textbf{dk}}{\textbf{p + cd}} = m$

11. $\dfrac{3k}{m} - \dfrac{d}{p} = \dfrac{l}{c}$
$3ckp - cdm = lmp$
$3ckp = lmp + cdm$
$\dfrac{\textbf{3ckp}}{\textbf{lp + cd}} = m$

12. $\dfrac{a}{c} + d = \dfrac{x}{m}$
$am + cdm = cx$
$am = cx - cdm$
$\dfrac{\textbf{am}}{\textbf{x − dm}} = c$

13. If $m\angle BAD = K°$, then $m\angle ABC = (180 - K)°$
(a) $4x + 2y = 8$
(b) $4x + y = 6$

$$
\begin{array}{ll}
\text{(a)} & 4x + 2y = 8 \\
-1\text{(b)} & -4x - y = -6 \\
\hline
& y = 2
\end{array}
$$

(a) $4x + 2(2) = 8$
$x = 1$

14. $24x = -x^3 + 10x^2$
$x^3 - 10x^2 + 24x = 0$
$x(x - 6)(x - 4) = 0$
$x = 0, 4, 6$

15. $25p^2 - 81 = 0$
$(5p - 9)(5p + 9) = 0$
$$p = \frac{9}{5}, -\frac{9}{5}$$

16.

$$
\begin{array}{r}
x^3 - 4x^2 + 16x - 64 + \frac{255}{x+4} \\
x + 4 \overline{)\; x^4 + 0x^3 + 0x^2 + 0x - 1} \\
\underline{x^4 + 4x^3} \\
-4x^3 + 0x^2 \\
\underline{-4x^3 - 16x^2} \\
16x^2 + 0x \\
\underline{16x^2 + 64x} \\
-64x - 1 \\
\underline{-64x - 256} \\
255
\end{array}
$$

17. $\dfrac{x(x + 7)(x + 2)}{(x - 3)(x + 2)} \cdot \dfrac{(x + 5)(x + 5)}{x(x + 7)(x + 5)}$

$= \dfrac{x + 5}{x - 3}$

18. $\dfrac{1}{-32^{4/5}} = -\dfrac{1}{(32^{1/5})^4} = -\dfrac{1}{16}$

19. Height $= 4 \text{ ft} \times \dfrac{12 \text{ in.}}{1 \text{ ft}} = 48 \text{ in.}$

$V_{\text{Cone}} = \dfrac{1}{3}\Big(A_{\text{Base}} \times \text{height}\Big)$

$\phantom{V_{\text{Cone}}} = \dfrac{1}{3}\Big(\dfrac{1}{2}\pi(4)^2 + \dfrac{1}{2}(12)(8)$

$\phantom{V_{\text{Cone}} = \dfrac{1}{3}} - \dfrac{1}{2}\pi(2)^2 \Big)(48) \text{ in.}^3$

$\phantom{V_{\text{Cone}}} = \dfrac{1}{3}(6\pi + 48)(48) \text{ in.}^3 \approx 1069.44 \text{ in.}^3$

20. $\dfrac{5x^2 + \dfrac{1}{x}}{\dfrac{pm^2}{x} + 5} = \dfrac{\dfrac{5x^3 + 1}{x}}{\dfrac{pm^2 + 5x}{x}} \cdot \dfrac{\dfrac{x}{pm^2 + 5x}}{\dfrac{x}{pm^2 + 5x}}$

$= \dfrac{5x^3 + 1}{pm^2 + 5x}$

21. $3\sqrt{\dfrac{2}{17}} - 5\sqrt{\dfrac{17}{2}} = \dfrac{3\sqrt{2}}{\sqrt{17}}\dfrac{\sqrt{17}}{\sqrt{17}} - \dfrac{5\sqrt{17}}{\sqrt{2}}\dfrac{\sqrt{2}}{\sqrt{2}}$

$= \dfrac{3\sqrt{34}}{17} - \dfrac{5\sqrt{34}}{2} = \dfrac{6\sqrt{34}}{34} - \dfrac{85\sqrt{34}}{34}$

$= -\dfrac{79\sqrt{34}}{34}$

22. $3\sqrt{6}(2\sqrt{6} - \sqrt{12}) = 3\sqrt{2}\sqrt{3}(2\sqrt{2}\sqrt{3} - \sqrt{2}\sqrt{2}\sqrt{3})$
$= 36 - 18\sqrt{2}$

23. (a) Every point on this line is 2 units above the
x axis.
$y = 2$
(b) The y intercept is -4. The slope is positive, and
the rise over the run for any triangle drawn is $\frac{1}{2}$.

$y = \dfrac{1}{2}x - 4$

24. $\dfrac{5x - 2}{3} - \dfrac{2x - 4}{2} = 6$
$10x - 4 - 6x + 12 = 36$
$4x = 28$
$x = 7$

25. $3\dfrac{1}{5}k + \dfrac{2}{5} = \dfrac{1}{10}$

$\dfrac{16}{5}k = \dfrac{1}{10} - \dfrac{4}{10}$

$k = -\dfrac{3}{10} \cdot \dfrac{5}{16} = -\dfrac{3}{32}$

26. (a) $2x - y = 5$
$y = 2x - 5$
(b) $4x + 3y = 9$

$y = -\dfrac{4}{3}x + 3$

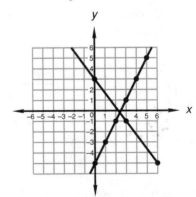

Solve the equations by elimination:

-2(a) $-4x + 2y = -10$

(b) $\dfrac{4x + 3y = \quad 9}{5y = \quad -1}$

$$y = -\dfrac{1}{5}$$

(a) $2x - \left(-\dfrac{1}{5}\right) = 5$

$$x = \dfrac{24}{5} \cdot \dfrac{1}{2} = \dfrac{12}{5}$$

$$\left(\dfrac{12}{5}, -\dfrac{1}{5}\right)$$

27. $\dfrac{x}{x+2} - \dfrac{2}{x^2 - 4}$

$= \dfrac{x(x-2)}{x^2 - 4} - \dfrac{2}{x^2 - 4} = \dfrac{x^2 - 2x - 2}{x^2 - 4}$

28. $\dfrac{(51{,}463 \times 10^{-14})(748{,}600 \times 10^{-21})}{7{,}861{,}523}$

$\approx \dfrac{(5 \times 10^{-10})(7 \times 10^{-16})}{8 \times 10^6} \approx \mathbf{4 \times 10^{-32}}$

29. $a^2 - ab^2 = \left(\dfrac{1}{2}\right)^2 - \dfrac{1}{2}\left(-\dfrac{1}{2}\right)^2 = \dfrac{1}{4} - \dfrac{1}{8} = \dfrac{1}{8}$

30. $-2^0 - 2^2 - (-2)^3 - |-2 - 3^0| - (-2)^3$

$= -1 - 4 + 8 - 3 + 8 = \mathbf{8}$

PROBLEM SET 45

1. $\dfrac{62}{100} \times S = 248$

$$S = 248 \cdot \dfrac{100}{62} = \mathbf{400}$$

2. Phosphorus: $1 \times 31 = 31$
Hydrogen: $3 \times 1 = 3$
Total: $\quad = 34$

$\dfrac{3}{34} = \dfrac{24}{T}$

$T = \mathbf{272\ grams}$

3. $\dfrac{64}{100} \times T = 5376$

$$T = 5376 \cdot \dfrac{100}{64} = 8400$$

Bromides $= 8400 - 5376 = \mathbf{3024}$

4. (a) $12N_R + 14N_C = 138$
(b) $N_R + N_C = 11$
Substitute $N_C = 11 - N_R$ into (a) and get:
(a') $12N_R + 14(11 - N_R) = 138$
$$N_R = \mathbf{8}$$
(b) $N_C = 11 - (8) = \mathbf{3}$

5. $R_B T_B + 80 = R_T T_T;$
$T_T = 8;\ T_B = 12;$
$R_T = R_B + 30$

$12R_B + 80 = 8(R_B + 30)$
$4R_B = 160$
$R_B = \mathbf{40\ mph}$
$R_T = (40) + 30 = \mathbf{70\ mph}$

6. $x^2 = 5$
$x = \pm\sqrt{5}$

7. $(x + 7)^2 = 11$
$x + 7 = \pm\sqrt{11}$
$x = \mathbf{-7 \pm \sqrt{11}}$

8. $(x + 12)^2 = 2$
$x + 12 = \pm\sqrt{2}$
$x = \mathbf{-12 \pm \sqrt{2}}$

9. $\left(x + \dfrac{3}{4}\right)^2 = 13$

$x + \dfrac{3}{4} = \pm\sqrt{13}$

$x = \mathbf{-\dfrac{3}{4} \pm \sqrt{13}}$

10. $\sin 33° = \dfrac{x}{10}$
$x = 10 \sin 33° \approx \mathbf{5.45}$

11. $\sin 38° = \dfrac{4}{p}$

$p = \dfrac{4}{\sin 38°} \approx \mathbf{6.50}$

12. $\cos A = \dfrac{8}{12}$
$A \approx \mathbf{48.19}$

$B \approx 180 - 90 - 48.19 \approx \mathbf{41.81}$

13. $100{,}000\ \text{mi}^2 \times \dfrac{5280\ \text{ft}}{1\ \text{mi}} \times \dfrac{5280\ \text{ft}}{1\ \text{mi}}$

$\times \dfrac{12\ \text{in.}}{1\ \text{ft}} \times \dfrac{12\ \text{in.}}{1\ \text{ft}}$

$= \mathbf{100{,}000(5280)(5280)(12)(12)\ in.^2}$

14. $\dfrac{(571{,}652)(40{,}316)}{214{,}000 \times 10^6} \approx \dfrac{(6 \times 10^5)(4 \times 10^4)}{2 \times 10^{11}}$

$= 1.2 \times 10^{-1} \approx \mathbf{1 \times 10^{-1}}$

15. $8 \times \overrightarrow{SF} = 12$

$\overrightarrow{SF} = \dfrac{3}{2}$

$x \times \dfrac{3}{2} = 5$

$x = \dfrac{10}{3}$

$c^2 = 8^2 + \left(\dfrac{10}{3}\right)^2$

$c^2 = \dfrac{676}{9}$

$c = \dfrac{26}{3}$

$\dfrac{26}{3} \cdot \dfrac{3}{2} = d + \dfrac{26}{3}$

$13 = d + \dfrac{26}{3}$

$\dfrac{39}{3} - \dfrac{26}{3} = d$

$\dfrac{13}{3} = d$

16. $\dfrac{mx}{4} + \dfrac{y}{p} = c$

$mxp + 4y = 4cp$

$4y = 4cp - mxp$

$\dfrac{4y}{4c - mx} = p$

17. $\dfrac{5x}{p} - k = \dfrac{m}{c}$

$5xc - ckp = mp$

$5xc = mp + ckp$

$\dfrac{5xc}{m + ck} = p$

18. $x^3 = -5x^2 + 50x$

$x^3 + 5x^2 - 50x = 0$

$x(x + 10)(x - 5) = 0$

$x = \mathbf{0, -10, 5}$

19. $36x^2 - 25 = 0$

$(6x - 5)(6x + 5) = 0$

$x = \dfrac{5}{6}, -\dfrac{5}{6}$

20. $\dfrac{x(x + 7)(x - 1)}{(x + 7)(x - 3)} \cdot \dfrac{(x + 5)(x - 3)}{x(x + 2)(x - 1)}$

$= \dfrac{x + 5}{x + 2}$

21. $-(-27)^{-5/3} = -\dfrac{1}{(-27)^{5/3}} = -\dfrac{1}{\left((-27)^{1/3}\right)^5}$

$= \dfrac{1}{243}$

22. Perimeter $= (24 + 4\pi)$ cm

Circumference$_{\text{Semicircle}} = 24 + 4\pi - 12 - 6$
$- 4 - 2 = 4\pi$ cm

Circumference$_{\text{Semicircle}} = \dfrac{2\pi r}{2} = 4\pi$ cm

$r = 4$ cm

Area$_{\text{Semicircle}} = \dfrac{\pi(4 \text{ cm})^2}{2} \approx \mathbf{25.12 \text{ cm}^2}$

23. $\dfrac{4xp - \dfrac{1}{x}}{6p - \dfrac{p^2}{x}} = \dfrac{\dfrac{4x^2p - 1}{x}}{\dfrac{6px - p^2}{x}} \cdot \dfrac{\dfrac{x}{6px - p^2}}{\dfrac{x}{6px - p^2}}$

$= \dfrac{\mathbf{4x^2p - 1}}{\mathbf{6px - p^2}}$

24. $5\sqrt{\dfrac{2}{5}} + 3\sqrt{\dfrac{5}{2}} = \dfrac{5\sqrt{2}}{\sqrt{5}}\dfrac{\sqrt{5}}{\sqrt{5}} + \dfrac{3\sqrt{5}}{\sqrt{2}}\dfrac{\sqrt{2}}{\sqrt{2}}$

$= \dfrac{5\sqrt{10}}{5} + \dfrac{3\sqrt{10}}{2} = \dfrac{10\sqrt{10}}{10} + \dfrac{15\sqrt{10}}{10}$

$= \dfrac{25\sqrt{10}}{10} = \dfrac{\mathbf{5\sqrt{10}}}{\mathbf{2}}$

25. $2\sqrt{5}(3\sqrt{15} - 2\sqrt{10}) = 2\sqrt{5}(3\sqrt{3}\sqrt{5} - 2\sqrt{2}\sqrt{5})$

$= \mathbf{30\sqrt{3} - 20\sqrt{2}}$

26. $y = -\dfrac{2}{3}x + b$

$-5 = -\dfrac{2}{3}(-2) + b$

$-\dfrac{19}{3} = b$

$y = -\dfrac{2}{3}x - \dfrac{19}{3}$

27. $\dfrac{5x - 2}{7} - \dfrac{x - 3}{5} = 4$

$25x - 10 - 7x + 21 = 140$

$18x = 129$

$x = \dfrac{43}{6}$

28. (a) $2x + 3y = 3$

$$y = -\frac{2}{3}x + 1$$

(b) $x - 5y = 20$

$$y = \frac{1}{5}x - 4$$

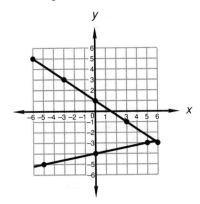

(a) $\quad\;\; 2x + 3y = 3$

-2(b) $\;\underline{-2x + 10y = -40}$

$\qquad\qquad 13y = -37$

$$y = -\frac{37}{13}$$

(b) $x - 5\left(-\dfrac{37}{13}\right) = 20$

$$x = \frac{260}{13} - \frac{185}{13} = \frac{75}{13}$$

$$\left(\frac{75}{13},\; -\frac{37}{13}\right)$$

29. $\dfrac{2}{x - 2} - \dfrac{3}{x + 2} - \dfrac{2x}{x^2 - 4}$

$$= \frac{2(x + 2)}{x^2 - 4} - \frac{3(x - 2)}{x^2 - 4} - \frac{2x}{x^2 - 4}$$

$$= \frac{2x + 4 - 3x + 6 - 2x}{x^2 - 4}$$

$$= \frac{-3x + 10}{x^2 - 4}$$

30. $a^2 - ab^3 = \left(-\dfrac{1}{2}\right)^2 - \left(-\dfrac{1}{2}\right)\left(\dfrac{1}{2}\right)^3 = \dfrac{1}{4} + \dfrac{1}{16}$

$$= \frac{4}{16} + \frac{1}{16} = \frac{5}{16}$$

PROBLEM SET 46

1. $\dfrac{240}{100} \times T = 1440$

$$T = 1440 \cdot \frac{100}{240} = 600$$

2. $\dfrac{1000}{4000} = \dfrac{B}{24,000}$

$$4000B = 24,000(1000)$$

$$B = \textbf{6000 kilograms}$$

3. Nitrogen: $\quad 1 \times 14 = 14$

Hydrogen: $\quad 4 \times 1 = 4$

Chlorine: $\quad 1 \times 35 = 35$

Total: $\qquad\qquad\;\; = 53$

$$\frac{35}{53} = \frac{140}{NH_4Cl}$$

$$35NH_4Cl = 53(140)$$

$$NH_4Cl = \textbf{212 grams}$$

4. (a) $\dfrac{N_F}{N_E} = \dfrac{7}{2} \;\longrightarrow\; 2N_F = 7N_E$

(b) $N_F = 3N_E + 11$

Substitute (b) into (a) and get:

(a') $2(3N_E + 11) = 7N_E$

$$N_E = \textbf{22}$$

(b) $N_F = 3(22) + 11 = \textbf{77}$

5. $R_M T_M + 16 = R_B T_B$;

$T_M = T_B = 4$; $R_M = 16$

$$16(4) + 16 = 4R_B$$

$$80 = 4R_B$$

$$\textbf{20 mph} = R_B$$

6. $5\sqrt{\dfrac{3}{2}} + 2\sqrt{\dfrac{2}{3}} - \sqrt{600}$

$$= \frac{5\sqrt{3}}{\sqrt{2}}\frac{\sqrt{2}}{\sqrt{2}} + \frac{2\sqrt{2}}{\sqrt{3}}\frac{\sqrt{3}}{\sqrt{3}} - 10\sqrt{6}$$

$$= \frac{5\sqrt{6}}{2} + \frac{2\sqrt{6}}{3} - 10\sqrt{6}$$

$$= \frac{15\sqrt{6}}{6} + \frac{4\sqrt{6}}{6} - \frac{60\sqrt{6}}{6} = -\frac{\textbf{41}\sqrt{\textbf{6}}}{\textbf{6}}$$

7. $3\sqrt{\dfrac{5}{7}} - 5\sqrt{\dfrac{7}{5}} + 3\sqrt{140}$

$$= \frac{3\sqrt{5}}{\sqrt{7}}\frac{\sqrt{7}}{\sqrt{7}} - \frac{5\sqrt{7}}{\sqrt{5}}\frac{\sqrt{5}}{\sqrt{5}} + 6\sqrt{35}$$

$$= \frac{3\sqrt{35}}{7} - \frac{5\sqrt{35}}{5} + 6\sqrt{35}$$

$$= \frac{15\sqrt{35}}{35} - \frac{35\sqrt{35}}{35} + \frac{210\sqrt{35}}{35} = \frac{\textbf{38}\sqrt{\textbf{35}}}{\textbf{7}}$$

8. $\sqrt{5\sqrt{5}} = \left[5(5^{1/2})\right]^{1/2} = 5^{1/2}5^{1/4} = \textbf{5}^{\textbf{3/4}}$

9. $\sqrt[3]{6\sqrt{6}} = \left[6(6^{1/2})\right]^{1/3} = 6^{1/3}6^{1/6} = \textbf{6}^{\textbf{1/2}}$

10. $\sqrt{x^3 y^3}\sqrt[4]{xy} = (x^3y^3)^{1/2}(xy)^{1/4} = x^{3/2}y^{3/2}x^{1/4}y^{1/4}$

$$= \textbf{x}^{\textbf{7/4}}\textbf{y}^{\textbf{7/4}}$$

11. $\sqrt[4]{x^5 y^3}\sqrt[3]{xy^5} = (x^5y^3)^{1/4}(xy^5)^{1/3}$

$$= x^{5/4}y^{3/4}x^{1/3}y^{5/3} = \textbf{x}^{\textbf{19/12}}\textbf{y}^{\textbf{29/12}}$$

12. $\left(x - \dfrac{2}{5}\right)^2 = 7$

$\qquad x - \dfrac{2}{5} = \pm\sqrt{7}$

$\qquad\qquad x = \dfrac{2}{5} \pm \sqrt{7}$

13. $\left(x - \dfrac{1}{4}\right)^2 = 5$

$\qquad x - \dfrac{1}{4} = \pm\sqrt{5}$

$\qquad\qquad x = \dfrac{1}{4} \pm \sqrt{5}$

14. $B + 24 + 90 = 180$

$\qquad\qquad B = 66$

$\cos 24° = \dfrac{7}{y}$

$\qquad y = \dfrac{7}{\cos 24°} \approx 7.66$

15. $\cos A = \dfrac{4}{7}$

$\qquad A \approx 55.15$

$\sin 55.15° \approx \dfrac{C}{7}$

$\qquad C \approx 7 \sin 55.15° \approx 5.74$

16. $\dfrac{(476{,}800)(9{,}016{,}423 \times 10^4)}{408 \times 10^{10}}$

$\approx \dfrac{(5 \times 10^5)(9 \times 10^{10})}{4 \times 10^{12}} \approx 1 \times 10^4$

17. $\dfrac{ax}{b} - c = \dfrac{k}{m}$

$axm - bcm = kb$

$\qquad m = \dfrac{kb}{ax - bc}$

18. $\dfrac{x}{m} - \dfrac{yb}{c} = p$

$cx - ybm = cmp$

$cx - cmp = ybm$

$\qquad c = \dfrac{ybm}{x - mp}$

19. $6 \times \overrightarrow{SF} = \dfrac{7}{2} \quad \rightarrow \quad \overrightarrow{SF} = \dfrac{7}{12}$

$4 \cdot \dfrac{7}{12} = x$

$\qquad \dfrac{7}{3} = x$

$5 \cdot \dfrac{7}{12} = y$

$\qquad \dfrac{35}{12} = y$

20. $\qquad -40x = 13x^2 + x^3$

$x^3 + 13x^2 + 40x = 0$

$x(x + 8)(x + 5) = 0$

$\qquad x = 0, -8, -5$

21. $\dfrac{(x + 9)(x + 5)}{x(x + 9)(x + 1)} \cdot \dfrac{x(x - 2)(x + 1)}{(x - 7)(x + 5)}$

$= \dfrac{x - 2}{x - 7}$

22. $8^{-4/3} = \dfrac{1}{8^{4/3}} = \dfrac{1}{(8^{1/3})^4} = \dfrac{1}{16}$

23. $x = 2(105) - 40 = \mathbf{170}$

$y = \dfrac{360 - 210}{2} = \mathbf{75}$

$z = \dfrac{360 - 2(65)}{2} = \mathbf{115}$

24. $\dfrac{\dfrac{m^2}{x} - x}{\dfrac{p^2}{x} + 2x} = \dfrac{\dfrac{m^2 - x^2}{x}}{\dfrac{p^2 + 2x^2}{x}} \cdot \dfrac{\dfrac{x}{p^2 + 2x^2}}{\dfrac{x}{p^2 + 2x^2}}$

$= \dfrac{m^2 - x^2}{p^2 + 2x^2}$

25. $3\sqrt{7}(2\sqrt{14} - \sqrt{7}) = 3\sqrt{7}(2\sqrt{2}\sqrt{7} - \sqrt{7})$

$= \mathbf{42\sqrt{2} - 21}$

26. Graph the line to find the slope.

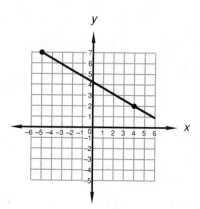

$m = \dfrac{-5}{+9} = -\dfrac{5}{9}$

Since the slopes of perpendicular lines are negative reciprocals of each other:

$m_\perp = \dfrac{9}{5}$

$y = \dfrac{9}{5}x + b$

$4 = \dfrac{9}{5}(-2) + b$

$\dfrac{38}{5} = b$

$y = \dfrac{9}{5}x + \dfrac{38}{5}$

27. $\dfrac{-x + 3}{2} - \dfrac{2x + 4}{3} = 5$

$\qquad -3x + 9 - 4x - 8 = 30$

$\qquad\qquad\qquad x = -\dfrac{29}{7}$

28. $\dfrac{-2x - 5}{4} - \dfrac{x - 2}{3} = 7$

$\qquad -6x - 15 - 4x + 8 = 84$

$\qquad\qquad\qquad x = -\dfrac{91}{10}$

29. (a) $x - y = -3$

$\qquad\quad y = x + 3$

(b) $x + 2y = -2$

$\qquad\quad y = -\dfrac{1}{2}x - 1$

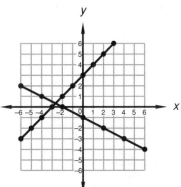

Substitute $y = x + 3$ into (b) and get:

(b′) $x + 2(x + 3) = -2$

$\qquad\qquad 3x = -8$

$\qquad\qquad\; x = -\dfrac{8}{3}$

(a) $y = \left(-\dfrac{8}{3}\right) + \dfrac{9}{3} = \dfrac{1}{3}$

$\left(-\dfrac{8}{3}, \dfrac{1}{3}\right)$

30. $a^2 - a^3b - ab$

$= \left(-\dfrac{1}{2}\right)^2 - \left(-\dfrac{1}{2}\right)^3\left(-\dfrac{1}{4}\right) - \left(-\dfrac{1}{2}\right)\left(-\dfrac{1}{4}\right)$

$= \dfrac{1}{4} - \dfrac{1}{32} - \dfrac{1}{8} = \dfrac{3}{32}$

PROBLEM SET 47

1. $\dfrac{4}{7} \times T = 2800$

$T = 2800 \cdot \dfrac{7}{4} = \mathbf{4900}$

2. $N \qquad N + 2 \qquad N + 4 \qquad N + 6$

$5(N + N + 2) = 7(N + 2 + N + 6) - 10$

$\qquad 10N + 10 = 14N + 46$

$\qquad\qquad\quad N = -9$

The desired integers are **−9, −7, −5, and −3.**

3. $\dfrac{70}{100} \times G = 1400$

$\qquad G = 1400 \cdot \dfrac{100}{70} = \mathbf{2000}$

4. Hydrogen: $\quad 2 \times 1 = 2$
Sulfur: $\qquad\; 1 \times 32 = 32$
Oxygen: $\quad\; 3 \times 16 = 48$
Total: $\qquad\qquad\quad = 82$

$\dfrac{48}{82} = \dfrac{192}{H_2SO_3}$

$48H_2SO_3 = 82(192)$

$\quad H_2SO_3 = \mathbf{328\ grams}$

5. $R_RT_R + R_WT_W = 48;$
$R_R = 8;\ R_W = 4;$
$T_R + T_W = 7$

$8(7 - T_W) + 4T_W = 48$

$\qquad\qquad -4T_W = -8$

$\qquad\qquad\quad T_W = 2$

$T_R = 7 - (2) = 5$

$D_W = 4(2) = \mathbf{8\ kilometers}$

$D_R = 8(5) = \mathbf{40\ kilometers}$

6. $\dfrac{52\ \text{in.}}{\text{sec}} \times \dfrac{1\ \text{ft}}{12\ \text{in.}} \times \dfrac{1\ \text{mi}}{5280\ \text{ft}} \times \dfrac{60\ \text{sec}}{1\ \text{min}}$

$\times \dfrac{60\ \text{min}}{1\ \text{hr}} = \dfrac{\mathbf{52(60)(60)}}{\mathbf{(12)(5280)}}\ \dfrac{\mathbf{mi}}{\mathbf{hr}}$

7. $\dfrac{805\ \text{mi}}{\text{hr}} \times \dfrac{5280\ \text{ft}}{1\ \text{mi}} \times \dfrac{1\ \text{hr}}{60\ \text{min}} \times \dfrac{1\ \text{min}}{60\ \text{sec}}$

$= \dfrac{\mathbf{805(5280)}}{\mathbf{(60)(60)}}\ \dfrac{\mathbf{ft}}{\mathbf{sec}}$

8. $\dfrac{13\ \text{in.}}{\text{sec}} \times \dfrac{1\ \text{ft}}{12\ \text{in.}} \times \dfrac{1\ \text{yd}}{3\ \text{ft}} \times \dfrac{60\ \text{sec}}{1\ \text{min}} \times \dfrac{60\ \text{min}}{1\ \text{hr}}$

$= \dfrac{\mathbf{13(60)(60)}}{\mathbf{(12)(3)}}\ \dfrac{\mathbf{yd}}{\mathbf{hr}}$

9. $\sqrt[3]{25\sqrt{5}} = \sqrt[3]{5^2\sqrt{5}} = \left[5^2(5^{1/2})\right]^{1/3} = 5^{2/3}5^{1/6}$

$= \mathbf{5^{5/6}}$

10. $\sqrt[7]{9\sqrt{3}} = \sqrt[7]{3^2\sqrt{3}} = \left[3^2(3^{1/2})\right]^{1/7} = 3^{2/7}3^{1/14}$

$= \mathbf{3^{5/14}}$

11. $\sqrt{x^2 y}\sqrt[3]{y^2 x} = (x^2y)^{1/2}(y^2x)^{1/3} = xy^{1/2}y^{2/3}x^{1/3}$

$= \mathbf{x^{4/3}y^{7/6}}$

12. $\sqrt{2\sqrt[3]{2}} = \left[2(2^{1/3})\right]^{1/2} = 2^{1/2}2^{1/6} = \mathbf{2^{2/3}}$

13. $3\sqrt{\dfrac{2}{3}} - 5\sqrt{\dfrac{3}{2}} + 2\sqrt{24}$

$= \dfrac{3\sqrt{2}}{\sqrt{3}}\dfrac{\sqrt{3}}{\sqrt{3}} - \dfrac{5\sqrt{3}}{\sqrt{2}}\dfrac{\sqrt{2}}{\sqrt{2}} + 4\sqrt{6}$

$= \dfrac{2\sqrt{6}}{2} - \dfrac{5\sqrt{6}}{2} + \dfrac{8\sqrt{6}}{2} = \dfrac{\mathbf{5\sqrt{6}}}{\mathbf{2}}$

14. $3\sqrt{\dfrac{5}{7}} - 2\sqrt{\dfrac{7}{5}} - \sqrt{315}$

$= \dfrac{3\sqrt{5}}{\sqrt{7}}\dfrac{\sqrt{7}}{\sqrt{7}} - \dfrac{2\sqrt{7}}{\sqrt{5}}\dfrac{\sqrt{5}}{\sqrt{5}} - 3\sqrt{35}$

$= \dfrac{15\sqrt{35}}{35} - \dfrac{14\sqrt{35}}{35} - \dfrac{105\sqrt{35}}{35} = -\dfrac{\mathbf{104\sqrt{35}}}{\mathbf{35}}$

15. $\sin 19° = \dfrac{x}{7}$

$x = 7 \sin 19° \approx \mathbf{2.28}$

16. $\cos 15° = \dfrac{9}{x}$

$x = \dfrac{9}{\cos 15°} \approx \mathbf{9.32}$

17. $(x - 3)^2 = 5$

$x - 3 = \pm\sqrt{5}$

$x = \mathbf{3 \pm \sqrt{5}}$

18. $\left(x + \dfrac{2}{7}\right)^2 = \dfrac{4}{49}$

$x + \dfrac{2}{7} = \pm\dfrac{2}{7}$

$x = -\dfrac{2}{7} \pm \dfrac{2}{7}$

$x = \mathbf{0, -\dfrac{4}{7}}$

19. $\dfrac{(36{,}421 \times 10^5)(493{,}025)}{40{,}216 \times 10^7}$

$\approx \dfrac{(4 \times 10^9)(5 \times 10^5)}{4 \times 10^{11}} = \mathbf{5 \times 10^3}$

20. $\dfrac{ay}{x} + p = \dfrac{m}{c}$

$acy + cpx = mx$

$acy = mx - cpx$

$\dfrac{\mathbf{acy}}{\mathbf{m - cp}} = x$

21. $\dfrac{a}{x} - \dfrac{c}{p} = b$

$ap - cx = bpx$

$ap - bpx = cx$

$p = \dfrac{\mathbf{cx}}{\mathbf{a - bx}}$

22. $B^2 = 4^2 + 3^2$

$B^2 = 25$

$B = 5$

$4 \times \overrightarrow{SF} = 2$

$\overrightarrow{SF} = \dfrac{1}{2}$

$5 \cdot \dfrac{1}{2} = C$

$\dfrac{5}{2} = C$

23. $x = 180 - 150 = \mathbf{30}$

$y = 180 - 90 - 30 = \mathbf{60}$

$z = y = \mathbf{60}$

$s = 180 - 90 - 60 = \mathbf{30}$

$p = 180 - 90 - 30 = \mathbf{60}$

$m = 180 - 90 - 60 = \mathbf{30}$

24. $\dfrac{\dfrac{x^2 p}{m} - m}{\dfrac{x}{m} - p} = \dfrac{\dfrac{x^2 p - m^2}{m}}{\dfrac{x - pm}{m}} \cdot \dfrac{\dfrac{m}{x - pm}}{\dfrac{m}{x - pm}}$

$= \dfrac{\mathbf{x^2 p - m^2}}{\mathbf{x - pm}}$

25. $\dfrac{x(x + 7)(x - 2)}{(x + 7)(x + 5)} \cdot \dfrac{(x + 5)(x + 3)}{x(x - 5)(x + 3)}$

$= \dfrac{\mathbf{x - 2}}{\mathbf{x - 5}}$

26. $-27^{-4/3} = -\dfrac{1}{27^{4/3}} = -\dfrac{1}{(27^{1/3})^4} = -\dfrac{\mathbf{1}}{\mathbf{81}}$

27. (a) $x - 4y = 8$

$y = \dfrac{1}{4}x - 2$

(b) $2x - 3y = 9$

$y = \dfrac{2}{3}x - 3$

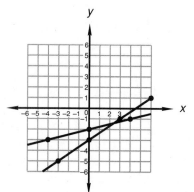

Solve the equations by elimination:

-2(a) $-2x + 8y = -16$

(b) $\underline{\quad 2x - 3y = \quad 9 \quad}$

$5y = -7$

$y = -\dfrac{7}{5}$

(a) $x - 4\left(-\dfrac{7}{5}\right) = 8$

$x = \dfrac{12}{5}$

$\left(\dfrac{12}{5}, \ -\dfrac{7}{5}\right)$

28. $y = \dfrac{2}{5}x + b$

$5 = \dfrac{2}{5}(-2) + b$

$\dfrac{29}{5} = b$

$y = \dfrac{2}{5}x + \dfrac{29}{5}$

29. $20x = -12x^2 - x^3$

$x^3 + 12x^2 + 20x = 0$

$x(x + 10)(x + 2) = 0$

$x = \mathbf{0, -2, -10}$

30. $-a - a^2 - a^3 - ab$

$= -\left(-\dfrac{1}{2}\right) - \left(-\dfrac{1}{2}\right)^2 - \left(-\dfrac{1}{2}\right)^3 - \left(-\dfrac{1}{2}\right)\left(\dfrac{1}{5}\right)$

$= \dfrac{1}{2} - \dfrac{1}{4} + \dfrac{1}{8} + \dfrac{1}{10} = \dfrac{\mathbf{19}}{\mathbf{40}}$

PROBLEM SET 48

1. $0.32 \times F = 512$

$F = \dfrac{512}{0.32} = \mathbf{1600}$

2. $\dfrac{1400}{1540} = \dfrac{NG}{6160}$

$1540NG = 1400(6160)$

$NG = \mathbf{5600 \ grams}$

3. Iron: $1 \times 56 = 56$

Bromine: $2 \times 80 = 160$

Total: $= 216$

$\dfrac{56}{216} = \dfrac{Fe}{972}$

$216Fe = 56(972)$

$Fe = \mathbf{252 \ grams}$

4. (a) $5N_B = 3N_G + 17$

(b) $6N_G = N_B + 2 \ \rightarrow \ N_B = 6N_G - 2$

Substitute $N_B = 6N_G - 2$ into (a) and get:

(a') $5(6N_G - 2) = 3N_G + 17$

$27N_G = 27$

$N_G = \mathbf{1}$

(b) $N_B = 6(1) - 2 = \mathbf{4}$

5. $R_F T_F = R_E T_E$; $T_F = 20$;

$T_E = 10$; $R_E = R_F + 30$

$20R_F = 10(R_F + 30)$

$10R_F = 300$

$R_F = \mathbf{30 \ mph}$

$R_E = (30) + 30 = \mathbf{60 \ mph}$

6. $\sqrt{x - 3} - 3 = 4$

$\sqrt{x - 3} = 7$

$x - 3 = 49$

$x = \mathbf{52}$

Check: $\sqrt{52 - 3} - 3 = 4$

$7 - 3 = 4$ check

7. $\sqrt{x + 5} + 3 = -4$

$\sqrt{x + 5} = -7$

$x + 5 = 49$

$x = 44$

Check: $\sqrt{44 + 5} + 3 = -4$

$7 + 3 = -4$ not true

No real number solution.

8. $\sqrt{x^2 + 8x + 15} - 5 = x$

$x^2 + 8x + 15 = x^2 + 10x + 25$

$-2x = 10$

$x = \mathbf{-5}$

Check: $\sqrt{(-5)^2 + 8(-5) + 15} - 5 = -5$

$-5 = -5$

9. $\dfrac{60 \ mi}{hr} \times \dfrac{5280 \ ft}{1 \ mi} \times \dfrac{1 \ hr}{60 \ min} \times \dfrac{1 \ min}{60 \ sec}$

$= \dfrac{\mathbf{60(5280)}}{\mathbf{(60)(60)}} \dfrac{ft}{sec}$

10. $\sqrt{2\sqrt[3]{2}} = \left[2\left(2^{1/3}\right)\right]^{1/2} = 2^{1/2}2^{1/6} = \mathbf{2^{2/3}}$

11. $\sqrt{m^2 y} \sqrt[3]{m^4 y} = \left(m^2 y\right)^{1/2}\left(m^4 y\right)^{1/3}$

$= my^{1/2}m^{4/3}y^{1/3} = \mathbf{m^{7/3}y^{5/6}}$

12. $\sqrt{8\sqrt[3]{2}} = \sqrt{2^3 \sqrt[3]{2}} = \left[2^3\left(2^{1/3}\right)\right]^{1/2} = 2^{3/2}2^{1/6}$

$= \mathbf{2^{5/3}}$

13. $3\sqrt{\dfrac{2}{5}} - \sqrt{\dfrac{5}{2}} + 2\sqrt{40}$

$= \dfrac{3\sqrt{2}}{\sqrt{5}} \dfrac{\sqrt{5}}{\sqrt{5}} - \dfrac{\sqrt{5}}{\sqrt{2}} \dfrac{\sqrt{2}}{\sqrt{2}} + 4\sqrt{10}$

$= \dfrac{6\sqrt{10}}{10} - \dfrac{5\sqrt{10}}{10} + \dfrac{40\sqrt{10}}{10} = \dfrac{\mathbf{41\sqrt{10}}}{\mathbf{10}}$

14. $2\sqrt{\dfrac{7}{11}} - \sqrt{\dfrac{11}{7}} + 2\sqrt{308}$

$= \dfrac{2\sqrt{7}}{\sqrt{11}} \dfrac{\sqrt{11}}{\sqrt{11}} - \dfrac{\sqrt{11}}{\sqrt{7}} \dfrac{\sqrt{7}}{\sqrt{7}} + 4\sqrt{77}$

$= \dfrac{14\sqrt{77}}{77} - \dfrac{11\sqrt{77}}{77} + \dfrac{308\sqrt{77}}{77} = \dfrac{\mathbf{311\sqrt{77}}}{\mathbf{77}}$

15. $\left(x - \dfrac{2}{7}\right)^2 = 4$

$x - \dfrac{2}{7} = \pm 2$

$x = \dfrac{2}{7} \pm \dfrac{14}{7}$

$x = \dfrac{\mathbf{16}}{\mathbf{7}}, -\dfrac{\mathbf{12}}{\mathbf{7}}$

16. $(x - 3)^2 = 16$

$x - 3 = \pm 4$

$x = 3 \pm 4$

$x = \mathbf{7, -1}$

17. $C = 180 - 90 - 63 = \mathbf{27}$

$\sin 63° = \dfrac{b}{5}$

$b = 5 \sin 63° \approx \mathbf{4.46}$

18. $\cos A = \dfrac{4}{6}$

$A \approx \mathbf{48.19}$

19. $\dfrac{(4{,}071{,}623)(51{,}642 \times 10^5)}{200{,}000 \times 10^{-13}}$

$\approx \dfrac{(4 \times 10^6)(5 \times 10^9)}{2 \times 10^{-8}} = \mathbf{1 \times 10^{24}}$

20. $\dfrac{b}{x} - \dfrac{c}{p} + k = \dfrac{m}{y}$

$bpy - cxy + kpxy = mxp$

$bpy + kpxy - mxp = cxy$

$p = \dfrac{\mathbf{cxy}}{\mathbf{by + kxy - mx}}$

21. $2(180 - A)° = 4(90 - A)° + 40°$

$360 - 2A = 400 - 4A$

$2A = 40$

$A = \mathbf{20°}$

22. Divide $3x^3 - 2$ by $x + 1$:

$$x + 1 \overline{\smash{)}\,3x^3 + 0x^2 + 0x - 2} \quad 3x^2 - 3x + 3 - \dfrac{5}{x+1}$$

$$\underline{3x^3 + 3x^2}$$
$$- 3x^2 + 0x$$
$$\underline{- 3x^2 - 3x}$$
$$3x - 2$$
$$\underline{3x + 3}$$
$$- 5$$

23. $D^2 = 5^2 + 4^2$

$D^2 = 25 + 16$

$D^2 = 41$

$D = \sqrt{41}$

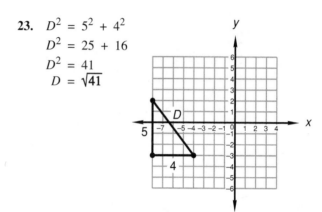

24. $\dfrac{(x - 2)(x - 1)}{x(x + 5)(x - 1)} \cdot \dfrac{x(x + 7)(x + 5)}{(x + 3)(x - 2)}$

$= \dfrac{x + 7}{x + 3}$

25. $-16^{-3/4} = -\dfrac{1}{16^{3/4}} = -\dfrac{1}{(16^{1/4})^3} = -\dfrac{\mathbf{1}}{\mathbf{8}}$

26. $V_{\text{Cone}} = \dfrac{1}{3}\left(A_{\text{Base}} \times \text{height}\right)$

$= \dfrac{1}{3}\left(12(8) - \dfrac{1}{2}\pi(4)^2\right)(4) \text{ in.}^3$

$= \dfrac{1}{3}(96 - 8\pi)(4) \text{ in.}^3 \approx \mathbf{94.51 \text{ in.}^3}$

27. $D^2 = 8^2 + 9^2$

$D^2 = 64 + 81$

$D^2 = 145$

$D = \sqrt{\mathbf{145}}$

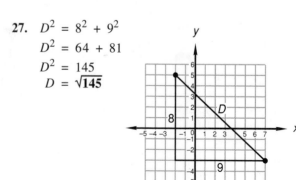

28.
$$\dfrac{\dfrac{x-1}{x}-\dfrac{1}{x^2}}{\dfrac{x+2}{x^2}-4}$$

$$=\dfrac{\dfrac{x^2-x-1}{x^2}}{\dfrac{x+2-4x^2}{x^2}}\cdot\dfrac{\dfrac{x^2}{x+2-4x^2}}{\dfrac{x^2}{x+2-4x^2}}$$

$$=\dfrac{x^2-x-1}{-4x^2+x+2}$$

29. Graph the line to find the slope.

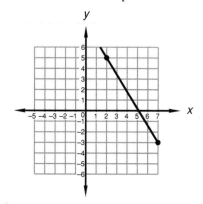

$$\text{Slope}=\dfrac{-8}{+5}=-\dfrac{8}{5}$$

$$y=-\dfrac{8}{5}x+b$$

$$5=-\dfrac{8}{5}(2)+b$$

$$\dfrac{41}{5}=b$$

$$y=-\dfrac{\mathbf{8}}{\mathbf{5}}x+\dfrac{\mathbf{41}}{\mathbf{5}}$$

30.
$$\dfrac{2x-3}{2}-\dfrac{x}{2}=\dfrac{5}{3}$$
$$6x-9-3x=10$$
$$3x=19$$
$$x=\dfrac{\mathbf{19}}{\mathbf{3}}$$

PROBLEM SET 49

1. $5(N-13)=4(-N)-92$
$$9N=-27$$
$$N=\mathbf{-3}$$

2. $\dfrac{260}{100}\times CB=1092$

$$CB=1092\cdot\dfrac{100}{260}=\mathbf{420}$$

3. Beryllium: $\quad 1\times 9=9$
Fluoride: $\quad 2\times 19=38$
Total: $\qquad\qquad =47$

$$\dfrac{38}{47}=\dfrac{95}{\text{BeF}_2}$$
$$38\text{BeF}_2=47(95)$$
$$\text{BeF}_2=\mathbf{117.5\ grams}$$

4. (a) $20N_R+N_Q=205$
(b) $N_R+N_Q=15$
Substitute $N_Q=15-N_R$ into (a) and get:
(a') $20N_R+(15-N_R)=205$
$$19N_R=190$$
$$N_R=\mathbf{10}$$
(b) $N_Q=15-(10)=\mathbf{5}$

5. $R_BT_B+200=R_GT_G;$
$T_B=T_G=10;\ R_G=240$

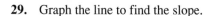

$$10R_B+200=10(240)$$
$$10R_B=2200$$
$$R_B=\mathbf{220\ yards\ per\ minute}$$

6. $y=-6x+b$
Use the point $(3,0)$ for x and y.
$$0=-6(3)+b$$
$$18=b$$
$$y=\mathbf{-6}x+\mathbf{18}$$

7. $B=180-160=20$
$A=B=20$
$$\sin 20°=\dfrac{N}{40}$$
$$N=40\sin 20°\approx\mathbf{13.68}$$

8. $\sqrt{x-4}-3=5$
$$x-4=64$$
$$x=\mathbf{68}$$

Check: $\sqrt{68-4}-3=5$
$$5=5\quad\text{check}$$

9. $\sqrt{x^2-2x+5}=x+1$
$$x^2-2x+5=x^2+2x+1$$
$$4x=4$$
$$x=\mathbf{1}$$

Check: $\sqrt{1-2+5}=2$
$$2=2\quad\text{check}$$

10. $\dfrac{200\ \text{in.}}{\text{min}}\times\dfrac{1\ \text{ft}}{12\ \text{in.}}\times\dfrac{1\ \text{min}}{60\ \text{sec}}=\dfrac{\mathbf{200}}{\mathbf{(12)(60)}}\dfrac{\text{ft}}{\text{sec}}$

11. $\sqrt[3]{9\sqrt{3}}=\sqrt[3]{3^2\sqrt{3}}=\left[3^2(3^{1/2})\right]^{1/3}=3^{2/3}3^{1/6}$
$$=\mathbf{3^{5/6}}$$

12. $\sqrt[4]{2}\sqrt[3]{2}=2^{1/4}2^{1/3}=\mathbf{2^{7/12}}$

13. $\sqrt{x^2 y^3}\sqrt{xy^4} = (x^2 y^3)^{1/2}(xy^4)^{1/2} = xy^{3/2}x^{1/2}y^2$
$= x^{3/2}y^{7/2}$

14. $\sqrt[3]{xy}\sqrt{xy} = (xy)^{1/3}(xy)^{1/2} = x^{1/3}y^{1/3}x^{1/2}y^{1/2}$
$= x^{5/6}y^{5/6}$

15. $3\sqrt{\dfrac{7}{2}} + 2\sqrt{\dfrac{2}{7}} - 3\sqrt{56}$

$= \dfrac{3\sqrt{7}}{\sqrt{2}}\dfrac{\sqrt{2}}{\sqrt{2}} + \dfrac{2\sqrt{2}}{\sqrt{7}}\dfrac{\sqrt{7}}{\sqrt{7}} - 6\sqrt{14}$

$= \dfrac{21\sqrt{14}}{14} + \dfrac{4\sqrt{14}}{14} - \dfrac{84\sqrt{14}}{14} = -\dfrac{59\sqrt{14}}{14}$

16. $5\sqrt{\dfrac{2}{9}} + 3\sqrt{\dfrac{9}{2}} + \sqrt{162}$

$= \dfrac{5\sqrt{2}}{\sqrt{9}}\dfrac{\sqrt{9}}{\sqrt{9}} + \dfrac{3\sqrt{9}}{\sqrt{2}}\dfrac{\sqrt{2}}{\sqrt{2}} + 3\sqrt{18}$

$= \dfrac{10\sqrt{18}}{18} + \dfrac{27\sqrt{18}}{18} + \dfrac{54\sqrt{18}}{18}$

$= \dfrac{91\sqrt{18}}{18} = \dfrac{91\sqrt{2}}{6}$

17. $\left(x - \dfrac{5}{3}\right)^2 = 5$

$x - \dfrac{5}{3} = \pm\sqrt{5}$

$x = \dfrac{5}{3} \pm \sqrt{5}$

18. $(x + 2)^2 = 16$
$x + 2 = \pm 4$
$x = -2 \pm 4$
$x = 2, -6$

19. $\dfrac{(517{,}832 \times 10^{-14})(80{,}123)}{200{,}000 \times 10^{-42}}$

$\approx \dfrac{(5 \times 10^{-9})(8 \times 10^4)}{2 \times 10^{-37}} = 2 \times 10^{33}$

20. $\dfrac{x}{y} + c = \dfrac{m}{x} - d$

$x^2 + cxy = my - dxy$
$x^2 = my - dxy - cxy$

$\dfrac{x^2}{m - dx - cx} = y$

21. $\dfrac{2}{x} - \dfrac{3}{y} = \dfrac{m}{p}$
$2py - 3px = mxy$

$p = \dfrac{mxy}{2y - 3x}$

22. $(x + 1)^3 = (x + 1)(x + 1)(x + 1)$
$= (x^2 + 2x + 1)(x + 1)$
$= x^3 + 2x^2 + x + x^2 + 2x + 1$
$= x^3 + 3x^2 + 3x + 1$

23. $\dfrac{(x - 5)(x - 1)}{x(x - 5)(x - 4)} \cdot \dfrac{x(x - 4)(x - 3)}{(x - 2)(x - 1)}$

$= \dfrac{x - 3}{x - 2}$

24. $D^2 = 9^2 + 1^2$
$D^2 = 81 + 1$
$D^2 = 82$
$D = \sqrt{82}$

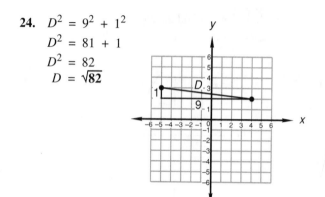

25. Use the graph in the solution for Problem 24 to find the slope:

Slope $= \dfrac{-1}{+9} = -\dfrac{1}{9}$

$y = -\dfrac{1}{9}x + b$

$2 = -\dfrac{1}{9}(4) + b$

$\dfrac{22}{9} = b$

$y = -\dfrac{1}{9}x + \dfrac{22}{9}$

26. Number of sides $= \dfrac{360°}{12°} = 30$

27. $\dfrac{x}{ay^2} - \dfrac{3x}{a^2 y^3} - \dfrac{5}{y^4}$

$= \dfrac{xay^2}{a^2 y^4} - \dfrac{3xy}{a^2 y^4} - \dfrac{5a^2}{a^2 y^4}$

$= \dfrac{xay^2 - 3xy - 5a^2}{a^2 y^4}$

28. $\dfrac{-3}{x + 3} - \dfrac{3x + 2}{x^2 - 9}$

$= \dfrac{-3(x - 3)}{x^2 - 9} - \dfrac{3x + 2}{x^2 - 9}$

$= \dfrac{-3x + 9 - 3x - 2}{x^2 - 9} = \dfrac{-6x + 7}{x^2 - 9}$

29. $\dfrac{x-2}{2} - \dfrac{x-3}{3} = 2$

$3x - 6 - 2x + 6 = 12$

$x = \mathbf{12}$

30. $\dfrac{x}{2} - \dfrac{x+2}{5} = 1$

$5x - 2x - 4 = 10$

$3x = 14$

$x = \mathbf{\dfrac{14}{3}}$

PROBLEM SET 50

1. (a) $N_T = 5N_W + 10$

(b) $N_T = 10N_W$

Substitute (b) into (a) and get:

(a′) $10N_W = 5N_W + 10$

$5N_W = 10$

$N_W = \mathbf{2}$

(b) $N_T = 10(2) = \mathbf{20}$

2. $R_S T_S + R_B T_B = 104;\ R_S = 8;$

$R_B = 12;\ T_S + T_B = 10$

$8T_S + 12(10 - T_S) = 104$

$-4T_S = -16$

$T_S = 4$

$T_B = 10 - (4) = 6$

$D_S = 8(4) = \mathbf{32\ miles}$

$D_B = 12(6) = \mathbf{72\ miles}$

$\overleftarrow{\underset{104}{D_SD_B}}$

3. Sodium: $1 \times 23 = 23$

Chlorine: $1 \times 35 = 35$

Total: $= 58$

$\dfrac{23}{58} = \dfrac{\text{Na}}{348}$

$58\text{Na} = 23(348)$

$\text{Na} = \mathbf{138\ grams}$

4. $N \qquad N+2 \qquad N+4$

$7(N + N + 4) = 10(N + 2) - 48$

$14N + 28 = 10N - 28$

$4N = -56$

$N = -14$

The desired integers are $\mathbf{-14}, \mathbf{-12},$ and $\mathbf{-10}$.

5. $\dfrac{240}{100} \times OS = 432$

$OS = 432 \cdot \dfrac{100}{240} = \mathbf{180\ inches\ per\ day}$

6. $x^2 + 8x - 4 = 0$

$(x^2 + 8x + \quad) = 4$

$x^2 + 8x + 16 = 4 + 16$

$(x + 4)^2 = 20$

$x + 4 = \pm 2\sqrt{5}$

$x = \mathbf{-4 \pm 2\sqrt{5}}$

7. $12x + x^2 - 5 = 0$

$(x^2 + 12x + \quad) = 5$

$x^2 + 12x + 36 = 5 + 36$

$(x + 6)^2 = 41$

$x + 6 = \pm\sqrt{41}$

$x = \mathbf{-6 \pm \sqrt{41}}$

8. $x^2 = 7x - 3$

$(x^2 - 7x + \quad) = -3$

$x^2 - 7x + \dfrac{49}{4} = -3 + \dfrac{49}{4}$

$\left(x - \dfrac{7}{2}\right)^2 = \dfrac{37}{4}$

$x - \dfrac{7}{2} = \pm\dfrac{\sqrt{37}}{2}$

$x = \mathbf{\dfrac{7}{2} \pm \dfrac{\sqrt{37}}{2}}$

9. $y = 3x + b$

Use the point $(-3, 0)$ for x and y.

$0 = 3(-3) + b$

$9 = b$

$\mathbf{y = 3x + 9}$

10. $B = 180 - 120 = 60$

$C = 180 - 90 - 60 = \mathbf{30}$

$\tan 60° = \dfrac{M}{10}$

$M = 10 \tan 60° \approx \mathbf{17.32}$

11. $\sqrt{x^2 - 4x + 20} = x + 2$

$x^2 - 4x + 20 = x^2 + 4x + 4$

$8x = 16$

$x = \mathbf{2}$

Check: $\sqrt{4 - 8 + 20} = 4$

$4 = 4$ check

12. $-5 = -\sqrt{x + 5} + 1$

$6 = \sqrt{x + 5}$

$36 = x + 5$

$\mathbf{31} = x$

Check: $-5 = -\sqrt{31 + 5} + 1$

$-5 = -5$ check

13. $\dfrac{400\ \text{yd}}{\text{sec}} \times \dfrac{3\ \text{ft}}{1\ \text{yd}} \times \dfrac{1\ \text{mi}}{5280\ \text{ft}} \times \dfrac{60\ \text{sec}}{1\ \text{min}}$

$\times\ \dfrac{60\ \text{min}}{1\ \text{hr}} = \dfrac{\mathbf{400(3)(60)(60)}}{\mathbf{5280}}\ \dfrac{\mathbf{mi}}{\mathbf{hr}}$

14. $\sqrt[5]{2\sqrt[3]{2}} = \left[2\left(2^{1/3}\right)\right]^{1/5} = 2^{1/5}2^{1/15} = \mathbf{2^{4/15}}$

15. $\sqrt{9\sqrt{3}} = \sqrt{3^2\sqrt{3}} = \left[3^2(3^{1/2})\right]^{1/2} = 3 \cdot 3^{1/4}$

$= \mathbf{3^{5/4}}$

16. $\sqrt{m^3 y^5}\sqrt[3]{m^2 y^2} = \left(m^3 y^5\right)^{1/2}\left(m^2 y^2\right)^{1/3}$

$= m^{3/2}y^{5/2}m^{2/3}y^{2/3} = \mathbf{m^{13/6}y^{19/6}}$

17. $5\sqrt{\dfrac{3}{11}} + 2\sqrt{\dfrac{11}{3}} - \sqrt{297}$

$= \dfrac{5\sqrt{3}}{\sqrt{11}}\dfrac{\sqrt{11}}{\sqrt{11}} + \dfrac{2\sqrt{11}}{\sqrt{3}}\dfrac{\sqrt{3}}{\sqrt{3}} - 3\sqrt{33}$

$= \dfrac{15\sqrt{33}}{33} + \dfrac{22\sqrt{33}}{33} - \dfrac{99\sqrt{33}}{33} = \mathbf{-\dfrac{62\sqrt{33}}{33}}$

18. $\dfrac{(746{,}800 \times 10^{14})(703{,}916 \times 10^4)}{500{,}000}$

$\approx \dfrac{(7 \times 10^{19})(7 \times 10^9)}{5 \times 10^5} = 9.8 \times 10^{23}$

$\approx \mathbf{1 \times 10^{24}}$

19. $\dfrac{mxc}{p} - k = \dfrac{2}{r}$

$mxcr - kpr = 2p$

$mxcr = 2p + kpr$

$c = \mathbf{\dfrac{2p + kpr}{mxr}}$

20. $\dfrac{4}{x} - \dfrac{3x}{p} = \dfrac{c}{m}$

$4pm - 3x^2m = cpx$

$4pm - cpx = 3x^2m$

$p = \mathbf{\dfrac{3x^2 m}{4m - cx}}$

21. $a^2 = 24^2 + 7^2$

$a^2 = 576 + 49$

$a^2 = 625$

$a = \sqrt{625} = 25$

$24 \times \overrightarrow{SF} = 6$

$\overrightarrow{SF} = \dfrac{1}{4}$

$25 \cdot \dfrac{1}{4} = c$

$\dfrac{25}{4} = c$

22. Divide $4x^3 + 3x + 5$ by $2x - 3$:

$$
\begin{array}{r}
2x^2 + 3x + 6 + \frac{23}{2x-3} \\
2x - 3 \overline{)\,4x^3 + 0x^2 + 3x + 5} \\
\underline{4x^3 - 6x^2} \\
6x^2 + 3x \\
\underline{6x^2 - 9x} \\
12x + 5 \\
\underline{12x - 18} \\
23
\end{array}
$$

23. $x^3 - 28x = 3x^2$

$x^3 - 3x^2 - 28x = 0$

$x(x - 7)(x + 4) = 0$

$x = \mathbf{0, -4, 7}$

24. $\dfrac{(x - 7)(x - 7)}{x(x - 7)(x - 6)} \cdot \dfrac{x(x - 6)(x + 2)}{(x - 7)(x + 5)}$

$= \dfrac{x + 2}{x + 5}$

25. $-49^{3/2} = -\left(49^{1/2}\right)^3 = \mathbf{-343}$

26. $V_{\text{Prism}} = A_{\text{Base}} \times \text{Height}$

$\text{Height} = \dfrac{V_{\text{Prism}}}{A_{\text{Base}}}$

$= \dfrac{(600 - 50\pi)\ \text{cm}^3}{\left[\frac{1}{2}(10)(10) + (10)(10) - \frac{1}{2}\pi(5)^2\right]\text{cm}^2}$

$= \dfrac{(600 - 50\pi)\ \text{cm}^3}{\left(150 - \frac{25}{2}\pi\right)\text{cm}^2} = \mathbf{4\ cm}$

27. Write the equation of the given line in slope-intercept form.

$-5y = -3x + 2$

$y = \dfrac{3}{5}x - \dfrac{2}{5}$

Since the slopes of perpendicular lines are negative reciprocals of each other:

$m_{\perp} = -\dfrac{5}{3}$

$y = -\dfrac{5}{3}x + b$

$5 = -\dfrac{5}{3}(-2) + b$

$\dfrac{5}{3} = b$

$y = \mathbf{-\dfrac{5}{3}x + \dfrac{5}{3}}$

28. $D^2 = 2^2 + 0^2$

$D^2 = 4$

$D = \sqrt{4} = 2$

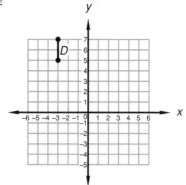

29. $-2(x^0 - x - 3) - |-2^0 - 3| = x - 2(-2 - 4)$

$-2(-x - 2) - 4 = x + 4 + 8$

$2x + 4 - 4 = x + 12$

$x = \mathbf{12}$

30. $\dfrac{x^2 y^{-2}}{z^2 p^0}\left(\dfrac{4p^0 x^{-5} z^2}{y^{-2}} - \dfrac{3x^{-4} y^2}{z^{-2} p}\right)$

$= \mathbf{4x^{-3} - 3x^{-2}p^{-1}}$

PROBLEM SET 51

1. $\dfrac{3}{17} \times K = 72$

$K = 72 \cdot \dfrac{17}{3} = \mathbf{408}$

2. $\dfrac{22}{100} \times S = 528$

$S = 528 \cdot \dfrac{100}{22} = \mathbf{2400}$

3. Sodium: $2 \times 23 = 46$
Sulfur: $1 \times 32 = 32$
Oxygen: $4 \times 16 = 64$
Total: $= 142$

$\dfrac{46}{142} = \dfrac{115}{Na_2SO_4}$

$46Na_2SO_4 = 115(142)$

$Na_2SO_4 = \mathbf{355\ grams}$

4. (a) $10N_D + 25N_Q = 650$
(b) $N_D + N_Q = 35$
Substitute $N_Q = 35 - N_D$ into (a) and get:
(a') $10N_D + 25(35 - N_D) = 650$

$-15N_D = -225$

$N_D = \mathbf{15}$

(b) $N_Q = 35 - (15) = \mathbf{20}$

5. $R_B T_B = R_J T_J;\ R_B = 40;$
$R_J = 50;\ T_B = T_J + 2$

$40(T_J + 2) = 50T_J$

$80 = 10T_J$

$8 = T_J$

$D_J = D_B = 50(8) = \mathbf{400\ miles}$

6. $5 + 6i - 3i - 2 = \mathbf{3 + 3i}$

7. $5ii - 8iii + 2i - 4 - \sqrt{-5}$
$= 5(ii) - 8(ii)i + 2i - 4 - \sqrt{5}i$
$= 5(-1) - 8(-1)i + 2i - 4 - \sqrt{5}i$
$= -5 + 8i + 2i - 4 - \sqrt{5}i = \mathbf{-9 + \left(10 - \sqrt{5}\right)i}$

8. $5i^3 + 3i^2 + 7ii + 4 + 2i^7$
$= 5(ii)i + 3(ii) + 7(ii) + 4 + 2(ii)(ii)(ii)i$
$= 5(-1)i + 3(-1) + 7(-1) + 4 + 2(-1)(-1)(-1)i$
$= -5i - 3 - 7 + 4 - 2i$
$= \mathbf{-6 - 7i}$

9. $-3i^2 - 2i + i^3 - 3 = -3(ii) - 2i + i(ii) - 3$
$= -3(-1) - 2i + i(-1) - 3 = 3 - 2i - i - 3$
$= \mathbf{-3i}$

10. $x^2 = -x + 1$

$(x^2 + x + \quad) = 1$

$x^2 + x + \dfrac{1}{4} = 1 + \dfrac{1}{4}$

$\left(x + \dfrac{1}{2}\right)^2 = \dfrac{5}{4}$

$x + \dfrac{1}{2} = \pm\dfrac{\sqrt{5}}{2}$

$x = \mathbf{-\dfrac{1}{2} \pm \dfrac{\sqrt{5}}{2}}$

11. $-4 = -x^2 - 3x$

$(x^2 + 3x + \quad) = 4$

$x^2 + 3x + \dfrac{9}{4} = 4 + \dfrac{9}{4}$

$\left(x + \dfrac{3}{2}\right)^2 = \dfrac{25}{4}$

$x + \dfrac{3}{2} = \pm\dfrac{5}{2}$

$x = -\dfrac{3}{2} \pm \dfrac{5}{2}$

$x = \mathbf{1, -4}$

12. $y = -4x + b$
Use the point $(2, 2)$ for x and y.
$2 = -4(2) + b$
$10 = b$
$\mathbf{y = -4x + 10}$

13. $\cos 30° = \dfrac{2}{m}$

$m = \dfrac{2}{\cos 30°} \approx \mathbf{2.31}$

14. $\sqrt{x-3} + 5 = 2$

$\sqrt{x-3} = -3$

$x - 3 = 9$

$x = 12$

Check: $\sqrt{12-3} + 5 = 2$

$8 = 2$

No real number solution.

15. $\dfrac{20 \text{ in.}}{\text{hr}} \times \dfrac{1 \text{ ft}}{12 \text{ in.}} \times \dfrac{1 \text{ mi}}{5280 \text{ ft}} \times \dfrac{1 \text{ hr}}{60 \text{ min}}$

$= \dfrac{20}{(12)(5280)(60)} \dfrac{\text{mi}}{\text{min}}$

16. $\sqrt[5]{3\sqrt{3}} = [3(3^{1/2})]^{1/5} = 3^{1/5}3^{1/10} = \mathbf{3^{3/10}}$

17. $\sqrt{9\sqrt[3]{3}} = \sqrt{3^2\sqrt[3]{3}} = [3^2(3^{1/3})]^{1/2}$

$= 3 \cdot 3^{1/6} = \mathbf{3^{7/6}}$

18. $\sqrt{x^2y^2m}\sqrt[4]{xym^2} = (x^2y^2m)^{1/2}(xym^2)^{1/4}$

$= xym^{1/2}x^{1/4}y^{1/4}m^{1/2} = \mathbf{x^{5/4}y^{5/4}m}$

19. $3\sqrt{\dfrac{2}{13}} + 3\sqrt{\dfrac{13}{2}} + 3\sqrt{104}$

$= \dfrac{3\sqrt{2}}{\sqrt{13}}\dfrac{\sqrt{13}}{\sqrt{13}} + \dfrac{3\sqrt{13}}{\sqrt{2}}\dfrac{\sqrt{2}}{\sqrt{2}} + 6\sqrt{26}$

$= \dfrac{6\sqrt{26}}{26} + \dfrac{39\sqrt{26}}{26} + \dfrac{156\sqrt{26}}{26} = \dfrac{\mathbf{201\sqrt{26}}}{\mathbf{26}}$

20. $(-27)^{-5/3} = \dfrac{1}{(-27)^{5/3}} = \dfrac{1}{\left((-27)^{1/3}\right)^5} = -\dfrac{\mathbf{1}}{\mathbf{243}}$

21. $\dfrac{(4{,}941{,}625)(7{,}041{,}683)}{0.00007142 \times 10^{-5}}$

$\approx \dfrac{(5 \times 10^6)(7 \times 10^6)}{7 \times 10^{-10}} = \mathbf{5 \times 10^{22}}$

22. $\dfrac{x}{p} - c = \dfrac{k}{m}$

$mx - cmp = kp$

$mx = kp + cmp$

$\dfrac{\mathbf{mx}}{\mathbf{k + cm}} = p$

23. $\dfrac{3}{p} - \dfrac{x}{R_1} = \dfrac{1}{R_2}$

$3R_1R_2 - xpR_2 = pR_1$

$R_2 = \dfrac{pR_1}{3R_1 - xp}$

24. $x = \dfrac{50}{2} = \mathbf{25}$

$y = 180 - 120 - 25 = \mathbf{35}$

$m = x = \mathbf{25}$

$z = y = \mathbf{35}$

25. Write the equation of the given line in slope-intercept form.

$y = -\dfrac{3}{2}x + \dfrac{5}{2}$

Since the slopes of perpendicular lines are negative reciprocals of each other:

$m_\perp = \dfrac{2}{3}$

$y = \dfrac{2}{3}x + b$

$-2 = \dfrac{2}{3}(-4) + b$

$\dfrac{2}{3} = b$

$y = \dfrac{2}{3}x + \dfrac{2}{3}$

26. $D^2 = 3^2 + 0^2$

$D^2 = 9$

$D = \sqrt{9} = 3$

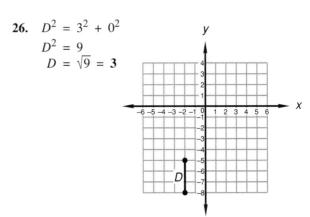

27. $\dfrac{x^{-2}y}{p^0}\left(\dfrac{4p^2}{x^{-2}y} - \dfrac{2px^2y}{y^2}\right) = 4p^2 - 2p$

28. Divide $2x^3 - 2x + 4$ by $x + 2$:

$$
\begin{array}{r}
2x^2 - 4x + 6 - \frac{8}{x+2} \\
x + 2 \overline{\smash{\big)}\ 2x^3 + 0x^2 - 2x + 4} \\
\underline{2x^3 + 4x^2} \\
-4x^2 - 2x \\
\underline{-4x^2 - 8x} \\
6x + 4 \\
\underline{6x + 12} \\
-8
\end{array}
$$

29. $\dfrac{-x - 2}{4} - \dfrac{x + 2}{3} = 3$

$-3x - 6 - 4x - 8 = 36$

$-7x = 50$

$x = -\dfrac{50}{7}$

30. $-2(x^0 - x - 2) - |-2| + 30(-x - x^0) = -x$

$-2(-x - 1) - 2 + 30(-x - 1) = -x$

$2x + 2 - 2 - 30x - 30 = -x$

$-27x = 30$

$x = -\dfrac{10}{9}$

PROBLEM SET 52

1. Iodine P_N + Iodine D_N = Iodine Total

$0.1(P_N) + 0.4(D_N) = 0.25(100)$

(a) $0.1P_N + 0.4D_N = 25$

(b) $P_N + D_N = 100$

Substitute $D_N = 100 - P_N$ into (a) and get:

(a′) $0.1P_N + 0.4(100 - P_N) = 25$

$-0.3P_N = -15$

$P_N = 50$

(b) $(50) + D_N = 100$

$D_N = 50$

50 ml 10%, 50 ml 40%

2. Salt P_N + Salt D_N = Salt Total

$0.25(P_N) + 0.05(D_N) = 0.1(1400)$

(a) $0.25P_N + 0.05D_N = 140$

(b) $P_N + D_N = 1400$

Substitute $D_N = 1400 - P_N$ into (a) and get:

(a′) $0.25P_N + 0.05(1400 - P_N) = 140$

$0.2P_N = 70$

$P_N = 350$

(b) $(350) + D_N = 1400$

$D_N = 1050$

350 ml 25%, 1050 ml 5%

3. Sodium: $2 \times 23 = 46$

Sulfur: $2 \times 32 = 64$

Oxygen: $3 \times 16 = 48$

Total: $= 158$

$\dfrac{46}{158} = \dfrac{Na}{1580}$

$158Na = 46(1580)$

$Na = \textbf{460 grams}$

4. (a) $4N_Y = 76 - 6N_R$

(b) $N_Y = N_R + 4$

Substitute (b) into (a) and get:

(a′) $4(N_R + 4) = 76 - 6N_R$

$10N_R = 60$

$N_R = \textbf{6}$

(b) $N_Y = (6) + 4 = \textbf{10}$

5. $R_L T_L + R_C T_C = 28$;

$R_L = 2$; $R_C = 8$;

$T_L + T_C{}' = 8$

$2T_L + 8(8 - T_L) = 28$

$-6T_L = -36$

$T_L = 6$

$T_C = 8 - (6) = 2$

$D_L = 2(6) = \textbf{12 kilometers}$

$D_C = 8(2) = \textbf{16 kilometers}$

6. $4i^2 - 3i + 2 = 4(ii) - 3i + 2$

$= 4(-1) - 3i + 2 = -4 - 3i + 2 = \textbf{-2 - 3}\textbf{\textit{i}}$

7. $3i^5 - i + 5 - \sqrt{-9} = 3i(ii)(ii) - i + 5 - 3i$

$= 3i - i + 5 - 3i = \textbf{5 - \textit{i}}$

8. $\sqrt{-16} - 2i^2 - 2i = 4i - 2(ii) - 2i$

$= 4i + 2 - 2i = \textbf{2 + 2\textit{i}}$

9. $2i^4 + \sqrt{-9} - 3i^3 = 2(ii)(ii) + 3i - 3i(ii)$

$= 2 + 3i + 3i = \textbf{2 + 6\textit{i}}$

10. $2x + 5 = x^2$

$(x^2 - 2x + \quad) = 5$

$x^2 - 2x + 1 = 5 + 1$

$(x - 1)^2 = 6$

$x - 1 = \pm\sqrt{6}$

$x = \textbf{1} \pm \sqrt{\textbf{6}}$

11. $x^2 - 5x = 2$

$x^2 - 5x + \dfrac{25}{4} = 2 + \dfrac{25}{4}$

$\left(x - \dfrac{5}{2}\right)^2 = \dfrac{33}{4}$

$x - \dfrac{5}{2} = \pm\dfrac{\sqrt{33}}{2}$

$x = \dfrac{\textbf{5}}{\textbf{2}} \pm \dfrac{\sqrt{\textbf{33}}}{\textbf{2}}$

12. $y = 2x + b$

Use the point $(4, 0)$ for x and y.

$0 = 2(4) + b$

$-8 = b$

$\textbf{\textit{y} = 2\textit{x} - 8}$

13. $B = 180 - 165 = 15$

$\sin 15° = \dfrac{7}{P}$

$P = \dfrac{7}{\sin 15°} \approx \textbf{27.05}$

14. $\sqrt{x^2 + 2x + 10} = x + 2$

$x^2 + 2x + 10 = x^2 + 4x + 4$

$2x = 6$

$x = \textbf{3}$

Check: $\sqrt{9 + 6 + 10} = 5$

$5 = 5$ check

15. $3 = -5 + \sqrt{x - 3}$

$8 = \sqrt{x - 3}$

$64 = x - 3$

$\textbf{67} = x$

Check: $3 = -5 + \sqrt{67 - 3}$

$3 = -5 + 8$ check

16. $\dfrac{200 \text{ in.}}{\text{hr}} \times \dfrac{1 \text{ ft}}{12 \text{ in.}} \times \dfrac{1 \text{ mi}}{5280 \text{ ft}} \times \dfrac{1 \text{ hr}}{60 \text{ min}}$

$= \dfrac{\textbf{200}}{\textbf{(12)(5280)(60)}} \dfrac{\text{mi}}{\text{min}}$

17. $2\sqrt{2\sqrt[4]{2}} = 2[2(2^{1/4})]^{1/2} = 2 \cdot 2^{1/2} \cdot 2^{1/8} = \textbf{2}^{\textbf{13/8}}$

18. $3\sqrt{9\sqrt[4]{3}} = 3\sqrt{3^2 \sqrt[4]{3}} = 3[3^2(3^{1/4})]^{1/2}$

$= 3 \cdot 3 \cdot 3^{1/8} = \textbf{3}^{\textbf{17/8}}$

19. $\sqrt{4x^3 y^5}\sqrt[3]{8xy^2} = (2^2 x^3 y^5)^{1/2}(2^3 xy^2)^{1/3}$

$= 2x^{3/2}y^{5/2}2x^{1/3}y^{2/3} = \textbf{4}x^{\textbf{11/6}}y^{\textbf{19/6}}$

20. $3\sqrt{\dfrac{2}{13}} - 5\sqrt{\dfrac{13}{2}} + \sqrt{104}$

$= \dfrac{3\sqrt{2}}{\sqrt{13}}\dfrac{\sqrt{13}}{\sqrt{13}} - \dfrac{5\sqrt{13}}{\sqrt{2}}\dfrac{\sqrt{2}}{\sqrt{2}} + 2\sqrt{26}$

$= \dfrac{6\sqrt{26}}{26} - \dfrac{65\sqrt{26}}{26} + \dfrac{52\sqrt{26}}{26} = -\dfrac{\textbf{7}\sqrt{\textbf{26}}}{\textbf{26}}$

21. $\dfrac{(987,612 \times 10^5)(413,280)}{(74,630)(400)}$

$\approx \dfrac{(1 \times 10^{11})(4 \times 10^5)}{(7 \times 10^4)(4 \times 10^2)} \approx \textbf{1} \times \textbf{10}^{\textbf{9}}$

22. $\dfrac{x}{p} - c + \dfrac{a}{b} = m$

$bx - bcp + ap = bmp$

$bx = bmp + bcp - ap$

$x = \dfrac{\textbf{bmp} + \textbf{bcp} - \textbf{ap}}{\textbf{b}}$

23. $\dfrac{x^2 + 3x - 28}{21x + 10x^2 + x^3} = \dfrac{(x + 7)(x - 4)}{x(x + 7)(x + 3)}$

$= \dfrac{x - 4}{x(x + 3)}$

24. $-16^{5/4} = -(16^{1/4})^5 = \textbf{-32}$

25. $\dfrac{\dfrac{x^2 y}{p^2} - p}{\dfrac{m}{p} - \dfrac{1}{p^2}} = \dfrac{\dfrac{x^2 y - p^3}{p^2}}{\dfrac{mp - 1}{p^2}} \cdot \dfrac{\dfrac{p^2}{mp - 1}}{\dfrac{p^2}{mp - 1}}$

$= \dfrac{x^2 y - p^3}{mp - 1}$

26. $z = \dfrac{60}{2} = \textbf{30}$

$p = z = \textbf{30}$

$x = 180 - 100 - 30 = \textbf{50}$

$y = x = \textbf{50}$

27. $-\dfrac{3}{x - 3} + \dfrac{2x + 4}{x^2 - 9}$

$= \dfrac{-3(x + 3)}{x^2 - 9} + \dfrac{2x + 4}{x^2 - 9}$

$= \dfrac{-3x - 9 + 2x + 4}{x^2 - 9} = \dfrac{-x - 5}{x^2 - 9}$

28. Write the equation of the given line in slope-intercept form.

$y = -\dfrac{2}{3}x + \dfrac{4}{3}$

Since parallel lines have the same slope:

$y = -\dfrac{2}{3}x + b$

$-7 = -\dfrac{2}{3}(-2) + b$

$-\dfrac{25}{3} = b$

$y = -\dfrac{2}{3}x - \dfrac{25}{3}$

29.

$$x - 4 \overline{)\,3x^3 + 0x^2 - x + 0\,} \quad 3x^2 + 12x + 47 + \dfrac{188}{x-4}$$

$\underline{3x^3 - 12x^2}$

$12x^2 - x$

$\underline{12x^2 - 48x}$

$47x + 0$

$\underline{47x - 188}$

188

30. $\dfrac{3x + 4}{2} - \dfrac{2x - 5}{3} = 4$

$9x + 12 - 4x + 10 = 24$

$5x = 2$

$x = \dfrac{2}{5}$

PROBLEM SET 53

1. Sodium: $2 \times 23 = 46$
Sulfur: $2 \times 32 = 64$
Oxygen: $3 \times 16 = 48$
Total: $ = 158$

Sodium $= \dfrac{46}{158} \approx 0.29 = \mathbf{29\%}$

2. Alcohol P_N + Alcohol D_N = Alcohol Total
$0.2P_N + 0.6D_N = 0.52(100)$
(a) $0.2P_N + 0.6D_N = 52$
(b) $P_N + D_N = 100$

Substitute $D_N = 100 - P_N$ into (a) and get:
(a') $0.2P_N + 0.6(100 - P_N) = 52$
$-0.4P_N = -8$
$P_N = 20$

(b) $(20) + D_N = 100$
$D_N = 80$

20 ml 20%, 80 ml 60%

3. Iodine P_N + Iodine D_N = Iodine Total
$0.4(P_N) + 0.8(D_N) = 0.72(250)$
(a) $0.4P_N + 0.8D_N = 180$
(b) $P_N + D_N = 250$

Substitute $D_N = 250 - P_N$ into (a) and get:
(a') $0.4P_N + 0.8(250 - P_N) = 180$
$-0.4P_N = -20$
$P_N = 50$

(b) $(50) + D_N = 250$
$D_N = 200$

50 ml 40%, 200 ml 80%

4. (a) $N_G + N_B = 80$
(b) $N_G = 5N_B + 8$
Substitute (b) into (a) and get:
(a') $(5N_B + 8) + N_B = 80$
$6N_B = 72$
$N_B = \mathbf{12}$
(b) $N_G = 5(12) + 8 = \mathbf{68}$

5. $R_M T_M + 200 = R_P T_P$;
$R_M = 50$; $T_M = T_P = 4$

$50(4) + 200 = 4R_P$
$400 = 4R_P$
100 mph $= R_P$

6. $9350 \text{ cm} \times \dfrac{1 \text{ m}}{100 \text{ cm}} \times \dfrac{1 \text{ km}}{1000 \text{ m}}$

$= \dfrac{9350}{(100)(1000)} \text{ km}$

7. $32 \text{ m} \times \dfrac{100 \text{ cm}}{1 \text{ m}} \times \dfrac{1 \text{ in.}}{2.54 \text{ cm}} \times \dfrac{1 \text{ ft}}{12 \text{ in.}}$

$\times \dfrac{1 \text{ yd}}{3 \text{ ft}} = \dfrac{32(100)}{(2.54)(12)(3)} \text{ yd}$

8. $16{,}480{,}000 \text{ mi}^2 \times \dfrac{5280 \text{ ft}}{1 \text{ mi}} \times \dfrac{5280 \text{ ft}}{1 \text{ mi}} \times \dfrac{12 \text{ in.}}{1 \text{ ft}}$

$\times \dfrac{12 \text{ in.}}{1 \text{ ft}} \times \dfrac{2.54 \text{ cm}}{1 \text{ in.}} \times \dfrac{2.54 \text{ cm}}{1 \text{ in.}}$

$= \mathbf{16{,}480{,}000(5280)(5280)(12)(12)(2.54)(2.54) \text{ cm}^2}$

9. $0.063 \text{ km}^2 \times \dfrac{1000 \text{ m}}{1 \text{ km}} \times \dfrac{1000 \text{ m}}{1 \text{ km}} \times \dfrac{100 \text{ cm}}{1 \text{ m}}$

$\times \dfrac{100 \text{ cm}}{1 \text{ m}} \times \dfrac{1 \text{ in.}}{2.54 \text{ cm}} \times \dfrac{1 \text{ in.}}{2.54 \text{ cm}} \times \dfrac{1 \text{ ft}}{12 \text{ in.}}$

$\times \dfrac{1 \text{ ft}}{12 \text{ in.}} \times \dfrac{1 \text{ mi}}{5280 \text{ ft}} \times \dfrac{1 \text{ mi}}{5280 \text{ ft}}$

$= \dfrac{(0.063)(1000)(1000)(100)(100)}{(2.54)(2.54)(12)(12)(5280)(5280)} \text{ mi}^2$

10. $-\sqrt{-4} + 2 + 2i^5 = -2i + 2 + 2i(ii)(ii)$
$= -2i + 2 + 2i = \mathbf{2}$

11. $2i^2 + 5i + 4 + \sqrt{-9} = 2(ii) + 5i + 4 + 3i$
$= \mathbf{2 + 8i}$

12. $-4i^5 + 2\sqrt{-16} = -4i(ii)(ii) + 2(4i)$
$= -4i + 8i = \mathbf{4i}$

13. $2i^3 - i^4 + 3i^2 = 2i(ii) - (ii)(ii) + 3(ii)$
$= -2i - 1 - 3 = \mathbf{-4 - 2i}$

14. $x^2 - 5 = 5x$
$(x^2 - 5x +) = 5$
$x^2 - 5x + \dfrac{25}{4} = 5 + \dfrac{25}{4}$
$\left(x - \dfrac{5}{2}\right)^2 = \dfrac{45}{4}$
$x - \dfrac{5}{2} = \pm\dfrac{3\sqrt{5}}{2}$
$x = \dfrac{5}{2} \pm \dfrac{3\sqrt{5}}{2}$

15. $-x^2 = -6x - 6$
$(x^2 - 6x +) = 6$
$x^2 - 6x + 9 = 6 + 9$
$(x - 3)^2 = 15$
$x - 3 = \pm\sqrt{15}$
$x = \mathbf{3 \pm \sqrt{15}}$

16. $A = 180 - 120 = 60$

$\cos 60° = \dfrac{10}{m}$

$m = \dfrac{10}{\cos 60°} = \mathbf{20}$

17. $\sqrt{x - 11} - 1 = 6$

$\sqrt{x - 11} = 7$

$x - 11 = 49$

$x = \mathbf{60}$

Check: $\sqrt{60 - 11} - 1 = 6$

$7 - 1 = 6$ check

18. $\sqrt{x^2 + 2x + 5} - 3 = x$

$x^2 + 2x + 5 = x^2 + 6x + 9$

$-4x = 4$

$x = \mathbf{-1}$

Check: $\sqrt{1 - 2 + 5} - 3 = -1$

$2 - 3 = -1$ check

19. $\sqrt[5]{2\sqrt[3]{2}} = [2(2^{1/3})]^{1/5} = 2^{1/5}2^{1/15} = \mathbf{2^{4/15}}$

20. $\sqrt{81\sqrt[4]{3}} = \sqrt{3^4\sqrt[4]{3}} = [3^4(3^{1/4})]^{1/2}$

$= 3^2 3^{1/8} = \mathbf{3^{17/8}}$

21. $\sqrt[5]{x^2 y}\sqrt[3]{xy^2} = (x^2y)^{1/5}(xy^2)^{1/3} = x^{2/5}y^{1/5}x^{1/3}y^{2/3}$

$= \mathbf{x^{11/15}y^{13/15}}$

22. $-4^{-5/2} = -\dfrac{1}{(4^{1/2})^5} = -\dfrac{1}{\mathbf{32}}$

23. $3\sqrt{\dfrac{2}{9}} - 2\sqrt{\dfrac{9}{2}} - 2\sqrt{50}$

$= \dfrac{3\sqrt{2}}{\sqrt{9}}\dfrac{\sqrt{9}}{\sqrt{9}} - \dfrac{2\sqrt{9}}{\sqrt{2}}\dfrac{\sqrt{2}}{\sqrt{2}} - 10\sqrt{2}$

$= \sqrt{2} - 3\sqrt{2} - 10\sqrt{2} = \mathbf{-12\sqrt{2}}$

24. $\dfrac{(2,135,820)(4,913,562)}{801,394,026}$

$\approx \dfrac{(2 \times 10^6)(5 \times 10^6)}{8 \times 10^8} \approx \mathbf{1 \times 10^4}$

25. $\dfrac{x}{y} - \dfrac{m}{p} + \dfrac{k}{c} = 0$

$cpx - cmy + kpy = 0$

$cpx + kpy = cmy$

$p = \dfrac{\mathbf{cmy}}{\mathbf{cx + ky}}$

26. $\dfrac{p}{x} + c = d$

$p + cx = dx$

$p = dx - cx$

$\dfrac{p}{d - c} = x$

27. $5 \times \overrightarrow{SF} = 10$

$\overrightarrow{SF} = 2$

$C \times 2 = 12$

$C = 6$

$A^2 = 5^2 + 6^2$

$A^2 = 25 + 36$

$A = \sqrt{61}$

$\sqrt{61} \times 2 = \sqrt{61} + B$

$2\sqrt{61} = \sqrt{61} + B$

$\mathbf{\sqrt{61}} = B$

28. $y = \dfrac{80}{2} = \mathbf{40}$

$x = 180 - 90 - 40 = \mathbf{50}$

$P = y = \mathbf{40}$

$Q = x = \mathbf{50}$

$R = 2(50) = \mathbf{100}$

29.

$$x + 1 \overline{\smash{\big)}\ x^3 + 0x^2 - 2x + 2}$$

$$\begin{array}{r} x^2 - x - 1 + \dfrac{3}{x+1} \\ \underline{x^3 + x^2} \\ -x^2 - 2x \\ \underline{-x^2 - x} \\ -x + 2 \\ \underline{-x - 1} \\ 3 \end{array}$$

30. (a) $2x - 3y = -9$

$-3y = -2x - 9$

$y = \dfrac{2}{3}x + 3$

(b) $5x + 3y = 3$

$3y = -5x + 3$

$y = -\dfrac{5}{3}x + 1$

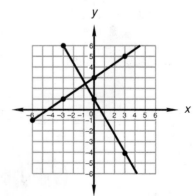

Solve the equations by elimination:

(a) $2x - 3y = -9$

(b) $\dfrac{5x + 3y = 3}{7x = -6}$

$x = -\dfrac{6}{7}$

(a) $y = \dfrac{2}{3}\left(-\dfrac{6}{7}\right) + 3 = \dfrac{17}{7}$

$\left(-\dfrac{6}{7},\ \dfrac{17}{7}\right)$

PROBLEM SET 54

1. Iodine P_N + Iodine D_N = Iodine Total

$0.2P_N + 0.7D_N = 0.575(400)$

(a) $0.2P_N + 0.7D_N = 230$

(b) $P_N + D_N = 400$

Substitute $D_N = 400 - P_N$ into (a) and get:

(a′) $0.2P_N + 0.7(400 - P_N) = 230$

$-0.5P_N = -50$

$P_N = 100$

(b) $(100) + D_N = 400$

$D_N = 300$

100 liters 20%, 300 liters 70%

2. Glycerine P_N + Glycerine D_N = Glycerine Total

$0.1(P_N) + 0.4(D_N) = 0.3(150)$

(a) $0.1P_N + 0.4D_N = 45$

(b) $P_N + D_N = 150$

Substitute $D_N = 150 - P_N$ into (a) and get:

(a′) $0.1P_N + 0.4(150 - P_N) = 45$

$-0.3P_N = -15$

$P_N = 50$

(b) $(50) + D_N = 150$

$D_N = 100$

50 ml 10%, 100 ml 40%

3. Potassium: $1 \times 39 = 39$

Chrome: $1 \times 35 = 35$

Oxygen: $3 \times 16 = 48$

Total: $= 122$

$\dfrac{39}{122} \approx 0.32 = \mathbf{32\%}$

4. (a) $\dfrac{N_D}{N_G} = \dfrac{5}{4} \;\rightarrow\; 4N_D = 5N_G$

(b) $4N_D = 3N_G + 40$

$-1(a)\ -4N_D + 5N_G = 0$

$(b)\ \dfrac{4N_D - 3N_G = 40}{2N_G = 40}$

$N_G = \mathbf{20}$

(b) $4N_D = 3(20) + 40$

$N_D = \mathbf{25}$

5. $R_C T_C = R_T T_T;\ R_C = 2R_T;$

$T_C = T_T - 3;\ R_T = 50$

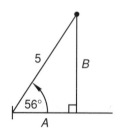

$100(T_T - 3) = 50T_T$

$50T_T = 300$

$T_T = 6$

$D_T = D_C = 50(6) = \mathbf{300\ miles}$

6. $A = 5\cos 56° \approx 2.80$

$B = 5\sin 56° \approx 4.15$

$2.80R + 4.15U$

7. $A = 8\cos 32° \approx 6.78$

$B = 8\sin 32° \approx 4.24$

$-6.78R - 4.24U$

8. $\dfrac{T}{V} = \dfrac{R}{S} \;\rightarrow\; ST = RV \;\rightarrow\; T = \dfrac{RV}{S}$

9. $100\ \text{km}^2 \times \dfrac{1000\ \text{m}}{1\ \text{km}} \times \dfrac{1000\ \text{m}}{1\ \text{km}} \times \dfrac{100\ \text{cm}}{1\ \text{m}}$

$\times \dfrac{100\ \text{cm}}{1\ \text{m}} = \mathbf{100(1000)(1000)(100)(100)\ cm^2}$

10. $100\ \text{ft}^2 \times \dfrac{12\ \text{in.}}{1\ \text{ft}} \times \dfrac{12\ \text{in.}}{1\ \text{ft}} \times \dfrac{2.54\ \text{cm}}{1\ \text{in.}}$

$\times \dfrac{2.54\ \text{cm}}{1\ \text{in.}} = \mathbf{100(12)(12)(2.54)(2.54)\ cm^2}$

11. $\dfrac{60\ \text{mi}}{\text{hr}} \times \dfrac{5280\ \text{ft}}{1\ \text{mi}} \times \dfrac{12\ \text{in.}}{1\ \text{ft}} \times \dfrac{2.54\ \text{cm}}{1\ \text{in.}}$

$\times \dfrac{1\ \text{m}}{100\ \text{cm}} \times \dfrac{1\ \text{km}}{1000\ \text{m}} \times \dfrac{1\ \text{hr}}{60\ \text{min}} \times \dfrac{1\ \text{min}}{60\ \text{sec}}$

$= \mathbf{\dfrac{60(5280)(12)(2.54)}{(100)(1000)(60)(60)}\ \dfrac{km}{sec}}$

12. $-\sqrt{-4} - 2i^3 + 4i^4 = -2i - 2i(ii) + 4(ii)(ii)$
$= -2i + 2i + 4 = \mathbf{4}$

13. $3i^3 + 2i - 4i^2 + \sqrt{-9} = 3i(ii) + 2i - 4(ii) + 3i$
$= -3i + 2i + 4 + 3i = \mathbf{4 + 2i}$

14. $-3i + 2i^2 - 2 + i = -3i + 2(ii) - 2 + i$
$= -3i - 2 - 2 + i = \mathbf{-4 - 2i}$

15. $2i^2 + 2i - 2\sqrt{-25} = 2(ii) + 2i - 10i = \mathbf{-2 - 8i}$

16.
$$x^2 = 7 + 3x$$
$$(x^2 - 3x + \quad) = 7$$
$$x^2 - 3x + \frac{9}{4} = 7 + \frac{9}{4}$$
$$\left(x - \frac{3}{2}\right)^2 = \frac{37}{4}$$
$$x - \frac{3}{2} = \pm\frac{\sqrt{37}}{2}$$
$$x = \frac{\mathbf{3}}{\mathbf{2}} \pm \frac{\sqrt{\mathbf{37}}}{\mathbf{2}}$$

17.
$$-7x = -x^2 + 3$$
$$(x^2 - 7x + \quad) = 3$$
$$x^2 - 7x + \frac{49}{4} = 3 + \frac{49}{4}$$
$$\left(x - \frac{7}{2}\right)^2 = \frac{61}{4}$$
$$x - \frac{7}{2} = \pm\frac{\sqrt{61}}{2}$$
$$x = \frac{\mathbf{7}}{\mathbf{2}} \pm \frac{\sqrt{\mathbf{61}}}{\mathbf{2}}$$

18. $y = \frac{3}{2}x + b$

Use the point $(6, 0)$ for x and y.

$$0 = \frac{3}{2}(6) + b$$
$$-9 = b$$
$$y = \frac{\mathbf{3}}{\mathbf{2}}\mathbf{x} - \mathbf{9}$$

19.
$$\frac{4x - 2}{5} - \frac{x - 3}{2} = 7$$
$$8x - 4 - 5x + 15 = 70$$
$$3x = 59$$
$$x = \frac{\mathbf{59}}{\mathbf{3}}$$

20. $\sqrt{x^2 - x - 5} + 1 = x$
$$x^2 - x - 5 = x^2 - 2x + 1$$
$$x = \mathbf{6}$$

Check: $\sqrt{36 - 6 - 5} + 1 = 6$
$$5 + 1 = 6 \quad \text{check}$$

21. $\sqrt{x - 2} - 11 = 1$
$$x - 2 = 144$$
$$x = \mathbf{146}$$

Check: $\sqrt{146 - 2} - 11 = 1$
$$12 - 11 = 1 \quad \text{check}$$

22.
$$\frac{x}{k} - cm = \frac{p}{c}$$
$$cx - c^2km = kp$$
$$cx = kp + c^2km$$
$$\frac{\mathbf{cx}}{\mathbf{p} + \mathbf{c^2m}} = k$$

23.
$$\frac{a}{p} - x + \frac{c}{m} = y$$
$$am - mpx + cp = mpy$$
$$cp = mpy + mpx - am$$
$$\frac{\mathbf{cp}}{\mathbf{py} + \mathbf{px} - \mathbf{a}} = m$$

24. $\sqrt[3]{4\sqrt{2}} = \sqrt[3]{2^2\sqrt{2}} = [2^2(2^{1/2})]^{1/3}$
$= 2^{2/3}2^{1/6} = \mathbf{2^{5/6}}$

25. $\sqrt[5]{9\sqrt[3]{3}} = \sqrt[5]{3^2\sqrt[3]{3}} = [3^2(3^{1/3})]^{1/5}$
$= 3^{2/5}3^{1/15} = \mathbf{3^{7/15}}$

26. $\sqrt[6]{xy}\sqrt[3]{xy^2} = (xy)^{1/6}(xy^2)^{1/3}$
$= x^{1/6}y^{1/6}x^{1/3}y^{2/3} = \mathbf{x^{1/2}y^{5/6}}$

27. $-81^{1/4} = \mathbf{-3}$

28. $2\sqrt{\frac{3}{7}} - 5\sqrt{\frac{7}{3}} + 2\sqrt{84}$

$$= \frac{2\sqrt{3}}{\sqrt{7}}\frac{\sqrt{7}}{\sqrt{7}} - \frac{5\sqrt{7}}{\sqrt{3}}\frac{\sqrt{3}}{\sqrt{3}} + 4\sqrt{21}$$

$$= \frac{6\sqrt{21}}{21} - \frac{35\sqrt{21}}{21} + \frac{84\sqrt{21}}{21} = \frac{\mathbf{55\sqrt{21}}}{\mathbf{21}}$$

29.
$$\frac{x^2p - \dfrac{x}{p^2}}{\dfrac{x^2y}{p^2} - x} = \frac{\dfrac{x^2p^3 - x}{p^2}}{\dfrac{x^2y - xp^2}{p^2}} \cdot \frac{\dfrac{p^2}{x^2y - xp^2}}{\dfrac{p^2}{x^2y - xp^2}}$$

$$= \frac{x^2p^3 - x}{x^2y - xp^2} = \frac{xp^3 - 1}{xy - p^2}$$

30. $\dfrac{4x^{-2}yp}{m^2y^{-1}}\left(\dfrac{p^{-1}m^2y}{16x^{-2}y} - \dfrac{2x^2y^0p}{m^{-2}}\right) = \dfrac{y^2}{4} - 8y^2p^2$

PROBLEM SET 55

1. $N \quad N + 1 \quad N + 2$

$N(N + 1) = -6(N + 2)$

$N^2 + N = -6N - 12$

$N^2 + 7N + 12 = 0$

$(N + 4)(N + 3) = 0$

$N = -4, -3$

The desired integers are **-4, -3, -2** and **-3, -2, -1**.

2. $N \quad N + 2 \quad N + 4$

$N(N + 4) = 9(N + 2) - 24$

$N^2 + 4N = 9N - 6$

$N^2 - 5N + 6 = 0$

$(N - 3)(N - 2) = 0$

$N = 3, 2$

Since the problem asks only for even integers, the desired integers are **2, 4,** and **6**.

3. Bromide P_N + Bromide D_N = Bromide Total

$0.05(P_N) + 0.4(D_N) = 0.12(60)$

(a) $0.05P_N + 0.4D_N = 7.2$

(b) $P_N + D_N = 60$

Substitute $D_N = 60 - P_N$ into (a) and get:

(a') $0.05P_N + 0.4(60 - P_N) = 7.2$

$-0.35P_N = -16.8$

$P_N = 48$

(b) $(48) + D_N = 60$

$D_N = 12$

48 ml 5%, 12 ml 40%

4. Carbon: $\quad 1 \times 12 = 12$

Chlorine: $\quad 4 \times 35 = 140$

Total: $\qquad\qquad = 152$

$\dfrac{12}{152} = \dfrac{C}{1368}$

$152C = 12(1368)$

$C = \textbf{108 grams}$

5. $3N \quad 3N + 3 \quad 3N + 6$

$6(3N) = 4(3N + 6) + 48$

$18N = 12N + 72$

$6N = 72$

$N = 12$

The desired integers are **36, 39,** and **42**.

6. $\dfrac{a + b}{x} + \dfrac{y}{m} = k$

$am + bm + xy = kmx$

$am + bm = kmx - xy$

$\dfrac{am + bm}{km - y} = x$

7. $\dfrac{mp}{c} + \dfrac{d + e}{x} = d$

$mpx + cd + ce = cdx$

$ce = cdx - mpx - cd$

$e = \dfrac{cdx - mpx - cd}{c}$

8. $\dfrac{mx}{y} + \dfrac{d}{a + b} = p$

$amx + bmx + dy = apy + bpy$

$amx + dy - apy = bpy - bmx$

$\dfrac{amx + dy - apy}{py - mx} = b$

9. $A = 40 \cos 35°$

$A \approx 32.77$

$B = 40 \sin 35°$

$B \approx 22.94$

32.77R − 22.94U

10. $A = 10 \cos 20°$

$A \approx 9.40$

$B = 10 \sin 20°$

$B \approx 3.42$

−9.40R − 3.42U

11. $4 \text{ yd}^3 \times \dfrac{3 \text{ ft}}{1 \text{ yd}} \times \dfrac{3 \text{ ft}}{1 \text{ yd}} \times \dfrac{3 \text{ ft}}{1 \text{ yd}} \times \dfrac{12 \text{ in.}}{1 \text{ ft}}$

$\times \dfrac{12 \text{ in.}}{1 \text{ ft}} \times \dfrac{12 \text{ in.}}{1 \text{ ft}} \times \dfrac{2.54 \text{ cm}}{1 \text{ in.}} \times \dfrac{2.54 \text{ cm}}{1 \text{ in.}}$

$\times \dfrac{2.54 \text{ cm}}{1 \text{ in.}} \times \dfrac{1 \text{ m}}{100 \text{ cm}} \times \dfrac{1 \text{ m}}{100 \text{ cm}} \times \dfrac{1 \text{ m}}{100 \text{ cm}}$

$= \dfrac{4(3)(3)(3)(12)(12)(12)(2.54)(2.54)(2.54)}{(100)(100)(100)} \text{ m}^3$

12. $\dfrac{1000 \text{ ft}}{\text{sec}} \times \dfrac{12 \text{ in.}}{1 \text{ ft}} \times \dfrac{2.54 \text{ cm}}{1 \text{ in.}} \times \dfrac{1 \text{ m}}{100 \text{ cm}}$

$\times \dfrac{1 \text{ km}}{1000 \text{ m}} \times \dfrac{60 \text{ sec}}{1 \text{ min}}$

$= \dfrac{1000(12)(2.54)(60)}{(100)(1000)} \dfrac{\text{km}}{\text{min}}$

13. $4i^5 - 2\sqrt{-9} - 2i^4 = 4i(ii)(ii) - 6i - 2(ii)(ii)$

$= 4i - 6i - 2 = \textbf{−2 − 2i}$

14. $4 + 2i^2 + 3i - \sqrt{-4} = 4 + 2(ii) + 3i - 2i$

$= 4 - 2 + 3i - 2i = \textbf{2 + i}$

15. $2i^3 + 2i^4 + 2 - 2i = 2i(ii) + 2(ii)(ii) + 2 - 2i$

$= -2i + 2 + 2 - 2i = \textbf{4 − 4i}$

16. $-3i^6 - 2i - 2 - 2i^2$
$= -3(ii)(ii)(ii) - 2i - 2 - 2(ii)$
$= 3 - 2i - 2 + 2 = \mathbf{3 - 2i}$

17.
$$x^2 = 5x + 5$$
$$(x^2 - 5x + \quad) = 5$$
$$x^2 - 5x + \frac{25}{4} = 5 + \frac{25}{4}$$
$$\left(x - \frac{5}{2}\right)^2 = \frac{45}{4}$$
$$x - \frac{5}{2} = \pm\frac{3\sqrt{5}}{2}$$
$$x = \frac{5}{2} \pm \frac{3\sqrt{5}}{2}$$

18.
$$x^2 - 6 = 6x$$
$$(x^2 - 6x + \quad) = 6$$
$$x^2 - 6x + 9 = 6 + 9$$
$$(x - 3)^2 = 15$$
$$x - 3 = \pm\sqrt{15}$$
$$x = \mathbf{3 \pm \sqrt{15}}$$

19. $\sqrt{x - 2} + 4 = 2$
$$x - 2 = 4$$
$$x = 6$$

Check: $\sqrt{6 - 2} + 4 = 2$
$2 + 4 = 2$ not true
No real number solution.

20. $\sqrt{x^2 - 2x + 14} - 12 = x$
$$x^2 - 2x + 14 = x^2 + 24x + 144$$
$$-26x = 130$$
$$x = \mathbf{-5}$$

Check: $\sqrt{25 + 10 + 14} - 12 = -5$
$7 - 12 = -5$ check

21. $\sqrt{16\sqrt{2}} = \sqrt{2^4\sqrt{2}} = [2^4(2^{1/2})]^{1/2}$
$= 2^2 2^{1/4} = \mathbf{2^{9/4}}$

22. $\sqrt[4]{27\sqrt[3]{3}} = \sqrt[4]{3^3\sqrt[3]{3}} = [3^3(3^{1/3})]^{1/4}$
$= 3^{3/4}3^{1/12} = \mathbf{3^{5/6}}$

23. $\sqrt[4]{x^2y}\sqrt{x^5y^2} = (x^2y)^{1/4}(x^5y^2)^{1/2}$
$= x^{1/2}y^{1/4}x^{5/2}y = \mathbf{x^3y^{5/4}}$

24. $-81^{5/4} = -(81^{1/4})^5 = \mathbf{-243}$

25. $4\sqrt{\dfrac{2}{11}} + 2\sqrt{\dfrac{11}{2}} - 4\sqrt{198}$

$= \dfrac{4\sqrt{2}}{\sqrt{11}}\dfrac{\sqrt{11}}{\sqrt{11}} + \dfrac{2\sqrt{11}}{\sqrt{2}}\dfrac{\sqrt{2}}{\sqrt{2}} - 12\sqrt{22}$

$= \dfrac{8\sqrt{22}}{22} + \dfrac{22\sqrt{22}}{22} - \dfrac{264\sqrt{22}}{22} = \mathbf{-\dfrac{117\sqrt{22}}{11}}$

26. $\dfrac{(4,183,256)(704,185 \times 10^{-42})}{802,164 \times 10^{30}}$

$\approx \dfrac{(4 \times 10^6)(7 \times 10^{-37})}{8 \times 10^{35}} \approx \mathbf{4 \times 10^{-66}}$

27. $4 \times \overrightarrow{SF} = 8$
$$\overrightarrow{SF} = 2$$
$$C \times 2 = 6$$
$$C = 3$$

$$A^2 = 4^2 + 3^2$$
$$A^2 = 25$$
$$A = 5$$

$$5 \cdot 2 = 5 + B$$
$$10 = 5 + B$$
$$\mathbf{5 = B}$$

28. $\dfrac{x}{y} = \dfrac{m}{p}$
$$xp = my$$
$$x = \mathbf{\dfrac{my}{p}}$$

29. $V_{\text{Prism}} = A_{\text{Base}} \times \text{height}$

$= \left[(8)(6) + \dfrac{1}{2}(8)(6) + \dfrac{1}{2}\pi(5)^2 \right.$

$\left. + \dfrac{1}{2}\pi(6)^2 \right] \text{ft}^2 \times 8 \text{ ft}$

$= \left(72 + \dfrac{61}{2}\pi\right)(8) \text{ ft}^3 \approx \mathbf{1342.16 \text{ ft}^3}$

30. $\dfrac{ap(x + 5)(x - 2)}{(x + 7)(x - 2)} \cdot \dfrac{(x + 7)(x - 3)}{ap(x + 5)(x - 4)}$

$= \dfrac{x - 3}{x - 4}$

PROBLEM SET 56

1. $N \qquad N + 2 \qquad N + 4$

$N(N + 2) = -10(N + 4) + 8$

$N^2 + 2N = -10N - 32$

$N^2 + 12N + 32 = 0$

$(N + 8)(N + 4) = 0$

$N = -8, -4$

The desired integers are $-8, -6, -4$ and $-4, -2, 0$.

2. Carbon: $1 \times 12 = 12$

Hydrogen: $2 \times 1 = 2$

Bromine: $2 \times 80 = 160$

Total: $= 174$

$\dfrac{160}{174} = \dfrac{320}{MB}$

$160 \, MB = 174(320)$

$MB = \mathbf{348 \ grams}$

$\dfrac{160}{174} \approx 0.92 = \mathbf{92\%}$

3. $R_W T_W = R_R T_R$;

$R_W = 4; \ R_R = 20;$

$T_W + T_R = 12$

$4T_W = 20(12 - T_W)$

$24T_W = 240$

$T_W = 10$

$D_W = 4(10) = \mathbf{40 \ miles}$

4. Fluorine P_N + Fluorine D_N = Fluorine Total

$0.2(P_N) + 0.8(D_N) = 0.56(1000)$

(a) $0.2P_N + 0.8D_N = 560$

(b) $P_N + D_N = 1000$

Substitute $D_N = 1000 - P_N$ into (a) and get:

(a') $0.2P_N + 0.8(1000 - P_N) = 560$

$-0.6P_N = -240$

$P_N = \mathbf{400}$

(b) $(400) + D_N = 1000$

$D_N = 600$

400 gallons 20%, 600 gallons 80%

5. $3\dfrac{3}{5} \times D = 1440$

$D = 1440 \cdot \dfrac{5}{18} = \mathbf{400}$

6. $x + 10 = 46$

$x = \mathbf{36}$

$y + 20 = 30$

$y = \mathbf{10}$

7. (a) $x = \dfrac{120 + 100}{2} = \mathbf{110}$

(b) $x = \dfrac{180 - 60}{2} = \mathbf{60}$

8. $\dfrac{x + 1}{y} = \dfrac{m}{p}$

$px + p = my$

$px = my - p$

$x = \dfrac{\boldsymbol{my - p}}{\boldsymbol{p}}$

9. $\dfrac{a + x}{b} - \dfrac{c}{m} = \dfrac{p}{k}$

$akm + kmx - bck = bmp$

$kmx = bmp + bck - akm$

$x = \dfrac{\boldsymbol{bmp + bck - akm}}{\boldsymbol{km}}$

10. $\dfrac{m}{a + c} - \dfrac{x}{m} = p$

$m^2 - ax - cx = amp + cmp$

$m^2 - ax - amp = cmp + cx$

$\dfrac{\boldsymbol{m^2 - ax - amp}}{\boldsymbol{mp + x}} = c$

11. $A = 10 \cos 30°$

$A \approx 8.66$

$B = 10 \sin 30°$

$B = 5$

$\mathbf{-8.66R - 5U}$

12. $A = 20 \cos 60°$

$A = 10$

$B = 20 \sin 60°$

$B \approx 17.32$

$\mathbf{10R + 17.32U}$

13. $\dfrac{60 \text{ km}}{\text{hr}} \times \dfrac{1000 \text{ m}}{1 \text{ km}} \times \dfrac{100 \text{ cm}}{1 \text{ m}} \times \dfrac{1 \text{ in.}}{2.54 \text{ cm}}$

$\times \dfrac{1 \text{ hr}}{60 \text{ min}} \times \dfrac{1 \text{ min}}{60 \text{ sec}} = \dfrac{\mathbf{60(1000)(100)}}{\mathbf{(2.54)(60)(60)}} \dfrac{\textbf{in.}}{\textbf{sec}}$

14. $400 \text{ yd}^3 \times \dfrac{3 \text{ ft}}{1 \text{ yd}} \times \dfrac{3 \text{ ft}}{1 \text{ yd}} \times \dfrac{3 \text{ ft}}{1 \text{ yd}} \times \dfrac{12 \text{ in.}}{1 \text{ ft}}$

$\times \dfrac{12 \text{ in.}}{1 \text{ ft}} \times \dfrac{12 \text{ in.}}{1 \text{ ft}} \times \dfrac{2.54 \text{ cm}}{1 \text{ in.}} \times \dfrac{2.54 \text{ cm}}{1 \text{ in.}}$

$\times \dfrac{2.54 \text{ cm}}{1 \text{ in.}}$

$= \mathbf{400(3)(3)(3)(12)(12)(12)(2.54)(2.54)(2.54) \ cm^3}$

15. $3i^5 + 2\sqrt{-25} - 3i^2 = 3i(ii)(ii) + 10i - 3(ii)$
$= 3i + 10i + 3 = \mathbf{3 + 13i}$

16. $2i^4 - 3i^3 + 2i + 4 = 2(ii)(ii) - 3i(ii) + 2i + 4$
$= 2 + 3i + 2i + 4 = \mathbf{6 + 5i}$

17. $\sqrt[6]{4\sqrt[5]{2}} = \sqrt[6]{2^2\sqrt[5]{2}} = [2^2(2^{1/5})]^{1/6}$
$= 2^{1/3}2^{1/30} = \mathbf{2^{11/30}}$

18. $\sqrt{y^4}\sqrt{xy^2} = (y^4)^{1/2}(xy^2)^{1/2} = y^2 x^{1/2} y = \mathbf{x^{1/2}y^3}$

19. $\sqrt{25\sqrt[3]{5}} = \sqrt{5^2\sqrt[3]{5}} = [5^2(5^{1/3})]^{1/2}$
$= 5 \cdot 5^{1/6} = \mathbf{5^{7/6}}$

20. $4\sqrt{\dfrac{3}{4}} - 2\sqrt{\dfrac{4}{3}} - 2\sqrt{27}$

$= \dfrac{4\sqrt{3}}{2} - \dfrac{4}{\sqrt{3}}\dfrac{\sqrt{3}}{\sqrt{3}} - 6\sqrt{3}$

$= \dfrac{12\sqrt{3}}{6} - \dfrac{8\sqrt{3}}{6} - \dfrac{36\sqrt{3}}{6} = \mathbf{-\dfrac{16\sqrt{3}}{3}}$

21.
$$x^2 = 7x + 7$$
$$(x^2 - 7x + \quad) = 7$$
$$x^2 - 7x + \frac{49}{4} = 7 + \frac{49}{4}$$
$$\left(x - \frac{7}{2}\right)^2 = \frac{77}{4}$$
$$x - \frac{7}{2} = \pm\frac{\sqrt{77}}{2}$$
$$x = \mathbf{\frac{7}{2} - \frac{\sqrt{77}}{2}}$$

22.
$$-8x - 8 = -x^2$$
$$(x^2 - 8x + \quad) = 8$$
$$x^2 - 8x + 16 = 8 + 16$$
$$(x - 4)^2 = 24$$
$$x - 4 = \pm 2\sqrt{6}$$
$$x = \mathbf{4 \pm 2\sqrt{6}}$$

23. $\sqrt{x + 1} + 1 = 1$
$$x + 1 = 0$$
$$x = \mathbf{-1}$$

Check: $\sqrt{-1 + 1} + 1 = 1$
$$0 + 1 = 1 \quad \text{check}$$

24. $\sqrt{x^2 - 2x + 21} - 1 = x$
$$x^2 - 2x + 21 = x^2 + 2x + 1$$
$$-4x = -20$$
$$x = \mathbf{5}$$

Check: $\sqrt{25 - 10 + 21} - 1 = 5$
$$6 - 1 = 5 \quad \text{check}$$

25.
$$
\begin{array}{r}
4x^3 + 12x^2 + 36x + 108 + \frac{323}{x-3} \\
x - 3 \overline{) 4x^4 + 0x^3 + 0x^2 + 0x - 1} \\
\underline{4x^4 - 12x^3} \\
12x^3 + 0x^2 \\
\underline{12x^3 - 36x^2} \\
36x^2 + 0x \\
\underline{36x^2 - 108x} \\
108x - 1 \\
\underline{108x - 324} \\
323
\end{array}
$$

26. $\dfrac{\dfrac{ax}{y^2} - \dfrac{yp}{x}}{\dfrac{yp}{xy} - \dfrac{1}{y^2}} = \dfrac{\dfrac{ax^2 - y^3 p}{xy^2}}{\dfrac{y^2 p - x}{xy^2}} \cdot \dfrac{\dfrac{xy^2}{y^2 p - x}}{\dfrac{xy^2}{y^2 p - x}}$

$= \mathbf{\dfrac{ax^2 - y^3 p}{y^2 p - x}}$

27. Graph the line to find the slope.

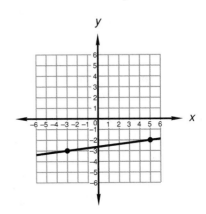

$m = \dfrac{+1}{+8} = \dfrac{1}{8}$

Since the slopes of perpendicular lines are negative reciprocals of each other:
$m_\perp = -8$
$$y = -8x + b$$
$$-5 = -8(-5) + b$$
$$-4 = b$$
$$y = \mathbf{-8x - 45}$$

28. $\dfrac{x - 3}{4} - \dfrac{2 - x}{8} = 6$
$$2x - 6 - 2 + x = 48$$
$$3x = 56$$
$$x = \mathbf{\frac{56}{3}}$$

29. $\dfrac{a - 5}{2} - \dfrac{3 - a}{4} = 1$

$2a - 10 - 3 + a = 4$

$3a = 17$

$a = \dfrac{17}{3}$

30. $\dfrac{x}{a^2 y} - \dfrac{3x + 2}{a^2 y(x - 1)} - \dfrac{4}{x^2 - 1}$

$= \dfrac{x(x^2 - 1)}{a^2 y(x^2 - 1)} - \dfrac{(3x + 2)(x + 1)}{a^2 y(x^2 - 1)}$

$\quad - \dfrac{4a^2 y}{a^2 y(x^2 - 1)}$

$= \dfrac{x^3 - x - 3x^2 - 5x - 2 - 4a^2 y}{a^2 y(x^2 - 1)}$

$= \dfrac{x^3 - 3x^2 - 6x - 4a^2 y - 2}{a^2 y(x^2 - 1)}$

PROBLEM SET 57

1. $\dfrac{P_1 V_1}{T_1} = \dfrac{P_2 V_2}{T_2}$

$\dfrac{100(4)}{800} = \dfrac{P_2(12)}{600}$

$P_2 = \mathbf{25\ N/m^2}$

2. $P_1 V_1 = P_2 V_2$

$7(42) = P_2(49)$

$P_2 = \mathbf{6\ atm}$

3. $\dfrac{P_1}{T_1} = \dfrac{P_2}{T_2}$

$\dfrac{400}{1200} = \dfrac{P_2}{300}$

$P_2 = \mathbf{100\ N/m^2}$

4. Alcohol P_N + Alcohol D_N = Alcohol Total

$0.2(P_N) + 0.4(D_N) = 0.352(1000)$

(a) $0.2P_N + 0.4D_N = 352$

(b) $P_N + D_N = 1000$

Substitute $D_N = 1000 - P_N$ into (a) and get:

(a′) $0.2P_N + 0.4(1000 - P_N) = 352$

$-0.2P_N = -48$

$P_N = 240$

(b) $(240) + D_N = 1000$

$D_N = 760$

240 gallons 20%, 760 gallons 40%

5. $N \qquad N + 2 \qquad N + 4 \qquad N + 6$

$(N + 4)(N + 6) = 10N + 49$

$N^2 + 10N + 24 = 10N + 49$

$N^2 - 25 = 0$

$(N - 5)(N + 5) = 0$

$N = 5, -5$

The desired integers are **5**, **7**, **9**, **11** and **−5**, **−3**, **−1**, **1**.

6. (a) $x = \dfrac{80 - 15}{2} = \mathbf{32.5}$

(b) $x = \dfrac{210 + 48}{2} = \mathbf{129}$

7. $\dfrac{a}{x - y} - \dfrac{c}{p} = m$

$ap - cx + cy = mpx - mpy$

$mpy + cy = mpx + cx - ap$

$y = \dfrac{mpx + cx - ap}{mp + c}$

8. $\dfrac{x - a}{p} - c = \dfrac{k}{d}$

$dx - ad - cdp = kp$

$dx - cdp - kp = ad$

$\dfrac{dx - cdp - kp}{d} = a$

9. $A = 4 \cos 40°$

$A \approx 3.06$

$B = 4 \sin 40°$

$B \approx 2.57$

3.06R + 2.57U

10. $A = 40 \cos 30°$

$A \approx 34.64$

$B = 40 \sin 30°$

$B = 20$

34.64R − 20U

11. $\dfrac{40\ cm}{sec} \times \dfrac{1\ in.}{2.54\ cm} \times \dfrac{1\ ft}{12\ in.} \times \dfrac{1\ mi}{5280\ ft}$

$\times \dfrac{60\ sec}{1\ min} \times \dfrac{60\ min}{1\ hr} = \dfrac{\mathbf{40(60)(60)}}{\mathbf{(2.54)(12)(5280)}}\ \dfrac{\mathbf{mi}}{\mathbf{hr}}$

12. $1000\ cm^3 \times \dfrac{1\ in.}{2.54\ cm} \times \dfrac{1\ in.}{2.54\ cm} \times \dfrac{1\ in.}{2.54\ cm}$

$\times \dfrac{1\ ft}{12\ in.} \times \dfrac{1\ ft}{12\ in.} \times \dfrac{1\ ft}{12\ in.}$

$= \dfrac{\mathbf{1000}}{\mathbf{(2.54)(2.54)(2.54)(12)(12)(12)}}\ ft^3$

13. $3i^3 - 2i^2 + i^4 - 5 = 3i(ii) - 2(ii) + (ii)(ii) - 5$

$= -3i + 2 + 1 - 5 = \mathbf{-2 - 3i}$

14. $-2\sqrt{-9} - 3i^2 + 2i - 2 = -6i - 3(ii) + 2i - 2$
$= -6i + 3 + 2i - 2 = \mathbf{1 - 4i}$

15. $\quad -5x - 6 = -x^2$
$(x^2 - 5x + \quad) = 6$
$x^2 - 5x + \dfrac{25}{4} = 6 + \dfrac{25}{4}$
$\left(x - \dfrac{5}{2}\right)^2 = \dfrac{49}{4}$
$x - \dfrac{5}{2} = \pm\dfrac{7}{2}$
$x = \dfrac{5}{2} \pm \dfrac{7}{2}$
$x = \mathbf{6, -1}$

16. $\quad -6x + x^2 = -5$
$(x^2 - 6x + \quad) = -5$
$x^2 - 6x + 9 = -5 + 9$
$(x - 3)^2 = 4$
$x - 3 = \pm 2$
$x = 3 \pm 2$
$x = \mathbf{5, 1}$

17. Use the graph to find the slope.
$y = \dfrac{11}{5}x + b$
Use the point $(-1, 6)$ for x and y.
$6 = \dfrac{11}{5}(-1) + b$
$\dfrac{41}{5} = b$
$y = \dfrac{\mathbf{11}}{\mathbf{5}}x + \dfrac{\mathbf{41}}{\mathbf{5}}$

18. $\sqrt{x - 2} + 2 = 3$
$x - 2 = 1$
$x = \mathbf{3}$

Check: $\sqrt{3 - 2} + 2 = 3$
$1 + 2 = 3$ check

19. $\sqrt{x^2 - x + 13} - 1 = x$
$x^2 - x + 13 = x^2 + 2x + 1$
$-3x = -12$
$x = \mathbf{4}$

Check: $\sqrt{16 - 4 + 13} - 1 = 4$
$5 - 1 = 4$ check

20. $\dfrac{x - 2}{4} - \dfrac{x}{3} = 5$
$3x - 6 - 4x = 60$
$-x = 66$
$x = \mathbf{-66}$

21. $\dfrac{x}{4} - \dfrac{3x + 1}{2} = 3$
$x - 6x - 2 = 12$
$-5x = 14$
$x = -\dfrac{\mathbf{14}}{\mathbf{5}}$

22. $\sqrt{9\sqrt[3]{3}} = \sqrt{3^2\sqrt[3]{3}} = [3^2(3^{1/3})]^{1/2}$
$= 3 \cdot 3^{1/6} = \mathbf{3^{7/6}}$

23. $\sqrt{x^4\sqrt[3]{x^2 y}} = [x^4(x^2 y)^{1/3}]^{1/2}$
$= x^2 x^{1/3} y^{1/6} = \mathbf{x^{7/3} y^{1/6}}$

24. $\sqrt[3]{4}\sqrt[5]{2} = \sqrt[3]{2^2}\sqrt[5]{2} = 2^{2/3}2^{1/5} = \mathbf{2^{13/15}}$

25. $3\sqrt{\dfrac{2}{5}} + 7\sqrt{\dfrac{5}{2}} - 2\sqrt{40}$
$= \dfrac{3\sqrt{2}}{\sqrt{5}}\dfrac{\sqrt{5}}{\sqrt{5}} + \dfrac{7\sqrt{5}}{\sqrt{2}}\dfrac{\sqrt{2}}{\sqrt{2}} - 4\sqrt{10}$
$= \dfrac{6\sqrt{10}}{10} + \dfrac{35\sqrt{10}}{10} - \dfrac{40\sqrt{10}}{10} = \dfrac{\sqrt{\mathbf{10}}}{\mathbf{10}}$

26. $16^{-5/4} = \dfrac{1}{16^{5/4}} = \dfrac{1}{(16^{1/4})^5} = \dfrac{\mathbf{1}}{\mathbf{32}}$

27. $\dfrac{\dfrac{x^2 y}{p^5 z} - 1}{\dfrac{x}{p^5} - \dfrac{4}{z}} = \dfrac{\dfrac{x^2 y - p^5 z}{p^5 z}}{\dfrac{xz - 4p^5}{p^5 z}} \cdot \dfrac{\dfrac{p^5 z}{xz - 4p^5}}{\dfrac{p^5 z}{xz - 4p^5}}$

$= \dfrac{x^2 y - p^5 z}{xz - 4p^5}$

28. $-2(-x^0 - 4^0) - 3x(2 - 6^0)$
$= (x)(-2 - 3^2 - 2) - x(-2 - 2^0)$
$4 - 6x + 3x = -13x + 3x$
$4 = -7x$
$-\dfrac{\mathbf{4}}{\mathbf{7}} = x$

29. $\quad -28x + x^3 = 3x^2$
$x^3 - 3x^2 - 28x = 0$
$x(x - 7)(x + 4) = 0$
$x = \mathbf{0, 7, -4}$

30. $\dfrac{x}{a^2} - \dfrac{x + 2}{a(a + 2)}$
$= \dfrac{x(a + 2)}{a^2(a + 2)} - \dfrac{a(x + 2)}{a^2(a + 2)}$
$= \dfrac{ax + 2x - ax - 2a}{a^2(a + 2)} = \dfrac{\mathbf{2x - 2a}}{\mathbf{a^2(a + 2)}}$

Problem Set 58

1. $\dfrac{P_1V_1}{T_1} = \dfrac{P_2V_2}{T_2}$

$\dfrac{4(6)}{600} = \dfrac{3(8)}{T_2}$

$T_2 = \textbf{600 K}$

2. Carbon: $6 \times 12 = 72$
Hydrogen: $8 \times 1 = 8$
Nitrogen: $1 \times 14 = 14$
Chlorine: $1 \times 35 = 35$
Total: $= 129$

$\dfrac{72}{129} = \dfrac{360}{CP}$

$72CP = 129(360)$

$CP = \textbf{645 grams}$

3. $\dfrac{Cl}{CP} = \dfrac{35}{129} \approx 0.27 = \textbf{27\%}$

4. Bromine P_N + Bromine D_N = Bromine Total

$0.1(P_N) + 0.4(D_N) = 0.16(50)$

(a) $0.1P_N + 0.4D_N = 8$

(b) $P_N + D_N = 50$

Substitute $D_N = 50 - P_N$ into (a) and get:

(a′) $0.1P_N + 0.4(50 - P_N) = 8$

$-0.3P_N = -12$

$P_N = 40$

(b) $(40) + D_N = 50$

$D_N = 10$

40 ml 10%, 10 ml 40%

5. $R_TT_T + 15 = R_RT_R$;
$T_T = T_R = 3$;
$R_R = 20$

$3R_T + 15 = 20(3)$

$3R_T = 45$

$R_T = 15$

$D_T = 15(3) = \textbf{45 miles}$

6. $3x^2 + 4x - 3 = 0$

$x^2 + \dfrac{4}{3}x - 1 = 0$

$\left(x^2 + \dfrac{4}{3}x + \quad\right) = 1$

$x^2 + \dfrac{4}{3}x + \dfrac{4}{9} = 1 + \dfrac{4}{9}$

$\left(x + \dfrac{2}{3}\right)^2 = \dfrac{13}{9}$

$x + \dfrac{2}{3} = \pm\dfrac{\sqrt{13}}{3}$

$x = -\dfrac{2}{3} \pm \dfrac{\sqrt{13}}{3}$

7. $4x^2 - x - 5 = 0$

$x^2 - \dfrac{1}{4}x - \dfrac{5}{4} = 0$

$\left(x^2 - \dfrac{1}{4}x + \quad\right) = \dfrac{5}{4}$

$x^2 - \dfrac{1}{4} + \dfrac{1}{64} = \dfrac{5}{4} + \dfrac{1}{64}$

$\left(x - \dfrac{1}{8}\right)^2 = \dfrac{81}{64}$

$x - \dfrac{1}{8} = \pm\dfrac{9}{8}$

$x = \dfrac{1}{8} \pm \dfrac{9}{8}$

$x = \dfrac{\textbf{5}}{\textbf{4}}, \textbf{-1}$

8. $3x^2 - 4 = -2x$

$x^2 + \dfrac{2}{3}x - \dfrac{4}{3} = 0$

$\left(x^2 + \dfrac{2}{3}x + \quad\right) = \dfrac{4}{3}$

$x^2 + \dfrac{2}{3}x + \dfrac{1}{9} = \dfrac{4}{3} + \dfrac{1}{9}$

$\left(x + \dfrac{1}{3}\right)^2 = \dfrac{13}{9}$

$x + \dfrac{1}{3} = \pm\dfrac{\sqrt{13}}{3}$

$x = -\dfrac{\textbf{1}}{\textbf{3}} \pm \dfrac{\sqrt{\textbf{13}}}{\textbf{3}}$

9. $x = 176 - 100 = \textbf{76}$

$y = \dfrac{360 - 176}{2} = \textbf{92}$

$z = \dfrac{100 + 76}{2} = \textbf{88}$

10. $\dfrac{x - a}{p} - c = \dfrac{y}{k}$

$kx - ak - ckp = py$

$kx - ak = py + ckp$

$\dfrac{\textbf{kx} - \textbf{ak}}{\textbf{y} + \textbf{ck}} = p$

11. $\dfrac{a}{x - p} - c = \dfrac{y}{k}$

$ak - ckx + ckp = xy - py$

$ckp + py = xy + ckx - ak$

$p = \dfrac{\textbf{xy} + \textbf{ckx} - \textbf{ak}}{\textbf{ck} + \textbf{y}}$

12. $A = 4\cos 40°$
$A \approx 3.06$

$B = 4\sin 40°$
$B \approx 2.57$

−3.06R − 2.57U

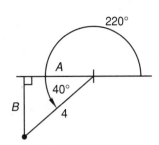

13. $A = 10\cos 45°$
$A \approx 7.07$

$B = 10\sin 45°$
$B \approx 7.07$

7.07R − 7.07U

14. $\dfrac{70 \text{ m}}{\text{sec}} \times \dfrac{100 \text{ cm}}{1 \text{ m}} \times \dfrac{1 \text{ in.}}{2.54 \text{ cm}} \times \dfrac{1 \text{ ft}}{12 \text{ in.}}$

$\times \dfrac{1 \text{ mi}}{5280 \text{ ft}} \times \dfrac{60 \text{ sec}}{1 \text{ min}} \times \dfrac{60 \text{ min}}{1 \text{ hr}}$

$= \dfrac{70(100)(60)(60)}{(2.54)(12)(5280)} \dfrac{\text{mi}}{\text{hr}}$

15. $40 \text{ ft}^3 \times \dfrac{12 \text{ in.}}{1 \text{ ft}} \times \dfrac{12 \text{ in.}}{1 \text{ ft}} \times \dfrac{12 \text{ in.}}{1 \text{ ft}} \times \dfrac{2.54 \text{ cm}}{1 \text{ in.}}$

$\times \dfrac{2.54 \text{ cm}}{1 \text{ in.}} \times \dfrac{2.54 \text{ cm}}{1 \text{ in.}}$

$= \mathbf{40(12)(12)(12)(2.54)(2.54)(2.54) \text{ cm}^3}$

16. $\sqrt[4]{9}\sqrt[5]{3} = \sqrt[4]{3^2}\sqrt[5]{3} = (3^2)^{1/4}3^{1/5}$
$= 3^{1/2}3^{1/5} = \mathbf{3^{7/10}}$

17. $\sqrt[7]{4\sqrt{2}} = \sqrt[7]{2^2\sqrt{2}} = [2^2(2^{1/2})]^{1/7}$
$= 2^{2/7}2^{1/14} = \mathbf{2^{5/14}}$

18. $\sqrt{x^2 y^3}\sqrt[3]{xy^5} = (x^2 y^3)^{1/2}(xy^5)^{1/3}$
$= xy^{3/2}x^{1/3}y^{5/3} = \mathbf{x^{4/3}y^{19/6}}$

19. $-16^{-3/4} = -\dfrac{1}{16^{3/4}} = -\dfrac{1}{(16^{1/4})^3} = \mathbf{-\dfrac{1}{8}}$

20. $5\sqrt{\dfrac{5}{11}} - 2\sqrt{\dfrac{11}{5}} + 3\sqrt{220}$

$= \dfrac{5\sqrt{5}}{\sqrt{11}}\dfrac{\sqrt{11}}{\sqrt{11}} - \dfrac{2\sqrt{11}}{\sqrt{5}}\dfrac{\sqrt{5}}{\sqrt{5}} + 6\sqrt{55}$

$= \dfrac{25\sqrt{55}}{55} - \dfrac{22\sqrt{55}}{55} + \dfrac{330\sqrt{55}}{55} = \mathbf{\dfrac{333\sqrt{55}}{55}}$

21. $5i^3 - 6i^8 + 2\sqrt{-25} - 4i^2$
$= 5i(ii) - 6(ii)(ii)(ii)(ii) + 10i - 4(ii)$
$= -5i - 6 + 10i + 4 = \mathbf{-2 + 5i}$

22. $\dfrac{(40,621,857)(6,031,824)}{19,610 \times 10^{-24}}$

$\approx \dfrac{(4 \times 10^7)(6 \times 10^6)}{2 \times 10^{-20}} \approx \mathbf{1 \times 10^{34}}$

23. $\dfrac{(x+8)(x+2)}{(x+8)(x+3)} \cdot \dfrac{(x+5)(x+3)}{(x+5)(x-5)}$

$= \mathbf{\dfrac{x+2}{x-5}}$

24. $V_{\text{Cylinder}} = A_{\text{Base}} \times \text{height} = (72 - 6\pi) \text{ cm}^3$

$A_{\text{Base}} = \dfrac{(72 - 6\pi) \text{ cm}^3}{3 \text{ cm}}$

$(6)(4) - \dfrac{1}{2}\pi r^2 = 24 - 2\pi \text{ cm}^2$

$\dfrac{1}{2}\pi r^2 = 2\pi \text{ cm}^2$

$r^2 = 4 \text{ cm}^2$

$r = \mathbf{2 \text{ cm}}$

25. $\dfrac{\dfrac{x^2 a^2}{m} - \dfrac{4}{p^3}}{\dfrac{xa}{mp^3} - 5}$

$= \dfrac{\dfrac{x^2 a^2 p^3 - 4m}{mp^3}}{\dfrac{xa - 5mp^3}{mp^3}} \cdot \dfrac{\dfrac{mp^3}{xa - 5mp^3}}{\dfrac{mp^3}{xa - 5mp^3}}$

$= \mathbf{\dfrac{x^2 a^2 p^3 - 4m}{xa - 5mp^3}}$

26. (a) $3x + 4y = -4$
$4y = -3x - 4$
$y = -\dfrac{3}{4}x - 1$

(b) $x - 5y = 10$
$-5y = -x + 10$
$y = \dfrac{1}{5}x - 2$

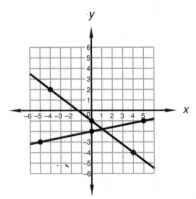

Solve the equations by elimination:

(a) $\quad 3x + 4y = -4$

-3(b) $-3x + 15y = -30$

$\overline{\qquad\qquad 19y = -34}$

$\qquad\qquad y = -\dfrac{34}{19}$

(b) $x = 5\left(-\dfrac{34}{19}\right) + 10 = \dfrac{20}{19}$

$\left(\dfrac{20}{19},\ -\dfrac{34}{19}\right)$

27.

$$x + 2 \overline{)\,4x^3 + 0x^2 + 0x - 5} \qquad 4x^2 - 8x + 16 - \dfrac{37}{x+2}$$

$\qquad\quad \underline{4x^3 + 8x^2}$

$\qquad\qquad\quad -8x^2 + 0x$

$\qquad\qquad\quad \underline{-8x^2 - 16x}$

$\qquad\qquad\qquad\quad 16x - 5$

$\qquad\qquad\qquad\quad \underline{16x + 32}$

$\qquad\qquad\qquad\qquad\quad -37$

28. $\quad \dfrac{3 - x}{2} - \dfrac{4x}{3} = 7$

$\qquad 9 - 3x - 8x = 42$

$\qquad\qquad -11x = 33$

$\qquad\qquad\quad x = -3$

29. $\quad x^0 - 2x - 5(x - 3^0) = -2x^0 - 7$

$\qquad 1 - 2x - 5x + 5 = -2 - 7$

$\qquad\qquad\qquad -7x = -15$

$\qquad\qquad\qquad\quad x = \dfrac{15}{7}$

30. $\quad \dfrac{b}{a(y + 1)} + \dfrac{cx}{a^2(y + 1)}$

$= \dfrac{ab}{a^2(y + 1)} + \dfrac{cx}{a^2(y + 1)} = \dfrac{ab + cx}{a^2(y + 1)}$

PROBLEM SET 59

1. $\quad N \qquad N + 1 \qquad N + 2 \qquad N + 3$

$\qquad\qquad N(N + 3) = 10(-N - 2) - 22$

$\qquad\qquad N^2 + 3N = -10N - 42$

$N^2 + 13N + 42 = 0$

$(N + 7)(N + 6) = 0$

$\qquad\qquad\qquad N = -7, -6$

The desired integers are $-7, -6, -5, -4$
and $-6, -5, -4, -3$.

2. Carbon: $\quad 2 \times 12 = 24$
   Hydrogen: $\ 6 \times 1 = 6$
   Oxygen: $\quad 1 \times 16 = 16$
   Total: $\qquad\qquad = 46$

$\qquad \dfrac{16}{46} = \dfrac{Ox}{460}$

$46Ox = 16(460)$

$\qquad Ox = \textbf{160 grams}$

3. $\quad R_R T_R + R_W T_W = 41;$
   $\quad R_R = 8;\ R_W = 3;$
   $\quad T_R + T_W = 7$

$\qquad 8T_R + 3(7 - T_R) = 41$

$\qquad\qquad\qquad 5T_R = 20$

$\qquad\qquad\qquad\ T_R = 4\ \text{hr}$

$\qquad T_W = 7 - (4) = 3\ \text{hr}$

$\qquad D_R = 4(8) = \textbf{32 miles}$

$\qquad D_W = 3(3) = \textbf{9 miles}$

4. Iodine P_N + Iodine D_N = Iodine Total
   $\qquad 0.9(P_N) + 0.7(D_N) = 0.78(100)$

(a) $0.9P_N + 0.7D_N = 78$

(b) $P_N + D_N = 100$

Substitute $D_N = 100 - P_N$ into (a) and get:

(a′) $0.9P_N + 0.7(100 - P_N) = 78$

$\qquad\qquad\qquad 0.2P_N = 8$

$\qquad\qquad\qquad\ P_N = 40$

(b) $(40) + D_N = 100$

$\qquad\qquad D_N = 60$

40 liters 90%, 60 liters 70%

5. $\qquad P_1 V_1 = P_2 V_2$
   $\qquad (14)(10) = (20)V_2$
   $\qquad\qquad V_2 = \textbf{7 liters}$

6. $\quad Zr = mCa + b$
   Use the graph to find the slope.

Slope $= \dfrac{-10}{7} = -1.43$

$Zr = -1.43Ca + b$
Use the point $(100, 18)$ for Ca and Zr.

$\qquad 18 = -1.43(100) + b$

$161 = b$

$\textbf{Zr} = \textbf{--1.43Ca + 161}$

7. (a) $x = \dfrac{78 + 42}{2} = \textbf{60}$

(b) $x = \dfrac{258 - 102}{2} = \textbf{78}$

8. (a) $\dfrac{x}{3} + \dfrac{3y}{4} = -\dfrac{1}{4}$

    (b) $0.04x - 0.2y = 1.13$

    (a′) $4x + 9y = -3$
    (b′) $4x - 20y = 113$

    $\begin{array}{r} \text{(a′)} \quad 4x + 9y = -3 \\ -1\text{(b′)} \; -4x + 20y = -113 \\ \hline 29y = -116 \\ y = -4 \end{array}$

    (a′) $4x + 9(-4) = -3$
    $4x = 33$
    $x = \dfrac{33}{4}$

    $\left(\dfrac{33}{4}, -4\right)$

9. $\tan\theta = \dfrac{2}{4}$
    $\theta = 26.57°$
    Since θ is a third-quadrant angle:
    $\theta = 26.578 + 180 = 206.57°$
    $H = \sqrt{2^2 + 4^2} = 2\sqrt{5}$

    $\underline{2\sqrt{5}/206.57°}$

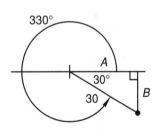

10. $A = 30\cos 30°$
    $A \approx 25.98$

    $B = 30\sin 30°$
    $B = 15$

    $\mathbf{25.98R - 15U}$

11. $2x^2 = x + 5$
    $\left(x^2 - \dfrac{1}{2}x + \quad\right) = \dfrac{5}{2}$
    $x^2 - \dfrac{1}{2}x + \dfrac{1}{16} = \dfrac{5}{2} + \dfrac{1}{16}$
    $\left(x - \dfrac{1}{4}\right)^2 = \dfrac{41}{16}$
    $x - \dfrac{1}{4} = \pm\dfrac{\sqrt{41}}{4}$
    $x = \dfrac{1}{4} \pm \dfrac{\sqrt{41}}{4}$

12. $3x^2 - 5 = 2x$
    $\left(x^2 - \dfrac{2}{3}x + \quad\right) = \dfrac{5}{3}$
    $x^2 - \dfrac{2}{3}x + \dfrac{1}{9} = \dfrac{5}{3} + \dfrac{1}{9}$

$\left(x - \dfrac{1}{3}\right)^2 = \dfrac{16}{9}$
$x - \dfrac{1}{3} = \pm\dfrac{4}{3}$
$x = \dfrac{1}{3} \pm \dfrac{4}{3}$
$x = \dfrac{5}{3}, -1$

13. $3x^2 - 4 = x$
    $\left(x^2 - \dfrac{1}{3}x + \quad\right) = \dfrac{4}{3}$
    $x^2 - \dfrac{1}{3}x + \dfrac{1}{36} = \dfrac{4}{3} + \dfrac{1}{36}$
    $\left(x - \dfrac{1}{6}\right)^2 = \dfrac{49}{36}$
    $x - \dfrac{1}{6} = \pm\dfrac{7}{6}$
    $x = \dfrac{1}{6} \pm \dfrac{7}{6}$
    $x = \dfrac{4}{3}, -1$

14. $3^2 = \left(\dfrac{\sqrt{11}}{2}\right)^2 + H^2$
    $9 = \dfrac{11}{4} + H^2$
    $\dfrac{25}{4} = H^2$
    $\dfrac{5}{2} = H$
    $\text{Area} = \dfrac{(b \times H)}{2}$
    $= \dfrac{\sqrt{11} \times \dfrac{5}{2}}{2} \text{ cm}^2 = \dfrac{5\sqrt{11}}{4} \text{ cm}^2$

15. $\dfrac{p - s}{m} - \dfrac{c}{4} + x = 0$
    $4p - 4s - cm + 4mx = 0$
    $4p - 4s = cm - 4mx$
    $\dfrac{4p - 4s}{c - 4x} = m$

16. $\dfrac{m}{p + s} - \dfrac{c}{x} + 4 = 0$
    $mx - cp - cs + 4px + 4sx = 0$
    $4px - cp = cs - mx - 4sx$
    $p = \dfrac{cs - mx - 4sx}{4x - c}$

17. $\dfrac{40 \text{ in.}}{\text{sec}} \times \dfrac{2.54 \text{ cm}}{1 \text{ in.}} \times \dfrac{1 \text{ m}}{100 \text{ cm}} \times \dfrac{60 \text{ sec}}{1 \text{ min}}$
    $\times \dfrac{60 \text{ min}}{1 \text{ hr}} = \dfrac{40(2.54)(60)(60)}{100} \dfrac{\text{m}}{\text{hr}}$

18. $18{,}000 \text{ cm}^3 \times \dfrac{1 \text{ in.}}{2.54 \text{ cm}} \times \dfrac{1 \text{ in.}}{2.54 \text{ cm}} \times \dfrac{1 \text{ in.}}{2.54 \text{ cm}}$

$\times \dfrac{1 \text{ ft}}{12 \text{ in.}} \times \dfrac{1 \text{ ft}}{12 \text{ in.}} \times \dfrac{1 \text{ ft}}{12 \text{ in.}}$

$= \dfrac{18{,}000}{(2.54)(2.54)(2.54)(12)(12)(12)} \text{ ft}^3$

19. $i^4 + 5 + 3\sqrt{-9} - 2\sqrt{-4} = (ii)(ii) + 5 + 9i - 4i$
$= 1 + 5 + 9i - 4i = \mathbf{6 + 5i}$

20. $3i^5 - 2i^2 - 4 - i = 3i(ii)(ii) - 2(ii) - 4 - i$
$= 3i + 2 - 4 - i = \mathbf{-2 + 2i}$

21. $\sqrt{x^2 - 4x + 39} + 1 = x + 4$
$\qquad x^2 - 4x + 39 = x^2 + 6x + 9$
$\qquad\qquad -10x = -30$
$\qquad\qquad\qquad x = 3$

Check: $\sqrt{9 - 12 + 39} + 1 = 3 + 4$
$\qquad\qquad\qquad 6 + 1 = 3 + 4 \quad$ check

22. $\sqrt{x - 3} + 2 = -4$
$\qquad x - 3 = 36$
$\qquad\quad x = 39$

Check: $\sqrt{39 - 3} + 2 = -4$
$\qquad\qquad 6 + 2 = -4 \quad$ not true
No real number solution.

23. $\sqrt[5]{3\sqrt{3}} = [3(3^{1/2})]^{1/5} = 3^{1/5}3^{1/10} = \mathbf{3^{3/10}}$

24. $\sqrt[4]{4\sqrt{2}} = \sqrt[4]{2^2 \sqrt{2}} = [2^2(2^{1/2})]^{1/4}$
$= 2^{1/2}2^{1/8} = \mathbf{2^{5/8}}$

25. $-16^{-3/4} = -\dfrac{1}{16^{3/4}} = -\dfrac{1}{\left(16^{1/4}\right)^3} = \mathbf{-\dfrac{1}{8}}$

26. $\sqrt{x^2 y}\,\sqrt[3]{y^5 x^4} = (x^2 y)^{1/2}(y^5 x^4)^{1/3}$
$= xy^{1/2}y^{5/3}x^{4/3} = \mathbf{x^{7/3}y^{13/6}}$

27. $\sqrt{\dfrac{2}{13}} - 4\sqrt{\dfrac{13}{2}} + 3\sqrt{234}$

$= \dfrac{\sqrt{2}}{\sqrt{13}}\dfrac{\sqrt{13}}{\sqrt{13}} - \dfrac{4\sqrt{13}}{\sqrt{2}}\dfrac{\sqrt{2}}{\sqrt{2}} + 9\sqrt{26}$

$= \dfrac{2\sqrt{26}}{26} - \dfrac{52\sqrt{26}}{26} + \dfrac{234\sqrt{26}}{26} = \mathbf{\dfrac{92\sqrt{26}}{13}}$

28. $\dfrac{x - 4}{2} - \dfrac{x}{3} - 2 = \dfrac{5}{2}$
$3x - 12 - 2x - 12 = 15$
$\qquad\qquad\qquad x = \mathbf{39}$

29. $\dfrac{x}{y} + \dfrac{x^2 + 2}{y^2 a} + \dfrac{x^3}{y(a + y)}$

$= \dfrac{axy(a + y)}{y^2 a(a + y)} + \dfrac{(x^2 + 2)(a + y)}{y^2 a(a + y)}$

$\quad + \dfrac{ax^3 y}{y^2 a(a + y)}$

$= \dfrac{axy(a + y) + (x^2 + 2)(a + y) + ax^3 y}{y^2 a(a + y)}$

30. $\dfrac{(x^0 yp)^{-3} x^3 yp^0 p^{-2}}{x^2 p^2 y^0 (y^{-3} p)^2 x^{-3} y^0 p^2} = \dfrac{y^{-3} p^{-3} x^3 yp^{-2}}{x^2 p^2 y^{-6} p^2 x^{-3} p^2}$

$= \dfrac{p^{-5} x^3 y^{-2}}{x^{-1} p^6 y^{-6}} = \mathbf{p^{-11}x^4 y^4}$

PROBLEM SET 60

1. $B = kG$
$\quad (8) = k(2)$
$\quad\;\; 4 = k$

$\quad B = 4(7) = \mathbf{28}$

2. $\text{RPM} = \dfrac{k}{N_t}$

$\quad\; 100 = \dfrac{k}{40}$

$\quad 4000 = k$

$\quad \text{RPM} = \dfrac{4000}{25} = \mathbf{160}$

3. $\quad C = kP$
$\quad\; 40 = k(20{,}000)$
$\; 0.002 = k$

$\quad C = 0.002(8000) = \mathbf{16}$

4. $\dfrac{P_1}{T_1} = \dfrac{P_2}{T_2}$

$\dfrac{400}{500} = \dfrac{P_2}{1000}$
$P_2(500) = 400(1000)$
$\quad P_2 = \mathbf{800 \text{ N/m}^2}$

5. (a) $N_P = N_T + 50$
(b) $2N_P = 3N_T + 60$
Substitute (a) into (b) and get:
(b′) $2(N_T + 50) = 3N_T + 60$
$\qquad\qquad N_T = \mathbf{40}$
(a) $N_P = (40) + 50 = \mathbf{90}$

6. $A = 4 \cos 45°$
$A \approx 2.83$

$B = 4 \sin 45°$
$B \approx 2.83$

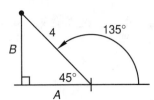

$-2.83R + 2.83U$

7. $\tan \theta = \dfrac{4}{2}$
$\theta = 63.43°$
Since θ is a second-quadrant angle:
$\theta = 180 - 63.43 = 116.57°$
$H = \sqrt{2^2 + 4^2} = 2\sqrt{5}$

$\underline{2\sqrt{5}/116.57°}$

8. (a) $\dfrac{2}{5}x + \dfrac{3}{2}y = 34$
(b) $0.02x + 0.3y = 6.2$

(a') $4x + 15y = 340$
(b') $2x + 30y = 620$

$\begin{array}{r} -2(a') \quad -8x - 30y = -680 \\ (b') \quad \underline{2x + 30y = 620} \\ -6x = -60 \\ x = 10 \end{array}$

(a') $4(10) + 15y = 340$
$y = 20$
(10, 20)

9. $4x^2 - 3 = x$

$\left(x^2 - \dfrac{1}{4}x + \right) = \dfrac{3}{4}$

$x^2 - \dfrac{1}{4}x + \dfrac{1}{64} = \dfrac{3}{4} + \dfrac{1}{64}$

$\left(x - \dfrac{1}{8}\right)^2 = \dfrac{49}{64}$

$x - \dfrac{1}{8} = \pm\dfrac{7}{8}$

$x = \dfrac{1}{8} \pm \dfrac{7}{8}$

$x = 1, -\dfrac{3}{4}$

10. $3x^2 = 2x + 1$

$\left(x^2 - \dfrac{2}{3}x + \right) = \dfrac{1}{3}$

$x^2 - \dfrac{2}{3}x + \dfrac{1}{9} = \dfrac{1}{3} + \dfrac{1}{9}$

$\left(x - \dfrac{1}{3}\right)^2 = \dfrac{4}{9}$

$x - \dfrac{1}{3} = \pm\dfrac{2}{3}$

$x = \dfrac{1}{3} \pm \dfrac{2}{3}$

$x = 1, -\dfrac{1}{3}$

11. $Na = mC + b$
Use the graph to find the slope.

$m = \dfrac{3}{80} = 0.037$

$Na = 0.037C + b$
By inspection $b = 3.9$.
$Na = 0.037C + 3.9$

12. (a) $6x + 10y + 30 = 110$
(b) $4x + 10y + 10 = 70$

$\begin{array}{r} 1(a) \quad 6x + 10y = 80 \\ -1(b) \quad \underline{-4x - 10y = -60} \\ 2x = 20 \\ x = 10 \end{array}$

(a) $6(10) + 10y = 80$
$y = 2$

13.
$\dfrac{a}{x + y} = c + \dfrac{m}{d}$

$ad = cdx + cdy + mx + my$

$ad - cdy - my = cdx + mx$

$\dfrac{ad - cdy - my}{cd + m} = x$

14.
$\dfrac{xy + c}{m} + d = \dfrac{k}{z}$

$xyz + cz + dmz = km$

$cz = km - xyz - dmz$

$c = \dfrac{km - xyz - dmz}{z}$

15. $3i^3 - \sqrt{-4} + 3\sqrt{-9} - 2 + i^2$
$= 3i(ii) - 2i + 9i - 2 + (ii)$
$= -3i - 2i + 9i - 2 - 1 = \mathbf{-3 + 4i}$

16. $-3i^4 - 2i^2 + 2 - 3\sqrt{-25}$
$= -3(ii)(ii) - 2(ii) + 2 - 15i$
$= -3 + 2 + 2 - 15i = \mathbf{1 - 15i}$

17. $\dfrac{60 \text{ cm}}{\text{sec}} \times \dfrac{1 \text{ in.}}{2.54 \text{ cm}} \times \dfrac{1 \text{ ft}}{12 \text{ in.}} \times \dfrac{1 \text{ mi}}{5280 \text{ ft}}$

$\times \dfrac{60 \text{ sec}}{1 \text{ min}} \times \dfrac{60 \text{ min}}{1 \text{ hr}} = \dfrac{60(60)(60)}{(2.54)(12)(5280)} \dfrac{\text{mi}}{\text{hr}}$

18. $1,400,000 \text{ cm}^3 \times \dfrac{1 \text{ in.}}{2.54 \text{ cm}} \times \dfrac{1 \text{ in.}}{2.54 \text{ cm}}$

$\times \dfrac{1 \text{ in.}}{2.54 \text{ cm}} \times \dfrac{1 \text{ ft}}{12 \text{ in.}} \times \dfrac{1 \text{ ft}}{12 \text{ in.}} \times \dfrac{1 \text{ ft}}{12 \text{ in.}}$

$\times \dfrac{1 \text{ yd}}{3 \text{ ft}} \times \dfrac{1 \text{ yd}}{3 \text{ ft}} \times \dfrac{1 \text{ yd}}{3 \text{ ft}}$

$= \dfrac{1,400,000}{(2.54)(2.54)(2.54)(12)(12)(12)(3)(3)(3)} \text{ yd}^3$

19. $\sqrt[3]{27\sqrt{3}} = \sqrt[3]{3^3 \sqrt{3}} = [3^3(3^{1/2})]^{1/3}$

$= 3 \cdot 3^{1/6} = \mathbf{3^{7/6}}$

20. $\sqrt[4]{81}\sqrt[3]{3} = \sqrt[4]{3^4}\sqrt[3]{3} = (3^4)^{1/4}(3^{1/3})$

$= 3 \cdot 3^{1/3} = \mathbf{3^{4/3}}$

21. $81^{-3/4} = \dfrac{1}{81^{3/4}} = \dfrac{1}{\left(81^{1/4}\right)^3} = \mathbf{\dfrac{1}{27}}$

22. $\sqrt{m^2 p}\sqrt[3]{m^5 p^4} = (m^2 p)^{1/2}(m^5 p^4)^{1/3}$

$= mp^{1/2}m^{5/3}p^{4/3} = \mathbf{m^{8/3}p^{11/6}}$

23. $5\sqrt{\dfrac{2}{9}} + 3\sqrt{\dfrac{9}{2}} - 5\sqrt{8}$

$= \dfrac{5\sqrt{2}}{3} + \dfrac{9}{\sqrt{2}}\dfrac{\sqrt{2}}{\sqrt{2}} - 10\sqrt{2}$

$= \dfrac{10\sqrt{2}}{6} + \dfrac{27\sqrt{2}}{6} - \dfrac{60\sqrt{2}}{6} = \mathbf{-\dfrac{23\sqrt{2}}{6}}$

24. $\dfrac{(5,162,348)(0.0000165)}{0.003217642}$

$\approx \dfrac{(5 \times 10^6)(2 \times 10^{-5})}{3 \times 10^{-3}} \approx \mathbf{3 \times 10^4}$

25. $15 \times \overrightarrow{SF} = 20$

$\overrightarrow{SF} = \dfrac{4}{3}$

$8 \cdot \dfrac{4}{3} = K$

$\dfrac{32}{3} = K$

26. $x = 180 - 90 - 40 = \mathbf{50}$

$k = 180 - 90 - 50 = \mathbf{40}$

$s = 180 - 90 - 40 = \mathbf{50}$

27. (a) $2x - 3y = -3$

(b) $2x + y = 8$

Substitute $y = -2x + 8$ into (a) and get:

(a') $2x - 3(-2x + 8) = -3$

$8x = 21$

$x = \dfrac{21}{8}$

(b) $y = -2\left(\dfrac{21}{8}\right) + 8 = \dfrac{11}{4}$

$\left(\dfrac{21}{8}, \dfrac{11}{4}\right)$

28. $x^3 + 50x = 15x^2$

$x^3 - 15x^2 + 50x = 0$

$x(x - 10)(x - 5) = 0$

$x = \mathbf{0, 5, 10}$

29. $y = \dfrac{2}{5}x + b$

Use the point $(-40, 2)$ for x and y.

$2 = \dfrac{2}{5}(-40) + b$

$18 = b$

$y = \dfrac{2}{5}x + 18$

30. $\dfrac{3x^{-4}y^0 y^2}{x^2 x}\left(\dfrac{2x^2 y^2}{y^4} - \dfrac{y^0 x^0 p^{-2}}{x^4 y^4}\right)$

$= \dfrac{6x^{-2}y^4}{x^3 y^4} - \dfrac{3x^{-4}p^{-2}y^2}{x^7 y^4} = \mathbf{6x^{-5} - 3x^{-11}p^{-2}y^{-2}}$

PROBLEM SET 61

1. $R_D = kS$

$0.005 = k(5)$

$0.001 = k$

$R_D = 0.001(0.3) = \mathbf{0.0003 \text{ kg per second}}$

2. $\dfrac{P_1}{T_1} = \dfrac{P_2}{T_2}$

$\dfrac{P_1}{1000} = \dfrac{300}{600}$

$P_1 = \dfrac{300(1000)}{600}$

$P_1 = \mathbf{500 \text{ N/m}^2}$

3. Water one − water out = water final

$W_1 - W_0 = W_F$

$(100) - (E) = (100 - E)$

$0.9(100) - (E) = 0.8(100 - E)$

$90 - E = 80 - 0.8E$

$-0.2E = -10$

$E = \mathbf{50 \text{ gallons}}$

4. Butterfat one + butterfat added = butterfat final

$B_1 + B_A = B_F$

$(900) + (P_N) = (900 + P_N)$

$0.02(900) + (P_N) = 0.1(900 + P_N)$

$18 + P_N = 90 + 0.1P_N$

$0.9P_N = 72$

$P_N = \mathbf{80 \text{ pounds}}$

5. Glycol$_1$ + glycol added = glycol final
$$(100) + (P_N) = (100 + P_N)$$
$$0.2(100) + 0.3(P_N) = 0.25(100 + P_N)$$
$$20 + 0.3P_N = 25 + 0.25P_N$$
$$0.05P_N = 5$$
$$P_N = \textbf{100 kilograms}$$

6. $A = 20 \cos 15°$
$A \approx 19.32$

$B = 20 \sin 15°$
$B \approx 5.18$

$\textbf{–19.32R + 5.18U}$

7. $\tan \theta = \dfrac{2}{6}$

$\theta \approx 18.43°$
Since θ is a fourth-quadrant angle:
$\theta = 360 - 18.43 = 341.57°$

$H = \sqrt{6^2 + 2^2} = 2\sqrt{10}$

$\textbf{2}\sqrt{\textbf{10}}/\textbf{341.57°}$

8. (a) $\dfrac{2}{3}x - \dfrac{2}{5}y = -4$
(b) $0.2x + 0.9y = 19.2$

(a′) $10x - 6y = -60$
(b′) $2x + 9y = 192$

(a′)$$ $10x - 6y = -60$
$-5(b')$ $-10x - 45y = -960$
$- 51y = -1020$
$y = 20$

(a′) $10x - 6(20) = -60$
$x = 6$

(6, 20)

9. $$-4x - 4 = -5x^2$$
$$\left(x^2 - \dfrac{4}{5}x + \right) = \dfrac{4}{5}$$
$$x^2 - \dfrac{4}{5}x + \dfrac{4}{25} = \dfrac{4}{5} + \dfrac{4}{25}$$
$$\left(x - \dfrac{2}{5}\right)^2 = \dfrac{24}{25}$$
$$x - \dfrac{2}{5} = \pm\dfrac{2\sqrt{6}}{5}$$
$$x = \dfrac{\textbf{2}}{\textbf{5}} \pm \dfrac{\textbf{2}\sqrt{\textbf{6}}}{\textbf{5}}$$

10. $$-x = 7 - 2x^2$$
$$\left(x^2 - \dfrac{1}{2}x + \right) = \dfrac{7}{2}$$
$$x^2 - \dfrac{1}{2}x + \dfrac{1}{16} = \dfrac{7}{2} + \dfrac{1}{16}$$
$$\left(x - \dfrac{1}{4}\right)^2 = \dfrac{57}{16}$$
$$x - \dfrac{1}{4} = \pm\dfrac{\sqrt{57}}{4}$$
$$x = \dfrac{\textbf{1}}{\textbf{4}} \pm \dfrac{\sqrt{\textbf{57}}}{\textbf{4}}$$

11. $Pb = mSb + b$
Use the graph to find the slope.
$m = \dfrac{100}{2.35} = 42.6$
$Pb = 42.6Sb + b$
Use the point (0.75, 0) for Sb and Pb.
$0 = 42.6(0.75) + b$
$-31.9 = b$
$\textbf{Pb = 42.6Sb – 31.9}$

12. $x + 3 = 68$
$x = \textbf{65}$
$y - 12 + 30 + 68 + 62 + 78 + 32 + 54$
$= 360$
$y = \textbf{48}$

13. $$\dfrac{(m + c + x)b}{k} + \dfrac{a}{d} = p$$
$$bdm + bcd + bdx + ka = dkp$$
$$dkp - bdm - bdx - ka = bcd$$
$$\dfrac{\textbf{dkp – bdm – bdx – ka}}{\textbf{bd}} = c$$

14. $$\dfrac{4y}{2a + x} + \dfrac{m}{c} = d$$
$$4cy + 2am + mx = 2acd + cdx$$
$$2am - 2acd = cdx - 4cy - mx$$
$$a = \dfrac{\textbf{cdx – 4cy – mx}}{\textbf{2m – 2cd}}$$

15. $-2i^2 - 3i - 4 - 2\sqrt{-4} = -2(ii) - 3i - 4 - 4i$
$= 2 - 3i - 4 - 4i = \textbf{–2 – 7i}$

16. $8i^4 - 2i^3 - 2i - 6 - 4\sqrt{-16}$
$= 8(ii)(ii) - 2i(ii) - 2i - 6 - 16i$
$= 8 + 2i - 2i - 6 - 16i = \textbf{2 – 16i}$

17. $\dfrac{40 \text{ cm}^3}{\text{sec}} \times \dfrac{1 \text{ in.}}{2.54 \text{ cm}} \times \dfrac{1 \text{ in.}}{2.54 \text{ cm}} \times \dfrac{1 \text{ in.}}{2.54 \text{ cm}}$

$ \times \dfrac{60 \text{ sec}}{1 \text{ min}} \times \dfrac{60 \text{ min}}{1 \text{ hr}}$

$= \dfrac{\textbf{40(60)(60)}}{\textbf{(2.54)(2.54)(2.54)}} \dfrac{\text{in.}^3}{\text{hr}}$

18. $4 \text{ ft}^3 \times \dfrac{12 \text{ in.}}{1 \text{ ft}} \times \dfrac{12 \text{ in.}}{1 \text{ ft}} \times \dfrac{12 \text{ in.}}{1 \text{ ft}} \times \dfrac{2.54 \text{ cm}}{1 \text{ in.}}$

$\times \dfrac{2.54 \text{ cm}}{1 \text{ in.}} \times \dfrac{2.54 \text{ cm}}{1 \text{ in.}}$

$= \mathbf{4(12)(12)(12)(2.54)(2.54)(2.54) \text{ cm}^3}$

19. $\sqrt{32\sqrt{2}} = \sqrt{2^5\sqrt{2}} = [2^5(2^{1/2})]^{1/2}$
$= 2^{5/2}2^{1/4} = \mathbf{2^{11/4}}$

20. $\sqrt[3]{8\sqrt[3]{2}} = \sqrt[3]{2^3\sqrt[3]{2}} = [2^3(2^{1/3})]^{1/3}$
$= 2 \cdot 2^{1/9} = \mathbf{2^{10/9}}$

21. $-8^{-5/3} = -\dfrac{1}{8^{5/3}} = -\dfrac{1}{\left(8^{1/3}\right)^5} = -\dfrac{\mathbf{1}}{\mathbf{32}}$

22. $\sqrt{x^5 y}\sqrt[4]{y^2 x} = (x^5 y)^{1/2}(y^2 x)^{1/4}$
$= x^{5/2}y^{1/2}y^{1/2}x^{1/4} = \mathbf{x^{11/4}y}$

23. $3\sqrt{\dfrac{2}{3}} - 4\sqrt{\dfrac{3}{2}} + 8\sqrt{24}$

$= \dfrac{3\sqrt{2}}{\sqrt{3}}\dfrac{\sqrt{3}}{\sqrt{3}} - \dfrac{4\sqrt{3}}{\sqrt{2}}\dfrac{\sqrt{2}}{\sqrt{2}} + 16\sqrt{6}$

$= \dfrac{6\sqrt{6}}{6} - \dfrac{12\sqrt{6}}{6} + \dfrac{96\sqrt{6}}{6} = \mathbf{15\sqrt{6}}$

24. $\dfrac{(41,685,231)(0.0012846 \times 10^{-14})}{0.001998 \times 10^{-10}}$

$\approx \dfrac{(4 \times 10^7)(1 \times 10^{-17})}{2 \times 10^{-13}} = \mathbf{2 \times 10^3}$

25. $A^2 = 5^2 + 12^2$
$A^2 = 169$
$A = 13$

$5 \times \overrightarrow{SF} = 4$
$\overrightarrow{SF} = \dfrac{4}{5}$

$13 \cdot \dfrac{4}{5} = C$
$\dfrac{\mathbf{52}}{\mathbf{5}} = C$

26. $\dfrac{r + 1}{s} = \dfrac{t - 1}{v}$
$rv + v = st - s$
$rv = st - s - v$
$r = \dfrac{\mathbf{st - s - v}}{\mathbf{v}}$

27. (a) $x - 2y = -6$
$-2y = -x - 6$
$y = \dfrac{1}{2}x + 3$

(b) $x + y = -1$
$y = -x - 1$

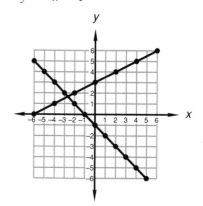

Substitute $y = -x - 1$ into (a) and get:
(a') $x - 2(-x - 1) = -6$
$x = -\dfrac{8}{3}$

(b) $\left(-\dfrac{8}{3}\right) + y = -1$
$y = \dfrac{5}{3}$

$\left(-\dfrac{\mathbf{8}}{\mathbf{3}}, \dfrac{\mathbf{5}}{\mathbf{3}}\right)$

28. $-15x = -x^3 + 2x^2$
$x^3 - 2x^2 - 15x = 0$
$x(x - 5)(x + 3) = 0$
$x = \mathbf{0, 5, -3}$

29. Write the equation of the given line in slope-intercept form.
$2x + 3y = 5$
$y = -\dfrac{2}{3}x + \dfrac{5}{3}$

Since the slopes of perpendicular lines are negative reciprocals of each other:

$m_\perp = \dfrac{3}{2}$

$y = \dfrac{3}{2}x + b$

$4 = \dfrac{3}{2}(-2) + b$

$7 = b$

$y = \dfrac{\mathbf{3}}{\mathbf{2}}x + \mathbf{7}$

30. $-2^2 - 3^2 - (2^0)^2 - (-2)^0 = -x(-2x^0 - 5^0)2^2$
$-4 - 9 - 1 - 1 = (2x + x)4$
$-15 = 12x$
$-\dfrac{\mathbf{5}}{\mathbf{4}} = x$

PROBLEM SET 62

1. $N_V = \dfrac{k}{S}$

$8 = \dfrac{k}{2}$

$16 = k$

$N_V = \dfrac{16}{8} = \mathbf{2}$

2. $\dfrac{P_1 V_1}{T_1} = \dfrac{P_2 V_2}{T_2}$

$\dfrac{(600)(2)}{(300)} = \dfrac{(400)(4)}{T_2}$

$T_2 = \mathbf{400\ K}$

3. $S_1 + S_A = S_T$

$(40) + (P_N) = (40 + P_N)$

$0.05(40) + 0.2(P_N) = 0.1(40 + P_N)$

$2 + 0.2P_N = 4 + 0.1P_N$

$0.1P_N = 2$

$P_N = \mathbf{20\ liters}$

4. $R_S T_S + R_T T_T = 152;$

$R_S = 8;\ R_T = 20;$

$T_S + T_T = 10$

$8T_S + 20(10 - T_S) = 152$

$-12T_S = -48$

$T_S = 4$

$D_S = 8(4) = \mathbf{32\ miles}$

5.
Carbon: $1 \times 12 = 12$
Hydrogen: $4 \times 1 = 4$
Oxygen: $1 \times 16 = 16$
Nitrogen: $2 \times 14 = 28$
Total: $= 60$

$\dfrac{N}{Tot} = \dfrac{28}{60} \approx 0.467 = \mathbf{46.7\%}$

6. $A = 20 \cos 20°$

$A \approx 18.79$

$B = 20 \sin 20°$

$B \approx 6.84$

$\mathbf{18.79R - 6.84U}$

7. $\tan \theta = \dfrac{5}{2}$

$\theta = 68.20°$

Since θ is a second-quadrant angle:

$\theta = 180 - 68.20 = 111.80°$

$H = \sqrt{5^2 + 2^2} = \sqrt{29}$

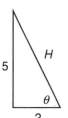

$\mathbf{\sqrt{29}\underline{/111.80°}}$

8. (a) $\dfrac{1}{3}x - \dfrac{2}{3}y = -1$

(b) $0.02x + 0.4y = 2.58$

(a′) $x - 2y = -3$

(b′) $2x + 40y = 258$

$-2(a′)\ -2x + 4y = 6$

$\ \ (b′)\ \ \underline{2x + 40y = 258}$

$44y = 264$

$y = 6$

(a′) $x - 2(6) = -3$

$x = 9$

$\mathbf{(9, 6)}$

9. $-x + 2x^2 + 3 = 0$

$\left(x^2 - \dfrac{1}{2}x + \quad\right) = -\dfrac{3}{2}$

$x^2 - \dfrac{1}{2}x + \dfrac{1}{16} = -\dfrac{3}{2} + \dfrac{1}{16}$

$\left(x - \dfrac{1}{4}\right)^2 = -\dfrac{23}{16}$

$x - \dfrac{1}{4} = \pm\dfrac{\sqrt{23}}{4}i$

$x = \dfrac{1}{4} \pm \dfrac{\sqrt{23}}{4}i$

10. $-5x + 6x^2 = -3$

$\left(x^2 - \dfrac{5}{6}x + \quad\right) = -\dfrac{1}{2}$

$x^2 - \dfrac{5}{6}x + \dfrac{25}{144} = -\dfrac{1}{2} + \dfrac{25}{144}$

$\left(x - \dfrac{5}{12}\right)^2 = -\dfrac{47}{144}$

$x - \dfrac{5}{12} = \pm\dfrac{\sqrt{47}}{12}i$

$x = \dfrac{5}{12} \pm \dfrac{\sqrt{47}}{12}i$

11. $Bi = mHg + b$
Use the graph to find the slope.

$m = \dfrac{24}{1.5} = 16$
$Bi = 16Hg + b$
Use the point (6, 24) for Hg and Bi.
$24 = 16(6) + b$
$-72 = b$
$Bi = 16Hg - 72$

12. Volume $= \pi r^2 h = 11,520\pi$ in.^{3}

$r^2 = \dfrac{11,520\pi \text{ in.}^3}{20\pi \text{ in.}}$

$r = \sqrt{576 \text{ in.}^2} = \textbf{24 in.}$

13. $\dfrac{(a + b)m}{c} - k = \dfrac{p}{r}$
$amr + bmr - ckr = cp$
$bmr = cp + ckr - amr$
$b = \dfrac{\textbf{cp + ckr - amr}}{\textbf{mr}}$

14. $\dfrac{6x}{2y + 4a} - c = \dfrac{p}{r}$
$6xr - 2cry - 4acr = 2py + 4ap$
$6xr - 4acr - 4ap = 2py + 2cry$
$\dfrac{\textbf{3xr - 2acr - 2ap}}{\textbf{p + cr}} = y$

15. $3i^3 - \sqrt{-4} - 2 - \sqrt{9} + 3$
$= 3i(ii) - 2i - 2 - 3 + 3$
$= \textbf{-2 - 5i}$

16. $-3i^3 + 2i - 4 - 3i^2 - 2\sqrt{9}$
$= -3i(ii) + 2i - 4 - 3(ii) - 6$
$= \textbf{-7 + 5i}$

17. $\dfrac{600 \text{ cm}^3}{\text{min}} \times \dfrac{1 \text{ in.}}{2.54 \text{ cm}} \times \dfrac{1 \text{ in.}}{2.54 \text{ cm}} \times \dfrac{1 \text{ in.}}{2.54 \text{ cm}}$

$\times \dfrac{1 \text{ ft}}{12 \text{ in.}} \times \dfrac{1 \text{ ft}}{12 \text{ in.}} \times \dfrac{1 \text{ ft}}{12 \text{ in.}} \times \dfrac{1 \text{ min}}{60 \text{ sec}}$

$= \dfrac{\textbf{600}}{\textbf{(2.54)(2.54)(2.54)(12)(12)(12)(60)}} \dfrac{\textbf{ft}^3}{\textbf{sec}}$

18. $20 \text{ yd}^3 \times \dfrac{3 \text{ ft}}{1 \text{ yd}} \times \dfrac{3 \text{ ft}}{1 \text{ yd}} \times \dfrac{3 \text{ ft}}{1 \text{ yd}} \times \dfrac{12 \text{ in.}}{1 \text{ ft}}$

$\times \dfrac{12 \text{ in.}}{1 \text{ ft}} \times \dfrac{12 \text{ in.}}{1 \text{ ft}} \times \dfrac{2.54 \text{ cm}}{1 \text{ in.}} \times \dfrac{2.54 \text{ cm}}{1 \text{ in.}}$

$\times \dfrac{2.54 \text{ cm}}{1 \text{ in.}}$

$= \textbf{20(3)(3)(3)(12)(12)(12)(2.54)(2.54)(2.54) cm}^3$

19. $\sqrt{16\sqrt{2}} = \sqrt{2^4\sqrt{2}} = [2^4(2^{1/2})]^{1/2}$
$= 2^2(2^{1/4}) = \textbf{2}^{\textbf{9/4}}$

20. $\sqrt[4]{4}\sqrt[5]{2} = \sqrt[4]{2^2}\sqrt[5]{2} = (2^2)^{1/4}2^{1/5}$
$= 2^{1/2}2^{1/5} = \textbf{2}^{\textbf{7/10}}$

21. $-4^{5/2} = -(4^{1/2})^5 = \textbf{-32}$

22. $\sqrt{xy^7}\sqrt[3]{x^5y} = (xy^7)^{1/2}(x^5y)^{1/3}$
$= x^{1/2}y^{7/2}x^{5/3}y^{1/3} = \textbf{x}^{\textbf{13/6}}\textbf{y}^{\textbf{23/6}}$

23. $3\sqrt{\dfrac{2}{5}} - 5\sqrt{\dfrac{5}{2}} - 3\sqrt{40}$

$= \dfrac{3\sqrt{2}}{\sqrt{5}}\dfrac{\sqrt{5}}{\sqrt{5}} - \dfrac{5\sqrt{5}}{\sqrt{2}}\dfrac{\sqrt{2}}{\sqrt{2}} - 6\sqrt{10}$

$= \dfrac{6\sqrt{10}}{10} - \dfrac{25\sqrt{10}}{10} - \dfrac{60\sqrt{10}}{10} = \textbf{-}\dfrac{\textbf{79}\sqrt{\textbf{10}}}{\textbf{10}}$

24. $\dfrac{(0.000618427 \times 10^{14})(7,891,642)}{3,728,196,842}$

$\approx \dfrac{(6 \times 10^{10})(8 \times 10^6)}{4 \times 10^9} \approx \textbf{1} \times \textbf{10}^\textbf{8}$

25. $C^2 = 24^2 + 7^2$
$C^2 = 625$
$C = 25$

$24 \times \overrightarrow{SF} = 6$
$\overrightarrow{SF} = \dfrac{1}{4}$

$25 \times \overrightarrow{SF} = B$
$25 \cdot \dfrac{1}{4} = B$
$\dfrac{25}{4} = B$

26. (a) $x = \dfrac{68 + 58}{2} = \textbf{63}°$

(b) $x = \dfrac{160 - 52}{2} = \textbf{54}°$

27. (a) $2x - 3y = -9$

$-3y = -2x - 9$

$y = \dfrac{2}{3}x + 3$

(b) $2x + 3y = -3$

$3y = -2x - 3$

$y = -\dfrac{2}{3}x - 1$

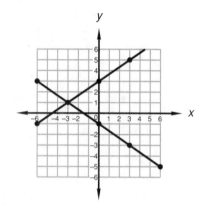

Substitute $y = -\dfrac{2}{3}x - 1$ into (a) and get:

(a') $2x - 3\left(-\dfrac{2}{3}x - 1\right) = -9$

$4x = -12$

$x = -3$

(b) $y = -\dfrac{2}{3}(-3) - 1 = 1$

$(-3, 1)$

28. $50x + x^3 = 15x^2$

$x^3 - 15x^2 + 50x = 0$

$x(x - 10)(x - 5) = 0$

$x = \mathbf{0, 5, 10}$

29. $y = -\dfrac{2}{7}x + b$

$-7 = -\dfrac{2}{7}(-5) + b$

$-\dfrac{59}{7} = b$

$y = -\dfrac{2}{7}x - \dfrac{59}{7}$

30. $\dfrac{3 + x}{2} - \dfrac{2}{7} = 4$

$21 + 7x - 4 = 56$

$7x = 39$

$x = \dfrac{39}{7}$

1. $T = kD$

$20 = k(400)$

$\dfrac{1}{20} = k$

$T = \dfrac{1}{20}(60) = \mathbf{3}$

2. $\dfrac{V_1}{T_1} = \dfrac{V_2}{T_2}$

$\dfrac{20}{800} = \dfrac{12}{T_2}$

$T_2 = \mathbf{480\ K}$

3. $W_1 + W_A = W_T$

$(50) + (P_N) = (50 + P_N)$

$0.96(50) + (P_N) = 0.99(50 + P_N)$

$48 + P_N = 49.5 + 0.99P_N$

$0.01P_N = 1.5$

$P_N = \mathbf{150\ gallons}$

4. $R_R T_R + 4000 = R_O T_O;$

$R_R = 20;\ R_O = 40;$

$T_R = T_O$

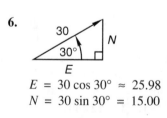

$20T + 4000 = 40T$

$20T = 4000$

$T = 200$ seconds

Length $= D_O = 40(200) = \mathbf{8000\ feet}$

5. Carbon: $\quad 3 \times 12 = 36$

Hydrogen: $\ 7 \times 1 = 7$

Chlorine: $\ \ 1 \times 35 = 35$

Total: $\qquad\qquad = 78$

$\dfrac{36}{78} = \dfrac{48}{TW}$

$36TW = 48(78)$

$TW = \mathbf{104\ grams}$

6.

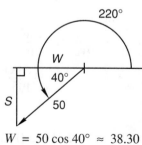

$E = 30 \cos 30° \approx 25.98$

$N = 30 \sin 30° = 15.00$

$W = 50 \cos 40° \approx 38.30$

$S = 50 \sin 40° \approx 32.14$

$$\begin{array}{r} 25.98R + 15.00U \\ -38.30R - 32.14U \\ \hline \mathbf{-12.32R - 17.14U} \end{array}$$

7.

$R = 30 \cos 45° \approx 21.21$
$U = 30 \sin 45° \approx 21.21$

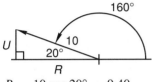

$R = 10 \cos 20° \approx 9.40$
$U = 10 \sin 20° \approx 3.42$

$$\begin{array}{r} 21.21R + 21.21U \\ -9.40R + 3.42U \\ \hline \mathbf{11.81R + 24.63U} \end{array}$$

8. $H = \sqrt{(7.3)^2 + (26.34)^2} = 27.33$

$\tan \theta = \dfrac{26.34}{7.3}$

$\theta \approx 74.51°$

27.33$\underline{/74.51°}$

9. (a) $\dfrac{1}{5}x - \dfrac{5}{2}y = -48$
(b) $0.4x + 0.05y = 5$

(a') $2x - 25y = -480$
(b') $40x + 5y = 500$

$$\begin{array}{r} \text{(a')} \quad 2x - 25y = -480 \\ 5\text{(b')} \quad 200x + 25y = 2500 \\ \hline 202x \qquad\quad = 2020 \\ x = 10 \end{array}$$

(a') $2(10) - 25y = -480$
$y = 20$

(10, 20)

10.
$$-x = -2x^2 - 5$$
$$\left(x^2 - \tfrac{1}{2}x + \quad\right) = -\tfrac{5}{2}$$
$$x^2 - \tfrac{1}{2}x + \tfrac{1}{16} = -\tfrac{5}{2} + \tfrac{1}{16}$$
$$\left(x - \tfrac{1}{4}\right)^2 = -\tfrac{39}{16}$$
$$x - \tfrac{1}{4} = \pm\tfrac{\sqrt{39}}{4}i$$
$$\mathbf{x = \tfrac{1}{4} \pm \tfrac{\sqrt{39}}{4}i}$$

11.
$$3x^2 = -4 + 2x$$
$$\left(x^2 - \tfrac{2}{3}x + \quad\right) = -\tfrac{4}{3}$$
$$x^2 - \tfrac{2}{3}x + \tfrac{1}{9} = -\tfrac{4}{3} + \tfrac{1}{9}$$
$$\left(x - \tfrac{1}{3}\right)^2 = -\tfrac{11}{9}$$
$$x - \tfrac{1}{3} = \pm\tfrac{\sqrt{11}}{3}i$$
$$\mathbf{x = \tfrac{1}{3} \pm \tfrac{\sqrt{11}}{3}i}$$

12. $Mo = mZr + b$
Use the graph to find the slope.

$m = \dfrac{-150}{20} = -7.5$

$Mo = -7.5Zr + b$
Use the point (116, 500) for Zr and Mo.
$500 = -7.5(116) + b$
$1370 = b$
$Mo = -7.5Zr + 1370$

13. $H^2 = 4^2 + 3^2$
$H^2 = 25$
$H = 5$

S.A. $= \dfrac{2(6 \times 4)}{2} + 2(20 \times 5) + (20 \times 6)$

$= 24 + 200 + 120 = 344 \text{ m}^2$

$344 \text{ m}^2 \times \dfrac{100 \text{ cm}}{1 \text{ m}} \times \dfrac{100 \text{ cm}}{1 \text{ m}} = \mathbf{3{,}440{,}000 \text{ cm}^2}$

14.
$$\dfrac{p + zy}{m} - c = \dfrac{a}{b}$$
$$bp + bzy - bcm = am$$
$$bzy = am + bcm - bp$$
$$z = \dfrac{\mathbf{am + bcm - bp}}{\mathbf{by}}$$

15.
$$\frac{p + zy}{m} - c = \frac{a}{b}$$
$$bp + bzy - bcm = am$$
$$bp + bzy = am + bcm$$
$$\frac{\bm{bp + bzy}}{\bm{a + bc}} = m$$

16. $4i^3 - i^5 + 2i^2 - \sqrt{-16}$
$$= 4i(ii) - i(ii)(ii) + 2(ii) - 4i = \bm{-2 - 9i}$$

17. $3 - 2i^5 - 3i^4 + \sqrt{-4} - i$
$$= 3 - 2i(ii)(ii) - 3(ii)(ii) + 2i - i = \bm{-i}$$

18. $\dfrac{400 \text{ cm}}{\text{min}} \times \dfrac{1 \text{ in.}}{2.54 \text{ cm}} \times \dfrac{1 \text{ ft}}{12 \text{ in.}} \times \dfrac{1 \text{ yd}}{3 \text{ ft}}$
$$\times \dfrac{1 \text{ min}}{60 \text{ sec}} = \dfrac{\bm{400}}{\bm{(2.54)(12)(3)(60)}} \dfrac{\bm{yd}}{\bm{sec}}$$

19. $4 \text{ mi}^3 \times \dfrac{5280 \text{ ft}}{1 \text{ mi}} \times \dfrac{5280 \text{ ft}}{1 \text{ mi}} \times \dfrac{5280 \text{ ft}}{1 \text{ mi}}$
$$\times \dfrac{12 \text{ in.}}{1 \text{ ft}} \times \dfrac{12 \text{ in.}}{1 \text{ ft}} \times \dfrac{12 \text{ in.}}{1 \text{ ft}} \times \dfrac{2.54 \text{ cm}}{1 \text{ in.}}$$
$$\times \dfrac{2.54 \text{ cm}}{1 \text{ in.}} \times \dfrac{2.54 \text{ cm}}{1 \text{ in.}} \times \dfrac{1 \text{ m}}{100 \text{ cm}}$$
$$\times \dfrac{1 \text{ m}}{100 \text{ cm}} \times \dfrac{1 \text{ m}}{100 \text{ cm}} \times \dfrac{1 \text{ km}}{1000 \text{ m}} \times \dfrac{1 \text{ km}}{1000 \text{ m}}$$
$$\times \dfrac{1 \text{ km}}{1000 \text{ m}}$$
$$= \dfrac{\bm{4(5280)^3 (12)^3 (2.54)^3}}{\bm{(100)(100)(100)(1000)(1000)(1000)}} \ \bm{km^3}$$

20. $\sqrt[5]{3\sqrt[4]{3}} = \left[3(3^{1/4})\right]^{1/5} = 3^{1/5}3^{1/20} = \bm{3^{1/4}}$

21. $\sqrt[5]{4\sqrt[4]{2}} = \sqrt[5]{2^2 \sqrt[4]{2}} = \left[2^2(2^{1/4})\right]^{1/5}$
$$= 2^{2/5}2^{1/20} = \bm{2^{9/20}}$$

22. $\dfrac{4}{(-27)^{-2/3}} = 4\left((-27)^{2/3}\right) = 4\left((-27)^{1/3}\right)^2$
$$= \bm{36}$$

23. $\sqrt[4]{xy^2} \sqrt{x^3 y} = (xy^2)^{1/4}(x^3 y)^{1/2}$
$$= x^{1/4}y^{1/2}x^{3/2}y^{1/2} = \bm{x^{7/4}y}$$

24. $3\sqrt{\dfrac{2}{7}} + 5\sqrt{\dfrac{7}{2}} - 3\sqrt{126}$
$$= \dfrac{3\sqrt{2}}{\sqrt{7}}\dfrac{\sqrt{7}}{\sqrt{7}} + \dfrac{5\sqrt{7}}{\sqrt{2}}\dfrac{\sqrt{2}}{\sqrt{2}} - 9\sqrt{14}$$
$$= \dfrac{6\sqrt{14}}{14} + \dfrac{35\sqrt{14}}{14} - \dfrac{126\sqrt{14}}{14} = -\dfrac{\bm{85\sqrt{14}}}{\bm{14}}$$

25. $\dfrac{(476,158 \times 10^{22})(79,318,642)}{(983,704)(514.0 \times 10^{-14})}$
$$\approx \dfrac{(5 \times 10^{27})(8 \times 10^7)}{(1 \times 10^6)(5 \times 10^{-12})} = \bm{8 \times 10^{40}}$$

26. $\text{Volume}_{\text{Cone}} = \dfrac{1}{3}(\text{Area}_{\text{Base}})(\text{Height}) = 12\pi \text{ m}^3$
$$\dfrac{1}{3}\pi r^2 h = 12\pi \text{ m}^3$$
$$r^2 = 6 \text{ m}^2$$
$$r = \bm{\sqrt{6} \text{ m}}$$

27. $12(90 - A) = (180 - A) + 20$
$$1080 - 12A = 200 - A$$
$$-11A = -880$$
$$A = \bm{80°}$$

28. Write the equation of the given line in slope-intercept form.
$$5x + 4y = 7$$
$$y = -\dfrac{5}{4}x + \dfrac{7}{4}$$
Since parallel lines have the same slope:
$$y = -\dfrac{5}{4}x + b$$
$$-7 = -\dfrac{5}{4}(2) + b$$
$$-\dfrac{9}{2} = b$$
$$y = -\dfrac{\bm{5}}{\bm{4}}x - \dfrac{\bm{9}}{\bm{2}}$$

29. $-2^0 - 2^2 - 2^2(-2 - 1^0)x - 3x$
$$- 7x^0y^0 - 4 = 2$$
$$-1 - 4 + 12x - 3x - 7 - 4 = 2$$
$$9x = 18$$
$$x = \bm{2}$$

30. $\dfrac{4x + 5}{3} - \dfrac{x}{7} = 2$
$$28x + 35 - 3x = 42$$
$$25x = 7$$
$$x = \dfrac{\bm{7}}{\bm{25}}$$

PROBLEM SET 64

1. $C = \dfrac{k}{F}$
$$300 = \dfrac{k}{2}$$
$$k = 600$$
$$C = \dfrac{600}{0.5} = \bm{1200 \text{ grams}}$$

2. $\dfrac{P_1}{T_1} = \dfrac{P_2}{T_2}$
$$\dfrac{700}{400} = \dfrac{2800}{T_2}$$
$$T_2 = \bm{1600 \text{ K}}$$

3. Iodine P_N + Iodine D_N = Iodine Total
$$0.3(P_N) + 0.8(D_N) = 0.4(50)$$
(a) $0.3P_N + 0.8D_N = 20$
(b) $P_N + D_N = 50$

Substitute $D_N = 50 - P_N$ into (a) and get:
(a′) $0.3P_N + 0.8(50 - P_N) = 20$
$$-0.5P_N = -20$$
$$P_N = 40$$

(b) $(40) + D_N = 50$
$$D_N = 10$$

40 liters 30%, 10 liters 80%

4. $R_I T_I = R_O T_O$;
$R_I = 400$; $R_O = 100$;
$T_I + T_O = 40$

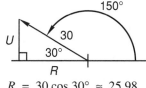

$400T_I = 100(40 - T_I)$
$500T_I = 4000$
$T_I = 8$ hr
$D_I = 400(8) =$ **3200 kilometers**

5. Carbon: $1 \times 12 = 12$
Hydrogen: $3 \times 1 = 3$
Iodine: $1 \times 127 = 127$
Total: $= 142$
$$\frac{I}{T} = \frac{127}{142} \approx 0.894 = \mathbf{89.4\%}$$

6. $m + \dfrac{2}{\dfrac{2}{c} + s} = m + \dfrac{2}{\dfrac{2 + cs}{c}}$
$$= m + \frac{2c}{2 + cs} = \frac{2m + cms + 2c}{2 + cs}$$

7. $\dfrac{m}{a} + \dfrac{3}{2 + \dfrac{s}{a}} = \dfrac{m}{a} + \dfrac{3}{\dfrac{2a + s}{a}}$
$$= \frac{m}{a} + \frac{3a}{2a + s} = \frac{2am + ms + 3a^2}{a(2a + s)}$$

8. $\dfrac{m}{c} + \dfrac{8}{2 + \dfrac{m}{c}} = \dfrac{m}{c} + \dfrac{8}{\dfrac{2c + m}{c}}$
$$= \frac{m}{c} + \frac{8c}{2c + m} = \frac{2cm + m^2 + 8c^2}{c(2c + m)}$$

9. $9i^3 - 3i^4 + 2\sqrt{-4} + \sqrt{-2}\sqrt{-2}$
$= 9i(ii) - 3(ii)(ii) + 4i + \sqrt{2}i\sqrt{2}i$
$= -9i - 3 + 4i - 2 = \mathbf{-5 - 5i}$

10. $\sqrt{-4} + \sqrt{-2}\sqrt{-2} - 4i^3$
$= 2i + \sqrt{2}i\sqrt{2}i - 4i(ii) = \mathbf{-2 + 6i}$

11. $i^4 - 3i^2 - 2\sqrt{-2}\sqrt{-3}$
$= (ii)(ii) - 3(ii) - 2\sqrt{2}i\sqrt{3}i = \mathbf{4 + 2\sqrt{6}}$

12. $2\sqrt{-9} - 3i^4 + 2\sqrt{3}\sqrt{-3} + i$
$= 6i - 3(ii)(ii) + 2\sqrt{3}\sqrt{3}i + i = \mathbf{-3 + 13i}$

13. $(2 + 3i)(5 - 3i)$
$= 10 - 6i + 15i - 9i^2 = \mathbf{19 + 9i}$

14. $(3i - 5)(2 + 4i)$
$= 6i + 12i^2 - 10 - 20i = \mathbf{-22 - 14i}$

15. $(2i - 4)(i + 2) = 2i^2 + 4i - 4i - 8 = \mathbf{-10}$

16.

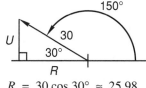

$R = 10 \cos 10° \approx 9.85$
$U = 10 \sin 10° \approx 1.74$

$R = 30 \cos 30° \approx 25.98$
$U = 30 \sin 30° = 15.00$

$$\begin{array}{r} 9.85R + 1.74U \\ -25.98R + 15.00U \\ \hline \mathbf{-16.13R + 16.74U} \end{array}$$

17. $H = \sqrt{4^2 + 6^2}$
$H = \sqrt{52} = 2\sqrt{13}$

$\tan \theta = \dfrac{6}{4}$
$\theta \approx 56.31°$

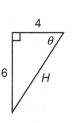

Since θ is a third-quadrant angle:
$\theta = 56.31 + 180 = 236.31°$

$2\sqrt{13}\underline{/236.31°}$

18. (a) $\dfrac{3}{2}x - \dfrac{1}{5}y = 28$

(b) $0.02x + 0.4y = 4.4$

(a′) $15x - 2y = 280$

(b′) $2x + 40y = 440$

20(a′) $300x - 40y = 5600$

(b′) $\underline{2x + 40y = 440}$

$302x = 6040$

$x = 20$

(b′) $2(20) + 40y = 440$

$40y = 400$

$y = 10$

(20, 10)

19. $3x^2 = -2x - 5$

$\left(x^2 + \dfrac{2}{3}x + \right) = -\dfrac{5}{3}$

$x^2 + \dfrac{2}{3}x + \dfrac{1}{9} = -\dfrac{5}{3} + \dfrac{1}{9}$

$\left(x + \dfrac{1}{3}\right)^2 = -\dfrac{14}{9}$

$x + \dfrac{1}{3} = \pm\dfrac{\sqrt{14}}{3}i$

$x = -\dfrac{1}{3} \pm \dfrac{\sqrt{14}}{3}i$

20. $-3x + 2x^2 = -7$

$\left(x^2 - \dfrac{3}{2}x + \right) = -\dfrac{7}{2}$

$x^2 - \dfrac{3}{2}x + \dfrac{9}{16} = -\dfrac{7}{2} + \dfrac{9}{16}$

$\left(x - \dfrac{3}{4}\right)^2 = -\dfrac{47}{16}$

$x - \dfrac{3}{4} = \pm\dfrac{\sqrt{47}}{4}i$

$x = \dfrac{3}{4} \pm \dfrac{\sqrt{47}}{4}i$

21. $W = mIr + b$

Use the graph to find the slope.

$m = \dfrac{-34}{250} = -0.13$

$W = -0.13Ir + b$

Use the point (1850, 50) for Ir and W.

$50 = -0.13(1850) + b$

$290 = b$

$W = -0.13Ir + 290$

22. Area = Area$_{\text{Triangle}}$ + Area$_{\text{Semicircle}}$

$\phantom{\text{Area}} = \dfrac{1}{2}bH + \dfrac{1}{2}\pi r^2$

$H^2 = 6^2 - \left(\dfrac{5}{2}\right)^2$

$H^2 = 36 - \dfrac{25}{4}$

$H = \dfrac{\sqrt{119}}{2}$

Area $= \left[\dfrac{1}{2}(5)\left(\dfrac{\sqrt{119}}{2}\right) + \dfrac{1}{2}\pi(3)^2\right] m^2$

$\phantom{\text{Area }} = \left(\dfrac{5\sqrt{119}}{4} + \dfrac{9}{2}\pi\right) m^2$

23. $\dfrac{400 \text{ cm}^3}{\text{sec}} \times \dfrac{1 \text{ in.}}{2.54 \text{ cm}} \times \dfrac{1 \text{ in.}}{2.54 \text{ cm}} \times \dfrac{1 \text{ in.}}{2.54 \text{ cm}}$

$ \times \dfrac{60 \text{ sec}}{\text{min}} \times \dfrac{60 \text{ min}}{1 \text{ hr}}$

$= \dfrac{\mathbf{400(60)(60)}}{\mathbf{(2.54)(2.54)(2.54)}} \dfrac{\textbf{in.}^3}{\textbf{hr}}$

24. $\sqrt[3]{9\sqrt[3]{3}} = \sqrt[3]{3^2\sqrt[3]{3}} = \left[3^2(3^{1/3})\right]^{1/3}$

$ = 3^{2/3}3^{1/9} = 3^{7/9}$

25. $2\sqrt{\dfrac{7}{5}} - 3\sqrt{\dfrac{5}{7}} + 2\sqrt{140}$

$= \dfrac{2\sqrt{7}}{\sqrt{5}}\dfrac{\sqrt{5}}{\sqrt{5}} - \dfrac{3\sqrt{5}}{\sqrt{7}}\dfrac{\sqrt{7}}{\sqrt{7}} + 4\sqrt{35}$

$= \dfrac{14\sqrt{35}}{35} - \dfrac{15\sqrt{35}}{35} + \dfrac{140\sqrt{35}}{35} = \dfrac{\mathbf{139\sqrt{35}}}{\mathbf{35}}$

26. $\dfrac{a(b + c)}{x} - m = \dfrac{d}{f}$

$abf + acf - fmx = dx$

$abf + acf = dx + fmx$

$\dfrac{\textbf{abf + acf}}{\textbf{d + fm}} = x$

27. $\dfrac{a(b + c)}{x} - m = \dfrac{d}{f}$

$abf + acf - fmx = dx$

$acf = dx + fmx - abf$

$c = \dfrac{\textbf{dx + fmx - abf}}{\textbf{af}}$

28. (a) $x - 3y = -6$

$$-3y = -x - 6$$

$$y = \frac{1}{3}x + 2$$

(b) $2x + 5y = 15$

$$5y = -2x + 15$$

$$y = -\frac{2}{5}x + 3$$

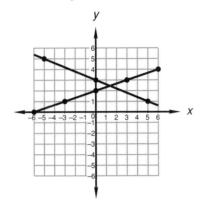

Substitute $x = 3y - 6$ into (b) and get:

(b') $2(3y - 6) + 5y = 15$

$$11y = 27$$

$$y = \frac{27}{11}$$

(a) $x = 3\left(\frac{27}{11}\right) - 6 = \frac{15}{11}$

$$\left(\frac{15}{11}, \frac{27}{11}\right)$$

29.
$$56x = -15x^2 - x^3$$

$$x^3 + 15x^2 + 56x = 0$$

$$x(x + 7)(x + 8) = 0$$

$$x = \mathbf{0, -7, -8}$$

30.
$$\frac{(146,842 \times 10^2)(0.0007892)}{(96,478 \times 10^{14})(0.000712 \times 10^{42})}$$

$$\approx \frac{(1 \times 10^7)(8 \times 10^{-4})}{(1 \times 10^{19})(7 \times 10^{38})} \approx \mathbf{1 \times 10^{-54}}$$

PROBLEM SET 65

1. $S = kP$

$$400 = k(100)$$

$$4 = k$$

$$S = 4(12) = \mathbf{48\ kilograms}$$

2. $P_1V_1 = P_2V_2$

$$800(2) = P_2(0.1)$$

$$P_2 = \mathbf{16,000\ N/m^2}$$

3. $0.84(500) - (E) = 0.8(500 - E)$

$$420 - E = 400 - 0.8E$$

$$-0.2E = -20$$

$$E = \mathbf{100\ milliliters}$$

4.   (a) $N_I = N_E + 400$

(b) $N_I = 4N_E + 100$

Substitute (b) into (a) and get:

(a') $(4N_E + 100) = N_E + 400$

$$3N_E = 300$$

$$N_E = \mathbf{100}$$

(a) $N_I = (100) + 400 = \mathbf{500}$

5. $\dfrac{96}{100} \times T = 1920$

$$T = 1920 \cdot \frac{100}{96} = 2000$$

$$\text{Phosgene}_{\text{Combined}} = 2000 - 1920$$

$$= \mathbf{80\ kilograms}$$

6.   (a) $R_WT_W = 8$

(b) $R_BT_B = 8$

(c) $R_B = 4R_W$

(d) $T_B = 2 - T_W$

Substitute (c) and (d) into (b) and get:

(b') $4R_W(2 - T_W) = 8$

Substitute (a) into (b') and get:

(b'') $8R_W - 4(8) = 8$

$$8R_W = 40$$

$$R_W = \mathbf{5}$$

$$R_B = \mathbf{20};\ T_W = \mathbf{\frac{8}{5}};\ T_B = \mathbf{\frac{2}{5}}$$

7.   (a) $R_PT_P = 600$

(b) $R_CT_C = 105$

(c) $R_P = 5R_C$

(d) $T_P + T_C = 15$

Substitute (c) and (d) into (a) and get:

(a') $5R_C(15 - T_C) = 600$

Substitute (b) into (a') and get:

(a'') $75R_C - 5(105) = 600$

$$75R_C = 1125$$

$$R_C = \mathbf{15}$$

$$R_P = \mathbf{75};\ T_C = \mathbf{7};\ T_P = \mathbf{8}$$

8.   (a) $R_1T_1 = 120$

(b) $R_2T_2 = 120$

(c) $R_1 = 2R_2$

(d) $T_1 + T_2 = 6$

Substitute (c) and (d) into (a) and get:

(a') $2R_2(6 - T_2) = 120$

Substitute (b) into (a') and get:

(a'') $12R_2 - 2(120) = 120$

$$12R_2 = 360$$

$$R_2 = \mathbf{30}$$

$$R_1 = \mathbf{60};\ T_2 = \mathbf{4};\ T_1 = \mathbf{2}$$

9. $x + \dfrac{1}{a + \dfrac{b}{c}} = x + \dfrac{1}{\dfrac{ac + b}{c}}$

$= x + \dfrac{c}{ac + b} = \dfrac{acx + bx + c}{ac + b}$

10. $\dfrac{4}{c} + \dfrac{1}{a + \dfrac{1}{b}} = \dfrac{4}{c} + \dfrac{1}{\dfrac{ab + 1}{b}}$

$= \dfrac{4}{c} + \dfrac{b}{ab + 1} = \dfrac{4ab + 4 + bc}{c(ab + 1)}$

11. $x + \dfrac{a}{1 + \dfrac{1}{a}} = x + \dfrac{a}{\dfrac{a + 1}{a}} = x + \dfrac{a^2}{a + 1}$

$= \dfrac{ax + x + a^2}{a + 1}$

12. $\sqrt{-4} - \sqrt{-3}\sqrt{-3} + 2i^5 - 4$
$= 2i - \sqrt{3}i\sqrt{3}i + 2i(ii)(ii) - 4$
$= 2i + 3 + 2i - 4 = \mathbf{-1 + 4i}$

13. $(5i - 2)(2i - 3)$
$= 10i^2 - 15i - 4i + 6 = \mathbf{-4 - 19i}$

14. $(-i - 3)(-2i + 4)$
$= 2i^2 - 4i + 6i - 12 = \mathbf{-14 + 2i}$

15.

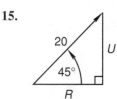

$R = 20 \cos 45° \approx 14.14$
$U = 20 \sin 45° \approx 14.14$

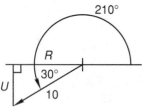

$R = 10 \cos 30° \approx 8.66$
$U = 10 \sin 30° \approx 5.00$

$\begin{array}{r} 14.14R + 14.14U \\ -8.66R - 5.00U \\ \hline \mathbf{5.48R + 9.14U} \end{array}$

16. $H = \sqrt{3^2 + 5^2}$
$H = \sqrt{34}$

$\tan \theta = \dfrac{5}{3}$

$\theta \approx 59.04°$

Since θ is a fourth-quadrant angle:
$\theta = 360 - 59.04 = 300.96°$

$\mathbf{\sqrt{34}/300.96°}$

17. (a) $\dfrac{2}{7}x - \dfrac{1}{6}y = 1$
(b) $0.3x + 0.07y = 0.84$

(a′) $12x - 7y = 42$
(b′) $\underline{30x + 7y = 84}$
$\ 42x = 126$
$\ x = 3$

(a′) $12(3) - 7y = 42$
$-7y = 6$
$y = -\dfrac{6}{7}$

$\left(3, -\dfrac{6}{7}\right)$

18. $-2 = -3x^2 - 7x$

$\left(x^2 + \dfrac{7}{3}x + \right) = \dfrac{2}{3}$

$x^2 + \dfrac{7}{3}x + \dfrac{49}{36} = \dfrac{2}{3} + \dfrac{49}{36}$

$\left(x + \dfrac{7}{6}\right)^2 = \dfrac{73}{36}$

$x + \dfrac{7}{6} = \pm\dfrac{\sqrt{73}}{6}$

$x = -\dfrac{7}{6} \pm \dfrac{\sqrt{73}}{6}$

19. $2x^2 - 4 = -5x$

$\left(x^2 + \dfrac{5}{2}x + \right) = 2$

$x^2 + \dfrac{5}{2}x + \dfrac{25}{16} = 2 + \dfrac{25}{16}$

$\left(x + \dfrac{5}{4}\right)^2 = \dfrac{57}{16}$

$x + \dfrac{5}{4} = \pm\dfrac{\sqrt{57}}{4}$

$x = -\dfrac{5}{4} \pm \dfrac{\sqrt{57}}{4}$

20. $K = mRa + b$
Use the graph to find the slope.

$m = \dfrac{500}{560} = 0.89$

$K = 0.89Ra + b$
Use the point (2100, 500) for Ra and K.
$500 = 0.89(2100) + b$
$-1369 = b$
$\mathbf{K = 0.89Ra - 1369}$

21. Since the measure of an inscribed angle equals half the measure of the intercepted arc:

$$\frac{m\widehat{AB}}{2} = y°$$

$$m\widehat{AB} = (2y)°$$

Since the measure of an arc of a circle is the same as the measure of the central angle:

$$x° = m\widehat{AB}$$
$$x° = (2y)°$$

22. $\dfrac{600 \text{ ft}^3}{\text{hr}} \times \dfrac{12 \text{ in.}}{1 \text{ ft}} \times \dfrac{12 \text{ in.}}{1 \text{ ft}} \times \dfrac{12 \text{ in.}}{1 \text{ ft}} \times \dfrac{1 \text{ hr}}{60 \text{ min}}$

$$= \frac{600(12)(12)(12)}{60} \frac{\text{in.}^3}{\text{min}}$$

23. $\sqrt{x^2\sqrt{y^3}} = [x^2 y^{3/2}]^{1/2} = xy^{3/4}$

24. $\sqrt[6]{4\sqrt[5]{2}} = \sqrt[6]{2^2\sqrt[5]{2}} = [2^2(2^{1/5})]^{1/6}$

$$= 2^{1/3}2^{1/30} = 2^{11/30}$$

25. $\sqrt{\dfrac{2}{9}} - 3\sqrt{\dfrac{9}{2}} - 2\sqrt{50} = \dfrac{\sqrt{2}}{3} - \dfrac{9}{\sqrt{2}}\dfrac{\sqrt{2}}{\sqrt{2}} - 10\sqrt{2}$

$$= \frac{2\sqrt{2}}{6} - \frac{27\sqrt{2}}{6} - \frac{60\sqrt{2}}{6} = -\frac{85\sqrt{2}}{6}$$

26. $\dfrac{x}{a(b+c)} - m = \dfrac{d}{f}$

$fx - abfm - acfm = abd + acd$

$fx = abd + acd + abfm + acfm$

$$x = \frac{abd + acd + abfm + acfm}{f}$$

27. $\dfrac{x}{a(b+c)} - m = \dfrac{d}{f}$

$fx - abfm - acfm = abd + acd$

$fx - acfm - acd = abd + abfm$

$$\frac{fx - acfm - acd}{ad + afm} = b$$

28. $\dfrac{(x-8)(x-4)}{x(x+5)(x-4)} \cdot \dfrac{x(x+5)(x+9)}{(x-8)(x+7)}$

$$= \frac{x+9}{x+7}$$

29. $\dfrac{(47{,}123 \times 10^5)(980)(476)}{(0.00134)(576 \times 10^5)}$

$$\approx \frac{(5 \times 10^9)(1 \times 10^3)(5 \times 10^2)}{(1 \times 10^{-3})(6 \times 10^7)}$$

$$\approx 4 \times 10^{10}$$

30. $\dfrac{4}{x+2} - \dfrac{3}{x^2-4}$

$$= \frac{4(x-2)}{x^2-4} - \frac{3}{x^2-4} = \frac{4x-11}{x^2-4}$$

Problem Set 66

1. $C = \dfrac{k}{U}$

$$5 = \frac{k}{20}$$
$$100 = k$$
$$C = \frac{100}{2} = \textbf{50 grams}$$

2. $\dfrac{P_1 V_1}{T_1} = \dfrac{P_2 V_2}{T_2}$

$$\frac{(400)(6)}{200} = \frac{(800)(60)}{T_2}$$
$$T_2 = \textbf{4000 K}$$

3. Arsenic P_N + Arsenic D_N = Arsenic Total

$0.6(P_N) + 0.2(D_N) = 0.36(200)$

(a) $0.6P_N + 0.2D_N = 72$

(b) $P_N + D_N = 200$

Substitute $D_N = 200 - P_N$ into (a) and get:

(a') $0.6P_N + 0.2(200 - P_N) = 72$
$$0.4P_N = 32$$
$$P_N = 80$$

(b) $(80) + D_N = 200$
$$D_N = 120$$

80 liters 60%, 120 liters 20%

4. $R_C T_C + 10 = R_D T_D$;

$R_C = 4$; $R_D = 6$;

$T_C = T_D$

$4T_D + 10 = 6T_D$
$10 = 2T_D$
5 hr $= T_D$

5. Carbon: $\quad 1 \times 12 = 12$
Hydrogen: $\quad 3 \times 1 = 3$
Bromine: $\quad 1 \times 80 = 80$
Total: $\qquad\qquad = 95$

$$\frac{80}{95} = \frac{Br}{950}$$
$$Br = \textbf{800 grams}$$

6. $\dfrac{1}{x+3} - \dfrac{7a}{-x-3} = \dfrac{1}{x+3} + \dfrac{7a}{x+3}$

$= \dfrac{1+7a}{x+3}$

7. $\dfrac{x+5}{x+3} + \dfrac{2x-3}{-3-x} = \dfrac{x+5}{x+3} - \dfrac{2x-3}{x+3}$

$= \dfrac{-x+8}{x+3}$

8. $\dfrac{4}{x^2-4} - \dfrac{2x}{x-2}$

$= \dfrac{4}{x^2-4} - \dfrac{2x(x+2)}{x^2-4}$

$= \dfrac{-2x^2-4x+4}{x^2-4}$

9. $\text{Area}_Q = \pi r^2 = 9\pi \text{ in.}^2$

$r = 3 \text{ in.}$

Since the radius of circle Q is 3 in., the radius of each semicircle is 1 in.

$A_{\text{Shaded}} = \dfrac{1}{2} A_Q + \dfrac{\pi r^2}{2} - \dfrac{2\pi r^2}{2}$

$= \dfrac{1}{2} A_Q - \dfrac{\pi r^2}{2}$

$= \dfrac{9\pi \text{ in.}^2}{2} - \dfrac{\pi (1 \text{ in.})^2}{2}$

$= \dfrac{9}{2}\pi \text{ in.}^2 - \dfrac{1}{2}\pi \text{ in.}^2 = \mathbf{4\pi \text{ in.}^2}$

10. (a) $R_T T_T = 300$

(b) $R_P T_P = 1200$

(c) $R_P = 8R_T$

(d) $T_T = T_P + 3$

Substitute (c) and (d) into (b) and get:

(b′) $8R_T(T_T - 3) = 1200$

Substitute (a) into (b′) and get:

(b″) $8(300) - 24R_T = 1200$

$-24R_T = -1200$

$R_T = 50$

$R_P = \mathbf{400}; \; T_T = \mathbf{6}; \; T_P = \mathbf{3}$

11. (a) $R_P T_P = 624$

(b) $R_T T_T = 364$

(c) $T_P = T_T - 4$

(d) $R_P = 4R_T$

Substitute (c) and (d) into (a) and get:

(a′) $4R_T(T_T - 4) = 624$

Substitute (b) into (a′) and get:

(a″) $4(364) - 16R_T = 624$

$-16R_T = -832$

$R_T = 52$

$R_P = \mathbf{208}; \; T_T = \mathbf{7}; \; T_P = \mathbf{3}$

12. $\text{Area} = \pi(3r)^2 - \pi r^2$

$= 9\pi r^2 - \pi r^2$

$= \mathbf{8\pi r^2 \text{ m}^2}$

13. $ax + \dfrac{1}{a + \dfrac{1}{x}} = ax + \dfrac{1}{\dfrac{ax+1}{x}}$

$= ax + \dfrac{x}{ax+1} = \dfrac{a^2x^2 + ax + x}{ax+1}$

14. $\dfrac{m}{x} + \dfrac{1}{x + \dfrac{1}{x}} = \dfrac{m}{x} + \dfrac{1}{\dfrac{x^2+1}{x}}$

$= \dfrac{m}{x} + \dfrac{x}{x^2+1} = \dfrac{mx^2 + m + x^2}{x(x^2+1)}$

15. $-\sqrt{-3}\sqrt{-2} + \sqrt{-4} - \sqrt{-3}\sqrt{-3} - 2i^3$

$= -\sqrt{3}i\sqrt{2}i + 2i - \sqrt{3}i\sqrt{3}i - 2i(ii)$

$= \sqrt{6} + 2i + 3 + 2i = \mathbf{3 + \sqrt{6} + 4i}$

16. $(4i - 2)(3 + 5i)$

$= 12i + 20i^2 - 6 - 10i = \mathbf{-26 + 2i}$

17.

$R = 4\cos 28° \approx 3.53$

$U = 4\sin 28° \approx 1.88$

$R = 10\cos 35° \approx 8.19$

$U = 10\sin 35° \approx 5.74$

$ 3.53R + 1.88U$

$\underline{ 8.19R + 5.74U}$

$ \mathbf{11.72R + 7.62U}$

18. $H = \sqrt{8^2 + 4^2}$

$H = \sqrt{80} = 4\sqrt{5}$

$\tan\theta = \dfrac{4}{8}$

$\theta \approx 26.57°$

$\mathbf{4\sqrt{5} \underline{/26.57°}}$

19. (a) $\dfrac{3}{8}x - \dfrac{1}{4}y = -2$

(b) $0.012x + 0.02y = 0.496$

(a′) $3x - 2y = -16$

(b') $12x + 20y = 496$

10(a') $30x - 20y = -160$

(b') $\underline{12x + 20y = 496}$

$42x = 336$

$x = 8$

(a') $3(8) - 2y = -16$

$2y = 40$

$y = 20$

(8, 20)

20.
$$-3 = -2x^2 - 6x$$
$$(x^2 + 3x +) = \frac{3}{2}$$
$$x^2 + 3x + \frac{9}{4} = \frac{3}{2} + \frac{9}{4}$$
$$\left(x + \frac{3}{2}\right)^2 = \frac{15}{4}$$
$$x + \frac{3}{2} = \pm\frac{\sqrt{15}}{2}$$
$$x = -\frac{3}{2} \pm \frac{\sqrt{15}}{2}$$

21.
$$5x^2 - 4 = -5x$$
$$(x^2 + x +) = \frac{4}{5}$$
$$x^2 + x + \frac{1}{4} = \frac{4}{5} + \frac{1}{4}$$
$$\left(x + \frac{1}{2}\right)^2 = \frac{21}{20}$$
$$x + \frac{1}{2} = \pm\frac{\sqrt{21}}{2\sqrt{5}}$$
$$x = -\frac{1}{2} \pm \frac{\sqrt{105}}{10}$$

22. Since this is a 30-60-90 triangle:

$2\overrightarrow{SF} = 7$

$\overrightarrow{SF} = \dfrac{7}{2}$

$x = \dfrac{7}{2}(\sqrt{3}) = \dfrac{7\sqrt{3}}{2}$

$y = \dfrac{7}{2}(1) = \dfrac{7}{2}$

23. $Mg = mCa + b$

Use the graph to find the slope.

$m = \dfrac{1000}{15} = 66.67$

$Mg = 66.67Ca + b$

Use the point (36, 2400) for Ca and Mg.

$2400 = 36(66.67) + b$

$0 = b$

$Mg = 66.67Ca$

24. $\dfrac{10 \text{ in.}^3}{\text{hr}} \times \dfrac{2.54 \text{ cm}}{1 \text{ in.}} \times \dfrac{2.54 \text{ cm}}{1 \text{ in.}} \times \dfrac{2.54 \text{ cm}}{1 \text{ in.}}$

$\times \dfrac{1 \text{ hr}}{60 \text{ min}} = \dfrac{10(2.54)(2.54)(2.54)}{60} \dfrac{\text{cm}^3}{\text{min}}$

25. $\sqrt[3]{x^{1/2}\sqrt{x^2}} = [x^{1/2}(x)]^{1/3} = x^{1/6}x^{1/3} = \boldsymbol{x^{1/2}}$

26. $\sqrt[3]{x^{1/5}\sqrt[4]{x}} = [x^{1/5}(x^{1/4})]^{1/3} = x^{1/15}x^{1/12} = \boldsymbol{x^{3/20}}$

27. $3\sqrt{\dfrac{2}{5}} + 7\sqrt{\dfrac{5}{2}} - 6\sqrt{40}$

$= \dfrac{3\sqrt{2}}{\sqrt{5}}\dfrac{\sqrt{5}}{\sqrt{5}} + \dfrac{7\sqrt{5}}{\sqrt{2}}\dfrac{\sqrt{2}}{\sqrt{2}} - 12\sqrt{10}$

$= \dfrac{6\sqrt{10}}{10} + \dfrac{35\sqrt{10}}{10} - \dfrac{120\sqrt{10}}{10} = \boldsymbol{-\dfrac{79\sqrt{10}}{10}}$

28.
$$\frac{px - y}{m} - c = \frac{k}{d}$$
$$dpx - dy - cdm = km$$
$$dpx - cdm - km = dy$$
$$\boldsymbol{\frac{dpx - cdm - km}{d} = y}$$

29.
$$\frac{m}{px - y} + \frac{k}{d} = -c$$
$$dm + kpx - ky = -cdpx + cdy$$
$$kpx + cdpx = cdy + ky - dm$$
$$\boldsymbol{x = \frac{cdy + ky - dm}{kp + cdp}}$$

30. $\dfrac{(40,213 \times 10^5)(748,609 \times 10^{-30})}{(0.164289)(506,217 \times 10^2)}$

$\approx \dfrac{(4 \times 10^9)(7 \times 10^{-25})}{(2 \times 10^{-1})(5 \times 10^7)} \approx \boldsymbol{3 \times 10^{-22}}$

PROBLEM SET 67

1. $R = kA$

$2000 = k(12,400)$

$\dfrac{5}{31} = k$

$3000 = \left(\dfrac{5}{31}\right)A$

$18,600 = A$

2. $P_1V_1 = P_2V_2$

$(800)(300) = (1200)V_2$

$V_2 = \boldsymbol{200 \text{ cm}^3}$

3. $0.2(400) + 0.8(D_N) = 0.32(400 + D_N)$

$80 + 0.8D_N = 128 + 0.32D_N$

$0.48D_N = 48$

$D_N = \boldsymbol{100 \text{ pounds}}$

4. (a) $N_N = N_D + 75$

(b) $8N_N = 10N_D + 140$

Substitute (a) into (b) and get:

(b') $8(N_D + 75) = 10N_D + 140$

$$460 = 2N_D$$

$$\mathbf{230} = N_D$$

(a) $N_N = (230) + 75 = \mathbf{305}$

5. $\dfrac{60}{100} \times T = 240$

$$T = 240 \cdot \dfrac{100}{60} = 400$$

$NR = 400 - 240 = \mathbf{160\ grams}$

6. $\dfrac{2}{-4 + \sqrt{5}} \cdot \dfrac{-4 - \sqrt{5}}{-4 - \sqrt{5}}$

$$= \dfrac{-8 - 2\sqrt{5}}{16 - 5} = \dfrac{\mathbf{-8 - 2\sqrt{5}}}{\mathbf{11}}$$

7. $\dfrac{1}{2\sqrt{2} + \sqrt{3}} \cdot \dfrac{2\sqrt{2} - \sqrt{3}}{2\sqrt{2} - \sqrt{3}}$

$$= \dfrac{2\sqrt{2} - \sqrt{3}}{8 - 3} = \dfrac{\mathbf{2\sqrt{2} - \sqrt{3}}}{\mathbf{5}}$$

8. $\dfrac{2}{3\sqrt{5} - 3} \cdot \dfrac{3\sqrt{5} + 3}{3\sqrt{5} + 3} = \dfrac{6\sqrt{5} + 6}{45 - 9}$

$$= \dfrac{\mathbf{\sqrt{5} + 1}}{\mathbf{6}}$$

9. $\dfrac{4x + 2}{x - 2} - \dfrac{3}{2 - x} = \dfrac{4x + 2}{x - 2} + \dfrac{3}{x - 2}$

$$= \dfrac{\mathbf{4x + 5}}{\mathbf{x - 2}}$$

10. $\dfrac{4}{x^2 - 9} + \dfrac{2}{x - 3}$

$$= \dfrac{4}{x^2 - 9} + \dfrac{2(x + 3)}{x^2 - 9} = \dfrac{\mathbf{2x + 10}}{\mathbf{x^2 - 9}}$$

11. (a) $R_P T_P = 1062$

(b) $R_T T_T = 295$

(c) $T_P = T_T - 2$

(d) $R_P = 6R_T$

Substitute (c) and (d) into (a) and get:

(a') $6R_T(T_T - 2) = 1062$

Substitute (b) into (a') and get:

(a") $6(295) - 12R_T = 1062$

$$-12R_T = -708$$

$$R_T = \mathbf{59}$$

$R_P = \mathbf{354};\ T_T = \mathbf{5};\ T_P = \mathbf{3}$

12. $4x + \dfrac{a}{x + \dfrac{a}{b}} = 4x + \dfrac{a}{\dfrac{bx + a}{b}}$

$$= 4x + \dfrac{ab}{bx + a} = \dfrac{\mathbf{4bx^2 + 4ax + ab}}{\mathbf{bx + a}}$$

13. $a + \dfrac{x}{m + \dfrac{1}{x}} = a + \dfrac{x}{\dfrac{mx + 1}{x}}$

$$= a + \dfrac{x^2}{mx + 1} = \dfrac{\mathbf{amx + a + x^2}}{\mathbf{mx + 1}}$$

14. $(5 - i)(6 + 2i)$

$$= 30 + 10i - 6i - 2i^2 = \mathbf{32 + 4i}$$

15. $\sqrt{-4} - \sqrt{-9} + \sqrt{-2}\sqrt{2} - 4i^2$

$$= 2i - 3i + \sqrt{2}i\sqrt{2} - 4(ii)$$

$$= 2i - 3i + 2i + 4 = \mathbf{4 + i}$$

16.

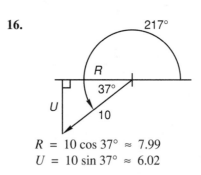

$R = 10 \cos 37° \approx 7.99$

$U = 10 \sin 37° \approx 6.02$

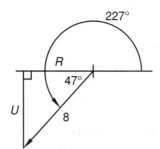

$R = 8 \cos 47° \approx 5.46$

$U = 8 \sin 47° \approx 5.85$

$$\begin{array}{r} -7.99R - 6.02U \\ -5.46R - 5.85U \\ \hline \mathbf{-13.45R - 11.87U} \end{array}$$

17. $H = \sqrt{4^2 + 3^2}$

$H = 5$

$\tan \theta = \dfrac{3}{4}$

$\theta \approx 36.87°$

Since θ is a third quadrant angle:

$\theta = 36.87 + 180 = 216.87°$

$\underline{5 / 216.87°}$

18. (a) $\dfrac{2}{7}x - \dfrac{2}{5}y = 0$

(b) $0.2x - 0.04y = 2.4$

(a') $10x - 14y = 0$

(b') $20x - 4y = 240$

$$-2(a') \quad -20x + 28y = 0$$
$$(b') \quad \underline{20x - 4y = 240}$$
$$24y = 240$$
$$y = 10$$

(b') $20x - 4(10) = 240$
$$20x = 280$$
$$x = 14$$

(14, 10)

19.
$$-x = -1 - 4x^2$$
$$\left(x^2 - \dfrac{1}{4}x + \right) = -\dfrac{1}{4}$$
$$x^2 - \dfrac{1}{4}x + \dfrac{1}{64} = -\dfrac{1}{4} + \dfrac{1}{64}$$
$$\left(x - \dfrac{1}{8}\right)^2 = -\dfrac{15}{64}$$
$$x - \dfrac{1}{8} = \pm\dfrac{\sqrt{15}}{8}i$$
$$x = \dfrac{1}{8} \pm \dfrac{\sqrt{15}}{8}i$$

20.
$$3x^2 + 5 = -2x$$
$$\left(x^2 + \dfrac{2}{3}x + \right) = -\dfrac{5}{3}$$
$$x^2 + \dfrac{2}{3}x + \dfrac{1}{9} = -\dfrac{5}{3} + \dfrac{1}{9}$$
$$\left(x + \dfrac{1}{3}\right)^2 = -\dfrac{14}{9}$$
$$x + \dfrac{1}{3} = \pm\dfrac{\sqrt{14}}{3}i$$
$$x = -\dfrac{1}{3} \pm \dfrac{\sqrt{14}}{3}i$$

21. Since this is a 30-60-90 triangle:
$$\sqrt{3} \times \overrightarrow{SF} = 5$$
$$\overrightarrow{SF} = \dfrac{5}{\sqrt{3}} = \dfrac{5\sqrt{3}}{3}$$
$$x = \dfrac{5\sqrt{3}}{3}(1) = \dfrac{5\sqrt{3}}{3}$$
$$y = \dfrac{5\sqrt{3}}{3}(2) = \dfrac{10\sqrt{3}}{3}$$

22.
$$\dfrac{400\ \text{cm}^3}{\text{hr}} \times \dfrac{1\ \text{in.}}{2.54\ \text{cm}} \times \dfrac{1\ \text{in.}}{2.54\ \text{cm}} \times \dfrac{1\ \text{in.}}{2.54\ \text{cm}}$$
$$\times \dfrac{1\ \text{hr}}{60\ \text{min}} = \dfrac{400}{(2.54)(2.54)(2.54)(60)}\ \dfrac{\text{in.}^3}{\text{min}}$$

23. $Ag = mAu + b$

Use the graph to find the slope.

$$m = -\dfrac{8}{40} = -0.2$$
$$Ag = -0.2Au + b$$

Use the point (120, 2) for Au and Ag.
$$2 = -0.2(120) + b$$
$$26 = b$$
$$Ag = -0.2Au + 26$$

24. $\sqrt{2\sqrt[6]{4}} = \sqrt{2\sqrt[6]{2^2}} = \left[2(2^{1/3})\right]^{1/2}$
$$= 2^{1/2}2^{1/6} = 2^{2/3}$$

25. $\sqrt[5]{x^2yp}\ \sqrt[3]{xy^2} = (x^2yp)^{1/5}(xy^2)^{1/3}$
$$= x^{2/5}y^{1/5}p^{1/5}x^{1/3}y^{2/3} = x^{11/15}y^{13/15}p^{1/5}$$

26. $4\sqrt{\dfrac{3}{11}} + 2\sqrt{\dfrac{11}{3}} - 2\sqrt{297}$

$$= \dfrac{4\sqrt{3}}{\sqrt{11}}\dfrac{\sqrt{11}}{\sqrt{11}} + \dfrac{2\sqrt{11}}{\sqrt{3}}\dfrac{\sqrt{3}}{\sqrt{3}} - 6\sqrt{33}$$

$$= \dfrac{12\sqrt{33}}{33} + \dfrac{22\sqrt{33}}{33} - \dfrac{198\sqrt{33}}{33} = -\dfrac{164\sqrt{33}}{33}$$

27. $\dfrac{a(b + c)}{x} - \dfrac{m}{y} = p$
$$aby + acy - mx = pxy$$
$$aby + acy = pxy + mx$$
$$\dfrac{aby + acy}{py + m} = x$$

28. $\dfrac{x}{a(b + c)} - \dfrac{y}{m} = p$
$$mx - aby - acy = abmp + acmp$$
$$mx - aby - abmp = acmp + acy$$
$$\dfrac{mx - aby - abmp}{amp + ay} = c$$

29. Graph the line to find the slope.

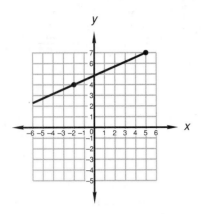

Slope $= \dfrac{3}{7}$

$y = \dfrac{3}{7}x + b$

Use the point $(-2, 4)$ for x and y.

$4 = \dfrac{3}{7}(-2) + b$

$\dfrac{34}{7} = b$

$y = \dfrac{3}{7}x + \dfrac{34}{7}$

30. $D^2 = 3^2 + 7^2$
$D^2 = 58$
$D = \sqrt{58}$

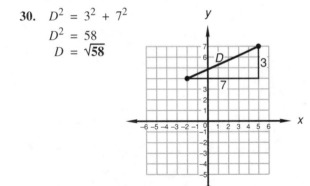

PROBLEM SET 68

1. $D = kP$

$800 = k(9600)$

$\dfrac{1}{12} = k$

$D = \dfrac{1}{12}(24,000) = \mathbf{2000}$

2. $P_1V_1 = P_2V_2$
$(1200)(200) = (1600)V_2$
$V_2 = \mathbf{150 \ ml}$

3. $0.4(800) + 0.2(D_N) = 0.36(800 + D_N)$
$320 + 0.2D_N = 288 + 0.36D_N$
$-0.16D_N = -32$
$D_N = \mathbf{200 \ pounds}$

4. (a) $N_F = 2N_E + 14$
(b) $2N_F = 20N_E - 100$
Substitute (a) into (b) and get:
(b′) $2(2N_E + 14) = 20N_E - 100$
$-16N_E = -128$
$N_E = \mathbf{8}$
(a) $N_F = 2(8) + 14 = \mathbf{30}$

5. $\dfrac{80}{100} \times T = 6720$

$T = 6720 \cdot \dfrac{100}{80}$

$T = \mathbf{8400}$

6. (a) $\dfrac{0.000418 \times 10^{-14}}{501,635 \times 10^6} \approx \dfrac{4 \times 10^{-18}}{5 \times 10^{11}}$
$= 8 \times 10^{-30}$
Calculator $= \mathbf{8.33 \times 10^{-30}}$

(b) $(0.00037 \times 10^{-13})(7231 \times 10^4)$
$\approx (4 \times 10^{-17})(7 \times 10^7) = 2.8 \times 10^{-9}$
Calculator $= \mathbf{2.68 \times 10^{-9}}$

7. (a) $\sqrt[3.8]{192} \rightarrow 3^4 = 81$ and $4^4 = 256$
Answer should be between 3 and 4.
Calculator $= \mathbf{3.99}$

(b) $(1.76)^{-3.42} \rightarrow 1^{-3} = 1$ and $2^{-3} = 0.125$
Answer should be between 1 and 0.125.
Calculator $= \mathbf{0.14}$

8. Since this is a 30-60-90 triangle:
$1 \times \overrightarrow{SF} = 2\sqrt{3}$
$\overrightarrow{SF} = 2\sqrt{3}$

$x = 2\sqrt{3}(2) = \mathbf{4\sqrt{3}}$
$y = 2\sqrt{3}(\sqrt{3}) = \mathbf{6}$

9. Since the area of a triangle is equal to half the product of the base and height, and all triangles in the parallelogram whose base is $\overline{AD}$ share the same height, **all have equal areas.**

10. $m\widehat{MN} = x$
$x + (2x + 10) + 54 + 89 = 360$
$3x = 207$
$x = 69°$
so $m\widehat{MN} = \mathbf{69°}$

11. $\dfrac{1}{-2 - \sqrt{2}} \cdot \dfrac{-2 + \sqrt{2}}{-2 + \sqrt{2}} = \dfrac{-2 + \sqrt{2}}{4 - 2}$

$= \dfrac{-2 + \sqrt{2}}{2}$

12. $\dfrac{4}{3\sqrt{2} - 1} \cdot \dfrac{3\sqrt{2} + 1}{3\sqrt{2} + 1} = \dfrac{12\sqrt{2} + 4}{18 - 1}$

$= \dfrac{12\sqrt{2} + 4}{17}$

13. $\dfrac{2}{3\sqrt{3} - 5} \cdot \dfrac{3\sqrt{3} + 5}{3\sqrt{3} + 5}$

$= \dfrac{6\sqrt{3} + 10}{27 - 25} = 3\sqrt{3} + 5$

14. $\dfrac{-7}{-x + 3} - \dfrac{2x}{x^2 - 9}$

$= \dfrac{7(x + 3)}{x^2 - 9} - \dfrac{2x}{x^2 - 9} = \dfrac{5x + 21}{x^2 - 9}$

15. $\dfrac{2}{4 - x} - \dfrac{3}{x^2 - 16}$

$= \dfrac{-2(x + 4)}{x^2 - 16} - \dfrac{3}{x^2 - 16} = \dfrac{-2x - 11}{x^2 - 16}$

16. (a) $R_G T_G = 140$
(b) $R_B T_B = 140$
(c) $R_B = 2R_G$
(d) $T_B = T_G - 7$

Substitute (c) and (d) into (b) and get:
(b') $2R_G(T_G - 7) = 140$
Substitute (a) into (b') and get:
(b'') $2(140) - 14R_G = 140$
$-14R_G = -140$
$R_G = 10$
$R_B = 20;\ T_G = 14;\ T_B = 7$

17. $4r + \dfrac{m}{x + \dfrac{m}{x}} = 4r + \dfrac{m}{\dfrac{x^2 + m}{x}}$

$= 4r + \dfrac{mx}{x^2 + m} = \dfrac{4rx^2 + 4mr + mx}{x^2 + m}$

18. $3a + \dfrac{ax}{a + \dfrac{x}{a}} = 3a + \dfrac{ax}{\dfrac{a^2 + x}{a}}$

$= 3a + \dfrac{a^2 x}{a^2 + x} = \dfrac{3a^3 + 3ax + a^2 x}{a^2 + x}$

19. $(5i - 2)(3i + 4)$
$= 15i^2 + 20i - 6i - 8 = -23 + 14i$

20. $\sqrt{-2}\sqrt{2} - 3i^2 - \sqrt{-9} + 2i^4 + 4$
$= \sqrt{2}i\sqrt{2} - 3(ii) - 3i + 2(ii)(ii) + 4$
$= 2i + 3 - 3i + 2 + 4 = 9 - i$

21.

$R = 40 \cos 45° \approx 28.28$
$U = 40 \sin 45° \approx 28.28$

$R = 10 \cos 24° \approx 9.14$
$U = 10 \sin 24° \approx 4.07$

$\begin{array}{r} 28.28R - 28.28U \\ 9.14R + 4.07U \\ \hline 37.42R - 24.21U \end{array}$

22. (a) $\dfrac{2}{5}x - \dfrac{2}{3}y = -2$
(b) $-0.06x - 0.4y = -7.2$

(a') $6x - 10y = -30$
(b') $\dfrac{-6x - 40y = -720}{- 50y = -750}$
$y = 15$

(a') $6x - 10(15) = -30$
$6x = 120$
$x = 20$

(20, 15)

23. $-2x = -1 - 4x^2$

$\left(x^2 - \dfrac{1}{2}x + \right) = -\dfrac{1}{4}$

$x^2 - \dfrac{1}{2}x + \dfrac{1}{16} = -\dfrac{1}{4} + \dfrac{1}{16}$

$\left(x - \dfrac{1}{4}\right)^2 = -\dfrac{3}{16}$

$x - \dfrac{1}{4} = \pm\dfrac{\sqrt{3}}{4}i$

$x = \dfrac{1}{4} \pm \dfrac{\sqrt{3}}{4}i$

24.
$$-3x^2 + 5 = -2x$$
$$\left(x^2 - \frac{2}{3}x + \quad\right) = \frac{5}{3}$$
$$x^2 - \frac{2}{3}x + \frac{1}{9} = \frac{5}{3} + \frac{1}{9}$$
$$\left(x - \frac{1}{3}\right)^2 = \frac{16}{9}$$
$$x - \frac{1}{3} = \pm\frac{4}{3}$$
$$x = \frac{1}{3} \pm \frac{4}{3}$$
$$x = \frac{5}{3}, -1$$

25. $\dfrac{700 \text{ cm}^3}{\text{min}} \times \dfrac{1 \text{ in.}}{2.54 \text{ cm}} \times \dfrac{1 \text{ in.}}{2.54 \text{ cm}} \times \dfrac{1 \text{ in.}}{2.54 \text{ cm}}$

$\times \dfrac{60 \text{ min}}{1 \text{ hr}} = \dfrac{\mathbf{700(60)}}{\mathbf{(2.54)(2.54)(2.54)}} \dfrac{\mathbf{in.^3}}{\mathbf{hr}}$

26. $\sqrt[7]{3\sqrt[3]{3}} = [3(3^{1/3})]^{1/7} = 3^{1/7}3^{1/21} = \mathbf{3^{4/21}}$

27. $\sqrt[3]{xy^5}\sqrt{x^5y} = (xy^5)^{1/3}(x^5y)^{1/2}$
$$= x^{1/3}y^{5/3}x^{5/2}y^{1/2} = \mathbf{x^{17/6}y^{13/6}}$$

28. $3\sqrt{\dfrac{2}{3}} - 5\sqrt{\dfrac{3}{2}} + 2\sqrt{24}$

$$= \dfrac{3\sqrt{2}}{\sqrt{3}}\dfrac{\sqrt{3}}{\sqrt{3}} - \dfrac{5\sqrt{3}}{\sqrt{2}}\dfrac{\sqrt{2}}{\sqrt{2}} + 4\sqrt{6}$$

$$= \dfrac{6\sqrt{6}}{6} - \dfrac{15\sqrt{6}}{6} + \dfrac{24\sqrt{6}}{6} = \mathbf{\dfrac{5\sqrt{6}}{2}}$$

29.
$$\dfrac{m}{x(a+b)} - \dfrac{p}{y} = c$$
$$my - apx - bpx = acxy + bcxy$$
$$my - apx - acxy = bcxy + bpx$$
$$\mathbf{\dfrac{my - apx - acxy}{cxy + px} = b}$$

30. Graph the line to find the slope.

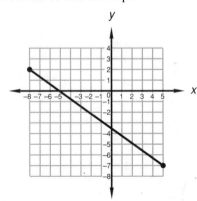

Slope $= -\dfrac{9}{13}$

$y = -\dfrac{9}{13}x + b$

Use the point $(-8, 2)$ for x and y.

$$2 = -\dfrac{9}{13}(-8) + b$$
$$-\dfrac{46}{13} = b$$
$$y = -\dfrac{9}{13}x - \dfrac{46}{13}$$

PROBLEM SET 69

1. $\dfrac{P_1V_1}{T_1} = \dfrac{P_2V_2}{T_2}$

$$T_2 = \dfrac{P_2V_2T_1}{P_1V_1}$$

$$= \dfrac{(0.0008 \times 10^{16})(0.013 \times 10^{-12})(800 \times 10^5)}{(80 \times 10^5)(0.0005 \times 10^{-7})}$$

$$= \dfrac{(8 \times 10^{12})(1.3 \times 10^{-14})(8 \times 10^7)}{(8 \times 10^6)(5 \times 10^{-11})}$$

$T_2 = \mathbf{2.08 \times 10^{10} \text{ K}}$

2. Length of side $= \dfrac{7}{\sqrt{2}} = \dfrac{7\sqrt{2}}{2}$ **units**

Area $= s^2 = \left(\dfrac{7\sqrt{2}}{2}\right)^2 = \dfrac{49}{2}$ **units²**

3. $B = \dfrac{k}{T}$

$400 = \dfrac{k}{5}$

$2000 = k$

$B = \dfrac{2000}{25} = \mathbf{80 \text{ tons}}$

4. (a) $0.8(P_N) + 0.4(D_N) = 1280$
(b) $P_N + D_N = 2000$
Substitute $D_N = 2000 - P_N$ into (a) and get:
(a') $0.8(P_N) + 0.4(2000 - P_N) = 1280$
$$0.4P_N = 480$$
$$P_N = 1200$$
(b) $(1200) + D_N = 2000$
$$D_N = 800$$
1200 ml 80%, 800 ml 40%

5. $R_TT_T + 40 = R_BT_B;$
$R_B = 50; R_T = 70;$
$T_B = T_T + 2$

$70T_T + 40 = 50(T_T + 2)$
$$20T_T = 60$$
$$T_T = 3 \text{ hr}$$
Since the train started at 2 p.m. and took 3 hours to get within 40 miles of the bus, it was **5 p.m.**

6. $\dfrac{1}{x-5} - \dfrac{2x-3}{5-x} = \dfrac{1}{x-5} + \dfrac{2x-3}{x-5}$

$= \dfrac{2x-2}{x-5}$

7. $\dfrac{2x+3}{x-2} + \dfrac{2x}{2-x} - \dfrac{3}{-x+2}$

$= \dfrac{2x+3}{x-2} - \dfrac{2x}{x-2} + \dfrac{3}{x-2} = \dfrac{6}{x-2}$

8. $\dfrac{3x+5}{x-5} + \dfrac{2}{x^2-25}$

$= \dfrac{(3x+5)(x+5)}{x^2-25} + \dfrac{2}{x^2-25}$

$= \dfrac{3x^2+15x+5x+25+2}{x^2-25}$

$= \dfrac{3x^2+20x+27}{x^2-25}$

9. $\dfrac{4}{3-\sqrt{2}} \cdot \dfrac{3+\sqrt{2}}{3+\sqrt{2}} = \dfrac{12+4\sqrt{2}}{9-2}$

$= \dfrac{12+4\sqrt{2}}{7}$

10. $\dfrac{2}{5-3\sqrt{2}} \cdot \dfrac{5+3\sqrt{2}}{5+3\sqrt{2}} = \dfrac{10+6\sqrt{2}}{25-18}$

$= \dfrac{10+6\sqrt{2}}{7}$

11. $\dfrac{2}{3-2\sqrt{8}} \cdot \dfrac{3+2\sqrt{8}}{3+2\sqrt{8}} = \dfrac{6+8\sqrt{2}}{9-32}$

$= \dfrac{-6-8\sqrt{2}}{23}$

12. (a) $R_G T_G = 171$
(b) $R_R T_R = 171$
(c) $R_R = 3R_G$
(d) $T_R = T_G - 6$

Substitute (c) and (d) into (b) and get:
(b') $3R_G(T_G - 6) = 171$
Substitute (a) into (b') and get:
(b'') $3(171) - 18R_G = 171$
$-18R_G = -342$
$R_G = 19$
$R_R = 57;\ T_G = 9;\ T_R = 3$

13. $m + \dfrac{1}{m + \dfrac{1}{m}} = m + \dfrac{1}{\dfrac{m^2+1}{m}}$

$= m + \dfrac{m}{m^2+1} = \dfrac{m^3+2m}{m^2+1}$

14. $x + \dfrac{a}{x + \dfrac{1}{a}} = x + \dfrac{a}{\dfrac{ax+1}{a}} = x + \dfrac{a^2}{ax+1}$

$= \dfrac{ax^2+x+a^2}{ax+1}$

15. $(4i-2)(2i-4) = 8i^2 - 16i - 4i + 8 = -20i$

16. $-\sqrt{-2}\sqrt{-2} + 2i^3 - i^2 = -\sqrt{2}i\sqrt{2}i + 2i(ii) - (ii)$
$= 2 - 2i + 1 = 3 - 2i$

17.

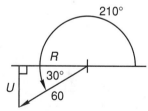

$R = 20\cos 30° \approx 17.32$
$U = 20\sin 30° = 10$

$R = 60\cos 30° \approx 51.96$
$U = 60\sin 30° = 30$

$\begin{array}{r} 17.32R + 10U \\ -51.96R - 30U \\ \hline -34.64R - 20U \end{array}$

18. $H = \sqrt{2^2 + 3^2}$
$H = \sqrt{13}$

$\tan\theta = \dfrac{3}{2}$
$\theta \approx 56.31°$

Since θ is a fourth-quadrant angle:
$\theta = -(56.31) + 360 = 303.69°$

$\sqrt{13}\underline{/303.69°}$

19. (a) $\frac{2}{3}x + \frac{2}{5}y = 28$

(b) $-0.05x - 0.2y = -5.5$

(a') $10x + 6y = 420$
(b') $-5x - 20y = -550$

$$\begin{array}{r} \text{(a')} \quad 10x + 6y = 420 \\ 2\text{(b')} \; -10x - 40y = -1100 \\ \hline -34y = -680 \\ y = 20 \end{array}$$

(a') $10x + 6(20) = 420$
$10x = 300$
$x = 30$

(30, 20)

20.
$$3x^2 + 8 = 5x$$
$$\left(x^2 - \frac{5}{3}x + \quad\right) = -\frac{8}{3}$$
$$x^2 - \frac{5}{3}x + \frac{25}{36} = -\frac{8}{3} + \frac{25}{36}$$
$$\left(x - \frac{5}{6}\right)^2 = -\frac{71}{36}$$
$$x - \frac{5}{6} = \pm\frac{\sqrt{71}}{6}i$$
$$x = \frac{5}{6} \pm \frac{\sqrt{71}}{6}i$$

21.
$$3x^2 + 8 = -5x$$
$$\left(x^2 + \frac{5}{3}x + \quad\right) = -\frac{8}{3}$$
$$x^2 + \frac{5}{3}x + \frac{25}{36} = -\frac{8}{3} + \frac{25}{36}$$
$$\left(x + \frac{5}{6}\right)^2 = -\frac{71}{36}$$
$$x + \frac{5}{6} = \pm\frac{\sqrt{71}}{6}i$$
$$x = -\frac{5}{6} \pm \frac{\sqrt{71}}{6}i$$

22. Since $\triangle ABC$ is equilateral:
$m\widehat{BC} = m\widehat{AB} = m\widehat{AC} = \mathbf{120°}$ and
$\overline{BC} = \overline{AB} = \overline{AC} = 10$

Use the Pythagorean theorem to find the height.
$10^2 = H^2 + 5^2$
$H^2 = 100 - 25$
$H = 5\sqrt{3}$

Area $= \dfrac{10 \times 5\sqrt{3}}{2} = \mathbf{25\sqrt{3} \text{ units}^2}$

23. $\dfrac{4 \text{ ft}^3}{\text{min}} \times \dfrac{12 \text{ in.}}{1 \text{ ft}} \times \dfrac{12 \text{ in.}}{1 \text{ ft}} \times \dfrac{12 \text{ in.}}{1 \text{ ft}} \times \dfrac{60 \text{ min}}{1 \text{ hr}}$

$= 4(12)(12)(12)(60)\dfrac{\text{in.}^3}{\text{hr}}$

24. $Cr = mV + b$
Use the graph to find the slope.

$m = \dfrac{900}{24} = 37$
$Cr = 37V + b$
Use the point (62, 200) for V and Cr.
$200 = 37(62) + b$
$-2094 = b$
$\mathbf{Cr = 37V - 2094}$

25. $\sqrt{27\sqrt[4]{3}} = \sqrt{3^3\sqrt[4]{3}} = \left[3^3(3^{1/4})\right]^{1/2}$
$= 3^{3/2}3^{1/8} = \mathbf{3^{13/8}}$

26. $\sqrt[3]{xm^5}\sqrt[4]{xm^2} = (xm^5)^{1/3}(xm^2)^{1/4}$
$= x^{1/3}m^{5/3}x^{1/4}m^{1/2} = \mathbf{x^{7/12}m^{13/6}}$

27. $2\sqrt{\dfrac{9}{2}} + 5\sqrt{\dfrac{2}{9}} - 5\sqrt{50}$

$= \dfrac{6}{\sqrt{2}}\dfrac{\sqrt{2}}{\sqrt{2}} + \dfrac{5\sqrt{2}}{3} - 25\sqrt{2}$

$= \dfrac{18\sqrt{2}}{6} + \dfrac{10\sqrt{2}}{6} - \dfrac{150\sqrt{2}}{6} = \mathbf{-\dfrac{61\sqrt{2}}{3}}$

28. $\dfrac{a(x + y)}{m} - \dfrac{c}{z} = k$
$axz + ayz - cm = kmz$
$axz = kmz + cm - ayz$
$x = \dfrac{kmz + cm - ayz}{az}$

29. $\dfrac{x + 2}{5} - \dfrac{3x - 3}{4} = 2$
$4x + 8 - 15x + 15 = 40$
$-11x = 17$
$x = -\dfrac{17}{11}$

30. (a) $\dfrac{472.2 \times 10^{-26}}{1658.27 \times 10^{10}}$

$\approx \dfrac{5 \times 10^{-24}}{2 \times 10^{13}} \approx 3 \times 10^{-37}$

Calculator $= \mathbf{2.85 \times 10^{-37}}$

(b) $(1.24)^{-2.73}$
Answer should be between 1 and 0.125.
Calculator $= \mathbf{0.56}$

PROBLEM SET 70

1. $\dfrac{P_1 V_1}{T_1} = \dfrac{P_2 V_2}{T_2}$

$T_2 = \dfrac{P_2 V_2 T_1}{P_1 V_1}$

$T_2 = \dfrac{(0.04 \times 10^5)(500)(4 \times 10^3)}{(0.001 \times 10^{-13})(0.04 \times 10^{14})}$

$T_2 = \dfrac{(4 \times 10^3)(5 \times 10^2)(4 \times 10^3)}{(1 \times 10^{-16})(4 \times 10^{12})}$

$T_2 = \mathbf{2 \times 10^{13}\ K}$

2. $0.2(400) + 0.5(D_N) = 0.26(400 + D_N)$

$80 + 0.5 D_N = 104 + 0.26 D_N$

$0.24 D_N = 24$

$D_N = \mathbf{100\ milliliters}$

3. (a) $50 N_T + 100 N_F = 5000$

(b) $N_T + N_F = 60$

Substitute $N_F = 60 - N_T$ into (a) and get:

(a') $50 N_T + 100(60 - N_T) = 5000$

$\qquad\qquad -50 N_T = -1000$

$\qquad\qquad\qquad N_T = \mathbf{20}$

(b) $N_F = 60 - (20) = \mathbf{40}$

4. Calcium: $\quad 1 \times 40 = 40$
Sulfur: $\qquad 1 \times 32 = 32$
Oxygen: $\quad 4 \times 16 = 64$
Total: $\qquad\qquad\ = 136$

$\dfrac{32}{136} = \dfrac{S}{680}$

$S = \mathbf{160\ grams}$

5. $N \qquad N + 2 \qquad N + 4$

$4(N)(N + 4) = -10(N + 2 + N + 4) + 28$

$4N^2 + 16N = -20N - 32$

$4N^2 + 36N + 32 = 0$

$N^2 + 9N + 8 = 0$

$(N + 8)(N + 1) = 0$

$\qquad\qquad N = -8, -1$

Since the problem asked for only the even integers, the desired integers are $\mathbf{-8, -6, -4.}$

6. $m = x\left(\dfrac{1}{2p} + \dfrac{3z}{b}\right)$

$m = \dfrac{x}{2p} + \dfrac{3xz}{b}$

$2bmp = bx + 6pxz$

$2bmp - bx = 6pxz$

$b = \dfrac{6pxz}{2mp - x}$

7. $ac = x\left(\dfrac{m}{r + s} + \dfrac{t}{z}\right)$

$ac = \dfrac{mx}{r + s} + \dfrac{tx}{z}$

$acrz + acsz = mxz + rtx + stx$

$acsz - stx = mxz + rtx - acrz$

$s = \dfrac{mxz + rtx - acrz}{acz - tx}$

8. $\dfrac{3x - 2}{x - 2} - \dfrac{3x}{2 - x} = \dfrac{3x - 2}{x - 2} + \dfrac{3x}{x - 2}$

$= \dfrac{6x - 2}{x - 2}$

9. $\dfrac{4}{x^2 + 8x + 12} + \dfrac{4x - 5}{x + 2}$

$= \dfrac{4}{(x + 6)(x + 2)} + \dfrac{(4x - 5)(x + 6)}{(x + 6)(x + 2)}$

$= \dfrac{4}{x^2 + 8x + 12} + \dfrac{4x^2 + 19x - 30}{x^2 + 8x + 12}$

$= \dfrac{\mathbf{4x^2 + 19x - 26}}{\mathbf{x^2 + 8x + 12}}$

10. $\dfrac{2}{\sqrt{2} - 4} \cdot \dfrac{\sqrt{2} + 4}{\sqrt{2} + 4} = \dfrac{2\sqrt{2} + 8}{2 - 16}$

$= \dfrac{\mathbf{-\sqrt{2} - 4}}{\mathbf{7}}$

11. $\dfrac{2}{3\sqrt{12} - 2} \cdot \dfrac{3\sqrt{12} + 2}{3\sqrt{12} + 2} = \dfrac{12\sqrt{3} + 4}{108 - 4}$

$= \dfrac{\mathbf{3\sqrt{3} + 1}}{\mathbf{26}}$

12. $\dfrac{2}{2\sqrt{3} - 2} \cdot \dfrac{2\sqrt{3} + 2}{2\sqrt{3} + 2} = \dfrac{4\sqrt{3} + 4}{12 - 4}$

$= \dfrac{\mathbf{\sqrt{3} + 1}}{\mathbf{2}}$

13. (a) $R_B T_B = 65$

(b) $R_X T_X = 104$

(c) $R_X = 2 R_B$

(d) $T_X = T_B - 1$

Substitute (c) and (d) into (b) and get:

(b') $2 R_B (T_B - 1) = 104$

Substitute (a) into (b') and get:

(b'') $2(65) - 2 R_B = 104$

$\qquad\qquad -2 R_B = -26$

$\qquad\qquad\quad R_B = \mathbf{13}$

$R_X = \mathbf{26}; \ T_B = \mathbf{5}; \ T_X = \mathbf{4}$

14. $xy + \dfrac{a}{1 + \dfrac{a}{b}} = xy + \dfrac{a}{\dfrac{b + a}{b}}$

$= xy + \dfrac{ab}{b + a} = \dfrac{xyb + xya + ab}{b + a}$

15. $\dfrac{m}{y} + \dfrac{x}{a + \dfrac{1}{y}} = \dfrac{m}{y} + \dfrac{x}{\dfrac{ay + 1}{y}}$

$= \dfrac{m}{y} + \dfrac{xy}{ay + 1} = \dfrac{amy + m + xy^2}{y(ay + 1)}$

16. $(2 - 3i)(5 - 6i)$

$= 10 - 12i - 15i + 18i^2 = \mathbf{-8 - 27i}$

17. $-\sqrt{-4} - \sqrt{-2}\sqrt{-3} + \sqrt{-9} = -2i - \sqrt{2}i\sqrt{3}i + 3i$

$= -2i + \sqrt{6} + 3i = \sqrt{6} + i$

18.

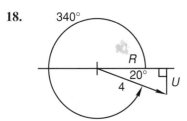

$R = 4\cos 20° \approx 3.76$
$U = 4\sin 20° \approx 1.37$

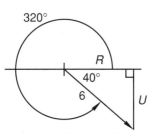

$R = 6\cos 40° \approx 4.60$
$U = 6\sin 40° \approx 3.86$

$3.76R - 1.37U$
$\underline{4.60R - 3.86U}$
$\mathbf{8.36R - 5.23U}$

19. $H = \sqrt{5^2 + 4^2}$
$H = \sqrt{41}$

$\tan\theta = \dfrac{5}{4}$
$\theta \approx 51.34°$
Since θ is a second-quadrant
angle:
$\theta = -(51.34) + 180 = 128.66°$

$\mathbf{\sqrt{41}/128.66°}$

20. (a) $\dfrac{1}{5}x - \dfrac{1}{4}y = -2$
(b) $0.07x + 0.3y = 5.5$
(a') $4x - 5y = -40$
(b') $7x + 30y = 550$

$6(a')$ $24x - 30y = -240$
(b') $\underline{7x + 30y = 550}$
$31x = 310$
$x = 10$
(a') $4(10) - 5y = -40$
$-5y = -80$
$y = 16$
(10, 16)

21. $3x^2 + 1 = 4x$
$\left(x^2 - \dfrac{4}{3}x + \right) = -\dfrac{1}{3}$
$x^2 - \dfrac{4}{3}x + \dfrac{4}{9} = -\dfrac{1}{3} + \dfrac{4}{9}$
$\left(x - \dfrac{2}{3}\right)^2 = \dfrac{1}{9}$
$x - \dfrac{2}{3} = \pm\dfrac{1}{3}$
$x = \dfrac{2}{3} \pm \dfrac{1}{3}$
$x = \mathbf{1, \dfrac{1}{3}}$

22. $-4x + 7 = -3x^2$
$\left(x^2 - \dfrac{4}{3}x + \right) = -\dfrac{7}{3}$
$x^2 - \dfrac{4}{3}x + \dfrac{4}{9} = -\dfrac{7}{3} + \dfrac{4}{9}$
$\left(x - \dfrac{2}{3}\right)^2 = -\dfrac{17}{9}$
$x - \dfrac{2}{3} = \pm\dfrac{\sqrt{17}}{3}i$
$x = \mathbf{\dfrac{2}{3} \pm \dfrac{\sqrt{17}}{3}i}$

23. Since this is a 30-60-90 triangle:
$1 \times \overrightarrow{SF} = 4$
$\overrightarrow{SF} = 4$
$a = 4(\sqrt{3}) = \mathbf{4\sqrt{3}}$
$b = 4(2) = \mathbf{8}$

24. $Co = mNi + b$
Use the graph to find the slope.
$m = \dfrac{-20}{13} = -1.5$
$Co = -1.5Ni + b$
Use the point (138, 50) for Ni and Co.
$50 = -1.5(138) + b$
$257 = b$
$Co = \mathbf{-1.5Ni + 257}$

25. (a) $\dfrac{46,831 \times 10^{-42}}{9140.26 \times 10^{-33}} \approx \dfrac{5 \times 10^{-38}}{9 \times 10^{-30}}$

$\approx 6 \times 10^{-9}$

Calculator = **5.12×10^{-9}**

(b) $\sqrt[2.7]{146}$

Answer should be between 5 and 7.

Calculator = **6.33**

26. $\sqrt[5]{x^4 y^3}\ \sqrt[4]{x^2 y} = (x^4 y^3)^{1/5}(x^2 y)^{1/4}$

$= x^{4/5} y^{3/5} x^{1/2} y^{1/4} = \mathbf{x^{13/10} y^{17/20}}$

27. $\sqrt{\dfrac{7}{2}} + 2\sqrt{\dfrac{2}{7}} - 2\sqrt{126}$

$= \dfrac{\sqrt{7}}{\sqrt{2}}\dfrac{\sqrt{2}}{\sqrt{2}} + \dfrac{2\sqrt{2}}{\sqrt{7}}\dfrac{\sqrt{7}}{\sqrt{7}} - 6\sqrt{14}$

$= \dfrac{7\sqrt{14}}{14} + \dfrac{4\sqrt{14}}{14} - \dfrac{84\sqrt{14}}{14} = \mathbf{-\dfrac{73\sqrt{14}}{14}}$

28.

$$x - 1 \overline{)\ 4x^3 + 0x^2 + 0x - 2}$$

quotient: $4x^2 + 4x + 4 + \dfrac{2}{x-1}$

$\underline{4x^3 - 4x^2}$

$4x^2 + 0x$

$\underline{4x^2 - 4x}$

$4x - 2$

$\underline{4x - 4}$

$+ 2$

29. Write the equation of the given line in slope-intercept form.

$4x + 3y = 5$

$y = -\dfrac{4}{3}x + \dfrac{5}{3}$

Since the slopes of perpendicular lines are negative reciprocals of each other:

$m_\perp = \dfrac{3}{4}$

$y = \dfrac{3}{4}x + b$

$5 = \dfrac{3}{4}(-2) + b$

$\dfrac{13}{2} = b$

$\mathbf{y = \dfrac{3}{4}x + \dfrac{13}{2}}$

30. $\dfrac{4x - 2}{5} - \dfrac{3x - 2}{4} = 10$

$16x - 8 - 15x + 10 = 200$

$x = \mathbf{198}$

PROBLEM SET 71

1. $P_1 V_1 = P_2 V_2$

$P_2 = \dfrac{P_1 V_1}{V_2}$

$P_2 = \dfrac{(40,000 \times 10^{-3})(4000 \times 10^{-2})}{8000 \times 10^{-2}}$

$P_2 = \dfrac{(4 \times 10^1)(4 \times 10^1)}{(8 \times 10^1)} = \mathbf{20\ N/m^2}$

2. $F = kP$

$240 = k(4800)$

$\dfrac{1}{20} = k$

$600 = \left(\dfrac{1}{20}\right)P$

$\mathbf{12,000} = P$

3. (a) $0.1P_N + 0.3D_N = 34$

(b) $P_N + D_N = 200$

Substitute $D_N = 200 - P_N$ into (a) and get:

(a′) $0.1P_N + 0.3(200 - P_N) = 34$

$-0.2P_N = -26$

$P_N = 130$

(b) $(130) + D_N = 200$

$D_N = 70$

130 ml 10%, 70 ml 30%

4. $R_R T_R = R_J T_J$;

$R_R = 8$; $R_J = 24$;

$T_R + T_J = 8$

$8T_R = 24(8 - T_R)$

$32T_R = 192$

$T_R = 6\ hr$

$D_R = D_J = 6(8) = \mathbf{48\ miles}$

5. Hydrogen: $2 \times 1 = 2$

Carbon: $1 \times 12 = 12$

Oxygen: $3 \times 16 = 48$

Total: $= 62$

$\dfrac{H}{T} = \dfrac{2}{62} \approx 0.032 = \mathbf{3.2\%}$

6. See Lesson 71.

7. $4x^2 - 2x - 6 = 0$

$x = \dfrac{-(-2) \pm \sqrt{(-2)^2 - 4(4)(-6)}}{2(4)}$

$x = \dfrac{2 \pm \sqrt{100}}{8} = \dfrac{2 \pm 10}{8} = \mathbf{\dfrac{3}{2}, -1}$

8.
$$2x^2 = -x - 4$$
$$2x^2 + x + 4 = 0$$
$$x = \frac{-1 \pm \sqrt{(1)^2 - 4(2)(4)}}{2(2)} = \frac{-1 \pm \sqrt{-31}}{4}$$
$$x = -\frac{1}{4} \pm \frac{\sqrt{31}}{4}i$$

9. $x = \dfrac{360 - 280}{2} = 40$

$y = \dfrac{180 - 80}{2} = 50$

Since the diagonals of a rhombus are perpendicular bisectors of each other: $z = 90$

10.
$$r = m\left(\frac{1}{x + c} + \frac{3}{y}\right)$$
$$r = \frac{m}{x + c} + \frac{3m}{y}$$
$$rxy + cry = my + 3mx + 3cm$$
$$rxy - 3mx = my + 3cm - cry$$
$$x = \frac{my + 3cm - cry}{ry - 3m}$$

11. $\dfrac{4x + 5}{x - 2} - \dfrac{4}{2 - x} = \dfrac{4x + 5}{x - 2} + \dfrac{4}{x - 2}$

$= \dfrac{4x + 9}{x - 2}$

12. $\dfrac{4}{2 - 3\sqrt{12}} \cdot \dfrac{2 + 3\sqrt{12}}{2 + 3\sqrt{12}} = \dfrac{8 + 24\sqrt{3}}{4 - 108}$

$= \dfrac{-1 - 3\sqrt{3}}{13}$

13. (a) $R_M T_M = 160$
(b) $R_P T_P = 400$
(c) $R_P = 2R_M$
(d) $T_P = T_M + 1$

Substitute (c) and (d) into (b) and get:
(b′) $2R_M(T_M + 1) = 400$
Substitute (a) into (b′) and get:
(b″) $2(160) + 2R_M = 400$
$$2R_M = 80$$
$$R_M = 40$$
$R_P = 80$; $T_M = 4$; $T_P = 5$

14. $x + \dfrac{x}{1 + \dfrac{1}{x}} = x + \dfrac{x}{\dfrac{x + 1}{x}} = x + \dfrac{x^2}{x + 1}$

$= \dfrac{2x^2 + x}{x + 1}$

15. $a + \dfrac{b}{a + \dfrac{a}{b}} = a + \dfrac{b}{\dfrac{ab + a}{b}}$

$= a + \dfrac{b^2}{ab + a} = \dfrac{a^2b + a^2 + b^2}{ab + a}$

16. $-3i^3 + 2\sqrt{-2}\sqrt{2} - \sqrt{-9} = -3i(ii) + 2\sqrt{2}\sqrt{2}i - 3i$
$= 3i + 4i - 3i = 4i$

17. $(-i - 1)(-3i + 2) = 3i^2 - 2i + 3i - 2$
$= -5 + i$

18.

$R = 20 \cos 70° \approx 6.84$
$U = 20 \sin 70° \approx 18.79$

$R = 10 \cos 40° \approx 7.66$
$U = 10 \sin 40° \approx 6.43$

$6.84R + 18.79U$
$\underline{7.66R + 6.43U}$
$\mathbf{14.50R + 25.22U}$

19. $H = \sqrt{4^2 + 4^2}$
$H = 4\sqrt{2}$

$\tan \theta = \dfrac{4}{4} = 1$
$$\theta = 45°$$
Since θ is a fourth-quadrant angle:
$\theta = -(45) + 360 = 315°$

$\mathbf{4\sqrt{2}\underline{/315°}}$

20. (a) $\dfrac{3}{8}x - \dfrac{1}{2}y = 2$
(b) $0.06x - 0.2y = -0.64$

(a′) $3x - 4y = 16$
(b′) $6x - 20y = -64$

$-2(a′) \quad -6x + 8y = -32$
$(b′) \quad \underline{6x - 20y = -64}$
$\quad\quad -12y = -96$
$\quad\quad\quad y = 8$

(a') $3x - 4(8) = 16$
$$3x = 48$$
$$x = 16$$
(16, 8)

21. $3x^2 - 2x + 5 = 0$

$(x^2 - \frac{2}{3}x + \quad) = -\frac{5}{3}$

$x^2 - \frac{2}{3}x + \frac{1}{9} = -\frac{5}{3} + \frac{1}{9}$

$\left(x - \frac{1}{3}\right)^2 = -\frac{14}{9}$

$x - \frac{1}{3} = \pm\frac{\sqrt{14}}{3}i$

$x = \frac{1}{3} \pm \frac{\sqrt{14}}{3}i$

22. Since this is a 30-60-90 triangle:
$2 \times \overrightarrow{SF} = 11$

$\overrightarrow{SF} = \frac{11}{2}$

$a = \frac{11}{2}(1) = \frac{11}{2}$

$b = \frac{11}{2}(\sqrt{3}) = \frac{11\sqrt{3}}{2}$

23. $Pb = mB + b$
Use the graph to find the slope.

$m = \frac{74}{2} = 37$

$Pb = 37B + b$
Use the point (4, 0) for B and Pb.
$$0 = 37(4) + b$$
$$-148 = b$$
$$Pb = 37B - 148$$

24. $\sqrt{81\sqrt{3}} = \sqrt{3^4\sqrt{3}} = [3^4(3^{1/2})]^{1/2}$
$= 3^2 3^{1/4} = 3^{9/4}$

25. $\sqrt[3]{x^5 y^6}\sqrt{xy^3} = (x^5 y^6)^{1/3}(xy^3)^{1/2}$
$= x^{5/3}y^2 x^{1/2}y^{3/2} = x^{13/6}y^{7/2}$

26. $3\sqrt{\frac{2}{5}} + 3\sqrt{\frac{5}{2}} - 6\sqrt{40}$

$= \frac{3\sqrt{2}}{\sqrt{5}}\frac{\sqrt{5}}{\sqrt{5}} + \frac{3\sqrt{5}}{\sqrt{2}}\frac{\sqrt{2}}{\sqrt{2}} - 12\sqrt{10}$

$= \frac{6\sqrt{10}}{10} + \frac{15\sqrt{10}}{10} - \frac{120\sqrt{10}}{10} = -\frac{99\sqrt{10}}{10}$

27. $(x - 2)^3 = (x - 2)(x - 2)(x - 2)$
$= (x^2 - 4x + 4)(x - 2)$
$= x^3 - 2x^2 - 4x^2 + 8x + 4x - 8$
$= x^3 - 6x^2 + 12x - 8$

28. $\dfrac{x^2 y - \dfrac{1}{y}}{\dfrac{x^2}{y} - 6} = \dfrac{\dfrac{x^2 y^2 - 1}{y}}{\dfrac{x^2 - 6y}{y}} \cdot \dfrac{\dfrac{y}{x^2 - 6y}}{\dfrac{y}{x^2 - 6y}}$

$= \dfrac{x^2 y^2 - 1}{x^2 - 6y}$

29. (a) $\dfrac{-471{,}635 \times 10^5}{0.0071893 \times 10^{-14}}$

$\approx \dfrac{-5 \times 10^{10}}{7 \times 10^{-17}} \approx -7 \times 10^{26}$

Calculator $= -\mathbf{6.56 \times 10^{26}}$

(b) $(2.4)^{-3.06}$
Answer should be between 0.125 and 0.037.
Calculator $= \mathbf{6.86 \times 10^{-2}}$

30. $\dfrac{b}{z} = \dfrac{x}{y}$
$by = xz$
$b = \dfrac{xz}{y}$

PROBLEM SET 72

1. $0.52(500) - (W_R) = 0.4(500 - W_R)$
$$260 - W_R = 200 - 0.4W_R$$
$$-0.6W_R = -60$$
$$W_R = \mathbf{100 \text{ ml}}$$

2. $3.07 \times AP = 31{,}314$
$$AP = \mathbf{10{,}200}$$

3. $\dfrac{66}{100} \times M = 5412$
$$M = \mathbf{8200}$$

4. (a) $N_F = N_J + 160$
(b) $6N_J = N_F + 40$
Substitute (a) into (b) and get:
(b') $6N_J = (N_J + 160) + 40$
$$5N_J = 200$$
$$N_J = \mathbf{40}$$
(a) $N_F = (40) + 160 = \mathbf{200}$

5. $\dfrac{P_1}{T_1} = \dfrac{P_2}{T_2}$

$P_2 = \dfrac{P_1 T_2}{T_1}$

$P_2 = \dfrac{(500 \times 10^5)(0.002 \times 10^5)}{0.0004 \times 10^7}$

$P_2 = \dfrac{(5 \times 10^7)(2 \times 10^2)}{4 \times 10^3}$

$P_2 = \mathbf{2.5 \times 10^6 \ N/m^2}$

6. There is never an exact solution to these problems. One possible solution is given here.
$S = mC + b$
Use the graph to find the slope.
$m = \dfrac{3}{40} = 0.075$
$S = 0.075C + b$
Use the point (40, 1) for C and S.
$1 = 0.075(40) + b$
$-2 = b$
$\mathbf{S = 0.075C - 2}$

7.

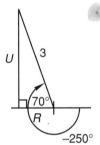

$R = 3 \cos 70° \approx 1.03$
$U = 3 \sin 70° \approx 2.82$

$\mathbf{-1.03R + 2.82U}$

8. $H = \sqrt{60^2 + 20^2}$
$H = 20\sqrt{10}$
$\tan \theta = \dfrac{20}{60} = \dfrac{1}{3}$
$\theta \approx 18.43°$
Since θ is a third-quadrant angle:
$\theta = 18.43 + 180 = 198.43°$

$\mathbf{20\sqrt{10}/198.43°}$

9. See Lesson 71.

10. $5x^2 = -7 - 2x$
$5x^2 + 2x + 7 = 0$
$x = \dfrac{-2 \pm \sqrt{(2)^2 - 4(5)(7)}}{2(5)}$
$x = \dfrac{-2 \pm \sqrt{-136}}{10} = \mathbf{-\dfrac{1}{5} \pm \dfrac{\sqrt{34}}{5} i}$

11. $3x^2 + 7x = -3$
$3x^2 + 7x + 3 = 0$
$x = \dfrac{-7 \pm \sqrt{7^2 - 4(3)(3)}}{2(3)}$
$x = \dfrac{-7 \pm \sqrt{13}}{6} = \mathbf{-\dfrac{7}{6} \pm \dfrac{\sqrt{13}}{6}}$

12. $a = x\left(\dfrac{1}{m_1} + \dfrac{y}{m_2} \right)$

$a = \dfrac{x}{m_1} + \dfrac{xy}{m_2}$

$a m_1 m_2 = m_2 x + m_1 xy$

$a m_1 m_2 - m_1 xy = m_2 x$

$m_1 = \dfrac{m_2 x}{a m_2 - xy}$

13. $\dfrac{2x + 3}{x^2 - 2x - 8} + \dfrac{3x - 2}{x - 4}$

$= \dfrac{2x + 3}{(x - 4)(x + 2)} + \dfrac{(3x - 2)(x + 2)}{(x - 4)(x + 2)}$

$= \dfrac{2x + 3 + 3x^2 + 4x - 4}{x^2 - 2x - 8}$

$= \dfrac{3x^2 + 6x - 1}{x^2 - 2x - 8}$

14. $\dfrac{3}{2 + 3\sqrt{20}} \cdot \dfrac{2 - 3\sqrt{20}}{2 - 3\sqrt{20}} = \dfrac{6 - 18\sqrt{5}}{4 - 180}$

$= \dfrac{-3 + 9\sqrt{5}}{88}$

15. (a) $R_P T_P = 693$
(b) $R_C T_C = 165$
(c) $R_P = 3R_C$
(d) $T_P = T_C + 2$

Substitute (c) and (d) into (a) and get:
(a') $3R_C(T_C + 2) = 693$
Substitute (b) into (a') and get:
(a'') $3(165) + 6R_C = 693$
$6R_C = 198$
$R_C = 33$
$R_P = \mathbf{99}; \ T_C = \mathbf{5}; \ T_P = \mathbf{7}$

16. $ab + \dfrac{b}{b + \dfrac{1}{b}} = ab + \dfrac{b}{\dfrac{b^2 + 1}{b}}$

$= ab + \dfrac{b^2}{b^2 + 1} = \dfrac{ab^3 + ab + b^2}{b^2 + 1}$

17. $x^2 + \dfrac{y}{y + \dfrac{1}{xy}} = x^2 + \dfrac{y}{\dfrac{xy^2 + 1}{xy}}$

$= x^2 + \dfrac{xy^2}{xy^2 + 1} = \dfrac{x^3y^2 + x^2 + xy^2}{xy^2 + 1}$

18. $(-3i - 5)(i + 5)$

$= -3i^2 - 15i - 5i - 25 = \textbf{-22 - 20}\boldsymbol{i}$

19. $-4i^3 - 3i^2 + \sqrt{-9} - \sqrt{-3}\sqrt{-3}$

$= -4i(ii) - 3(ii) + 3i - \sqrt{3}\sqrt{3}(ii)$

$= 4i + 3 + 3i + 3 = \textbf{6 + 7}\boldsymbol{i}$

20. (a) $\dfrac{1}{4}x - \dfrac{1}{5}y = 2$

(b) $0.03x - 0.4y = -1.64$

(a′) $5x - 4y = 40$
(b′) $3x - 40y = -164$

$\begin{array}{rcl} -10(\text{a}′) & -50x + 40y & = -400 \\ (\text{b}′) & 3x - 40y & = -164 \\ \hline & -47x & = -564 \\ & x & = 12 \end{array}$

(a′) $5(12) - 4y = 40$

$-4y = -20$

$y = 5$

$\textbf{(12, 5)}$

21. $\quad 2x^2 - x + 4 = 0$

$\left(x^2 - \dfrac{1}{2}x + \quad\right) = -2$

$x^2 - \dfrac{1}{2}x + \dfrac{1}{16} = -2 + \dfrac{1}{16}$

$\left(x - \dfrac{1}{4}\right)^2 = -\dfrac{31}{16}$

$x - \dfrac{1}{4} = \pm\dfrac{\sqrt{31}}{4}i$

$x = \dfrac{\textbf{1}}{\textbf{4}} \pm \dfrac{\sqrt{\textbf{31}}}{\textbf{4}}\boldsymbol{i}$

22. (a) $x = \dfrac{135 + 17}{2} = \textbf{76°}$

(b) $x = \dfrac{272 - 40}{2} = \textbf{116°}$

23. $\sqrt[5]{4\sqrt{2}} = \sqrt[5]{2^2\sqrt{2}} = \left[2^2(2^{1/2})\right]^{1/5}$

$= 2^{2/5}2^{1/10} = \textbf{2}^{\textbf{1/2}}$

24. $\dfrac{-2^0}{-8^{-4/3}} = -\left[-(8^{1/3})^4\right] = -[-(16)] = \textbf{16}$

25. $\sqrt[4]{a^5y}\sqrt{ay^4} = (a^5y)^{1/4}(ay^4)^{1/2}$

$= a^{5/4}y^{1/4}a^{1/2}y^2 = \boldsymbol{a}^{\textbf{7/4}}\boldsymbol{y}^{\textbf{9/4}}$

26. $2\sqrt{\dfrac{3}{5}} + 4\sqrt{\dfrac{5}{3}} - 2\sqrt{135}$

$= \dfrac{2\sqrt{3}}{\sqrt{5}}\dfrac{\sqrt{5}}{\sqrt{5}} + \dfrac{4\sqrt{5}}{\sqrt{3}}\dfrac{\sqrt{3}}{\sqrt{3}} - 6\sqrt{15}$

$= \dfrac{6\sqrt{15}}{15} + \dfrac{20\sqrt{15}}{15} - \dfrac{90\sqrt{15}}{15} = -\dfrac{\textbf{64}\sqrt{\textbf{15}}}{\textbf{15}}$

27. $A^2 = 6^2 + 4^2$

$A = 2\sqrt{13}$

$6 \times \overrightarrow{SF} = 4$

$\overrightarrow{SF} = \dfrac{2}{3}$

$2\sqrt{13} \cdot \dfrac{2}{3} = B$

$\dfrac{\textbf{4}\sqrt{\textbf{13}}}{\textbf{3}} = B$

28. $(-2)^0 - 2^2 - 2 - 2^0 - |-2 - 2| - 2^3$

$= -2(-2x - 2)$

$1 - 4 - 2 - 1 - 4 - 8 = 4x + 4$

$-22 = 4x$

$-\dfrac{\textbf{11}}{\textbf{2}} = x$

29. Write the equation of the given line in slope-intercept form.

$4y - 3x = 1$

$y = \dfrac{3}{4}x + \dfrac{1}{4}$

Since the slopes of perpendicular lines are negative reciprocals of each other:

$m_\perp = -\dfrac{4}{3}$

$y = -\dfrac{4}{3}x + b$

$0 = -\dfrac{4}{3}(-7) + b$

$-\dfrac{28}{3} = b$

$y = -\dfrac{\textbf{4}}{\textbf{3}}x - \dfrac{\textbf{28}}{\textbf{3}}$

30. (a) $\dfrac{-35,123 \times 10^4}{-798 \times 10^{-15}} \approx \dfrac{-4 \times 10^8}{-8 \times 10^{-13}}$

$= 5 \times 10^{20}$

Calculator $= \textbf{4.40} \times \textbf{10}^{\textbf{20}}$

(b) $\sqrt[3.8]{243}$

Answer should be between 3 and 5.

Calculator $= \textbf{4.24}$

PROBLEM SET 73

1. Carbon: $2 \times 12 = 24$
Hydrogen: $5 \times 1 = 5$
Bromine: $1 \times 80 = 80$
Total: $= 109$

$$\frac{24}{109} = \frac{48}{T}$$
$$T = \textbf{218 grams}$$

2. (a) $2N_P = 4N_D + 8$
(a') $N_P = 2N_D + 4$
(b) $7N_D = 3N_P - 4$
Substitute (a') into (b) and get:
(b') $7N_D = 3(2N_D + 4) - 4$
$$N_D = \textbf{8}$$
(a') $N_P = 2(8) + 4 = \textbf{20}$

3. $R_L T_L + R_T T_T = 256;$
$R_L = 16;\ R_T = 12;$
$T_L = T_T + 2$

$$16(T_T + 2) + 12T_T = 256$$
$$28T_T = 224$$
$$T_T = 8 \text{ hr}$$
$$D_T = 8(12) = \textbf{96 miles}$$

4. $N_F = \dfrac{k}{S_C}$

$$50 = \frac{k}{50}$$
$$2500 = k$$

$$N_F = \frac{2500}{25} = \textbf{100}$$

5. $\dfrac{V_1}{T_1} = \dfrac{V_2}{T_2}$

$$T_2 = \frac{V_2 T_1}{V_1}$$

$$T_2 = \frac{(0.08)(700 \times 10^5)}{0.0004}$$

$$T_2 = \frac{(8 \times 10^{-2})(7 \times 10^7)}{4 \times 10^{-4}}$$

$$T_2 = \textbf{1.4} \times \textbf{10}^{\textbf{10}}\ \textbf{K}$$

6. $\dfrac{3 + \sqrt{5}}{2 - 2\sqrt{5}} \cdot \dfrac{2 + 2\sqrt{5}}{2 + 2\sqrt{5}}$

$$= \frac{6 + 6\sqrt{5} + 2\sqrt{5} + 10}{4 - 20} = \frac{-2 - \sqrt{5}}{2}$$

7. $\dfrac{\sqrt{8} - 2\sqrt{2}}{3\sqrt{2} - 2\sqrt{3}} = \dfrac{0}{3\sqrt{2} - 2\sqrt{3}} = \textbf{0}$

8. Length of side $= \dfrac{7}{\sqrt{2}} = \dfrac{7\sqrt{2}}{2}$ m

Area $= s^2 = \left(\dfrac{7\sqrt{2}}{2}\right)^2 = \dfrac{49}{2}$ m^2

9. There is never an exact solution for these problems. One possible solution is given here.
$H = mC + b$
Use the graph to find the slope.

$$m = \frac{-45}{5} = -9$$
$$H = -9C + b$$
Use the point $(105, 80)$ for C and H.
$$80 = -9(105) + b$$
$$1025 = b$$
$$\boldsymbol{H = -9C + 1025}$$

10.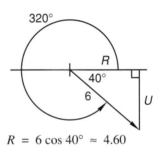

$R = 8 \cos 20° \approx 7.52$
$U = 8 \sin 20° \approx 2.74$

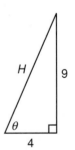

$R = 6 \cos 40° \approx 4.60$
$U = 6 \sin 40° \approx 3.86$

$$\begin{array}{r} 7.52R + 2.74U \\ 4.60R - 3.86U \\ \hline \textbf{12.12R} - \textbf{1.12U} \end{array}$$

11. $H = \sqrt{9^2 + 4^2}$
$H = \sqrt{97}$

$$\tan \theta = \frac{9}{4}$$
$$\theta \approx 66.04°$$

$$\boldsymbol{\sqrt{97}\,\underline{/66.04°}}$$

12. See Lesson 71.

13.
$$2x^2 + 5 = -5x$$
$$2x^2 + 5x + 5 = 0$$
$$x = \frac{-5 \pm \sqrt{5^2 - 4(2)(5)}}{2(2)} = \frac{-5 \pm \sqrt{-15}}{4}$$
$$x = -\frac{5}{4} \pm \frac{\sqrt{15}}{4}i$$

14.
$$2x^2 - 4 = -5x$$
$$2x^2 + 5x - 4 = 0$$
$$x = \frac{-5 \pm \sqrt{5^2 - 4(2)(-4)}}{2(2)} = \frac{-5 \pm \sqrt{57}}{4}$$
$$x = -\frac{5}{4} \pm \frac{\sqrt{57}}{4}$$

15.
$$xc = px\left(\frac{1}{km_1} - \frac{1}{m_2}\right)$$
$$xc = \frac{px}{km_1} - \frac{px}{m_2}$$
$$ckm_1m_2x = m_2px - km_1px$$
$$ckm_1m_2x + km_1px = m_2px$$
$$m_1 = \frac{m_2 p}{ckm_2 + kp}$$

16.
$$\frac{3x - 2}{x - 2} - \frac{4x - 3}{2 - x}$$
$$= \frac{3x - 2}{x - 2} + \frac{4x - 3}{x - 2} = \frac{7x - 5}{x - 2}$$

17.
(a) $R_A T_A = 160$
(b) $R_B T_B = 240$
(c) $R_B = 2R_A$
(d) $T_A + T_B = 7$

Substitute (c) and (d) into (b) and get:
(b') $2R_A(7 - T_A) = 240$
Substitute (a) into (b') and get:
(b'') $14R_A - 2(160) = 240$
$$14R_A = 560$$
$$R_A = \mathbf{40}$$
$$R_B = \mathbf{80}; \ T_A = \mathbf{4}; \ T_B = \mathbf{3}$$

18. $ax^2 - \dfrac{a}{a - \dfrac{1}{ax}} = ax^2 - \dfrac{a}{\dfrac{a^2x - 1}{ax}}$
$$= ax^2 - \frac{a^2x}{a^2x - 1} = \frac{a^3x^3 - ax^2 - a^2x}{a^2x - 1}$$

19. $(3i + 2)(i - 4) - \sqrt{-9}$
$$= 3i^2 - 12i + 2i - 8 - 3i = \mathbf{-11 - 13i}$$

20. (a) $2x + 3y = 6$
$$y = -\frac{2}{3}x + 2$$
(b) $x - 2y = 4$
$$y = \frac{1}{2}x - 2$$

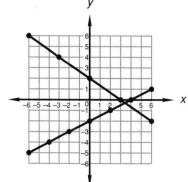

2(a) $4x + 6y = 12$
3(b) $\underline{3x - 6y = 12}$
$$7x \qquad = 24$$
$$x = \frac{24}{7}$$

(b) $\left(\dfrac{24}{7}\right) - 2y = 4$
$$-2y = \frac{4}{7}$$
$$y = -\frac{2}{7}$$
$$\left(\frac{24}{7}, -\frac{2}{7}\right)$$

21.
$$x^2 + 6 = 3x$$
$$(x^2 - 3x + \quad) = -6$$
$$x^2 - 3x + \frac{9}{4} = -6 + \frac{9}{4}$$
$$\left(x - \frac{3}{2}\right)^2 = -\frac{15}{4}$$
$$x - \frac{3}{2} = \pm\frac{\sqrt{15}}{2}i$$
$$x = \frac{3}{2} \pm \frac{\sqrt{15}}{2}i$$

22. $x = \dfrac{80}{2} = \mathbf{40}$
$$y = \frac{80}{2} = \mathbf{40}$$
$$a = \frac{50}{2} = \mathbf{25}$$
$$b = \frac{50}{2} = \mathbf{25}$$
$$c = 180 - 40 - 25 = \mathbf{115}$$

23. $3\sqrt{9\sqrt{3}} = 3\sqrt{3^2\sqrt{3}} = 3[3^2(3^{1/2})]^{1/2}$
$= 3 \cdot 3(3^{1/4}) = \mathbf{3^{9/4}}$

24. $\dfrac{-2^2}{-16^{-3/4}} = -[-(4)(16^{1/4})^3] = \mathbf{32}$

25. $\sqrt{a^2 x^0 yx^{1/2} y^2} = (a^2 x^0 yx^{1/2} y^2)^{1/2}$
$= ay^{1/2}x^{1/4}y = \mathbf{ax^{1/4}y^{3/2}}$

26. $2\sqrt{\dfrac{5}{8}} + 3\sqrt{\dfrac{8}{5}} - 2\sqrt{40}$

$= \dfrac{2\sqrt{5}}{\sqrt{8}}\dfrac{\sqrt{8}}{\sqrt{8}} + \dfrac{3\sqrt{8}}{\sqrt{5}}\dfrac{\sqrt{5}}{\sqrt{5}} - 2\sqrt{40}$

$= \dfrac{10\sqrt{40}}{40} + \dfrac{24\sqrt{40}}{40} - \dfrac{80\sqrt{40}}{40}$

$= -\dfrac{46\sqrt{40}}{40} = -\dfrac{\mathbf{23}\sqrt{\mathbf{10}}}{\mathbf{10}}$

27. $\dfrac{x(x - 8)(x + 2)}{(x - 10)(x + 2)} \cdot \dfrac{(x - 10)(x + 5)}{x(x - 8)(x + 3)}$

$= \dfrac{\mathbf{x + 5}}{\mathbf{x + 3}}$

28. $y = -\dfrac{2}{5}x + b$

$2 = -\dfrac{2}{5}(-5) + b$

$0 = b$

$y = -\dfrac{2}{5}x$

29. (a) $\dfrac{4,168,214 \times 10^{24}}{74.612 \times 10^{-5.34}} \approx \dfrac{4 \times 10^{30}}{7 \times 10^{-4}}$
$\approx 6 \times 10^{33}$
Calculator $= \mathbf{1.22 \times 10^{34}}$

(b) $(4.01)^{-5.34}$
Answer should be between 9×10^{-4} and 2×10^{-4}.
Calculator $= \mathbf{6.01 \times 10^{-4}}$

30. $\dfrac{3x - 5}{2} - \dfrac{x}{3} = 6$
$9x - 15 - 2x = 36$
$7x = 51$
$x = \dfrac{51}{7}$

PROBLEM SET 74

1. $R_G T_G = 28$; $R_B T_B = 28$;
$R_B = 2R_G$; $T_G + T_B = 12$

$2R_G(12 - T_G) = 28$
$24R_G - 2R_G T_G = 28$
$24R_G - 2(28) = 28$
$24R_G = 84$
$R_G = \mathbf{3.5}$ **mph**
$R_B = \mathbf{7}$ **mph**; $T_G = \mathbf{8}$ **hr**; $T_B = \mathbf{4}$ **hr**

2. $R_R T_R = 80$;
$R_C T_C = 30$;
$R_R = 4R_C$;
$T_R = T_C - 2$

$4R_C(T_C - 2) = 80$
$4R_C T_C - 8R_C = 80$
$4(30) - 8R_C = 80$
$-8R_C = -40$
$R_C = \mathbf{5}$ **mph**
$R_R = \mathbf{20}$ **mph**; $T_C = \mathbf{6}$ **hr**; $T_R = \mathbf{4}$ **hr**

3. $\dfrac{P_1 V_1}{T_1} = \dfrac{P_2 V_2}{T_2}$

$T_2 = \dfrac{P_2 V_2 T_1}{P_1 V_1}$

$T_2 = \dfrac{(400 \times 10^5)(500 \times 10^4)(0.06 \times 10^6)}{(0.004 \times 10^5)(0.02 \times 10^4)}$

$T_2 = \dfrac{(4 \times 10^7)(5 \times 10^6)(6 \times 10^4)}{(4 \times 10^2)(2 \times 10^2)}$

$T_2 = \mathbf{1.5 \times 10^{14}}$ **K**

4. $0.92(2000) - E_A = 0.8(2000 - E_A)$
$1840 - E_A = 1600 - 0.8E_A$
$-0.2E_A = -240$
$E_A = \mathbf{1200}$ **liters**

5. $N \qquad N + 2 \qquad N + 4$
$4(N + 2)(N + 4) = 20(N + N + 2) + 12$
$4N^2 + 24N + 32 = 40N + 52$
$4N^2 - 16N - 20 = 0$
$4(N^2 - 4N - 5) = 0$
$(N - 5)(N + 1) = 0$
$N = -1, 5$
The desired integers are $\mathbf{-1, 1, 3}$ and $\mathbf{5, 7, 9}$.

6. $\dfrac{2 - \sqrt{3}}{-\sqrt{3} - 2} \cdot \dfrac{-\sqrt{3} + 2}{-\sqrt{3} + 2}$

$= \dfrac{-2\sqrt{3} + 4 + 3 - 2\sqrt{3}}{3 - 4} = \mathbf{-7 + 4\sqrt{3}}$

7. $\dfrac{3\sqrt{2} - 4}{\sqrt{2} - 3} \cdot \dfrac{\sqrt{2} + 3}{\sqrt{2} + 3}$

$= \dfrac{6 + 9\sqrt{2} - 4\sqrt{2} - 12}{2 - 9} = \dfrac{6 - 5\sqrt{2}}{7}$

8. $\dfrac{4\sqrt{2} - 5}{2 - 3\sqrt{8}} \cdot \dfrac{2 + 3\sqrt{8}}{2 + 3\sqrt{8}}$

$= \dfrac{8\sqrt{2} + 48 - 10 - 30\sqrt{2}}{4 - 72}$

$= \dfrac{-19 + 11\sqrt{2}}{34}$

9. There is never an exact solution for these problems. One possible solution is given here.

$N = mF + b$

Use the graph to find the slope.

$m = \dfrac{350}{5} = 70$

$N = 70F + b$

Use the point (30, 50) for F and N.

$50 = 70(30) + b$

$-2050 = b$

$\mathbf{N = 70F - 2050}$

10.

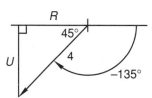

$R = 4 \cos 45° \approx 2.83$
$U = 4 \sin 45° \approx 2.83$

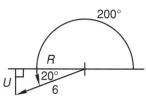

$R = 6 \cos 20° \approx 5.64$
$U = 6 \sin 20° \approx 2.05$

$-2.83R - 2.83U$
$\underline{-5.64R - 2.05U}$
$\mathbf{-8.47R - 4.88U}$

11. $H = \sqrt{5^2 + 5^2}$
$H = 5\sqrt{2}$

$\tan \theta = \dfrac{5}{5} = 1$

$\theta = 45°$

Since θ is a third-quadrant angle:
$\theta = 45 + 180 = 225°$

$\mathbf{5\sqrt{2}/225°}$

12. See Lesson 71.

13. $-3x^2 - x = 4$
$3x^2 + x + 4 = 0$

$x = \dfrac{-1 \pm \sqrt{1^2 - 4(3)(4)}}{2(3)} = \dfrac{-1 \pm \sqrt{-47}}{6}$

$= \mathbf{-\dfrac{1}{6} \pm \dfrac{\sqrt{47}}{6}\, i}$

14. $-x - 3x^2 = -4$
$3x^2 + x - 4 = 0$

$x = \dfrac{-1 \pm \sqrt{1^2 - 4(3)(-4)}}{2(3)} = \dfrac{-1 \pm \sqrt{49}}{6}$

$= -\dfrac{1}{6} \pm \dfrac{7}{6} = \mathbf{1, -\dfrac{4}{3}}$

15. $a = m\left(\dfrac{1}{pc} - \dfrac{k}{x}\right)$

$a = \dfrac{m}{pc} - \dfrac{km}{x}$

$acpx = mx - ckmp$

$ckmp = mx - acpx$

$\dfrac{\boldsymbol{ckmp}}{\boldsymbol{m - acp}} = \boldsymbol{x}$

16. $\dfrac{a}{m} = c\left(\dfrac{1}{x_1} + \dfrac{b}{x_2}\right)$

$\dfrac{a}{m} = \dfrac{c}{x_1} + \dfrac{bc}{x_2}$

$ax_1x_2 = cmx_2 + bcmx_1$

$ax_1x_2 - bcmx_1 = cmx_2$

$x_1 = \dfrac{\boldsymbol{cmx_2}}{\boldsymbol{ax_2 - bcm}}$

17. $ax - \dfrac{a}{x - \dfrac{x}{a}} = ax - \dfrac{a}{\dfrac{ax - x}{a}}$

$= ax - \dfrac{a^2}{ax - x} = \dfrac{a^2x^2 - ax^2 - a^2}{ax - x}$

18. $(2i - 3)(i - 3) + \sqrt{-9} + 3i^3$
$= 2i^2 - 6i - 3i + 9 + 3i + 3i(i i)$
$= -2 - 6i - 3i + 9 + 3i - 3i = \mathbf{7 - 9i}$

19. $\dfrac{3x - 2}{x^2 + 7x + 10} - \dfrac{1}{x + 5}$

$= \dfrac{3x - 2}{x^2 + 7x + 10} - \dfrac{x + 2}{x^2 + 7x + 10}$

$= \dfrac{\mathbf{2x - 4}}{\mathbf{x^2 + 7x + 10}}$

20. (a) $\dfrac{3}{2}x - \dfrac{2}{5}y = 2$

(b) $3x + 0.5y = 17$

(a') $15x - 4y = 20$

(b') $30x + 5y = 170$

-2(a') $-30x + 8y = -40$

(b') $\underline{30x + 5y = 170}$

$13y = 130$

$y = 10$

(a') $15x - 4(10) = 20$

$15x = 60$

$x = 4$

(4, 10)

21. $\qquad 3x^2 - x = -7$

$\left(x^2 - \dfrac{1}{3}x + \right) = -\dfrac{7}{3}$

$x^2 - \dfrac{1}{3}x + \dfrac{1}{36} = -\dfrac{7}{3} + \dfrac{1}{36}$

$\left(x - \dfrac{1}{6}\right)^2 = -\dfrac{83}{36}$

$x - \dfrac{1}{6} = \pm\dfrac{\sqrt{83}}{6}i$

$x = \dfrac{1}{6} \pm \dfrac{\sqrt{83}}{6}i$

22. (a) $4x - y = 45$

(b) $3x + 3y = 90$

3(a) $12x - 3y = 135$

(b) $\underline{3x + 3y = 90}$

$15x = 225$

$x = 15$

(b) $3(15) + 3y = 90$

$3y = 45$

$y = 15$

23. $2\sqrt{4\sqrt{2}} = 2\sqrt{2^2\sqrt{2}} = 2[2^2 2^{1/2}]^{1/2}$

$= 2(2)(2^{1/4}) = \mathbf{2^{9/4}}$

24. $\dfrac{-3^0(-3^0)^2}{-9^{-3/2}} = -[-(9^{1/2})^3] = \mathbf{27}$

25. $\sqrt{4x^2y^5}\,\sqrt[3]{8y^5x} = (4x^2y^5)^{1/2}(8y^5x)^{1/3}$

$= 2xy^{5/2}2y^{5/3}x^{1/3} = \mathbf{4x^{4/3}y^{25/6}}$

26. $2\sqrt{\dfrac{6}{7}} + 3\sqrt{\dfrac{7}{6}} - 3\sqrt{42}$

$= \dfrac{2\sqrt{6}}{\sqrt{7}}\dfrac{\sqrt{7}}{\sqrt{7}} + \dfrac{3\sqrt{7}}{\sqrt{6}}\dfrac{\sqrt{6}}{\sqrt{6}} - 3\sqrt{42}$

$= \dfrac{12\sqrt{42}}{42} + \dfrac{21\sqrt{42}}{42} - \dfrac{126\sqrt{42}}{42} = \mathbf{-\dfrac{31\sqrt{42}}{14}}$

27. Write the equation of the given line in slope-intercept form.

$4x - y = 7$

$y = 4x - 7$

Since parallel lines have the same slopes:

$y = 4x + b$

$0 = 4(-2) + b$

$8 = b$

$\mathbf{y = 4x + 8}$

28. $\sqrt{x - 5} - 2 = 7$

$\phantom{\sqrt{x -}}x - 5 = 81$

$\phantom{\sqrt{x -}}x = \mathbf{86}$

Check: $\sqrt{86 - 5} - 2 = 7$

$\phantom{Check: \sqrt{8}}9 - 2 = 7 \quad$ check

29. $\dfrac{4x - 3}{7} - \dfrac{x - 2}{3} = 5$

$12x - 9 - 7x + 14 = 105$

$5x = 100$

$x = \mathbf{20}$

30. Length of side $= \dfrac{9\sqrt{2}}{\sqrt{2}} = \mathbf{9\ m}$

Area $= s^2 = (9\ m)^2 = \mathbf{81\ m^2}$

Problem Set 75

1. $R_D T_D = 120$;

$R_E T_E = 360$;

$R_E = 2R_D$;

$T_E = T_D + 1$

$2R_D(T_D + 1) = 360$

$2R_D T_D + 2R_D = 360$

$2(120) + 2R_D = 360$

$2R_D = 120$

$R_D = \mathbf{60\ mph}$

$R_E = \mathbf{120\ mph};\ T_D = \mathbf{2\ hr};\ T_E = \mathbf{3\ hr}$

2. $\dfrac{P_1}{T_1} = \dfrac{P_2}{T_2}$

$P_2 = \dfrac{P_1 T_2}{T_1}$

$P_2 = \dfrac{(0.0036 \times 10^{-2})(40 \times 10^4)}{50 \times 10^7}$

$P_2 = \dfrac{(3.6 \times 10^{-5})(4 \times 10^5)}{5 \times 10^8}$

$P_2 = \mathbf{2.88 \times 10^{-8}\ N/m^2}$

3. $T = \dfrac{k}{N_M}$

$10 = \dfrac{k}{500}$

$5000 = k$

$T = \dfrac{5000}{200} = $ **25 days**

4. $R_B T_B + 20 = R_T T_T;$

$R_T = 60; \ R_B = 40;$

$T_B = T_T + 2$

D_T

D_B 20

$40(T_T + 2) + 20 = 60T_T$

$40T_T + 100 = 60T_T$

$100 = 20T_T$

5 hr $= T_T$

5. (a) $4N_L + 6N_P = 192$

(b) $N_L = N_P - 2$

Substitute (b) into (a) and get:

(a') $4(N_P - 2) + 6N_P = 192$

$10N_P = 200$

$N_P = $ **20**

(b) $N_L = (20) - 2 = $ **18**

6. $\dfrac{x + 4}{x^2 + 2x - 3} - \dfrac{2}{1 - x}$

$= \dfrac{x + 4}{(x + 3)(x - 1)} + \dfrac{2(x + 3)}{(x + 3)(x - 1)}$

$= \dfrac{3x + 10}{x^2 + 2x - 3}$

7. $\dfrac{x - 3}{x^2 - x - 12} - \dfrac{x + 2}{-3 - x}$

$= \dfrac{x - 3}{(x - 4)(x + 3)} + \dfrac{(x + 2)(x - 4)}{(x - 4)(x + 3)}$

$= \dfrac{x - 3 + x^2 - 2x - 8}{x^2 - x - 12}$

$= \dfrac{x^2 - x - 11}{x^2 - x - 12}$

8. $\dfrac{2 - \sqrt{2}}{3 + \sqrt{2}} \cdot \dfrac{3 - \sqrt{2}}{3 - \sqrt{2}}$

$= \dfrac{6 - 2\sqrt{2} - 3\sqrt{2} + 2}{9 - 2} = \dfrac{8 - 5\sqrt{2}}{7}$

9. $\dfrac{2 - 3\sqrt{3}}{3 - 2\sqrt{12}} = \dfrac{2 - 3\sqrt{3}}{3 - 4\sqrt{3}} \cdot \dfrac{3 + 4\sqrt{3}}{3 + 4\sqrt{3}}$

$= \dfrac{6 + 8\sqrt{3} - 9\sqrt{3} - 36}{9 - 48} = \dfrac{30 + \sqrt{3}}{39}$

10. $A_{\text{Circle}} = \pi r^2 = 9\pi \text{ cm}^2$

$r = $ **3 cm**

Length of diagonal $= 2r = 2(3 \text{ cm}) = $ **6 cm**

$A_{\text{Square}} = s^2 = \left(\dfrac{6\sqrt{2}}{2} \text{ cm}\right)^2 = $ **18 cm²**

11. There is never an exact solution for these problems. One possible solution is given here.

$V = mK + b$

Use the graph to find the slope.

$m = -\dfrac{5}{4} = -1.25$

$V = -1.25K + b$

Use the point (95, 27) for K and V.

$27 = -1.25(95) + b$

$146 \approx b$

$V = $ **−1.25K + 146**

12. See Lesson 71.

13.

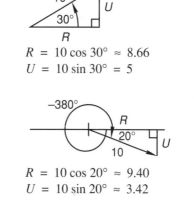

$R = 10 \cos 30° \approx 8.66$

$U = 10 \sin 30° = 5$

$R = 10 \cos 20° \approx 9.40$

$U = 10 \sin 20° \approx 3.42$

$8.66R + 5.00U$

$9.40R - 3.42U$

$\overline{\textbf{18.06R + 1.58U}}$

14. $H = \sqrt{8^2 + 2^2}$

$H = 2\sqrt{17}$

$\tan \theta = \dfrac{8}{2} = 4$

$\theta \approx 75.96°$

Since θ is a third-quadrant angle:

$\theta = 75.96 + 180 = 255.96°$

$2\sqrt{17}\underline{/255.96°}$

2

θ

8   H

15.
$$-2x + 4 = -5x^2$$
$$5x^2 - 2x + 4 = 0$$
$$x = \frac{-(-2) \pm \sqrt{(-2)^2 - 4(5)(4)}}{2(5)}$$
$$x = \frac{2 \pm \sqrt{-76}}{10} = \frac{1}{5} \pm \frac{\sqrt{19}}{5} i$$

16.
$$\frac{m}{x} = cm\left(\frac{a}{x} + \frac{b}{y}\right)$$
$$\frac{m}{x} = \frac{acm}{x} + \frac{bcm}{y}$$
$$my = acmy + bcmx$$
$$my - acmy = bcmx$$
$$y = \frac{bcmx}{m - acm} = \frac{bcx}{1 - ac}$$

17. Since this is a 30-60-90 triangle:
$$\sqrt{3} \times \vec{SF} = 8$$
$$\vec{SF} = \frac{8\sqrt{3}}{3}$$
$$x = \frac{8\sqrt{3}}{3} \cdot 2 = \frac{16\sqrt{3}}{3}$$
$$y = \frac{8\sqrt{3}}{3} \cdot 1 = \frac{8\sqrt{3}}{3}$$

18.
$$a^2y + \frac{a^2}{a + \dfrac{a}{y}} = a^2y + \frac{a^2}{\dfrac{ay + a}{y}}$$
$$= a^2y + \frac{a^2 y}{ay + a} = \frac{a^3 y^2 + a^3 y + a^2 y}{ay + a}$$
$$= \frac{a^2 y^2 + a^2 y + ay}{y + 1}$$

19.
$$(2 + i)(i - 4) - \sqrt{-16}$$
$$= 2i - 8 + i^2 - 4i - 4i$$
$$= 2i - 8 - 1 - 4i - 4i = -9 - 6i$$

20. (a) $\frac{1}{4}x + \frac{1}{3}y = 15$
(b) $0.02x + 0.2y = 6.4$

(a′) $3x + 4y = 180$
(b′) $2x + 20y = 640$

$$
\begin{array}{l}
-5(a′) \;\; -15x - 20y = -900 \\
(b′) \quad 2x + 20y = 640 \\
\hline
-13x = -260 \\
x = 20
\end{array}
$$

(a′) $3(20) + 4y = 180$
$$4y = 120$$
$$y = 30$$
(20, 30)

21.
$$3x^2 + 2 = -x$$
$$\left(x^2 + \frac{1}{3}x + \right) = -\frac{2}{3}$$
$$x^2 + \frac{1}{3}x + \frac{1}{36} = -\frac{2}{3} + \frac{1}{36}$$
$$\left(x + \frac{1}{6}\right)^2 = -\frac{23}{36}$$
$$x + \frac{1}{6} = \pm\frac{\sqrt{23}}{6} i$$
$$x = -\frac{1}{6} \pm \frac{\sqrt{23}}{6} i$$

22.
$$A_{\text{Triangle}} = \frac{1}{2}bH = 48 \text{ m}^2$$
$$H = \frac{48(2) \text{ m}^2}{12 \text{ m}}$$
$$H = \mathbf{8 \text{ m}}$$

$$A_{\text{Semicircle}} = \frac{1}{2}\pi r^2 = 2\pi$$
$$r^2 = 4 \text{ m}^2$$
$$r = \mathbf{2 \text{ m}}$$

23.
$$4\sqrt{2\sqrt[3]{2}} = 2^2[2(2^{1/3})]^{1/2} = 2^2 2^{1/2} 2^{1/6} = \mathbf{2^{8/3}}$$

24.
$$\frac{-2^0\,(-2^0)}{-4^{-3/2}} = -(4^{1/2})^3 = \mathbf{-8}$$

25.
$$\sqrt[4]{mp^5}\,\sqrt[3]{m^2 p^4} = (mp^5)^{1/4}(m^2 p^4)^{1/3}$$
$$= m^{1/4}p^{5/4}m^{2/3}p^{4/3} = \mathbf{m^{11/12}p^{31/12}}$$

26.
$$3\sqrt{\frac{7}{8}} + 2\sqrt{\frac{8}{7}} - 2\sqrt{56}$$
$$= \frac{3\sqrt{7}}{\sqrt{8}}\frac{\sqrt{8}}{\sqrt{8}} + \frac{2\sqrt{8}}{\sqrt{7}}\frac{\sqrt{7}}{\sqrt{7}} - 2\sqrt{56}$$
$$= \frac{21\sqrt{56}}{56} + \frac{16\sqrt{56}}{56} - \frac{112\sqrt{56}}{56}$$
$$= -\frac{75\sqrt{56}}{56} = \mathbf{-\frac{75\sqrt{14}}{28}}$$

27. (a)
$$\frac{4813 \times 10^{-14}}{0.01903 \times 10^{-22}} \approx \frac{5 \times 10^{-11}}{2 \times 10^{-24}}$$
$$\approx 3 \times 10^{13}$$
Calculator = $\mathbf{2.53 \times 10^{13}}$

(b) $\sqrt[3.6]{198}$
Answer should be between 3 and 5.
Calculator = **4.34**

28. Graph the line to find the slope.

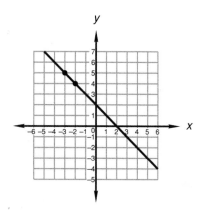

$$m = -\frac{1}{1} = -1$$

Since the slopes of perpendicular lines are negative reciprocals of each other:

$m_\perp = 1$

$y = x + b$

$(7) = (5) + b$

$2 = b$

$y = x + 2$

29. $\sqrt{3x - 5} - 2 = 7$

$3x - 5 = 81$

$3x = 86$

$$x = \frac{86}{3}$$

Check: $\sqrt{3\left(\frac{86}{3}\right) - 5} - 2 = 7$

$9 - 2 = 7$ check

30. $\dfrac{4x + 7}{2} - \dfrac{5x}{3} = 4$

$12x + 21 - 10x = 24$

$2x = 3$

$$x = \frac{3}{2}$$

PROBLEM SET 76

1. $R_C T_C = 32;$

$R_T T_T = 72;$

$R_T = 3R_C;$

$T_T = T_C - 2$

D_C |——→| D_T |——→|
  32 72

$3R_C(T_C - 2) = 72$

$3R_C T_C - 6R_C = 72$

$3(32) - 6R_C = 72$

$-6R_C = -24$

$R_C = 4$ mph

$R_T = 12$ mph; $T_C = 8$ hr; $T_T = 6$ hr

2. $\dfrac{P_1 V_1}{T_1} = \dfrac{P_2 V_2}{T_2}$

$V_2 = \dfrac{P_1 V_1 T_2}{P_2 T_1}$

$V_2 = \dfrac{(700 \times 10^5)(700 \times 10^{-7})(8000 \times 10^5)}{(3500 \times 10^4)(56 \times 10^4)}$

$V_2 = \dfrac{(7 \times 10^7)(7 \times 10^{-5})(8 \times 10^8)}{(3.5 \times 10^7)(5.6 \times 10^5)}$

$V_2 = 2 \times 10^{-1}$ liter

3. $0.64(500) - E_A = 0.4(500 - E_A)$

$320 - E_A = 200 - 0.4E_A$

$-0.6E_A = -120$

$E_A = 200$ ml

4. Sodium: $1 \times 23 = 23$

Chlorine: $1 \times 35 = 35$

Oxygen: $1 \times 16 = 16$

Total: $= 74$

$\dfrac{35}{39} = \dfrac{280}{\text{NaO}}$

NaO = 312 grams

5. (a) $4N_B = 3N_R + 14$

(b) $6N_R = N_B + 7$

Substitute $N_B = 6N_R - 7$ into (a) and get:

(a′) $4(6N_R - 7) = 3N_R + 14$

$21N_R = 42$

$N_R = 2$

(b) $N_B = 6(2) - 7 =$ **5**

6. (a) $x = 2y$

(b) $x + y + z = -198$

(c) $x - 3y - 2z = 16$

Substitute (a) into (b) and (c) and get:

(b′) $3y + z = -198$

(c′) $-y - 2z = 16$

2(b′) $6y + 2z = -396$

(c′) $\underline{-y - 2z = \quad 16}$

$5y \qquad = -380$

$y = -76$

(b′) $3(-76) + z = -198$

$z = 30$

(a) $x = 2(-76) = -152$

$(-152, -76, 30)$

7. (a) $2x + 2y - z = 81$
(b) $3x - y + 2z = 27$
(c) $x - 3z = 0$
$\qquad x = 3z$
Substitute (c) into (a) and (b) and get:
(a′) $2y + 5z = 81$
(b′) $-y + 11z = 27$

$\qquad$ (a′) $\quad 2y + 5z = 81$
2(b′) $\quad -2y + 22z = 54$
$\qquad\qquad\qquad 27z = 135$
$\qquad\qquad\qquad\quad z = 5$

$\qquad$ (a′) $2y + 5(5) = 81$
$\qquad\qquad\quad\ 2y = 56$
$\qquad\qquad\qquad\ y = 28$
(c) $x = 3(5) = 15$

(15, 28, 5)

8.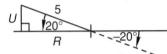

$R = 5 \cos 20° \approx 4.70$
$U = 5 \sin 20° \approx 1.71$

$R = 4 \cos 65° \approx 1.69$
$U = 4 \sin 65° \approx 3.63$

$-4.70R + 1.71U$
$-1.69R + 3.63U$
$\overline{-6.39R + 5.34U}$

9. $H = \sqrt{4^2 + 1^2}$
$H = \sqrt{17}$

$\tan \theta = \dfrac{1}{4}$
$\qquad \theta \approx 14.04°$
Since θ is a third-quadrant angle:
$\theta = 14.04 + 180 = 194.04°$
$\mathbf{\sqrt{17}/194.04°}$

10. $\dfrac{5x + 2}{x^2 + 3x - 10} - \dfrac{2x}{2 - x}$

$= \dfrac{5x + 2}{(x + 5)(x - 2)} + \dfrac{2x(x + 5)}{(x + 5)(x - 2)}$

$= \dfrac{5x + 2 + 2x^2 + 10x}{x^2 + 3x - 10}$

$= \dfrac{2x^2 + 15x + 2}{x^2 + 3x - 10}$

11. $\dfrac{2\sqrt{2} - 1}{1 - \sqrt{2}} \cdot \dfrac{1 + \sqrt{2}}{1 + \sqrt{2}}$

$= \dfrac{2\sqrt{2} + 4 - 1 - \sqrt{2}}{1 - 2} = -3 - \sqrt{2}$

12. $\dfrac{3 - \sqrt{2}}{4 + 2\sqrt{8}} = \dfrac{3 - \sqrt{2}}{4 + 4\sqrt{2}} \cdot \dfrac{4 - 4\sqrt{2}}{4 - 4\sqrt{2}}$

$= \dfrac{12 - 12\sqrt{2} - 4\sqrt{2} + 8}{16 - 32} = \dfrac{-5 + 4\sqrt{2}}{4}$

13. There is never an exact solution to these problems.
One possible solution is given here.
$Al = mB + b$
Use the graph to find the slope.

$m = \dfrac{42}{4} = 10.5$
$Al = 10.5B + b$
Use the point $(4, 180)$ for B and Al.
$180 = 10.5(4) + b$
$138 = b$
$\mathbf{Al = 10.5B + 138}$

14. See Lesson 71.

15. $\dfrac{a}{c + x} = m\left(\dfrac{1}{r} + \dfrac{1}{t}\right)$

$\qquad \dfrac{a}{c + x} = \dfrac{m}{r} + \dfrac{m}{t}$

$\qquad\qquad art = cmt + mtx + cmr + mrx$
$art - cmt - cmr = mtx + mrx$

$\qquad\qquad\qquad x = \dfrac{art - cmt - cmr}{mt + mr}$

16. $\dfrac{a}{c + x} = m\left(\dfrac{1}{r} + \dfrac{1}{t}\right)$

$\qquad \dfrac{a}{c + x} = \dfrac{m}{r} + \dfrac{m}{t}$

$\qquad\qquad art = cmt + mtx + cmr + mrx$
$art - cmt - mtx = cmr + mrx$

$\qquad\qquad\qquad t = \dfrac{cmr + mrx}{ar - cm - mx}$

17. $\qquad 7x^2 - x - 1 = 0$

$\qquad \left(x^2 - \dfrac{1}{7}x + \right) = \dfrac{1}{7}$

$\qquad x^2 - \dfrac{1}{7}x + \dfrac{1}{196} = \dfrac{1}{7} + \dfrac{1}{196}$

$\qquad \left(x - \dfrac{1}{14}\right)^2 = \dfrac{29}{196}$

$\qquad\qquad x - \dfrac{1}{14} = \pm\dfrac{\sqrt{29}}{14}$

$\qquad\qquad\qquad x = \dfrac{1}{14} \pm \dfrac{\sqrt{29}}{14}$

18. $x + \dfrac{a}{a + \dfrac{a}{x}} = x + \dfrac{a}{\dfrac{ax + a}{x}}$

$= x + \dfrac{ax}{ax + a} = \dfrac{ax^2 + ax + ax}{ax + a}$

$= \dfrac{x^2 + 2x}{x + 1}$

19. $(3 - i)(2 - i) - 2i^2 - \sqrt{-9}$

$= 6 - 3i - 2i + i^2 - 2i^2 - 3i$

$= 6 - 3i - 2i - 1 + 2 - 3i = \mathbf{7 - 8i}$

20. (a) $\dfrac{1}{3}x + \dfrac{2}{3}y = 31$

(b) $0.02x + 0.7y = 27.6$

(a′) $x + 2y = 93$

(b′) $2x + 70y = 2760$

$-2(a′)\ -2x - \ 4y = -186$

(b′) $\underline{\ \ 2x + 70y = 2760}$

$66y = 2574$

$y = 39$

(a′) $x + 2(39) = 93$

$x = 15$

(15, 39)

21. The radius of the circle is equal to the diagonal of the rectangle.

$r = \sqrt{\left(\sqrt{11}\right)^2 + 5^2}$

$r = \mathbf{6}$

22. $5\sqrt{25\sqrt{5}} = 5\sqrt{5^2\sqrt{5}} = 5[5^2 5^{1/2}]^{1/2}$

$= 5\left(5(5^{1/4})\right) = \mathbf{5^{9/4}}$

23. $\dfrac{-1^0\,(-1^0)}{-4^{-5/2}} = -(4^{1/2})^5 = \mathbf{-32}$

24. $\sqrt[6]{my^3}\,\sqrt[4]{m^3 y} = (my^3)^{1/6}(m^3 y)^{1/4}$

$= m^{1/6}y^{1/2}m^{3/4}y^{1/4} = \mathbf{m^{11/12}y^{3/4}}$

25. $2\sqrt{\dfrac{3}{8}} + 3\sqrt{\dfrac{8}{3}} - 2\sqrt{216}$

$= \dfrac{2\sqrt{3}}{\sqrt{8}}\dfrac{\sqrt{8}}{\sqrt{8}} + \dfrac{3\sqrt{8}}{\sqrt{3}}\dfrac{\sqrt{3}}{\sqrt{3}} - 6\sqrt{24}$

$= \dfrac{6\sqrt{24}}{24} + \dfrac{24\sqrt{24}}{24} - \dfrac{144\sqrt{24}}{24} = -\dfrac{114\sqrt{24}}{24}$

$= \mathbf{-\dfrac{19\sqrt{6}}{2}}$

26. (a) $\dfrac{0.00842 \times 10^{18}}{4,198,312 \times 10^{-13}} \approx \dfrac{8 \times 10^{15}}{4 \times 10^{-7}}$

$= 2 \times 10^{22}$

Calculator $= \mathbf{2.01 \times 10^{22}}$

(b) $(4.63)^{5.12}$

Answer should be between 1024 and 3125.

Calculator $= \mathbf{2557.26}$

27. Graph the line to find the slope.

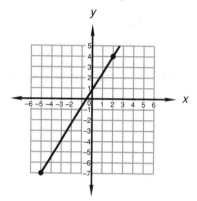

Slope $= \dfrac{11}{7}$

$y = \dfrac{11}{7}x + b$

$4 = \dfrac{11}{7}(2) + b$

$\dfrac{6}{7} = b$

$y = \mathbf{\dfrac{11}{7}x + \dfrac{6}{7}}$

28. $\sqrt{2x - 2} + 7 = 9$

$2x - 2 = 4$

$2x = 6$

$x = \mathbf{3}$

Check: $\sqrt{2(3) - 2} + 7 = 9$

$2 + 7 = 9$ check

29. $\dfrac{15x - 2}{3} - \dfrac{2x - 5}{4} = 3$

$60x - 8 - 6x + 15 = 36$

$54x = 29$

$x = \mathbf{\dfrac{29}{54}}$

30. $A_{\text{Figure}} = \dfrac{1}{2}bH + \dfrac{1}{2}\pi r^2$

$H = \sqrt{\sqrt{5}^2 - 1^2}$

$H = \sqrt{4} = 2$

$A_{\text{Figure}} = \left[\dfrac{1}{2}(2)(2) + \dfrac{1}{2}\pi\left(\dfrac{\sqrt{5}}{2}\right)^2\right]$ in.^{2}

$= \left(2 + \dfrac{5}{8}\pi\right)$ in.$^2 \approx \mathbf{3.96\ in.^2}$

PROBLEM SET 77

1. $R_H T_H = 320$;
$R_S T_S = 240$;
$R_S = R_H + 20$;
$T_H = 2T_S$

D_H 320 D_S 240

$(R_S - 20)2T_S = 320$
$2R_S T_S - 40T_S = 320$
$2(240) - 40T_S = 320$
$-40T_S = -160$
$T_S = \textbf{4 hr}$
$T_H = \textbf{8 hr}$; $R_S = \textbf{60 mph}$; $R_H = \textbf{40 mph}$

2. $E = \dfrac{k}{C}$

$500 = \dfrac{k}{10}$
$5000 = k$

$E = \dfrac{5000}{1} = \textbf{5000}$

3. $0.7(50) + W_A = 0.97(50 + W_A)$
$35 + W_A = 48.5 + 0.97W_A$
$0.03W_A = 13.5$
$W_A = \textbf{450 liters}$

4. $7N \qquad 7N + 7 \qquad 7N + 14$
$7N + 4(7N + 14) = 3(7N + 7) + 133$
$35N + 56 = 21N + 154$
$14N = 98$
$N = 7$
The desired integers are **49, 56, 63**.

5. $R_C T_C = R_Q T_Q$;
$T_C = 6$; $T_Q = 12$;
$R_C = R_Q + 4$

D_C D_Q

$R_C(6) = (R_C - 4)(12)$
$6R_C = 48$
$R_C = \textbf{8 mph}$

6. $\sqrt[3]{x^3 + 9x^2 - 27} = x + 3$
$x^3 + 9x^2 - 27 = x^3 + 9x^2 + 27x + 27$
$-27x = 54$
$x = \textbf{-2}$

Check: $\sqrt[3]{(-2)^3 + 9(-2)^2 - 27} - 1 = 0$
$1 - 1 = 0$ check

7. $\sqrt{m - 12} - \sqrt{m} + 2 = 0$
$\sqrt{m - 12} = \sqrt{m} - 2$
$m - 12 = m - 4\sqrt{m} + 4$
$-16 = -4\sqrt{m}$
$4 = \sqrt{m}$
$\textbf{16} = m$

Check: $\sqrt{16 - 12} - \sqrt{16} + 2 = 0$
$2 - 4 + 2 = 0$ check

8. (a) $2x + y - z = 7$
(b) $x - 2y + z = -2$
(c) $2y + z = 0$
$z = -2y$
Substitute (c) into (a) and (b) and get:
(a') $2x + 3y = 7$
(b') $x - 4y = -2$

$\begin{array}{r} \text{(a')} \quad 2x + 3y = 7 \\ -2\text{(b')} \; -2x + 8y = 4 \\ \hline 11y = 11 \\ y = 1 \end{array}$

(b') $x - 4(1) = -2$
$x = 2$

(c) $z = -2(1) = -2$

(2, 1, -2)

9. (a) $x + y + z = 7$
(b) $2x - y - z = -4$
(c) $z = 2y$
Substitute (c) into (a) and (b) and get:

$\begin{array}{r} \text{(a')} \quad x + 3y = 7 \\ \text{(b')} \; 2x - 3y = -4 \\ \hline 3x = 3 \\ x = 1 \end{array}$

(a') $(1) + 3y = 7$
$3y = 6$
$y = 2$

(c) $z = 2(2) = 4$

(1, 2, 4)

10. Since this is a 30-60-90 triangle:
$2 \times \overrightarrow{SF} = 8$
$\overrightarrow{SF} = 4$

$r = 4(\sqrt{3}) = \textbf{4}\sqrt{\textbf{3}}$
$t = 4(1) = \textbf{4}$

11.

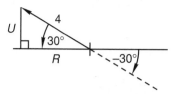

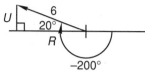

$R = 4 \cos 30° \approx 3.46$
$U = 4 \sin 30° = 2.00$

$R = 6 \cos 20° \approx 5.64$
$U = 6 \sin 20° \approx 2.05$

$$\begin{array}{r} -3.46R + 2.00U \\ -5.64R + 2.05U \\ \hline \mathbf{-9.10R + 4.05U} \end{array}$$

12. $H = \sqrt{7^2 + 2^2}$
$H = \sqrt{53}$

$\tan \theta = \dfrac{7}{2}$

$\theta \approx 74.05°$

Since θ is a third-quadrant angle:
$\theta = 74.05 + 180 = 254.05°$

$\mathbf{\sqrt{53}/254.05°}$

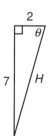

13. $\dfrac{7x + 2}{x^2 - 2x - 15} - \dfrac{2}{5 - x}$

$= \dfrac{7x + 2}{(x - 5)(x + 3)} + \dfrac{2(x + 3)}{(x - 5)(x + 3)}$

$= \dfrac{7x + 2 + 2x + 6}{x^2 - 2x - 15} = \dfrac{\mathbf{9x + 8}}{\mathbf{x^2 - 2x - 15}}$

14. $\dfrac{2 - \sqrt{2}}{2\sqrt{2} - 1} \cdot \dfrac{2\sqrt{2} + 1}{2\sqrt{2} + 1}$

$= \dfrac{4\sqrt{2} + 2 - 4 - \sqrt{2}}{8 - 1} = \dfrac{\mathbf{-2 + 3\sqrt{2}}}{\mathbf{7}}$

15. $\dfrac{3 + 2\sqrt{5}}{1 - \sqrt{5}} \cdot \dfrac{1 + \sqrt{5}}{1 + \sqrt{5}}$

$= \dfrac{3 + 3\sqrt{5} + 2\sqrt{5} + 10}{1 - 5} = \dfrac{\mathbf{-13 - 5\sqrt{5}}}{\mathbf{4}}$

16. $a + \dfrac{a}{a + \dfrac{a}{x}} = a + \dfrac{a}{\dfrac{ax + a}{x}}$

$= a + \dfrac{ax}{ax + a} = \dfrac{a^2 x + a^2 + ax}{ax + a}$

$= \dfrac{\mathbf{ax + a + x}}{\mathbf{x + 1}}$

17.
$$\dfrac{a}{x} = m\left(\dfrac{a}{R_1} + \dfrac{b}{R_2} \right)$$

$$\dfrac{a}{x} = \dfrac{am}{R_1} + \dfrac{bm}{R_2}$$

$$aR_1R_2 = amxR_2 + bmxR_1$$

$$aR_1R_2 - amxR_2 = bmxR_1$$

$$R_2 = \dfrac{\mathbf{bmxR_1}}{\mathbf{aR_1 - amx}}$$

18. $-\sqrt{-9} - 3i^3 - 2i^4 + 2$
$= -3i - 3i(ii) - 2(ii)(ii) + 2$
$= -3i + 3i - 2 + 2 = \mathbf{0}$

19. (a) $\dfrac{3}{7}x + \dfrac{2}{5}y = 10$

(b) $0.03x - 0.2y = -1.58$

(a′)$15x + 14y = 350$
(b′) $3x - 20y = -158$

$$\begin{array}{r} \text{(a′)} \quad 15x + 14y = 350 \\ -5\text{(b′)} \ -15x + 100y = 790 \\ \hline 114y = 1140 \\ y = 10 \end{array}$$

(b′) $3x - 20(10) = -158$
$x = 14$

$\mathbf{(14, 10)}$

20.
$$-7x - 1 = 2x^2$$

$$\left(x^2 + \dfrac{7}{2}x + \quad \right) = -\dfrac{1}{2}$$

$$x^2 + \dfrac{7}{2}x + \dfrac{49}{16} = -\dfrac{1}{2} + \dfrac{49}{16}$$

$$\left(x + \dfrac{7}{4} \right)^2 = \dfrac{41}{16}$$

$$x + \dfrac{7}{4} = \pm\dfrac{\sqrt{41}}{4}$$

$$x = -\dfrac{\mathbf{7}}{\mathbf{4}} \pm \dfrac{\sqrt{\mathbf{41}}}{\mathbf{4}}$$

21.
$$-8x - 1 = 2x^2$$
$$2x^2 + 8x + 1 = 0$$

$$x = \dfrac{-8 \pm \sqrt{8^2 - 4(2)(1)}}{2(2)} = \dfrac{-8 \pm \sqrt{56}}{4}$$

$$= \mathbf{-2} \pm \dfrac{\sqrt{\mathbf{14}}}{\mathbf{2}}$$

22. $m\angle AOC = 180 - 30 = \mathbf{150°}$

$m\angle OCA = m\angle OAC = \dfrac{180 - 150}{2} = \mathbf{15°}$

$\text{Area} = \dfrac{30}{360} \cdot \pi(3)^2 = \dfrac{\mathbf{3\pi}}{\mathbf{4}} \textbf{ units}^2$

23. $A_{\text{Square}} = s^2 = 4 \text{ cm}^2$

$\qquad s = \textbf{2 cm}$

Length of diagonal $= 2(\sqrt{2}) \text{ cm} = \textbf{2}\sqrt{\textbf{2}} \textbf{ cm}$

Radius of circle $= \dfrac{s}{2} = \dfrac{2 \text{ cm}}{2} = \textbf{1 cm}$

24. $\dfrac{4000 \text{ cm}^3}{\text{sec}} \times \dfrac{1 \text{ in.}}{2.54 \text{ cm}} \times \dfrac{1 \text{ in.}}{2.54 \text{ cm}} \times \dfrac{1 \text{ in.}}{2.54 \text{ cm}}$

$\qquad \times \dfrac{1 \text{ ft}}{12 \text{ in.}} \times \dfrac{1 \text{ ft}}{12 \text{ in.}} \times \dfrac{1 \text{ ft}}{12 \text{ in.}} \times \dfrac{60 \text{ sec}}{1 \text{ min}}$

$= \dfrac{4000(60)}{(2.54)(2.54)(2.54)(12)(12)(12)} \dfrac{\text{ft}^3}{\text{min}}$

25. $\sqrt[4]{4\sqrt[3]{2}} = \sqrt[4]{2^2 \sqrt[3]{2}} = [2^2 2^{1/3}]^{1/4}$

$= 2^{1/2} 2^{1/12} = \textbf{2}^{\textbf{7/12}}$

26. $\dfrac{(-3)^0 (-3^0)}{-9^{-3/2}} = -[-(9^{1/2})^3] = \textbf{27}$

27. $(2\sqrt{5} + 5)(5\sqrt{20} - 1) = (2\sqrt{5} + 5)(10\sqrt{5} - 1)$

$= 100 - 2\sqrt{5} + 50\sqrt{5} - 5 = \textbf{95} + \textbf{48}\sqrt{\textbf{5}}$

28. Write the equation of the given line in slope-intercept form.

$5x + 4y = 3$

$\qquad y = -\dfrac{5}{4}x + \dfrac{3}{4}$

Since the slopes of perpendicular lines are negative reciprocals of each other:

$m_\perp = \dfrac{4}{5}$

$y = \dfrac{4}{5}x + b$

$4 = \dfrac{4}{5}(-2) + b$

$\dfrac{28}{5} = b$

$y = \dfrac{4}{5}x + \dfrac{28}{5}$

29. $D^2 = 10^2 + 8^2$

$D^2 = 164$

$D = \textbf{2}\sqrt{\textbf{41}}$

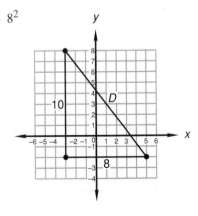

30. $\dfrac{3x - 1}{4} - \dfrac{x - 5}{7} = 1$

$21x - 7 - 4x + 20 = 28$

$17x = 15$

$x = \dfrac{15}{17}$

Problem Set 78

1. $R_R T_R = 135;$

$R_J T_J = 945;$

$T_R = T_J - 4; \; R_J = 3R_R$

$\qquad \overset{D_R}{\vert\!\!\xrightarrow{\hspace{1.5cm}}\!\!\vert} \qquad \overset{D_J}{\vert\!\!\xrightarrow{\hspace{1.5cm}}\!\!\vert}$

$\qquad\qquad 135 \qquad\qquad\qquad 945$

$3R_R(T_R + 4) = 945$

$3R_R T_R + 12R_R = 945$

$3(135) + 12R_R = 945$

$\qquad 12R_R = 540$

$\qquad R_R = \textbf{45 mph}$

$R_J = \textbf{135 mph}; \; T_R = \textbf{3 hr}; \; T_J = \textbf{7 hr}$

2. $N \qquad N + 2 \qquad N + 4 \qquad N + 6$

$(N + 2)(N + 6) = -3(N + N + 4) - 16$

$N^2 + 8N + 12 = -6N - 28$

$N^2 + 14N + 40 = 0$

$(N + 10)(N + 4) = 0$

$\qquad N = -10, -4$

The desired intergers are $\textbf{-10}, \textbf{-8}, \textbf{-6}, \textbf{-4}$

and $\textbf{-4}, \textbf{-2}, \textbf{0}, \textbf{2}$.

3. $\dfrac{86}{100} \times TC = 3440$

$\qquad TC = \textbf{4000 grams}$

4. $\dfrac{58}{100} \times C = 232$

$\qquad C = \textbf{400}$

5. (a) $N_F = 2N_S$

(b) $N_F \cdot N_S = 200$

Substitute (a) into (b) and get:

(b') $2N_S \cdot N_S = 200$

$\qquad N_S^2 = 100$

$\qquad N_S = \pm 10$

10, 20 and **-10, -20**

6. (a) $N = 5D$

(b) $2N = 7D + 6$

Substitute (a) into (b) and get:

(b') $2(5D) = 7D + 6$

$\qquad 3D = 6$

$\qquad D = 2$

(a) $N = 5(2) = 10$

Fraction $= \dfrac{N}{D} = \dfrac{\textbf{10}}{\textbf{2}}$

7. $\sqrt{x-9} + \sqrt{x} = 3$

$\sqrt{x-9} = 3 - \sqrt{x}$

$x - 9 = 9 - 6\sqrt{x} + x$

$-18 = -6\sqrt{x}$

$3 = \sqrt{x}$

$\mathbf{9} = x$

Check: $\sqrt{9-9} + \sqrt{9} = 3$

$0 + 3 = 3$ check

8. $\sqrt{x-8} + \sqrt{x} = 4$

$\sqrt{x-8} = 4 - \sqrt{x}$

$x - 8 = 16 - 8\sqrt{x} + x$

$-24 = -8\sqrt{x}$

$3 = \sqrt{x}$

$\mathbf{9} = x$

Check: $\sqrt{9-8} + \sqrt{9} = 4$

$1 + 3 = 4$ check

9. $\sqrt{k-24} = 6 - \sqrt{k}$

$k - 24 = 36 - 12\sqrt{k} + k$

$-60 = -12\sqrt{k}$

$5 = \sqrt{k}$

$\mathbf{25} = k$

Check: $\sqrt{25-24} = 6 - \sqrt{25}$

$1 = 6 - 5$ check

10. There is never an exact solution to these problems. One possible solution is given here.

$S = mI + b$

Use the graph to find the slope.

$m = \dfrac{95}{100} = 0.95$

$S = 0.95I + b$

Use the point (114, 20) for I and S.

$20 = 0.95(114) + b$

$-88 = b$

$\mathbf{S = 0.95I - 88}$

11. (a) $x - 2y - z = -9$

(b) $2x - y + 2z = 7$

(c) $3x - y = 0$

$y = 3x$

Substitute (c) into (a) and (b) and get:

(a′) $-5x - z = -9$

(b′) $-x + 2z = 7$

2(a′) $-10x - 2z = -18$

(b′) $\underline{\quad -x + 2z = \quad 7\quad}$

$-11x \qquad = -11$

$x = 1$

(b′) $-(1) + 2z = 7$

$z = 4$

(c) $y = 3(1) = 3$

$\mathbf{(1, 3, 4)}$

12.

$B = 10 \cos 27° \approx 8.91$

$C = 10 \sin 27° \approx 4.54$

$0.00R + \quad 8.00U$

$\underline{8.91R + \quad 4.54U}$

$\mathbf{8.91R + 12.54U}$

$\tan \theta = \dfrac{12.54}{8.91}$

$\theta \approx 54.61°$

$F = \sqrt{(8.91)^2 + (12.54)^2} \approx 15.38$

$\mathbf{15.38\underline{/54.61°}}$

13. $\dfrac{x-3}{x^2 + 5x - 14} + \dfrac{3x}{2 - x}$

$= \dfrac{x-3}{(x+7)(x-2)} - \dfrac{3x(x+7)}{(x+7)(x-2)}$

$= \dfrac{x - 3 - 3x^2 - 21x}{x^2 + 5x - 14}$

$= \dfrac{-3x^2 - 20x - 3}{x^2 + 5x - 14}$

14. $\dfrac{3 - \sqrt{5}}{\sqrt{5} + 2} \cdot \dfrac{\sqrt{5} - 2}{\sqrt{5} - 2}$

$= \dfrac{3\sqrt{5} - 6 - 5 + 2\sqrt{5}}{5 - 4} = \mathbf{-11 + 5\sqrt{5}}$

15. $\dfrac{2 + 2\sqrt{2}}{3 - 3\sqrt{2}} \cdot \dfrac{3 + 3\sqrt{2}}{3 + 3\sqrt{2}}$

$= \dfrac{6 + 6\sqrt{2} + 6\sqrt{2} + 12}{9 - 18} = \dfrac{\mathbf{-6 - 4\sqrt{2}}}{\mathbf{3}}$

16. $\dfrac{4 - 3\sqrt{2}}{1 - \sqrt{2}} \cdot \dfrac{1 + \sqrt{2}}{1 + \sqrt{2}}$

$= \dfrac{4 + 4\sqrt{2} - 3\sqrt{2} - 6}{1 - 2} = \mathbf{2 - \sqrt{2}}$

17. $x^2 = -x - 1$

$x^2 + x + 1 = 0$

$x = \dfrac{-1 \pm \sqrt{1^2 - 4(1)(1)}}{2(1)}$

$x = \dfrac{-1 \pm \sqrt{-3}}{2} = -\dfrac{1}{2} \pm \dfrac{\sqrt{3}}{2}i$

18.
$$mc = a\left(\frac{1}{x} + \frac{1}{R_1}\right)$$
$$mc = \frac{a}{x} + \frac{a}{R_1}$$
$$cmR_1x = aR_1 + ax$$
$$cmR_1x - aR_1 = ax$$
$$R_1 = \frac{ax}{cmx - a}$$

19.
$$mc = a\left(\frac{1}{x} + \frac{1}{R_1}\right)$$
$$mc = \frac{a}{x} + \frac{a}{R_1}$$
$$cmR_1x = aR_1 + ax$$
$$c = \frac{aR_1 + ax}{mR_1x}$$

20. $a + \dfrac{a}{a + \dfrac{x}{a}} = a + \dfrac{a}{\dfrac{a^2 + x}{a}}$

$= a + \dfrac{a^2}{a^2 + x} = \dfrac{a^3 + ax + a^2}{a^2 + x}$

21. $-5i^3 - \sqrt{-9} + \sqrt{-3}\sqrt{-3} = -5i(ii) - 3i + 3(ii)$
$= 5i - 3i - 3 = \mathbf{-3 + 2i}$

22. Since this is a 30-60-90 triangle:
$$2 \times \overrightarrow{SF} = \sqrt{5}$$
$$\overrightarrow{SF} = \frac{\sqrt{5}}{2}$$

$$a = \frac{\sqrt{5}}{2} \cdot 1 = \frac{\sqrt{5}}{2}$$
$$b = \frac{\sqrt{5}}{2} \cdot \sqrt{3} = \frac{\sqrt{15}}{2}$$

$$\text{Area} = \frac{bh}{2} = \frac{\frac{\sqrt{5}}{2}\left(\frac{\sqrt{15}}{2}\right)}{2} = \frac{5\sqrt{3}}{8} \text{ units}^2$$

23. $\dfrac{400 \text{ in.}^3}{\text{sec}} \times \dfrac{2.54 \text{ cm}}{1 \text{ in.}} \times \dfrac{2.54 \text{ cm}}{1 \text{ in.}} \times \dfrac{2.54 \text{ cm}}{1 \text{ in.}}$

$\times \dfrac{60 \text{ sec}}{1 \text{ min}} = \mathbf{400(2.54)(2.54)(2.54)(60)} \dfrac{\mathbf{cm^3}}{\mathbf{min}}$

24. $2\sqrt[5]{4\sqrt[6]{2}} = 2\sqrt[5]{2^2\sqrt[6]{2}} = 2[2^2(2^{1/6})]^{1/5}$
$= 2(2^{2/5}2^{1/30}) = \mathbf{2^{43/30}}$

25. $\dfrac{-2^0(2^{-2})}{4^{-3/2}} = \dfrac{-4^{3/2}}{4} = \dfrac{-(4^{1/2})^3}{4} = \mathbf{-2}$

26. $(2\sqrt{4} - 2)(3\sqrt{9} - 2) = (2)(7) = \mathbf{14}$

27. $4\sqrt{\dfrac{5}{8}} - 3\sqrt{\dfrac{8}{5}} + 2\sqrt{40}$

$= \dfrac{4\sqrt{5}}{\sqrt{8}}\dfrac{\sqrt{8}}{\sqrt{8}} - \dfrac{3\sqrt{8}}{\sqrt{5}}\dfrac{\sqrt{5}}{\sqrt{5}} + 2\sqrt{40}$

$= \dfrac{20\sqrt{40}}{40} - \dfrac{24\sqrt{40}}{40} + \dfrac{80\sqrt{40}}{40} = \dfrac{\mathbf{19\sqrt{10}}}{\mathbf{5}}$

28. (a) $\dfrac{70{,}218 \times 10^{-4}}{5062 \times 10^5} \approx \dfrac{7 \times 10^0}{5 \times 10^8} \approx 1 \times 10^{-8}$

Calculator $= \mathbf{1.39 \times 10^{-8}}$

(b) $5.\sqrt[4]{263}$

Answer should be between 2 and 4.
Calculator $= \mathbf{2.81}$

29. Solve for v:
$$\frac{\sqrt{mv}}{e} = p$$
$$\sqrt{mv} = ep$$
$$mv = e^2p^2$$
$$v = \frac{e^2p^2}{m} = \frac{(500)^2(100 \times 10^{-14})^2}{4 \times 10^7}$$
$$= \mathbf{6.25 \times 10^{-27}}$$

30.
$$\frac{S}{U} = \frac{T}{V}$$
$$SV = TU$$
$$S = \frac{TU}{V}$$

PROBLEM SET 79

1.
$$T_C = kN_P$$
$$2142 = k(3)$$
$$714 = k$$

$$T_C = 714(10) = \mathbf{\$7140}$$

2. $R_DT_D = 900;$
$T_D = 5T_E;$
$R_ET_E = 120;$
$R_D = R_E + 10$

$\overset{D_D}{\underset{900}{\longmapsto}}$ $\overset{D_E}{\underset{120}{\longmapsto}}$

$5T_E(R_E + 10) = 900$
$5R_ET_E + 50T_E = 900$
$5(120) + 50T_E = 900$
$50T_E = 300$
$T_E = 6 \text{ hr}$

$T_D = \mathbf{30 \text{ hr}}; \ R_E = \mathbf{20 \text{ kph}}; \ R_D = \mathbf{30 \text{ kph}}$

3. Carbon: $1 \times 12 = 12$
Oxygen: $2 \times 16 = 32$
Total: $\quad\quad = 44$

$$\frac{C}{T} = \frac{12}{44} \approx 0.273 = \mathbf{27.3\%}$$

4. (a) $0.6P_N + 0.9D_N = 39$
(b) $P_N + D_N = 50$
Substitute $D_N = 50 - P_N$ into (a) and get:
(a′) $0.6P_N + 0.9(50 - P_N) = 39$
$$-0.3P_N = -6$$
$$P_N = 20$$
(b) $(20) + D_N = 50$
$$D_N = 30$$

20 ml 60%, 30 ml 90%

5. (a) $\dfrac{N_W}{N_P} = \dfrac{2}{7}$
(a′) $7N_W = 2N_P$
(b) $5N_W = N_P + 120$
(b′) $N_P = 5N_W - 120$
Substitute (b′) into (a′) and get:
(a″) $7N_W = 2(5N_W - 120)$
$$3N_W = 240$$
$$N_W = \mathbf{80}$$
(b′) $N_P = 5(80) - 120 = \mathbf{280}$

6. $50{,}000 \text{ ml} \times \dfrac{1 \text{ liter}}{1000 \text{ ml}} = \mathbf{50 \text{ liters}}$

7. $20 \text{ ft}^3 \times \dfrac{12 \text{ in.}}{1 \text{ ft}} \times \dfrac{12 \text{ in.}}{1 \text{ ft}} \times \dfrac{12 \text{ in.}}{1 \text{ ft}} \times \dfrac{2.54 \text{ cm}}{1 \text{ in.}}$

$\times \dfrac{2.54 \text{ cm}}{1 \text{ in.}} \times \dfrac{2.54 \text{ cm}}{1 \text{ in.}} \times \dfrac{1 \text{ liter}}{1000 \text{ cm}^3}$

$= \dfrac{20(12)(12)(12)(2.54)(2.54)(2.54)}{1000}$ **liters**

8. $\sqrt{x^2 - 4x + 4} = x + 2$
$$x^2 - 4x + 4 = x^2 + 4x + 4$$
$$-8x = 0$$
$$x = \mathbf{0}$$

Check: $\sqrt{0^2 - 4(0) + 4} = 0 + 2$
$$2 = 2 \quad \text{check}$$

9. $\sqrt{s} = 4 - \sqrt{s + 8}$
$\sqrt{s + 8} = 4 - \sqrt{s}$
$$s + 8 = 16 - 8\sqrt{s} + s$$
$$8 = 8\sqrt{s}$$
$$1 = \sqrt{s}$$
$$\mathbf{1} = s$$

Check: $\sqrt{1} = 4 - \sqrt{1 + 8}$
$$1 = 4 - 3 \quad \text{check}$$

10. (a) $2x - y + 2z = 3$
(b) $x - y - 2z = -6$
(c) $3x - y = 0$
$$y = 3x$$

(a) $2x - y + 2z = 3$
(b) $\underline{x - y - 2z = -6}$
(d) $3x - 2y = -3$

Substitute (c) into (d) and get:
(d′) $3x - 2(3x) = -3$
$$-3x = -3$$
$$x = 1$$

(c) $y = 3(1) = 3$

(b) $(1) - (3) - 2z = -6$
$$-2z = -4$$
$$z = 2$$

(1, 3, 2)

11.

$A = 2 \cos 40° \approx 1.53$
$B = 2 \sin 40° \approx 1.29$

$C = 1 \cos 45° \approx 0.71$
$D = 1 \sin 45° \approx 0.71$

$1.53R - 1.29U$
$\underline{0.71R + 0.71U}$
$\mathbf{2.24R - 0.58U}$

$\tan \theta = -\dfrac{0.58}{2.24}$
$\theta \approx 345.48°$

$F = \sqrt{(2.24)^2 + (0.58)^2} \approx 2.31$

$\mathbf{2.31\underline{/345.48°}}$

12. $\dfrac{4x + 3}{x^2 - 9} - \dfrac{2x}{3 - x}$

$= \dfrac{4x + 3}{x^2 - 9} + \dfrac{2x(x + 3)}{x^2 - 9}$

$= \dfrac{4x + 3 + 2x^2 + 6x}{x^2 - 9}$

$= \dfrac{2x^2 + 10x + 3}{x^2 - 9}$

13. $\dfrac{-2 - \sqrt{3}}{2\sqrt{3} + 2} \cdot \dfrac{2\sqrt{3} - 2}{2\sqrt{3} - 2}$

$= \dfrac{-4\sqrt{3} + 4 - 6 + 2\sqrt{3}}{12 - 4} = \dfrac{-1 - \sqrt{3}}{4}$

14. $\dfrac{-1 - \sqrt{2}}{-5 - \sqrt{2}} \cdot \dfrac{-5 + \sqrt{2}}{-5 + \sqrt{2}}$

$= \dfrac{5 - \sqrt{2} + 5\sqrt{2} - 2}{25 - 2} = \dfrac{3 + 4\sqrt{2}}{23}$

15. See Lesson 71.

16.
$$\frac{a}{x} = m\left(\frac{a}{R_1} + \frac{b}{R_2}\right)$$
$$\frac{a}{x} = \frac{am}{R_1} + \frac{bm}{R_2}$$
$$aR_1R_2 = amxR_2 + bmxR_1$$
$$aR_1R_2 - amxR_2 = bmxR_1$$
$$a = \frac{bmxR_1}{R_1R_2 - mxR_2}$$

17.
$$\frac{a}{x} = m\left(\frac{a}{R_1} + \frac{b}{R_2}\right)$$
$$\frac{a}{x} = \frac{am}{R_1} + \frac{bm}{R_2}$$
$$aR_1R_2 = amxR_2 + bmxR_1$$
$$aR_1R_2 - amxR_2 = bmxR_1$$
$$\frac{aR_1R_2 - amxR_2}{mxR_1} = b$$

18. $ax - \dfrac{ax}{a - \dfrac{a}{x}} = ax - \dfrac{ax}{\dfrac{ax - a}{x}}$

$= ax - \dfrac{ax^2}{ax - a} = \dfrac{a^2x^2 - a^2x - ax^2}{ax - a}$

$= \dfrac{ax^2 - ax - x^2}{x - 1}$

19. $\sqrt{-4} - 3i^2 - 2i^4 + 2 - \sqrt{-2}\sqrt{-2}$
$= 2i + 3 - 2 + 2 + 2 = 5 + 2i$

20. (a) $\dfrac{1}{5}x - \dfrac{1}{4}y = -6$
(b) $0.2x + 0.2y = 12$

(a′) $4x - 5y = -120$
(b′) $2x + 2y = 120$

$\begin{array}{r} (a') \quad 4x - 5y = -120 \\ -2(b') \;\; -4x - 4y = -240 \\ \hline -9y = -360 \\ y = 40 \end{array}$

(b′) $2x + 2(40) = 120$
$2x = 40$
$x = 20$

(20, 40)

21. $3x^2 - 1 = 2x$
$3x^2 - 2x - 1 = 0$

$x = \dfrac{-(-2) \pm \sqrt{(-2)^2 - 4(3)(-1)}}{2(3)}$

$= \dfrac{2 \pm \sqrt{16}}{6} = \dfrac{1}{3} \pm \dfrac{2}{3} = 1, -\dfrac{1}{3}$

22. Since this is a 30-60-90 triangle:
$\sqrt{3} \times \overrightarrow{SF} = 3$
$\overrightarrow{SF} = \sqrt{3}$
$x = \sqrt{3}(1) = \sqrt{3}$
$y = \sqrt{3}(2) = 2\sqrt{3}$

23. Since this is a 45-45-90 triangle:
$1 \times \overrightarrow{SF} = \sqrt{3}$
$\overrightarrow{SF} = \sqrt{3}$
$a = \sqrt{3}(1) = \sqrt{3}$
$b = \sqrt{3}(\sqrt{2}) = \sqrt{6}$

24. $\dfrac{40 \text{ ml}}{\text{sec}} \times \dfrac{1 \text{ cm}^3}{1 \text{ ml}} \times \dfrac{1 \text{ in.}}{2.54 \text{ cm}} \times \dfrac{1 \text{ in.}}{2.54 \text{ cm}}$

$\times \dfrac{1 \text{ in.}}{2.54 \text{ cm}} \times \dfrac{60 \text{ sec}}{1 \text{ min}} \times \dfrac{60 \text{ min}}{1 \text{ hr}}$

$= \dfrac{40(60)(60)}{(2.54)(2.54)(2.54)} \dfrac{\text{in.}^3}{\text{hr}}$

25. $3\sqrt{9^2 \sqrt{3}} = 3\sqrt{3^4 \sqrt{3}} = 3[3^4 3^{1/2}]^{1/2}$
$= 3[3^2 3^{1/4}] = 3^{13/4}$

26. $2\sqrt{\dfrac{7}{2}} + 3\sqrt{\dfrac{2}{7}} - 2\sqrt{126}$

$= \dfrac{2\sqrt{7}}{\sqrt{2}} \dfrac{\sqrt{2}}{\sqrt{2}} + \dfrac{3\sqrt{2}}{\sqrt{7}} \dfrac{\sqrt{7}}{\sqrt{7}} - 6\sqrt{14}$

$= \dfrac{14\sqrt{14}}{14} + \dfrac{6\sqrt{14}}{14} - \dfrac{84\sqrt{14}}{14} = -\dfrac{32\sqrt{14}}{7}$

27. $(4\sqrt{2} + 3)(5\sqrt{2} - 4) = 40 - 16\sqrt{2} + 15\sqrt{2} - 12$
$= 28 - \sqrt{2}$

28. (a) $\dfrac{41,852 \times 10^{28}}{0.00492 \times 10^{-14}} \approx \dfrac{4 \times 10^{32}}{5 \times 10^{-17}}$

$= 8 \times 10^{48}$

Calculator $= 8.51 \times 10^{48}$

(b) $(194)^{-1.09}$

Answer should be between 5×10^{-3} and 3×10^{-5}.

Calculator $= 3.21 \times 10^{-3}$

29. $y = \dfrac{1}{5}x + b$

$7 = \dfrac{1}{5}(5) + b$

$6 = b$

$y = \dfrac{1}{5}x + 6$

30. $D^2 = 0^2 + 4^2$

$D^2 = 16$

$D = 4$

PROBLEM SET 80

1. $C = kN$

$119 = k(14)$

$k = 8.5$

$C = 8.5(32) = \$272$

2. $\dfrac{B_1}{B_2} = \dfrac{Y_2^2}{Y_1^2}$

$\dfrac{50}{B_2} = \dfrac{(10)^2}{(5)^2}$

$100B_2 = 1250$

$B_2 = 12.5$

3. (a) $\dfrac{N_B}{N_P} = \dfrac{4}{5} \longrightarrow 5N_B = 4N_P$

(b) $N_B = 2N_P - 1200$

Substitute (b) into (a) and get:

(a') $5(2N_P - 1200) = 4N_P$

$6N_P = 6000$

$N_P = 1000$

(b) $N_B = 2(1000) - 1200 = 800$

4. $3\dfrac{1}{4} \times P = 650$

$P = 650 \cdot \dfrac{4}{13} = 200$

5. (a) $40N_D + 2N_C = 820$

(b) $N_D + N_C = 30$

Substitute (b) into (a) and get:

(a') $40N_D + 2(30 - N_D) = 820$

$38N_D = 760$

$N_D = 20$

(b) $N_C = 30 - (20) = 10$

6. $\sqrt{x^2 - x - 2} - x + 2 = 0$

$x^2 - x - 2 = x^2 - 4x + 4$

$3x = 6$

$x = 2$

Check: $\sqrt{2^2 - 2 - 2} - 2 + 2 = 0$

$0 - 2 + 2 = 0$ check

7. $\sqrt{p + 20} + \sqrt{p} = 10$

$p + 20 = 100 - 20\sqrt{p} + p$

$20\sqrt{p} = 80$

$\sqrt{p} = 4$

$p = 16$

Check: $\sqrt{16 + 20} + \sqrt{16} = 10$

$6 + 4 = 10$ check

8. $\sqrt{s} - 18 + \sqrt{s - 36} = 0$

$s - 36 = 324 - 36\sqrt{s} + s$

$36\sqrt{s} = 360$

$\sqrt{s} = 10$

$s = 100$

Check: $\sqrt{100} - 18 + \sqrt{100 - 36} = 0$

$10 - 18 + 8 = 0$ check

9. (a) $x + y + z = 8$

(b) $2x - 3y - z = -6$

(c) $2x - z = 0$

$z = 2x$

Substitute (c) into (a) and (b) and get:

(a') $3x + y = 8$

(b') $-3y = -6$

(b') $-3y = -6$

$y = 2$

(a') $3x + (2) = 8$

$3x = 6$

$x = 2$

(c) $z = 2(2) = 4$

$(2, 2, 4)$

10.

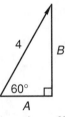

$A = 4 \cos 60° = 2.00$
$B = 4 \sin 60° ≈ 3.46$

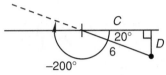

$C = 6 \cos 20° ≈ 5.64$
$D = 6 \sin 20° ≈ 2.05$

$2.00R + 3.46U$
$\underline{5.64R - 2.05U}$
$\mathbf{7.64R + 1.41U}$

11. $H = \sqrt{2^2 + 6^2}$
$H = 2\sqrt{10}$

$\tan \theta = \dfrac{6}{2} = 3$
$\theta ≈ 71.57°$
Since θ is a second-quadrant angle:
$\theta ≈ -(71.57) + 180 = 108.43°$

$\mathbf{2\sqrt{10}/108.43°}$

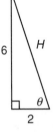

12. $\dfrac{4x + 2}{x^2 - 6x - 16} - \dfrac{3}{x - 8}$

$= \dfrac{4x + 2}{(x - 8)(x + 2)} - \dfrac{3(x + 2)}{(x - 8)(x + 2)}$

$= \dfrac{\mathbf{x - 4}}{\mathbf{x^2 - 6x - 16}}$

13. $\dfrac{3\sqrt{2} - 1}{1 + \sqrt{2}} \cdot \dfrac{1 - \sqrt{2}}{1 - \sqrt{2}}$

$= \dfrac{3\sqrt{2} - 6 - 1 + \sqrt{2}}{1 - 2} = \mathbf{7 - 4\sqrt{2}}$

14. $\dfrac{2 - 3\sqrt{2}}{3 - 2\sqrt{2}} \cdot \dfrac{3 + 2\sqrt{2}}{3 + 2\sqrt{2}}$

$= \dfrac{6 + 4\sqrt{2} - 9\sqrt{2} - 12}{9 - 8} = \mathbf{-6 - 5\sqrt{2}}$

15. $A^2 = 3^2 + 4^2$
$A = 5$
$4 \times \overrightarrow{SF} = 8$
$\overrightarrow{SF} = 2$
$(5)2 = B + 5$
$10 = B + 5$
$\mathbf{5 = B}$

16. $-3x^2 + 2 = -3x$

$(x^2 - x +) = \dfrac{2}{3}$

$x^2 - x + \dfrac{1}{4} = \dfrac{2}{3} + \dfrac{1}{4}$

$\left(x - \dfrac{1}{2}\right)^2 = \dfrac{11}{12}$

$x - \dfrac{1}{2} = \pm\dfrac{\sqrt{33}}{6}$

$\mathbf{x = \dfrac{1}{2} \pm \dfrac{\sqrt{33}}{6}}$

17. $a = xm\left(\dfrac{p}{y} + \dfrac{q}{c}\right)$

$a = \dfrac{mpx}{y} + \dfrac{mqx}{c}$

$acy = cmpx + mqxy$
$acy - mqxy = cmpx$

$\mathbf{y = \dfrac{cmpx}{ac - mqx}}$

18. $3a - \dfrac{3}{a - \dfrac{3}{a}} = 3a - \dfrac{3}{\dfrac{a^2 - 3}{a}}$

$= 3a - \dfrac{3a}{a^2 - 3} = \dfrac{3a^3 - 9a - 3a}{a^2 - 3}$

$= \mathbf{\dfrac{3a^3 - 12a}{a^2 - 3}}$

19. (a) Since this is a 45-45-90 triangle:
$\sqrt{2} \times \overrightarrow{SF} = 5$

$\overrightarrow{SF} = \dfrac{5\sqrt{2}}{2}$

$m = \dfrac{5\sqrt{2}}{2}(1) = \mathbf{\dfrac{5\sqrt{2}}{2}}$

$n = \dfrac{5\sqrt{2}}{2}(1) = \mathbf{\dfrac{5\sqrt{2}}{2}}$

(b) Since this is a 30-60-90 triangle:
$2 \times \overrightarrow{SF} = 5$

$\overrightarrow{SF} = \dfrac{5}{2}$

$c = \dfrac{5}{2}(1) = \mathbf{\dfrac{5}{2}}$

$d = \dfrac{5}{2}\left(\sqrt{3}\right) = \mathbf{\dfrac{5\sqrt{3}}{2}}$

20. $-\sqrt{-4} + \sqrt{-9} - i^3 + \sqrt{-2}\sqrt{-2} - 4i^4$
$= -2i + 3i - i(ii) + 2(ii) - 4(ii)(ii)$
$= -2i + 3i + i - 2 - 4 = \mathbf{-6 + 2i}$

21. (a) $\dfrac{2}{3}x - \dfrac{1}{4}y = 6$

(b) $0.07x + 0.06y = 1.32$

(a') $8x - 3y = 72$
(b') $7x + 6y = 132$

$\begin{aligned}
2(a')\ 16x - 6y &= 144 \\
(b')\ \underline{\ \ 7x + 6y = 132} \\
23x \ \ \ \ \ \ &= 276 \\
x &= 12
\end{aligned}$

$\begin{aligned}
(b')\ 7(12) + 6y &= 132 \\
6y &= 48 \\
y &= 8
\end{aligned}$

(12, 8)

22.
$$-x^2 = -x - 5$$
$$x^2 - x - 5 = 0$$
$$x = \frac{-(-1) \pm \sqrt{(-1)^2 - 4(1)(-5)}}{2(1)}$$
$$x = \frac{1 \pm \sqrt{21}}{2} = \mathbf{\frac{1}{2} \pm \frac{\sqrt{21}}{2}}$$

23. $\dfrac{600\ \text{cm}^3}{\text{min}} \times \dfrac{1\ \text{in.}}{2.54\ \text{cm}} \times \dfrac{1\ \text{in.}}{2.54\ \text{cm}} \times \dfrac{1\ \text{in.}}{2.54\ \text{cm}}$

$\times \dfrac{1\ \text{ft}}{12\ \text{in.}} \times \dfrac{1\ \text{ft}}{12\ \text{in.}} \times \dfrac{1\ \text{ft}}{12\ \text{in.}} \times \dfrac{60\ \text{min}}{1\ \text{hr}}$

$= \dfrac{\mathbf{600(60)}}{\mathbf{(2.54)(2.54)(2.54)(12)(12)(12)}}\ \dfrac{\mathbf{ft^3}}{\mathbf{hr}}$

24. Volume $= A_{\text{Base}} \times$ Height

$= 2\left(\dfrac{\pi r^2}{2}\right) \times H$

$= \pi(6\ \text{in.})^2\left(6\ \text{ft} \times \dfrac{12\ \text{in.}}{1\ \text{ft}}\right)$

$\approx$ **8138.88 1–in. sugar cubes**

25. $\sqrt[4]{x^5 y}\sqrt{xy^3} = (x^5 y)^{1/4}(xy^3)^{1/2}$

$= x^{5/4}y^{1/4}x^{1/2}y^{3/2} = \mathbf{x^{7/4}y^{7/4}}$

26. $\dfrac{-2^0\,(-2)^0}{-(4)^{-3/2}} = -\left[-(4^{1/2})^3\right] = \mathbf{8}$

27. (a) $x - 3y = 6$

$y = \dfrac{1}{3}x - 2$

(b) $2x + y = -1$

$y = -2x - 1$

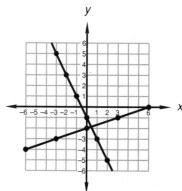

Substitute $y = -2x - 1$ into (a) and get:
(a') $x - 3(-2x - 1) = 6$
$7x = 3$
$x = \dfrac{3}{7}$

(b) $y = -2\left(\dfrac{3}{7}\right) - 1 = -\dfrac{13}{7}$

$\left(\dfrac{3}{7}, -\dfrac{13}{7}\right)$

28. $\dfrac{3}{ax} + \dfrac{3x}{a^2 x} + \dfrac{7x}{x + a}$

$= \dfrac{3a(x + a)}{a^2 x(x + a)} + \dfrac{3x(x + a)}{a^2 x(x + a)}$

$+ \dfrac{7a^2 x^2}{a^2 x(x + a)}$

$= \dfrac{3ax + 3a^2 + 3x^2 + 3ax + 7a^2 x^2}{a^2 x(x + a)}$

$= \dfrac{3a^2 + 3x^2 + 6ax + 7a^2 x^2}{a^2 x(x + a)}$

29.
$$x^3 = 4x^2 + 32x$$
$$x^3 - 4x^2 - 32x = 0$$
$$x(x - 8)(x + 4) = 0$$
$$x = \mathbf{0, -4, 8}$$

30. Write the equation of the given line in slope-intercept form.
$$-x - y - 1 = 0$$
$$y = -x - 1$$
Since the slopes of perpendicular lines are negative reciprocals of each other:
$$m_\perp = +1$$
$$y = x + b$$
$$-3 = -2 + b$$
$$-1 = b$$
$$\mathbf{y = x - 1}$$

PROBLEM SET 81

1. Variation method:

$M = kT^2$

$100 = k(2)^2$

$25 = k$

$M = 25(5^2) = \mathbf{625}$

Equal ratio method:

$$\frac{M_1}{M_2} = \frac{T_1^2}{T_2^2}$$

$$\frac{100}{M_2} = \frac{(2)^2}{(5)^2}$$

$4M_2 = 2500$

$M_2 = \mathbf{625}$

2. $M = \dfrac{k}{A^2}$

$4 = \dfrac{k}{(10)^2}$

$400 = k$

$M = \dfrac{400}{4} = \mathbf{100}$

3. $R_R T_R = 375$; $R_J T_J = 375$;

$R_R = 3R_J$; $T_R = T_J - 10$

$3R_J(T_J - 10) = 375$

$3R_J T_J - 30R_J = 375$

$3(375) - 30R_J = 375$

$-30R_J = -750$

$R_J = \mathbf{25\ mph}$

$R_R = \mathbf{75\ mph}$; $T_J = \mathbf{15\ hr}$; $T_R = \mathbf{5\ hr}$

4. (a) $0.1P_N + 0.4D_N = 38$

(b) $P_N + D_N = 200$

Substitute $D_N = 200 - P_N$ into (a) and get:

(a') $0.1P_N + 0.4(200 - P_N) = 38$

$-0.3P_N = -42$

$P_N = 140$

(b) $(140) + D_N = 200$

$D_N = 60$

140 liters 10%; 60 liters 40%

5. $\dfrac{40}{100} \times N_T = 1120$

$N_T = 1120 \cdot \dfrac{100}{40}$

$N_T = \mathbf{2800}$

6. $\dfrac{3 - i}{2 + 5i} \cdot \dfrac{2 - 5i}{2 - 5i}$

$= \dfrac{6 - 15i - 2i + 5i^2}{4 - 25i^2} = \dfrac{1}{29} - \dfrac{17}{29}i$

7. $\dfrac{3 - 2i}{2i - 4} \cdot \dfrac{2i + 4}{2i + 4}$

$= \dfrac{6i + 12 - 4i^2 - 8i}{4i^2 - 16} = -\dfrac{4}{5} + \dfrac{1}{10}i$

8. $(180 - A) = 4(90 - A) + 30$

$180 - A = 360 - 4A + 30$

$3A = 210$

$A = \mathbf{70°}$

9. $\sqrt{x^2 - x + 30} - 3 = x$

$x^2 - x + 30 = x^2 + 6x + 9$

$-7x = -21$

$x = \mathbf{3}$

Check: $\sqrt{3^2 - 3 + 30} - 3 = 3$

$6 - 3 = 3$ check

10. $\sqrt{p - 48} = 12 - \sqrt{p}$

$p - 48 = 144 - 24\sqrt{p} + p$

$24\sqrt{p} = 192$

$\sqrt{p} = 8$

$p = \mathbf{64}$

Check: $\sqrt{64 - 48} = 12 - \sqrt{64}$

$4 = 12 - 8$ check

11. (a) $x + 2y - 3z = 5$

(b) $2x - y - z = 0$

(c) $y - 3z = 0$

$y = 3z$

Substitute (c) into (a) and (b) and get:

(a') $x + 3z = 5$

(b') $2x - 4z = 0$

$-2(a')\ -2x - 6z = -10$

(b') $\underline{\ \ 2x - 4z = \quad 0}$

$-10z = -10$

$z = 1$

(a') $x + 3(1) = 5$

$x = 2$

(c) $y = 3(1) = 3$

(2, 3, 1)

12.

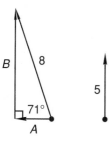

$A = 8 \cos 71° \approx 2.60$
$B = 8 \sin 71° \approx 7.56$

$$\begin{array}{r} -2.60R + 7.56U \\ 0.00R + 5.00U \\ \hline -2.60R + 12.56U \end{array}$$

$\tan \theta = -\dfrac{12.56}{2.60}$
$\quad\theta \approx 101.70°$

$F = \sqrt{(2.60)^2 + (12.56)^2} \approx 12.83$

12.83/101.70°

13. $\dfrac{7x - 2}{x^2 - 9} + \dfrac{3x}{3 - x}$

$= \dfrac{7x - 2}{(x - 3)(x + 3)} - \dfrac{3x(x + 3)}{(x - 3)(x + 3)}$

$= \dfrac{-3x^2 - 2x - 2}{x^2 - 9}$

14. $\dfrac{\sqrt{2} - 5}{\sqrt{2} - 2} \cdot \dfrac{\sqrt{2} + 2}{\sqrt{2} + 2}$

$= \dfrac{2 + 2\sqrt{2} - 5\sqrt{2} - 10}{2 - 4} = \dfrac{8 + 3\sqrt{2}}{2}$

15. $\dfrac{2\sqrt{3} - 1}{1 - 3\sqrt{3}} \cdot \dfrac{1 + 3\sqrt{3}}{1 + 3\sqrt{3}}$

$= \dfrac{2\sqrt{3} + 18 - 1 - 3\sqrt{3}}{1 - 27} = \dfrac{-17 + \sqrt{3}}{26}$

16. $\dfrac{1 + \sqrt{2}}{3 - \sqrt{2}} \cdot \dfrac{3 + \sqrt{2}}{3 + \sqrt{2}}$

$= \dfrac{3 + \sqrt{2} + 3\sqrt{2} + 2}{9 - 2} = \dfrac{5 + 4\sqrt{2}}{7}$

17. $x = \dfrac{40 + 50}{2} = 45$

$y = \dfrac{140 - 100}{2} = 20$

18. See Lesson 71.

19. $2a^2 - \dfrac{3a}{a + \dfrac{1}{a}} = 2a^2 - \dfrac{3a}{\dfrac{a^2 + 1}{a}}$

$= 2a^2 - \dfrac{3a^2}{a^2 + 1} = \dfrac{2a^4 - a^2}{a^2 + 1}$

20. $\sqrt{-9} - \sqrt{-2}\sqrt{-2} + \sqrt{-2}\sqrt{2} - 3i^3 - 2i^2$
$= 3i - 2(ii) + 2i - 3i(ii) - 2(ii)$
$= 3i + 2 + 2i + 3i + 2 = \mathbf{4 + 8i}$

21. (a) $\dfrac{2}{3}x - \dfrac{1}{3}y = 6$
(b) $0.15x + 0.01y = 0.84$

$$\begin{array}{ll} (a') & 2x - y = 18 \\ (b') & 15x + y = 84 \\ \hline & 17x = 102 \\ & x = 6 \end{array}$$

$$\begin{array}{ll} (a') & 2(6) - y = 18 \\ & y = -6 \end{array}$$

(6, −6)

22. $\qquad 2 = -2x^2 - 3x$
$2x^2 + 3x + 2 = 0$

$x = \dfrac{-3 \pm \sqrt{3^2 - 4(2)(2)}}{2(2)} = \dfrac{-3 \pm \sqrt{-7}}{4}$

$= -\dfrac{3}{4} \pm \dfrac{\sqrt{7}}{4}i$

23.
$$\require{enclose}
\begin{array}{r}
4x^2 + 16x + 63 + \frac{254}{x-4} \\
x - 4 \enclose{longdiv}{4x^3 + 0x^2 - x + 2} \\
\underline{4x^3 - 16x^2} \\
16x^2 - x \\
\underline{16x^2 - 64x} \\
63x + 2 \\
\underline{63x - 252} \\
254
\end{array}$$

24. Since this is a 45-45-90 triangle:
$\sqrt{2} \times \overrightarrow{SF} = 17$

$\overrightarrow{SF} = \dfrac{17\sqrt{2}}{2}$

$x = \dfrac{17\sqrt{2}}{2}(1) = \dfrac{17\sqrt{2}}{2}$

$y = \dfrac{17\sqrt{2}}{2}(1) = \dfrac{17\sqrt{2}}{2}$

25. $3\sqrt{9\sqrt[4]{3}} = 3\sqrt{3^2 \sqrt[4]{3}} = 3\left[3^2(3^{1/4})\right]^{1/2}$
$= 3 \cdot 3 \cdot 3^{1/8} = \mathbf{3^{17/8}}$

26. $2\sqrt{\dfrac{1}{5}} - 3\sqrt{5} + 3\sqrt{20} = \dfrac{2}{\sqrt{5}}\dfrac{\sqrt{5}}{\sqrt{5}} - 3\sqrt{5} + 6\sqrt{5}$

$= \dfrac{2\sqrt{5}}{5} - \dfrac{15\sqrt{5}}{5} + \dfrac{30\sqrt{5}}{5} = \dfrac{17\sqrt{5}}{5}$

27. $\dfrac{x}{x + y} + \dfrac{3}{x^2 y} + \dfrac{2}{xy}$

$= \dfrac{x^3 y}{x^2 y(x + y)} + \dfrac{3(x + y)}{x^2 y(x + y)}$

$\quad + \dfrac{2x(x + y)}{x^2 y(x + y)}$

$= \dfrac{x^3 y + 3x + 3y + 2x^2 + 2xy}{x^2 y(x + y)}$

28. $-4^2 - 3^0 - 2^0(x - x^0) - 3^0(-2x - 5) = 7$

$-16 - 1 - x + 1 + 2x + 5 = 7$

$x = 18$

29. $D^2 = 11^2 + 7^2$

$D^2 = 170$

$D = \sqrt{170}$

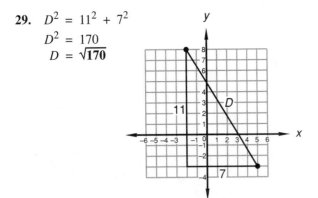

30. (a) $\dfrac{0.5061 \times 10^5}{0.0071643 \times 10^{-18}} \approx \dfrac{5 \times 10^4}{7 \times 10^{-21}}$

$\approx 7 \times 10^{24}$

Calculator $= \mathbf{7.06 \times 10^{24}}$

(b) $\sqrt[6.2]{594}$

Answer should be between 2 and 3.

Calculator $= \mathbf{2.80}$

PROBLEM SET 82

1. $\dfrac{Q_1}{Q_2} = \dfrac{S_2^2}{S_1^2}$

$\dfrac{300}{Q_2} = \dfrac{(10)^2}{(5)^2}$

$100Q_2 = 25(300)$

$Q_2 = \mathbf{75}$

2. $R_P T_P = 1920;$ $\overset{D_P}{\underset{1920}{\vdash\!\!\!\longrightarrow\!\!\!\dashv}}$ $\overset{D_R}{\underset{320}{\vdash\!\!\!\longrightarrow\!\!\!\dashv}}$

$R_R T_R = 320;$

$T_P = T_R + 4;$

$R_P = 3R_R$

$3R_R(T_R + 4) = 1920$

$3R_R T_R + 12R_R = 1920$

$3(320) + 12R_R = 1920$

$12R_R = 960$

$R_R = \mathbf{80\ mph}$

$R_P = \mathbf{240\ mph};\ T_R = \mathbf{4\ hr};\ T_P = \mathbf{8\ hr}$

3. $\dfrac{P_1 V_1}{T_1} = \dfrac{P_2 V_2}{T_2}$

$T_2 = \dfrac{P_2 V_2 T_1}{P_1 V_1}$

$T_2 = \dfrac{(50)(200)(540)}{(5)(250)} = \mathbf{4320\ K}$

4. $R_R T_R + R_W T_W = 60;$ $\overset{D_R \quad D_W}{\underset{60}{\vdash\!\!\longrightarrow\!\!\longrightarrow\!\!\dashv}}$

$R_R = 10;\ R_W = 5;$

$T_R + T_W = 8$

$10T_R + 5(8 - T_R) = 60$

$5T_R = 20$

$T_R = 4\ hr$

$T_W = 4\ hr$

$D_R = 10(4) = \mathbf{40\ miles}$

$D_W = 5(4) = \mathbf{20\ miles}$

5. Sodium: $2 \times 23 = 46$
Hydrogen: $1 \times 1 = 1$
Phosphorous: $1 \times 31 = 31$
Oxygen: $4 \times 16 = 64$
Total: $= 142$

$\dfrac{46}{142} = \dfrac{Na}{852}$

$142Na = 46(852)$

$Na = \mathbf{276\ grams}$

Percent $= \dfrac{Na}{T} = \dfrac{46}{142} \approx 0.324 = \mathbf{32.4\%}$

6. $\dfrac{m}{2 + \dfrac{m}{2 + \dfrac{2}{p}}} = \dfrac{m}{2 + \dfrac{m}{\dfrac{2p + 2}{p}}}$

$= \dfrac{m}{2 + \dfrac{mp}{2p + 2}} = \dfrac{m}{\dfrac{4p + 4 + mp}{2p + 2}}$

$= \dfrac{m(2p + 2)}{4p + 4 + mp}$

7. $\dfrac{p}{a + \dfrac{b}{m + \dfrac{3}{y}}} = \dfrac{p}{a + \dfrac{b}{\dfrac{my + 3}{y}}}$

$= \dfrac{p}{a + \dfrac{by}{my + 3}} = \dfrac{p}{\dfrac{amy + 3a + by}{my + 3}}$

$= \dfrac{p(my + 3)}{amy + 3a + by}$

8. $V_{\text{Cone}} = \dfrac{\pi r^2 h}{3} = 12\pi \text{ m}^3$

$h = \dfrac{12(3) \text{ m}^3}{9 \text{ m}^2} = \textbf{4 m}$

9. $\dfrac{2 - 3i}{4 + i} \cdot \dfrac{4 - i}{4 - i} = \dfrac{8 - 2i - 12i - 3}{16 + 1}$

$= \dfrac{5}{17} - \dfrac{14}{17}i$

10. $\dfrac{5 - i}{2 - 3i} \cdot \dfrac{2 + 3i}{2 + 3i} = \dfrac{10 + 15i - 2i + 3}{4 + 9}$

$= \textbf{1} + \textbf{i}$

11. $\sqrt{x^2 - x + 47} - 5 = x$

$\qquad x^2 - x + 47 = x^2 + 10x + 25$

$\qquad\qquad 11x = 22$

$\qquad\qquad\quad x = \textbf{2}$

Check: $\sqrt{4 - 2 + 47} - 5 = 2$

$\qquad\qquad 7 - 5 = 2 \quad$ check

12. $\sqrt{x + 24} + \sqrt{x} = 12$

$\qquad x + 24 = 144 - 24\sqrt{x} + x$

$\qquad 24\sqrt{x} = 120$

$\qquad \sqrt{x} = 5$

$\qquad x = \textbf{25}$

Check: $\sqrt{25 + 24} + \sqrt{25} = 12$

$\qquad\qquad 7 + 5 = 12 \quad$ check

13. (a) $2x - 2y - z = 16$

$\quad$ (b) $3x - y + 2z = 5$

$\quad$ (c) $-y + 3z = 0$

$\qquad\qquad y = 3z$

Substitute (c) into (a) and (b) and get:

(a′) $2x - 7z = 16$

(b′) $3x - z = 5$

$\qquad$ (a′) $\quad 2x - 7z = \quad 16$

-7(b′) $\dfrac{-21x + 7z = -35}{}$

$\qquad\qquad -19x \qquad\quad = -19$

$\qquad\qquad\qquad x = 1$

(b′) $3(1) - z = 5$

$\qquad\qquad z = -2$

(c) $y = 3(-2) = -6$

$\textbf{(1, -6, -2)}$

14.

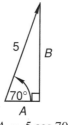

$A = 5 \cos 70° \approx 1.71$

$B = 5 \sin 70° \approx 4.70$

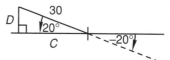

$C = 30 \cos 20° \approx 28.19$

$D = 30 \sin 20° \approx 10.26$

$\qquad 1.71R + \quad 4.70U$

$\dfrac{-28.19R + 10.26U}{}$

$\textbf{-26.48R + 14.96U}$

15. $H = \sqrt{7^2 + 4^2}$

$H = \sqrt{65}$

$\tan \theta = \dfrac{7}{4}$

$\quad \theta \approx 60.26°$

Since θ is a second-quadrant angle:

$\theta = -(60.26) + 180 = 119.74°$

$\underline{\sqrt{65}/119.74°}$

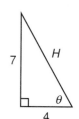

16. $\dfrac{3x - 2}{x^2 - 9} + \dfrac{x}{3 - x}$

$= \dfrac{3x - 2}{(x - 3)(x + 3)} - \dfrac{x(x + 3)}{(x + 3)(x - 3)}$

$= \dfrac{3x - 2 - x^2 - 3x}{x^2 - 9} = \dfrac{-x^2 - 2}{x^2 - 9}$

17. $\dfrac{3\sqrt{2} - 1}{1 + \sqrt{2}} \cdot \dfrac{1 - \sqrt{2}}{1 - \sqrt{2}}$

$= \dfrac{3\sqrt{2} - 6 - 1 + \sqrt{2}}{1 - 2} = \textbf{7} - \textbf{4}\sqrt{\textbf{2}}$

18. $\dfrac{-2 - \sqrt{5}}{2 + 2\sqrt{5}} \cdot \dfrac{2 - 2\sqrt{5}}{2 - 2\sqrt{5}}$

$= \dfrac{-4 + 4\sqrt{5} - 2\sqrt{5} + 10}{4 - 20} = \dfrac{\textbf{-3} - \sqrt{\textbf{5}}}{\textbf{8}}$

19. $5^2 = A^2 + 3^2$

$25 = A^2 + 9$

$16 = A^2$

$4 = A$

$5 \times \overrightarrow{SF} = 4$

$\overrightarrow{SF} = \dfrac{4}{5}$

$4 \cdot \dfrac{4}{5} = B$

$\dfrac{16}{5} = B$

20. $\dfrac{z}{m^2} = \dfrac{p}{m}\left(\dfrac{x}{a} + y\right)$

$\dfrac{z}{m^2} = \dfrac{px}{am} + \dfrac{py}{m}$

$az = mpx + ampy$

$az - ampy = mpx$

$a = \dfrac{mpx}{z - mpy}$

21. $\dfrac{z}{m^2} = \dfrac{p}{m}\left(\dfrac{x}{a} + y\right)$

$\dfrac{z}{m^2} = \dfrac{px}{am} + \dfrac{py}{m}$

$az = mpx + ampy$

$\dfrac{az}{mx + amy} = p$

22. Length of side $= \dfrac{\text{diagonal}}{\sqrt{2}} = \dfrac{6\text{ m}}{\sqrt{2}} = \mathbf{3\sqrt{2}\text{ m}}$

Area $= s^2 = (3\sqrt{2}\text{ m})^2 = \mathbf{18\ m^2}$

23. (a) $\dfrac{2}{5}x - \dfrac{1}{3}y = -1$

(b) $0.07x + 0.2y = 2.15$

(a') $6x - 5y = -15$

(b') $7x + 20y = 215$

$4(a')\ \ 24x - 20y = -60$

$\underline{(b')\ \ \ 7x + 20y = 215}$

$\ \ \ \ \ 31x\ \ \ \ \ \ = 155$

$\ \ \ \ \ \ \ \ \ \ \ x = 5$

(a') $6(5) - 5y = -15$

$-5y = -45$

$y = 9$

(5, 9)

24. See Lesson 71.

25. $-3x^2 - 2 = 5x$

$3x^2 + 5x + 2 = 0$

$x = \dfrac{-5 \pm \sqrt{5^2 - 4(3)(2)}}{2(3)} = \dfrac{-5 \pm \sqrt{1}}{6}$

$= -\dfrac{5}{6} \pm \dfrac{1}{6} = -\dfrac{2}{3}, -1$

26. $\sqrt{x^5 y}\ \sqrt[3]{x^2 y^5} = (x^5 y)^{1/2}(x^2 y^5)^{1/3}$

$= x^{5/2} y^{1/2} x^{2/3} y^{5/3} = \mathbf{x^{19/6} y^{13/6}}$

27. $3\sqrt{\dfrac{2}{9}} + 3\sqrt{\dfrac{9}{2}} - 4\sqrt{50}$

$= \dfrac{3\sqrt{2}}{3} + \dfrac{9}{\sqrt{2}}\dfrac{\sqrt{2}}{\sqrt{2}} - 20\sqrt{2}$

$= \dfrac{6\sqrt{2}}{6} + \dfrac{27\sqrt{2}}{6} - \dfrac{120\sqrt{2}}{6} = -\dfrac{\mathbf{29\sqrt{2}}}{\mathbf{2}}$

28. $\dfrac{-2^0}{-4^{-5/2}} = -\left[-(4^{1/2})^5\right] = \mathbf{32}$

29. (a) $4x - 3y = -3$

$y = \dfrac{4}{3}x + 1$

(b) $4x + 3y = 6$

$y = -\dfrac{4}{3}x + 2$

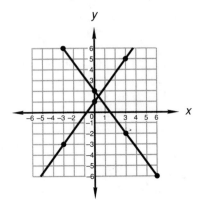

(a) $4x - 3y = -3$

(b) $\underline{4x + 3y = \ \ 6}$

$\ \ \ 8x\ \ \ \ \ \ \ = 3$

$\ \ \ \ \ \ \ \ x = \dfrac{3}{8}$

(b) $4\left(\dfrac{3}{8}\right) + 3y = 6$

$3y = \dfrac{9}{2}$

$y = \dfrac{3}{2}$

$\left(\dfrac{3}{8}, \dfrac{3}{2}\right)$

30. $\dfrac{400 \text{ cm}^3}{\text{sec}} \times \dfrac{1 \text{ in.}}{2.54 \text{ cm}} \times \dfrac{1 \text{ in.}}{2.54 \text{ cm}} \times \dfrac{1 \text{ in.}}{2.54 \text{ cm}}$

$\times \dfrac{1 \text{ ft}}{12 \text{ in.}} \times \dfrac{1 \text{ ft}}{12 \text{ in.}} \times \dfrac{1 \text{ ft}}{12 \text{ in.}} \times \dfrac{60 \text{ sec}}{1 \text{ min}}$

$= \dfrac{400(60)}{(2.54)(2.54)(2.54)(12)(12)(12)} \dfrac{\text{ft}^3}{\text{min}}$

PROBLEM SET 83

1. Equal ratio method:

$$\frac{R_1}{R_2} = \frac{C_1}{C_2}$$

$$\frac{1200}{R_2} = \frac{300}{100}$$

$$300R_2 = 100(1200)$$

$$R_2 = \mathbf{400}$$

Variation method:

$$R = kC$$
$$1200 = k(300)$$
$$4 = k$$

$$R = 4(100) = \mathbf{400}$$

2. $R_M T_M = 420;$

$R_T T_T = 270;$

$T_M = 2T_T;$

$R_T = R_M + 20$

$$(R_T - 20)2T_T = 420$$
$$2R_T T_T - 40T_T = 420$$
$$2(270) - 40T_T = 420$$
$$-40T_T = -120$$
$$T_T = \mathbf{3 \text{ hr}}$$

$T_M = \mathbf{6 \text{ hr}};\ R_T = \mathbf{90 \text{ kph}};\ R_M = \mathbf{70 \text{ kph}}$

3. $R_S T_S + 400 = R_M T_M;$

$R_M = 5;\ T_M = T_S = 400$

$$R_S(400) + 400 = 5(400)$$
$$400R_S = 1600$$
$$R_S = \mathbf{4 \text{ yards per second}}$$

4. (a) $4N_I = 8N_S + 80$

(a′) $N_I = 2N_S + 20$

(b) $10N_S = N_I + 140$

Substitute (a′) into (b) and get:

(b′) $10N_S = (2N_S + 20) + 140$

$$8N_S = 160$$
$$N_S = \mathbf{20}$$

(a′) $N_I = 2(20) + 20 = \mathbf{60}$

5. (a) $0.05P_N + 0.2D_N = 64$

(b) $P_N + D_N = 800$

Substitute $D_N = 800 - P_N$ into (a) and get:

(a′) $0.05P_N + 0.2(800 - P_N) = 64$

$$-0.15P_N = -96$$
$$P_N = 640$$

(b) $(640) + D_N = 800$

$$D_N = 160$$

640 ml 5%, 160 ml 20%

6. (a) $x + 6y = 40$

(b) $5x + 6y = 80$

$-1\text{(a)} \quad -x - 6y = -40$

$\underline{1\text{(b)} \quad 5x + 6y = 80}$

$\qquad\qquad 4x = 40$

$\qquad\qquad\qquad x = \mathbf{10}$

(a) $(10) + 6y = 40$

$$6y = 30$$
$$y = \mathbf{5}$$

7. $\dfrac{a^{x-3}\,m^{x+2}}{a^{x/3}\,m^{3x}} = a^{x-3-x/3}m^{x+2-3x} = a^{2x/3-3}m^{-2x+2}$

8. $\dfrac{(a^x)^m\,(b^x)^{m+3}}{a^{-m}} = a^{mx+m}b^{mx+3x}$

9. $\dfrac{a}{x + \dfrac{1}{a + \dfrac{1}{x}}} = \dfrac{a}{x + \dfrac{1}{\dfrac{ax + 1}{x}}}$

$= \dfrac{a}{x + \dfrac{x}{ax + 1}} = \dfrac{a}{\dfrac{ax^2 + x + x}{ax + 1}}$

$= \dfrac{a(ax + 1)}{ax^2 + 2x}$

10. $\dfrac{b}{a + \dfrac{b}{a + \dfrac{a}{b}}} = \dfrac{b}{a + \dfrac{b}{\dfrac{ab + a}{b}}}$

$= \dfrac{b}{a + \dfrac{b^2}{ab + a}} = \dfrac{b}{\dfrac{a^2b + a^2 + b^2}{ab + a}}$

$= \dfrac{b(ab + a)}{a^2b + a^2 + b^2}$

11. $\dfrac{2 - 4i}{1 + i} \cdot \dfrac{1 - i}{1 - i} = \dfrac{2 - 2i - 4i - 4}{1 + 1}$

$= \mathbf{-1 - 3i}$

12. $\dfrac{3 + 5i}{2 - 2i} \cdot \dfrac{2 + 2i}{2 + 2i} = \dfrac{6 + 6i + 10i - 10}{4 + 4}$

$= \mathbf{-\dfrac{1}{2} + 2i}$

13. (a) $2x + 3y - z = -3$
(b) $x + 2y = 0$
$$x = -2y$$
(c) $x - 2y + z = -2$
Substitute (b) into (a) and (c) and get:

(a') $-y - z = -3$
(c') $-4y + z = -2$
$$\overline{-5y \qquad = -5}$$
$$y = 1$$

(a') $-(1) - z = -3$
$$z = 2$$

(b) $x = -2(1) = -2$

(−2, 1, 2)

14. $\sqrt{k} + \sqrt{k + 32} = 8$
$$k + 32 = 64 - 16\sqrt{k} + k$$
$$16\sqrt{k} = 32$$
$$\sqrt{k} = 2$$
$$k = \mathbf{4}$$

Check: $\sqrt{4} + \sqrt{4 + 32} = 8$
$$2 + 6 = 8 \quad \text{check}$$

15.

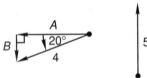

$A = 4 \cos 20° \approx 3.76$
$B = 4 \sin 20° \approx 1.37$

$$-3.76R - 1.37U$$
$$\underline{0.00R + 5.00U}$$
$$-3.76R + 3.63U$$

$$\tan \theta = -\frac{3.63}{3.76}$$
$$\theta \approx 136.01°$$

$$F = \sqrt{(3.76)^2 + (3.63)^2} \approx 5.23$$

5.23/136.01°

16. $\dfrac{5}{x} + \dfrac{6}{x^2 - 4} - \dfrac{3x}{2 - x}$

$$= \frac{5(x^2 - 4)}{x(x^2 - 4)} + \frac{6x}{x(x^2 - 4)} + \frac{3x^2(x + 2)}{x(x^2 - 4)}$$

$$= \frac{5x^2 - 20 + 6x + 3x^3 + 6x^2}{x(x^2 - 4)}$$

$$= \frac{3x^3 + 11x^2 + 6x - 20}{x(x^2 - 4)}$$

17. $\dfrac{-2\sqrt{2} - 2}{4 + \sqrt{2}} \cdot \dfrac{4 - \sqrt{2}}{4 - \sqrt{2}}$

$$= \frac{-8\sqrt{2} + 4 - 8 + 2\sqrt{2}}{16 - 2} = \frac{-2 - 3\sqrt{2}}{7}$$

18. $\dfrac{-\sqrt{3} - 3}{2 - \sqrt{3}} \cdot \dfrac{2 + \sqrt{3}}{2 + \sqrt{3}}$

$$= \frac{-2\sqrt{3} - 3 - 6 - 3\sqrt{3}}{4 - 3} = -9 - 5\sqrt{3}$$

19.
$$p = \frac{a}{x} - c\left(\frac{a}{m} - y\right)$$
$$p = \frac{a}{x} - \frac{ac}{m} + cy$$
$$mpx = am - acx + cmxy$$
$$am + cmxy - mpx = acx$$
$$m = \frac{acx}{a + cxy - px}$$

20.
$$p = \frac{a}{x} - c\left(\frac{a}{m} - y\right)$$
$$p = \frac{a}{x} - \frac{ac}{m} + cy$$
$$mpx = am - acx + cmxy$$
$$mpx - am = cmxy - acx$$
$$\frac{mxp - am}{mxy - ax} = c$$

21. Since this is a 30-60-90 triangle:
$$\sqrt{3} \times \overrightarrow{SF} = 4$$
$$\overrightarrow{SF} = \frac{4}{\sqrt{3}} = \frac{4\sqrt{3}}{3}$$
$$x = 2\left(\frac{4\sqrt{3}}{3}\right) = \frac{8\sqrt{3}}{3}$$
$$y = 1\left(\frac{4\sqrt{3}}{3}\right) = \frac{4\sqrt{3}}{3}$$

22. Since this is a 45-45-90 triangle:
$$1 \times \overrightarrow{SF} = 4$$
$$\overrightarrow{SF} = 4$$

$$m = \sqrt{2}(4) = \mathbf{4\sqrt{2}}$$
$$p = 1(4) = \mathbf{4}$$

23. (a) $x - \dfrac{2}{5}y = 11$
(b) $-0.05x - 0.2y = 1.65$

(a') $5x - 2y = 55$
(b') $-5x - 20y = 165$

Solve the equations by elimination:

$$(a')\quad 5x - 2y = 55$$
$$(b')\quad \underline{-5x - 20y = 165}$$
$$\quad -22y = 220$$
$$\quad y = -10$$

$$(a')\quad 5x - 2(-10) = 55$$
$$5x = 35$$
$$x = 7$$

(7, -10)

24.
$$2x^2 - 5x = 5$$
$$\left(x^2 - \frac{5}{2}x + \quad\right) = \frac{5}{2}$$
$$x^2 - \frac{5}{2}x + \frac{25}{16} = \frac{5}{2} + \frac{25}{16}$$
$$\left(x - \frac{5}{4}\right)^2 = \frac{65}{16}$$
$$x - \frac{5}{4} = \pm\frac{\sqrt{65}}{4}$$
$$\mathbf{x = \frac{5}{4} \pm \frac{\sqrt{65}}{4}}$$

25.
$$\sqrt{\frac{7}{4}} + 2\sqrt{\frac{4}{7}} - 5\sqrt{63} = \frac{\sqrt{7}}{2} + \frac{4}{\sqrt{7}}\frac{\sqrt{7}}{\sqrt{7}} - 15\sqrt{7}$$
$$= \frac{7\sqrt{7}}{14} + \frac{8\sqrt{7}}{14} - \frac{210\sqrt{7}}{14} = \mathbf{-\frac{195\sqrt{7}}{14}}$$

26.
$$\frac{-2^0(-3^0)}{-27^{-2/3}} = -\{-[-(27^{1/3})^2]\} = \mathbf{-9}$$

27.
$$\left(\sqrt[3]{x^2 y}\right)^4 = [(x^2y)^{1/3}]^4 = (x^{2/3}y^{1/3})^4 = \mathbf{x^{8/3}y^{4/3}}$$

28.
$$\frac{4 \text{ liters}}{\text{sec}} \times \frac{1000 \text{ cm}^3}{\text{liter}} \times \frac{60 \text{ sec}}{\text{min}} \times \frac{60 \text{ min}}{1 \text{ hr}}$$
$$= \mathbf{4(1000)(60)(60)\frac{cm^3}{hr}}$$

29.
$$D^2 = 9^2 + 8^2$$
$$D^2 = 145$$
$$\mathbf{D = \sqrt{145}}$$

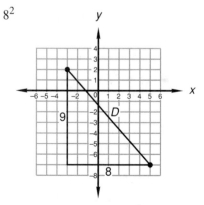

30. Graph the line to find the slope.

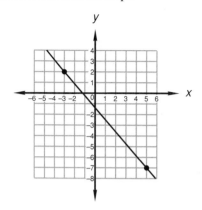

$$m = \frac{2 - (-7)}{-3 - 5} = -\frac{9}{8}$$

Use the point $(-3, 2)$ for x and y.

$$y = -\frac{9}{8}x + b$$
$$2 = -\frac{9}{8}(-3) + b$$
$$-\frac{27}{8} + \frac{16}{8} = b$$
$$-\frac{11}{8} = b$$
$$\mathbf{y = -\frac{9}{8}x - \frac{11}{8}}$$

PROBLEM SET 84

1. Equal ratio method:
$$\frac{A_1}{A_2} = \frac{B_2}{B_1}$$
$$\frac{20}{A_2} = \frac{10}{80}$$
$$1600 = 10A_2$$
$$\mathbf{160} = A_2$$

Variation method:
$$A = \frac{k}{B}$$
$$20 = \frac{k}{80}$$
$$1600 = k$$
$$A = \frac{1600}{10}$$
$$A = \mathbf{160}$$

2. $50 + 273 = 323$ kelvins
$$\frac{P_1}{T_1} = \frac{P_2}{T_2}$$
$$\frac{4}{323} = \frac{8}{T_2}$$
$$4T_2 = 2584$$
$$T_2 = 646 \text{ kelvins}$$
$$646 - 273 = \mathbf{373°C}$$

3. Lithium: $\quad 2 \times 7 = 14$
Calcium: $\quad 2 \times 40 = 80$
Oxygen: $\quad 7 \times 16 = 112$
Total: $\qquad\qquad = 206$

$$\frac{Li_1}{Li_2} = \frac{T_1}{T_2}$$

$$\frac{14}{56} = \frac{206}{T_2}$$

$$14T_2 = 11{,}536$$

$$T_2 = \textbf{824 grams}$$

$$\frac{Li_1}{T_1} \times 100 = \frac{14}{206} \cdot 100 \approx \textbf{6.8\%}$$

4. $0.1(160) + 0.3P_N = 0.22(160 + P_N)$
$$16 + 0.3P_N = 35.2 + 0.22P_N$$
$$0.08P_N = 19.2$$
$$P_N = \textbf{240 ml}$$

5. $\dfrac{250}{100} \cdot 140{,}000 = B_T$
$\quad \textbf{350,000 tons} = B_T$

6. $y^b x^{c+2} y^{b/3} x^{-2-p} = x^{c+2} x^{-2-p} y^b y^{b/3} = \boldsymbol{x^{c-p} y^{4b/3}}$

7. $\dfrac{y^b\, x^{c+3}}{y^{b/3}\, x^{2-p}} = x^{c+3} x^{-2+p} y^b y^{-b/3} = \boldsymbol{x^{c+p+1} y^{2b/3}}$

8. $(x^{a+3})^b x^{-2b+4} = x^{ab+3b} x^{-2b+4} = \boldsymbol{x^{b+ab+4}}$

9. $\dfrac{(y^a)^{b+2}\, y^{-ab}}{y^{-2+a}} = y^{ab+2a} y^{-ab} y^{2-a} = \boldsymbol{y^{a+2}}$

10. $\dfrac{p}{a + \dfrac{m}{1 + \dfrac{1}{am}}} = \dfrac{p}{a + \dfrac{am^2}{am + 1}}$

$\qquad = \dfrac{p(am + 1)}{a^2 m + a + am^2}$

11. $\dfrac{1}{p - \dfrac{b}{b - \dfrac{1}{x}}} = \dfrac{1}{p - \dfrac{bx}{bx - 1}}$

$\qquad = \dfrac{bx - 1}{bpx - p - bx}$

12. $\dfrac{2 - 3i}{1 + i} \cdot \dfrac{1 - i}{1 - i} = \dfrac{2 - 3i - 2i - 3}{1 + 1}$

$\qquad = \dfrac{-1 - 5i}{2} = -\dfrac{1}{2} - \dfrac{5}{2}i$

13. $\dfrac{3 + 4i}{3 - 3i} \cdot \dfrac{3 + 3i}{3 + 3i} = \dfrac{9 + 12i + 9i - 12}{9 + 9}$

$\qquad = \dfrac{-3 + 21i}{18} = -\dfrac{1}{6} + \dfrac{7}{6}i$

14. $\sqrt{s - 48} + \sqrt{s} = 8$
$$s - 48 = 64 - 16\sqrt{s} + s$$
$$16\sqrt{s} = 112$$
$$\sqrt{s} = 7$$
$$s = \textbf{49}$$

15. (a) $3x - y - 2z = -6$
(b) $2x - y + z = 2$
(c) $-y + z = 0$
$\qquad\quad z = y$
Substitute (c) into (a) and (b) and get:
(a′) $3x - 3y = -6$
(b′) $2x = 2$
$\qquad x = 1$

(a′) $3(1) - 3y = -6$
$\qquad -3y = -9$
$\qquad\quad y = 3$

(b) $2(1) - (3) + z = 2$
$\qquad -1 + z = 2$
$\qquad\qquad z = 3$

$(1, 3, 3)$

16.

$A = 6\cos 30° \approx 5.20$
$B = 6\sin 30° = 3.00$

$C = 4\cos 20° \approx 3.76$
$D = 4\sin 20° \approx 1.37$

$5.20R + 3.00U$
$\underline{3.76R + 1.37U}$
$\textbf{8.96R + 4.37U}$

17. $H = \sqrt{8^2 + 3^2}$
$\quad H = \sqrt{73}$

$\quad \tan \theta = \dfrac{8}{3}$

$\qquad \theta \approx 69.44°$

$\boldsymbol{\sqrt{73}\,\underline{/69.44°}}$

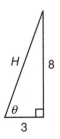

Algebra 2 Solutions Manual

18. $\dfrac{4}{x^2} - \dfrac{2}{-x(x-3)} + \dfrac{5x}{3-x}$

$= \dfrac{4(x-3)}{x^2(x-3)} + \dfrac{2x}{x^2(x-3)} - \dfrac{5x^3}{x^2(x-3)}$

$= \dfrac{4x - 12 + 2x - 5x^3}{x^2(x-3)}$

$= \dfrac{-5x^3 + 6x - 12}{x^2(x-3)}$

19. $\dfrac{4\sqrt{2}-5}{3\sqrt{2}+2} \cdot \dfrac{3\sqrt{2}-2}{3\sqrt{2}-2}$

$= \dfrac{24 - 15\sqrt{2} - 8\sqrt{2} + 10}{18 - 4}$

$= \dfrac{34 - 23\sqrt{2}}{14}$

20. $(2 + 3\sqrt{20})(4 - 5\sqrt{45}) = (2 + 6\sqrt{5})(4 - 15\sqrt{5})$
$= 8 + 24\sqrt{5} - 30\sqrt{5} - 450 = -442 - 6\sqrt{5}$

21.
$$\dfrac{m}{c} + x = p\left(\dfrac{1}{x} + \dfrac{b}{y}\right)$$
$$\dfrac{m}{c} + x = \dfrac{p}{x} + \dfrac{pb}{y}$$
$$mxy + x^2cy = cpy + cbpx$$
$$mxy = cpy + cbpx - x^2cy$$
$$\dfrac{mxy}{py + pbx - x^2y} = c$$

22. $\sqrt{-3}\sqrt{3} - \sqrt{2}\sqrt{2} - \sqrt{-4} + 3i^2 - 2i^5$
$= 3i - 2 - 2i - 3 - 2i = -5 - i$

23. $\quad -2x^2 - x = 3$
$-2x^2 - x - 3 = 0$

$x = \dfrac{1 \pm \sqrt{1 - 4(-3)(-2)}}{2(-2)} = \dfrac{1 \pm \sqrt{-23}}{-4}$

$= -\dfrac{1}{4} \pm \dfrac{\sqrt{23}}{4}i$

24. $A_{\text{Square}} = s^2 = 36\,\text{m}^2$
$\qquad\qquad s = 6\,\text{m}$

$r_{\text{Circle}} = \dfrac{s}{2} = 3\,\text{m}$

$A_{\text{Circle}} = \pi r^2 = \pi(3\,\text{m})^2 = 9\pi\,\text{m}^2$

25. There is never an exact solution for these problems. One possible solution is given here.
$Y = mB + b$
Use the graph to find the slope.

$m = \dfrac{-300}{6} = -50$

$Y = -50B + b$
Use the point (99, 100) for B and Y.
$\quad 100 = -50(99) + b$
$\quad 100 = -4950 + b$
$5050 = b$
$Y = -50B + 5050$

26. (a) $\dfrac{-2.065 \times 10^4}{-500 \times 10^6} \approx \dfrac{-2 \times 10^4}{-5 \times 10^8} = 4 \times 10^{-5}$
Calculator $= 4.13 \times 10^{-5}$

(b) $(84.9)^{-4.91}$
Answer should be between 3×10^{-10} and 4×10^{-10}.
Calculator $= 3.38 \times 10^{-10}$

27. $3\sqrt{\dfrac{5}{3}} + 2\sqrt{\dfrac{3}{5}} - 4\sqrt{60}$

$= \dfrac{3\sqrt{5}}{\sqrt{3}}\dfrac{\sqrt{3}}{\sqrt{3}} + \dfrac{2\sqrt{3}}{\sqrt{5}}\dfrac{\sqrt{5}}{\sqrt{5}} - 8\sqrt{15}$

$= \sqrt{15} + \dfrac{2\sqrt{15}}{5} - 8\sqrt{15}$

$= \dfrac{5\sqrt{15}}{5} + \dfrac{2\sqrt{15}}{5} - \dfrac{40\sqrt{15}}{5} = -\dfrac{33\sqrt{15}}{5}$

28. $\dfrac{40\,\text{cm}}{\text{sec}} \times \dfrac{60\,\text{sec}}{1\,\text{min}} \times \dfrac{60\,\text{min}}{1\,\text{hr}} \times \dfrac{1\,\text{in.}}{2.54\,\text{cm}}$

$\times \dfrac{1\,\text{ft}}{12\,\text{in.}} \times \dfrac{1\,\text{mi}}{5280\,\text{ft}} = \dfrac{40(60)(60)}{(2.54)(12)(5280)}\,\dfrac{\text{mi}}{\text{hr}}$

29. $-2^0[-2 - 3(x - 2^2)] - 3(2x - 5^0)$
$\quad = 7 - 2(2x + 2)$
$-[-2 - 3x + 12] - 6x + 3 = 7 - 4x - 4$
$\qquad\qquad 3x - 10 - 6x + 3 = -4x + 3$
$\qquad\qquad\qquad\qquad\qquad x = 10$

30. $-3\left\{[(-2 - 3) - 2] - 2[-4(-3 - 2^0)]\right\}$
$\quad + 2^0(-3)$
$= -3\{[-5 - 2] - 2[16]\} - 3$
$= -3(-7 - 32) - 3 = 114$

PROBLEM SET 85

1. Equal ratio method:

$$\frac{I_1}{I_2} = \frac{T_1}{T_2}$$

$$\frac{800}{I_2} = \frac{2400}{9}$$

$$I_2 = 3$$

Variation method:

$$I = kT$$
$$800 = k(2400)$$
$$\frac{1}{3} = k$$

$$N_I = \frac{1}{3}(9)$$
$$N_I = 3$$

2. $R_P T_P = 640;$
$R_B T_B = 280;$
$T_P = 2T_B;$
$R_P = R_B + 20$

$$\overset{D_P}{\underset{640}{\mid\!\longrightarrow\!\mid}} \quad \overset{D_B}{\underset{280}{\mid\!\longrightarrow\!\mid}}$$

$$(R_B + 20)2T_B = 640$$
$$2R_B T_B + 40T_B = 640$$
$$2(280) + 40T_B = 640$$
$$40T_B = 80$$
$$T_B = 2 \text{ hr}$$

$T_P = 4 \text{ hr};\ R_P = 160 \text{ mph};\ R_B = 140 \text{ mph}$

3. (a) $5N_H = N_D + 90$
   (b) $3N_D = N_H + 10$
   (b') $N_H = 3N_D - 10$
   Substitute (b') into (a) and get:
   (a') $5(3N_D - 10) = N_D + 90$
   $$15N_D - 50 = N_D + 90$$
   $$14N_D = 140$$
   $$N_D = 10$$
   (b') $N_H = 3(10) - 10 = 20$

4. (a) $0.1P_N + 0.5D_N = 136$
   (b) $P_N + D_N = 400$
   Substitute $D_N = 400 - P_N$ into (a) and get:
   (a') $0.1P_N + 0.5(400 - P_N) = 136$
   $$-0.4P_N = -64$$
   $$P_N = 160$$
   (b) $(160) + D_N = 400$
   $$D_N = 240$$
   $160 \text{ ft}^3\ 10\%,\ 240 \text{ ft}^3\ 50\%$

5. $N \quad N + 2 \quad N + 4$
   $$N(N + 4) = 10(-N - 2) - 25$$
   $$N^2 + 4N = -10N - 20 - 25$$
   $$N^2 + 14N + 45 = 0$$
   $$(N + 5)(N + 9) = 0$$
   $$N = -5, -9$$
   The desired integers are **–9, –7, –5 and –5, –3, –1.**

6. (a) $BT_D + 6T_D = 24$
   (b) $\underline{BT_D - 6T_D = 12}$
   $$2BT_D \qquad = 36$$
   $$BT_D = 18$$

   (a) $(18) + 6T_D = 24$
   $$6T_D = 6$$
   $$T_D = 1$$

   (b) $B(1) - 6(1) = 12$
   $$B = 18$$

7. (a) $x^2 + y^2 = 16$
   (b) $2x - y = 4$
   $$y = 2x - 4$$
   (b') $y^2 = 4x^2 - 16x + 16$
   Substitute (b') into (a) and get:
   (a') $x^2 + (4x^2 - 16x + 16) = 16$
   $$5x^2 - 16x + 16 = 16$$
   $$5x^2 - 16x = 0$$
   $$x(5x - 16) = 0$$
   $$x = 0 \quad \text{and} \quad x = \frac{16}{5}$$
   Substitute these values of x into (b) and solve for y.
   (b) $y = 2(0) - 4$
   $$y = -4$$

   (b) $y = 2\left(\frac{16}{5}\right) - 4$
   $$y = \frac{12}{5}$$

   $(0, -4)$ and $\left(\dfrac{16}{5},\ \dfrac{12}{5}\right)$

8. $$\frac{a^x (b^{y-2})^x}{a^{2x} (b^{-2})^x} = \frac{a^x b^{xy-2x}}{a^{2x} b^{-2x}}$$
   $$= a^{x-2x} b^{xy-2x+2x} = a^{-x} b^{xy}$$

9. $$\frac{a^x b^{x/3} b^{-2}}{a^{x/2}} = a^{x-x/2} b^{x/3-2} = a^{x/2} b^{x/3-2}$$

10. $$\frac{x}{a + \dfrac{b}{c + \dfrac{x}{m}}} = \frac{x}{a + \dfrac{bm}{cm + x}}$$
    $$= \frac{x(cm + x)}{acm + ax + bm}$$

11. $\dfrac{a}{2 + \dfrac{c}{c + \dfrac{b}{c}}} = \dfrac{a}{2 + \dfrac{c^2}{c^2 + b}}$

$= \dfrac{a(c^2 + b)}{2c^2 + 2b + c^2} = \dfrac{a(c^2 + b)}{3c^2 + 2b}$

12. $\dfrac{2 - 2i}{3 - 5i} \cdot \dfrac{3 + 5i}{3 + 5i} = \dfrac{6 - 6i + 10i + 10}{9 + 25}$

$= \dfrac{16 + 4i}{34} = \dfrac{8}{17} + \dfrac{2}{17}i$

13. $\dfrac{4 - \sqrt{2}i}{3 + \sqrt{2}i} \cdot \dfrac{3 - \sqrt{2}i}{3 - \sqrt{2}i}$

$= \dfrac{12 - 3\sqrt{2}i - 4\sqrt{2}i - 2}{9 + 2} = \dfrac{10}{11} - \dfrac{7\sqrt{2}}{11}i$

14. $\sqrt{k - 32} + \sqrt{k} = 8$

$k - 32 = 64 - 16\sqrt{k} + k$

$16\sqrt{k} = 96$

$\sqrt{k} = 6$

$k = 36$

15. (a) $x + 2y + 2z = 6$

(b) $2x - y + 3z = 6$

(c) $y - z = 0$

$y = z$

Substitute (c) into (a) and (b) and get:

(a′) $x + 4z = 6$

(b′) $2x + 2z = 6$

$\begin{array}{rr} \text{(a′)} & x + 4z = 6 \\ -2\text{(b′)} & \underline{-4x - 4z = -12} \\ & -3x = -6 \\ & x = 2 \end{array}$

(a′) $(2) + 4z = 6$

$4z = 4$

$z = 1$

(c) $y = 1$

(2, 1, 1)

16.

$B = 4\cos 32° \approx 3.39$

$C = 4\sin 32° \approx 2.12$

$0.00R + 6.00U$

$\underline{3.39R - 2.12U}$

$\mathbf{3.39R + 3.88U}$

$\tan \theta = \dfrac{3.88}{3.39}$

$\theta \approx 48.86°$

$F = \sqrt{(3.39)^2 + (3.88)^2} \approx 5.15$

5.15 $\underline{/48.86°}$

17. $\dfrac{2x + 3}{x - a} - \dfrac{4}{a - x} = \dfrac{2x + 3 + 4}{x - a}$

$= \dfrac{2x + 7}{x - a}$

18. $\dfrac{3 - 2\sqrt{2}}{5 - \sqrt{2}} \cdot \dfrac{5 + \sqrt{2}}{5 + \sqrt{2}}$

$= \dfrac{15 - 10\sqrt{2} + 3\sqrt{2} - 4}{25 - 2} = \dfrac{11 - 7\sqrt{2}}{23}$

19. $\dfrac{4 + \sqrt{3}}{2 - 2\sqrt{3}} \cdot \dfrac{2 + 2\sqrt{3}}{2 + 2\sqrt{3}}$

$= \dfrac{8 + 2\sqrt{3} + 8\sqrt{3} + 6}{4 - 12} = \dfrac{14 + 10\sqrt{3}}{-8}$

$= \dfrac{-7 - 5\sqrt{3}}{4}$

20. $c = m\left(\dfrac{d}{c} - p\right)$

$c = \dfrac{dm}{c} - mp$

$c^2 = dm - cmp$

$cmp = dm - c^2$

$p = \dfrac{dm - c^2}{cm}$

21. $-\sqrt{-2}\sqrt{2} - 3i^3 + 2i + \sqrt{-2}\sqrt{-2} - \sqrt{-9}$

$= -2i + 3i + 2i - 2 - 3i = \mathbf{-2}$

22. (a) $70 = \dfrac{120 + x}{2}$

$140 = 120 + x$

$\mathbf{20 = x}$

(b) $360 - 120 = 240$

$x = \dfrac{240 - 120}{2} = \mathbf{60}$

23. (a) $\dfrac{1}{9}x + \dfrac{1}{3}y = 3$

(b) $0.3x - 0.04y = 2.46$

(a′) $x + 3y = 27$

(b′) $30x - 4y = 246$

$\begin{array}{rr} 30\text{(a′)} & 30x + 90y = 810 \\ -1\text{(b′)} & \underline{-30x + 4y = -246} \\ & 94y = 564 \\ & y = 6 \end{array}$

(a′) $x + 3(6) = 27$

$x = 9$

(9, 6)

24. See Lesson 71.

25. There is never an exact solution to these problems. One possible solution is given here.

$Na = mMg + b$

Use the graph to find the slope.

$m = \dfrac{-10}{2} = -5$

$Na = -5Mg + b$

Use the point (54, 82) for Mg and Na.

$82 = -5(54) + b$

$352 = b$

$Na = -5Mg + 352$

26. (a) $x - 4y = -8$

$y = \dfrac{1}{4}x + 2$

(b) $3x + y = 6$

$y = -3x + 6$

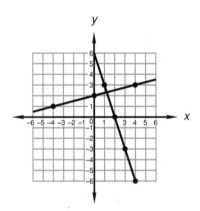

Substitute $y = -3x + 6$ into (a) and get:

(a') $x - 4(-3x + 6) = -8$

$x + 12x - 24 = -8$

$13x = 16$

$x = \dfrac{16}{13}$

(b) $y = -3\left(\dfrac{16}{13}\right) + 6$

$y = \dfrac{30}{13}$

$\left(\dfrac{16}{13}, \dfrac{30}{13}\right)$

27. $\dfrac{x}{r} = \dfrac{t}{s}$

$x = \dfrac{rt}{s}$

28. $\dfrac{100 \text{ ft}^3}{\text{sec}} \times \dfrac{60 \text{ sec}}{1 \text{ min}} \times \dfrac{12 \text{ in.}}{1 \text{ ft}} \times \dfrac{12 \text{ in.}}{1 \text{ ft}} \times \dfrac{12 \text{ in.}}{1 \text{ ft}}$

$\times \dfrac{2.54 \text{ cm}}{1 \text{ in.}} \times \dfrac{2.54 \text{ cm}}{1 \text{ in.}} \times \dfrac{2.54 \text{ cm}}{1 \text{ in.}}$

$= \mathbf{100(60)(12)(12)(12)(2.54)(2.54)(2.54)\ \dfrac{cm^3}{min}}$

29. $3\sqrt{\dfrac{5}{12}} + 3\sqrt{\dfrac{12}{5}} + 3\sqrt{240}$

$= \dfrac{3\sqrt{5}}{2\sqrt{3}}\dfrac{\sqrt{3}}{\sqrt{3}} + \dfrac{6\sqrt{3}}{\sqrt{5}}\dfrac{\sqrt{5}}{\sqrt{5}} + 12\sqrt{15}$

$= \dfrac{5\sqrt{15}}{10} + \dfrac{12\sqrt{15}}{10} + \dfrac{120\sqrt{15}}{10} = \mathbf{\dfrac{137\sqrt{15}}{10}}$

30. $\dfrac{-2^0}{-4^{-5/2}} = 4^{5/2} = (4^{1/2})^5 = \mathbf{32}$

PROBLEM SET 86

1. $\dfrac{V_1}{T_1} = \dfrac{V_2}{T_2}$

$\dfrac{400}{1273} = \dfrac{V_2}{2273}$

$V_2 = \dfrac{400(2273)}{1273} = \mathbf{714.22 \text{ liters}}$

2. $R_R T_R = R_P T_P;\ R_R = 20;$
$R_P = 2;\ T_R + T_P = 11$

$(20)T_R = 2(11 - T_R)$

$20T_R = 22 - 2T_R$

$22T_R = 22$

$T_R = 1$

$D_R = R_R T_R = 20(1) = 20 \text{ miles}$

Since D_R is only halfway to the swamp, the total distance to the swamp is $2D_R$ or **40 miles**.

3. $0.5(100) + 0.2(D_N) = 0.23(100 + D_N)$

$50 + 0.2D_N = 23 + 0.23D_N$

$27 = 0.03D_N$

900 liters $= D_N$

4. Hydrogen: $\quad 2 \times 1 = 2$

Carbon: $\quad 1 \times 12 = 12$

Oxygen: $\quad 3 \times 16 = 48$

Total: $\qquad\qquad = 62$

$\dfrac{12}{62} = \dfrac{C}{372}$

$62C = 12(372)$

$C = \mathbf{72 \text{ grams}}$

5. $0.14 \cdot 120,000 = RN$

$\mathbf{16,800 = RN}$

6. $A_{\text{Circle}} = \pi r^2 = 25\pi \text{ m}^2$

$r^2 = 25 \text{ m}^2$

$r = \mathbf{5 \text{ m}}$

Length of diagonal $= 2r = 2(5 \text{ m}) = \mathbf{10 \text{ m}}$

Length of side $= \dfrac{10 \text{ m}}{\sqrt{2}} \cdot \dfrac{\sqrt{2}}{\sqrt{2}} = \mathbf{5\sqrt{2} \text{ m}}$

Area of square $= s^2 = (5\sqrt{2} \text{ m})^2 = \mathbf{50 \text{ m}^2}$

7. $x \nleq -2$; $D = \{$Negative integers$\}$
$x > -2$

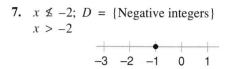

8. $-x + 3 \ngtr 2$; $D = \{$Reals$\}$
$-x + 3 \leq 2$
$-x \leq -1$
$x \geq 1$

9. $-x - 6 \ngeq -3$; $D = \{$Negative integers$\}$
$-x - 6 < -3$
$-x < 3$
$x > -3$

10. (a) $BT_D + 3T_D = 60$
(b) $BT_D - 3T_D = 36$
$\overline{\quad 2BT_D \qquad = 96}$
(c) $\qquad BT_D = 48$

(a) $(48) + 3T_D = 60$
$3T_D = 12$
$T_D = \mathbf{4}$

(c) $B(4) = 48$
$B = \mathbf{12}$

11. (a) $x^2 + y^2 = 4$
(b) $y - x = 1$
$y = x + 1$
(b′) $y^2 = (x + 1)^2 = x^2 + 2x + 1$
Substitute (b′) into (a) and get:
(a′) $x^2 + (x^2 + 2x + 1) = 4$
$2x^2 + 2x - 3 = 0$
Solve this equation by using the quadratic formula.

$$x = \frac{-2 \pm \sqrt{2^2 - 4(2)(-3)}}{2(2)} = -\frac{1}{2} \pm \frac{\sqrt{7}}{2}$$

Substitute these values of x into (b) and solve for y.

(b) $y = \left(-\dfrac{1}{2} + \dfrac{\sqrt{7}}{2}\right) + 1$

$y = \dfrac{1}{2} + \dfrac{\sqrt{7}}{2}$

(b) $y = \left(-\dfrac{1}{2} - \dfrac{\sqrt{7}}{2}\right) + 1$

$y = \dfrac{1}{2} - \dfrac{\sqrt{7}}{2}$

$\left(-\dfrac{1}{2} + \dfrac{\sqrt{7}}{2}, \dfrac{1}{2} + \dfrac{\sqrt{7}}{2}\right)$ and

$\left(-\dfrac{1}{2} - \dfrac{\sqrt{7}}{2}, \dfrac{1}{2} - \dfrac{\sqrt{7}}{2}\right)$

12. $\dfrac{x^{2a}(y^b)^{2a}x^{a/3}}{y^{ba/3}} = \dfrac{x^{2a}y^{2ab}x^{a/3}}{y^{ab/3}} = x^{7a/3}y^{5ab/3}$

13. $\dfrac{(x^{a+2})^2}{x^{2-a}} = \dfrac{x^{2a+4}}{x^{2-a}} = x^{3a+2}$

14. $\dfrac{1}{x + \dfrac{a}{x + \dfrac{1}{a}}} = \dfrac{1}{x + \dfrac{a^2}{ax + 1}}$

$= \dfrac{1}{\dfrac{ax^2 + x + a^2}{ax + 1}} = \dfrac{ax + 1}{ax^2 + x + a^2}$

15. $\dfrac{2 - 3i}{-5 + i} \cdot \dfrac{-5 - i}{-5 - i}$

$= \dfrac{-10 - 2i + 15i - 3}{25 + 1}$

$= \dfrac{-13 + 13i}{26} = -\dfrac{1}{2} + \dfrac{1}{2}i$

16. $\dfrac{3 + 2i}{5 - i} \cdot \dfrac{5 + i}{5 + i} = \dfrac{15 + 3i + 10i - 2}{25 + 1}$

$= \dfrac{13 + 13i}{26} = \dfrac{1}{2} + \dfrac{1}{2}i$

17. $\dfrac{2 + 1 + 6 + 13 + N}{5} = 9$

$N + 22 = 45$
$N = \mathbf{23}$

18. $\sqrt{p + 48} = 8 - \sqrt{p}$
$p + 48 = 64 - 16\sqrt{p} + p$
$16\sqrt{p} = 16$
$\sqrt{p} = 1$
$p = \mathbf{1}$

Check: $\sqrt{1 + 48} = 8 - \sqrt{1}$
$7 = 8 - 1$ check

19. (a) $x + 2y - z = 0$
(b) $3x + y - 2z = 3$
(c) $2x - z = 0$
$z = 2x$
Substitute (c) into (a) and (b) and get:
(a′) $-x + 2y = 0$
(b′) $-x + y = 3$

$\quad$ (a′) $-x + 2y = \;\; 0$
-1(b′) $\underline{\;\; x - \;\; y = -3}$
$y = -3$

(a′) $-x + 2(-3) = 0$
$x = -6$

(c) $z = 2(-6) = -12$

$(-6, -3, -12)$

20.

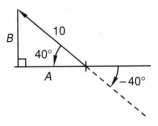

$A = 10 \cos 40° \approx 7.66$
$B = 10 \sin 40° \approx 6.43$

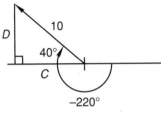

$C = 10 \cos 40° \approx 7.66$
$D = 10 \sin 40° \approx 6.43$

$$-7.66R + 6.43U$$
$$\underline{-7.66R + 6.43U}$$
$$-15.32R + 12.86U$$

21. $H = \sqrt{10^2 + 3^2}$
$H = \sqrt{109}$

$\tan \theta = \dfrac{10}{3}$

$\theta \approx 73.30°$

Since θ is a third-quadrant angle:
$\theta = 73.30 + 180 = 253.30°$

$\sqrt{109}\underline{/253.30°}$

22. $\dfrac{-3 - 2\sqrt{3}}{1 - 3\sqrt{3}} \cdot \dfrac{1 + 3\sqrt{3}}{1 + 3\sqrt{3}}$

$= \dfrac{-3 - 9\sqrt{3} - 2\sqrt{3} - 18}{1 - 27}$

$= \dfrac{-21 - 11\sqrt{3}}{-26} = \dfrac{21 + 11\sqrt{3}}{26}$

23.
$$\dfrac{x + 2}{y} - c = m\left(\dfrac{a}{b} + x\right)$$
$$\dfrac{x + 2}{y} - c = \dfrac{am}{b} + mx$$
$$bx + 2b - bcy = amy + bmxy$$
$$bx + 2b - bcy - bmxy = amy$$
$$\dfrac{amy}{x + 2 - cy - mxy} = b$$

24. $-\sqrt{-16} - \sqrt{3}\sqrt{-3} + \sqrt{-3}\sqrt{-3} = -4i - 3i - 3$
$= -3 - 7i$

25. $5\sqrt{\dfrac{2}{7}} + 3\sqrt{\dfrac{7}{2}} - 2\sqrt{56}$

$= \dfrac{5\sqrt{2}}{\sqrt{7}} \dfrac{\sqrt{7}}{\sqrt{7}} + \dfrac{3\sqrt{7}}{\sqrt{2}} \dfrac{\sqrt{2}}{\sqrt{2}} - 4\sqrt{14}$

$= \dfrac{10\sqrt{14}}{14} + \dfrac{21\sqrt{14}}{14} - \dfrac{56\sqrt{14}}{14} = -\dfrac{25\sqrt{14}}{14}$

26. $-3x^2 - 1 = 6x$

$x^2 + 2x + \dfrac{1}{3} = 0$

$(x^2 + 2x + \quad) = -\dfrac{1}{3}$

$x^2 + 2x + 1 = -\dfrac{1}{3} + 1$

$(x + 1)^2 = \dfrac{2}{3}$

$x + 1 = \pm\dfrac{\sqrt{6}}{3}$

$x = -1 \pm \dfrac{\sqrt{6}}{3}$

27. $\dfrac{40 \text{ in.}^2}{\text{min}} \times \dfrac{1 \text{ ft}}{12 \text{ in.}} \times \dfrac{1 \text{ ft}}{12 \text{ in.}} \times \dfrac{1 \text{ yd}}{3 \text{ ft}} \times \dfrac{1 \text{ yd}}{3 \text{ ft}}$

$\times \dfrac{60 \text{ min}}{1 \text{ hr}} = \dfrac{40(60)}{(12)(12)(3)(3)} \dfrac{\text{yd}^2}{\text{hr}}$

28. $-2x^2 - 1 = 6x$
$2x^2 + 6x + 1 = 0$

$x = \dfrac{-6 \pm \sqrt{6^2 - 4(2)(1)}}{2(2)}$

$x = -\dfrac{3}{2} \pm \dfrac{\sqrt{7}}{2}$

29. (a) $2x - 5y = -15$

$y = \dfrac{2}{5}x + 3$

(b) $3x + 4y = -4$

$y = -\dfrac{3}{4}x - 1$

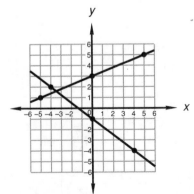

Solve the equations by elimination:

-3(a) $-6x + 15y = 45$

2(b) $\underline{6x + 8y = -8}$

$23y = 37$

$y = \dfrac{37}{23}$

(b) $3x + 4\left(\dfrac{37}{23}\right) = -4$

$3x = \dfrac{-240}{23}$

$x = -\dfrac{80}{23}$

$\left(-\dfrac{80}{23}, \dfrac{37}{23}\right)$

30. Write the equation of the given line in slope-intercept form.

$4x - 6y = 25$

$-6y = -4x + 25$

$y = \dfrac{2}{3}x - \dfrac{25}{6}$

Since the slopes of perpendicular lines are negative reciprocals of each other:

$m_\perp = -\dfrac{3}{2}$

$y = -\dfrac{3}{2}x + b$

$2 = -\dfrac{3}{2}(7) + b$

$\dfrac{25}{2} = b$

$y = -\dfrac{3}{2}x + \dfrac{25}{2}$

PROBLEM SET 87

1. $R_J T_J = 1440$;

$R_B T_B = 120$;

$R_J = 4R_B$;

$T_J = T_B + 4$

$4R_B(T_B + 4) = 1440$

$4(120) + 16R_B = 1440$

$16R_B = 960$

$R_B = \textbf{60 mph}$

$R_J = \textbf{240 mph}$; $T_B = \textbf{2 hr}$; $T_J = \textbf{6 hr}$

2. Equal ratio method:

$\dfrac{HS_1}{HS_2} = \dfrac{H_2}{H_1}$

$\dfrac{10}{HS_2} = \dfrac{200}{400}$

$HS_2 = \textbf{20 hours}$

Variation method:

$HS = \dfrac{k}{H}$

$10 = \dfrac{k}{400}$

$k = 4000$

$HS = \dfrac{4000}{200} = \textbf{20 hours}$

3. (a) $0.8P_N + 0.3D_N = 240$

(b) $P_N + D_N = 600$

Substitute $D_N = 600 - P_N$ into (a) and get:

(a$'$) $0.8P_N + 0.3(600 - P_N) = 240$

$0.5P_N = 60$

$P_N = 120$

(b) $(120) + D_N = 600$

$D_N = 480$

120 in.^{3} 80%, 480 in.^{3} 30%

4. (a) $10N_H = N_P - 102$

(b) $100N_H = N_P + 168$

(b$'$) $N_P = 100N_H - 168$

Substitute (b$'$) into (a) and get:

(a$'$) $10N_H = (100N_H - 168) - 102$

$-90N_H = -270$

$N_H = \textbf{3}$

(b$'$) $N_P = 100(3) - 168 = \textbf{132}$

5. Postassium (K): $1 \times 39 = 39$

Chlorine: $1 \times 35 = 35$

Oxygen: $3 \times 16 = 48$

Total: $= 122$

Percect Ox $= \dfrac{\text{Ox}}{\text{Tot}} \times 100$

$= \dfrac{48}{122} \cdot 100 \approx \textbf{39.34\%}$

$\dfrac{48}{122} = \dfrac{576}{\text{Tot}}$

$48\,\text{Tot} = 122(576)$

$\text{Tot} = \textbf{1464 grams}$

6. Refer to Lesson 87.

7. $(x_1, y_1) = (-2, 108)$

$(x_2, y_2) = (-21, 47)$

$m = \dfrac{y_2 - y_1}{x_2 - x_1} = \dfrac{47 - 108}{-21 - (-2)} = \dfrac{-61}{-19} = \dfrac{61}{19}$

8. $x \not< -4$; $D = \{\text{Integers}\}$

$x \geq -4$

9. $-x + 2 \leq -3$; $D = \{$Positive integers$\}$
$$-x \leq -5$$
$$x \geq 5$$

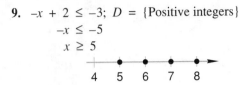

10. (a) $BT_D + 4T_D = 36$
(b) $\dfrac{BT_D - 4T_D = 12}{2BT_D \qquad = 48}$
(c) $BT_D = 24$

(a) $(24) + 4T_D = 36$
$$4T_D = 12$$
$$T_D = \mathbf{3}$$

(c) $B(3) = 24$
$$B = \mathbf{8}$$

11. (a) $x^2 + y^2 = 4$
(b) $x - 2y = 1$
$$x = 2y + 1$$
(b′) $x^2 = 4y^2 + 4y + 1$
Substitute (b′) into (a) and get:
(a′) $(4y^2 + 4y + 1) + y^2 = 4$
$$5y^2 + 4y - 3 = 0$$

Solve this equation by using the quadratic formula.

$$y = \frac{-4 \pm \sqrt{4^2 - 4(5)(-3)}}{2(5)} = -\frac{2}{5} \pm \frac{\sqrt{19}}{5}$$

Substitute these values of y into (b) and solve for x.

(b) $x = 2\left(-\dfrac{2}{5} + \dfrac{\sqrt{19}}{5}\right) + 1$

$$x = \frac{1}{5} + \frac{2\sqrt{19}}{5}$$

(b) $x = 2\left(-\dfrac{2}{5} - \dfrac{\sqrt{19}}{5}\right) + 1$

$$x = \frac{1}{5} - \frac{2\sqrt{19}}{5}$$

$\left(\dfrac{1}{5} + \dfrac{2\sqrt{19}}{5}, -\dfrac{2}{5} + \dfrac{\sqrt{19}}{5}\right)$ and

$\left(\dfrac{1}{5} - \dfrac{2\sqrt{19}}{5}, -\dfrac{2}{5} - \dfrac{\sqrt{19}}{5}\right)$

12. $\dfrac{m^a\, m^{2a+2}\, y^{-b}}{y^{2-b}} = m^{3a+2}y^{-2}$

13. $\dfrac{(m^{a-3})^2\, y}{m^a\, y^{b+1}} = \dfrac{m^{2a-6}\, y}{m^a\, y^{b+1}} = m^{a-6}y^{-b}$

14. $\dfrac{a}{a + \dfrac{a}{a + \dfrac{b}{a}}} = \dfrac{a}{a + \dfrac{a^2}{a^2 + b}}$

$$= \frac{a(a^2 + b)}{a^3 + ab + a^2} = \frac{a^2 + b}{a^2 + b + a}$$

15. $\dfrac{5i - 2}{-1 - i} \cdot \dfrac{-1 + i}{-1 + i} = \dfrac{-5i - 5 + 2 - 2i}{1 + 1}$
$$= -\frac{3}{2} - \frac{7}{2}i$$

16. $\dfrac{-3 + 2i}{-2 - i} \cdot \dfrac{-2 + i}{-2 + i} = \dfrac{6 - 3i - 4i - 2}{4 + 1}$
$$= \frac{4}{5} - \frac{7}{5}i$$

17. (a) $(9315 \times 10^3)(-2.065 \times 10^4)$
Estimate: $(9 \times 10^6)(-2 \times 10^4)$
$$= -1.8 \times 10^{11}$$
Calculator: $\mathbf{-1.92 \times 10^{11}}$

(b) $\sqrt[2.7]{1001.94}$
Answer should be between 10 and 31.
Calculator: **12.92**

18. (a) $x + y - 2z = 7$
(b) $3x - y - z = 3$
(c) $2x + z = 0$
$$z = -2x$$
Substitute (c) into (a) and (b) and get:
(a′) $5x + y = 7$
(b′) $\dfrac{5x - y = 3}{10x \qquad = 10}$
$$x = 1$$

(a′) $5(1) + y = 7$
$$y = 2$$

(c) $z = -2(1) = -2$

$(\mathbf{1,\ 2,\ -2})$

19. $\sqrt{z} - \sqrt{z - 45} = 5$
$$\sqrt{z} - 5 = \sqrt{z - 45}$$
$$z - 10\sqrt{z} + 25 = z - 45$$
$$10\sqrt{z} = 70$$
$$\sqrt{z} = 7$$
$$z = \mathbf{49}$$

Check: $\sqrt{49} - \sqrt{49 - 45} = 5$
$$7 - 2 = 5 \quad \text{check}$$

20.

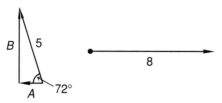

$A = 5 \cos 72° \approx 1.55$
$B = 5 \sin 72° \approx 4.76$

$\begin{array}{r} -1.55R + 4.76U \\ 8.00R + 0.00U \\ \hline 6.45R + 4.76U \end{array}$

$\tan \theta = \dfrac{4.76}{6.45}$
$\theta \approx 36.43°$
$F = \sqrt{(4.76)^2 + (6.45)^2} \approx 8.02$

8.02/36.43°

21. $\dfrac{2 - \sqrt{3}}{\sqrt{3} + 2} \cdot \dfrac{\sqrt{3} - 2}{\sqrt{3} - 2}$

$= \dfrac{2\sqrt{3} - 4 - 3 + 2\sqrt{3}}{3 - 4} = 7 - 4\sqrt{3}$

22. $-3i(ii) - \sqrt{-2}\sqrt{-3} = \sqrt{6} + 3i$

23. $3\sqrt{\dfrac{7}{3}} + 3\sqrt{\dfrac{3}{7}} - 4\sqrt{189}$

$= \dfrac{3\sqrt{7}}{\sqrt{3}}\dfrac{\sqrt{3}}{\sqrt{3}} + \dfrac{3\sqrt{3}}{\sqrt{7}}\dfrac{\sqrt{7}}{\sqrt{7}} - 12\sqrt{21}$

$= \dfrac{21\sqrt{21}}{21} + \dfrac{9\sqrt{21}}{21} - \dfrac{252\sqrt{21}}{21} = -\dfrac{74\sqrt{21}}{7}$

24. $\dfrac{y + 4}{m} = \dfrac{ap}{b} + \dfrac{p}{c}$
$bcy + 4bc = acmp + bmp$
$bcy + 4bc - acmp = bmp$

$c = \dfrac{bmp}{by + 4b - amp}$

25. $y = 2(70) - 80 = \mathbf{60}$

$2x = 360 - 2(75)$
$2x = 210$
$x = \mathbf{105}$

26. $-3x^2 - 4 = 2x$
$3x^2 + 2x + 4 = 0$

$x = \dfrac{-2 \pm \sqrt{2^2 - 4(3)(4)}}{2(3)} = -\dfrac{1}{3} \pm \dfrac{\sqrt{11}}{3}i$

27. $\dfrac{10 \text{ km}}{\text{hr}} \times \dfrac{1000 \text{ m}}{1 \text{ km}} \times \dfrac{100 \text{ cm}}{1 \text{ m}} \times \dfrac{1 \text{ in.}}{2.54 \text{ cm}}$

$\times \dfrac{1 \text{ hr}}{60 \text{ min}} \times \dfrac{1 \text{ min}}{60 \text{ sec}} = \dfrac{10(1000)(100)}{(2.54)(60)(60)} \dfrac{\text{in.}}{\text{sec}}$

28. See Lesson 71.

29. $3\sqrt{9\sqrt[4]{3}} = 3\sqrt{3^2 \sqrt[4]{3}} = 3(3)(3^{1/8}) = \mathbf{3^{17/8}}$

30. $\dfrac{-9^{-3/2}}{-(-27)^{-2/3}} = \dfrac{-\left((-27)^{1/3}\right)^2}{-(9^{1/2})^3} = \dfrac{-9}{-27} = \dfrac{1}{3}$

PROBLEM SET 88

1. $N \qquad N + 2 \qquad N + 4$
$N(N + 4) = 6(-N - 2) + 12$
$N^2 + 4N = -6N$
$N^2 + 10N = 0$
$N(N + 10) = 0$
$N = 0, -10$
The desired integers are **0, 2, 4** and **-10, -8, -6.**

2. Tin (Sn): $1 \times 119 = 119$
Chromate: $1 \times 52 = 52$
Oxygen: $4 \times 16 = 64$
Total: $= 235$

$\dfrac{119}{235} = \dfrac{595}{\text{Tot}}$
Total = **1175 grams**

3. $R_D T_D = 400;$
$R_G T_G = 1120;$
$R_D = R_G - 20;$
$T_G = 2T_D$

$\begin{array}{c} \overset{D_G}{\vdash\!\!\!-\!\!\!-\!\!\!\dashv} \\ 1120 \end{array} \qquad \begin{array}{c} \overset{D_D}{\vdash\!\!\!-\!\!\!-\!\!\!\dashv} \\ 400 \end{array}$

$(R_D + 20)2T_D = 1120$
$2(400) + 40T_D = 1120$
$40T_D = 320$
$T_D = \mathbf{8\ hr}$
$T_G = \mathbf{16\ hr};\ R_D = \mathbf{50\ mph};\ R_G = \mathbf{70\ mph}$

4. $\dfrac{P_1 V_1}{T_1} = \dfrac{P_2 V_2}{T_2}$

$V_2 = \dfrac{P_1 V_1 T_2}{T_1 P_2}$

$V_2 = \dfrac{(740)(10)(1473)}{(573)(1480)}$

$V_2 = \mathbf{12.85\ liters}$

5. $PV = nRT$

$\dfrac{PV}{RT} = n$

$n = \dfrac{(1)(5)}{(0.0821)(251)}$ mole $= \mathbf{0.243\ mole}$

6. Refer to Section 88.A

7. $D = \sqrt{(x_1 - x_2)^2 + (y_1 - y_2)^2}$

$D = \sqrt{(5 - (-3))^2 + ((-2) - 3)^2}$

$= \sqrt{8^2 + (-5)^2} = \sqrt{64 + 25} = \sqrt{89}$

8. $-x - 2 \le 4$; $D = \{$Negative integers$\}$

$-x \le 6$

$x \ge -6$

9. $-x + 2 \not> 3$; $D = \{$Negative integers$\}$

$-x + 2 \le 3$

$-x \le 1$

$x \ge -1$

10. (a) $BT_D + 6T_D = 22$

(b) $BT_D - 6T_D = 10$

$\overline{2BT_D \qquad = 32}$

(c) $BT_D = 16$

(a) $(16) + 6T_D = 22$

$6T_D = 6$

$T_D = 1$

(c) $B = 16$

11. (a) $x^2 + y^2 = 2$

(b) $x - y = 1$

$x = y + 1$

(b′) $x^2 = (y + 1)^2 = y^2 + 2y + 1$

Substitute (b′) into (a) and get:

(a′) $(y^2 + 2y + 1) + y^2 = 2$

$2y^2 + 2y - 1 = 0$

Solve this equation by using the quadratic formula.

$y = \dfrac{-2 \pm \sqrt{2^2 - 4(2)(-1)}}{2(2)} = -\dfrac{1}{2} \pm \dfrac{\sqrt{3}}{2}$

Substitute these values of y into (b) and solve for x.

(b) $x = \left(-\dfrac{1}{2} + \dfrac{\sqrt{3}}{2}\right) + 1$

$x = \dfrac{1}{2} + \dfrac{\sqrt{3}}{2}$

(b) $x = \left(-\dfrac{1}{2} - \dfrac{\sqrt{3}}{2}\right) + 1$

$x = \dfrac{1}{2} - \dfrac{\sqrt{3}}{2}$

$\left(\dfrac{1}{2} + \dfrac{\sqrt{3}}{2}, -\dfrac{1}{2} + \dfrac{\sqrt{3}}{2}\right)$ and

$\left(\dfrac{1}{2} - \dfrac{\sqrt{3}}{2}, -\dfrac{1}{2} - \dfrac{\sqrt{3}}{2}\right)$

12. $\dfrac{(x^2)^{a+b}\, x^{-2a+b}\, y^a}{y^{a/4}} = x^{2a+2b-2a+b} y^{a-a/4}$

$= x^{3b} y^{3a/4}$

13. $\dfrac{(y^{2a+2})^2\, y^{a/2} b}{y^a b^{2a}} = y^{4a+4+a/2-a} b^{1-2a}$

$= y^{7a/2+4} b^{1-2a}$

14. $\dfrac{r}{m + \dfrac{r}{\dfrac{1}{r} + m}} = \dfrac{r}{m + \dfrac{r^2}{1 + rm}}$

$= \dfrac{r(1 + rm)}{m + rm^2 + r^2}$

15. $\dfrac{pc}{p - \dfrac{c^2}{p - \dfrac{1}{pc}}} = \dfrac{cp}{p - \dfrac{c^3 p}{cp^2 - 1}}$

$= \dfrac{cp(cp^2 - 1)}{cp^3 - p - c^3 p} = \dfrac{cp(cp^2 - 1)}{p(cp^2 - c^3 - 1)}$

$= \dfrac{c(cp^2 - 1)}{cp^2 - c^3 - 1}$

16. $\dfrac{4i - 1}{3i - 2} \cdot \dfrac{3i + 2}{3i + 2} = \dfrac{-12 - 3i + 8i - 2}{-9 - 4}$

$= \dfrac{-14 + 5i}{-13} = \dfrac{14}{13} - \dfrac{5}{13}i$

17. (a) $\dfrac{5712 \times 10^{-2}}{0.0416 \times 10^3}$

Estimate: $\dfrac{(6 \times 10^1)}{(4 \times 10^1)} = 1.5$

Calculator: **1.37**

(b) $(184.3)^{-1.62}$

Answer should be between 5×10^{-3} and 2.5×10^{-5}.

Calculator: $\mathbf{2.14 \times 10^{-4}}$

18. $\sqrt{z - 35} + \sqrt{z} = 7$

$z - 35 = 49 - 14\sqrt{z} + z$

$14\sqrt{z} = 84$

$\sqrt{z} = 6$

$z = 36$

Check: $\sqrt{36 - 35} + \sqrt{36} = 7$

$1 + 6 = 7$

$7 = 7$ check

19. (a) $2x + 2y - z = 14$
(b) $3x + 3y + z = 16$
(c) $x - 2y = 0$
$x = 2y$

Substitute (c) into (a) and (b) and get:

(a') $6y - z = 14$
(b') $\underline{9y + z = 16}$
$15y \quad\quad = 30$
$y = 2$

(c) $x = 2(2) = 4$

(a) $2(4) + 2(2) - z = 14$
$12 - z = 14$
$z = -2$

(4, 2, –2)

20.

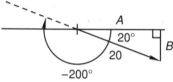

$A = 20 \cos 20° \approx 18.79$
$B = 20 \sin 20° \approx 6.84$

$C = 30 \cos 30° \approx 25.98$
$D = 30 \sin 30° = 15.00$

$18.79R - 6.84U$
$\underline{25.98R - 15.00U}$
$\mathbf{44.77R - 21.84U}$

21. $H = \sqrt{12^2 + 4^2}$
$H = \sqrt{160} = 4\sqrt{10}$

$\tan \theta = \dfrac{12}{4} = 3$
$\theta \approx 71.57°$
Since θ is a fourth-quadrant angle:
$\theta = 360 - 71.57 = 288.43°$

$4\sqrt{10}\underline{/288.43°}$

22. $4i^2 - \sqrt{-9} = \mathbf{-4 - 3i}$

23. $4\sqrt{\dfrac{9}{3}} + 3\sqrt{\dfrac{3}{9}} - 5\sqrt{27}$
$= 4\sqrt{3} + \sqrt{3} - 15\sqrt{3} = \mathbf{-10\sqrt{3}}$

24. $\dfrac{1 - \sqrt{2}}{3 - 2\sqrt{2}} \cdot \dfrac{3 + 2\sqrt{2}}{3 + 2\sqrt{2}}$

$= \dfrac{3 - 3\sqrt{2} + 2\sqrt{2} - 4}{9 - 8} = \mathbf{-1 - \sqrt{2}}$

25. $\dfrac{4 + \sqrt{3}}{1 - \sqrt{3}} \cdot \dfrac{1 + \sqrt{3}}{1 + \sqrt{3}}$

$= \dfrac{4 + \sqrt{3} + 4\sqrt{3} + 3}{1 - 3} = \dfrac{\mathbf{-7 - 5\sqrt{3}}}{\mathbf{2}}$

26.
$\dfrac{x + 2y}{c} = y\left(\dfrac{1}{x} - \dfrac{1}{r}\right)$

$\dfrac{x + 2y}{c} = \dfrac{y}{x} - \dfrac{y}{r}$

$x^2 r + 2xyr = cyr - cxy$

$cxy = cry - 2rxy - rx^2$

$\dfrac{cxy}{cy - 2xy - x^2} = r$

27. $-x^2 - 2x - 2 = 0$
$x^2 + 2x = -2$
$(x^2 + 2x + 1) = -2 + 1$
$(x + 1)^2 = -1$
$x + 1 = \pm i$
$\mathbf{x = -1 \pm i}$

28. $\dfrac{10 \text{ ml}}{\text{sec}} \times \dfrac{60 \text{ sec}}{1 \text{ min}} \times \dfrac{1 \text{ cm}^3}{1 \text{ ml}} \times \dfrac{1 \text{ in.}}{2.54 \text{ cm}}$

$\times \dfrac{1 \text{ in.}}{2.54 \text{ cm}} \times \dfrac{1 \text{ in.}}{2.54 \text{ cm}}$

$= \dfrac{(10)(60)}{(2.54)(2.54)(2.54)} \dfrac{\text{in.}^3}{\text{min}}$

29. $\dfrac{4}{3}\overrightarrow{SF} = \dfrac{7}{4}$

$\overrightarrow{SF} = \dfrac{21}{16}$

$x = \dfrac{7}{3}\overrightarrow{SF}$

$x = \dfrac{7}{3}\left(\dfrac{21}{16}\right)$

$x = \dfrac{49}{16}$

$y + 83 = 180$
$y = \mathbf{97}$

30. $\dfrac{-3x - 2}{2} - \dfrac{2x - 4}{3} = 7$
$-9x - 6 - 4x + 8 = 42$
$-13x = 40$
$x = -\dfrac{40}{13}$

PROBLEM SET 89

1. Equal ratio method:

$$\frac{A_1}{A_2} = \frac{B_1}{B_2}$$

$$\frac{500}{A_2} = \frac{10}{42}$$

$$A_2 = \mathbf{2100}$$

Variation method:

$$A = kB$$

$$500 = k(10)$$

$$50 = k$$

$$A = 50(42)$$

$$A = \mathbf{2100}$$

2. $R_RT_R + R_WT_W = 65$;

$R_R = 10$; $R_W = 5$;

$T_R = 9 - T_W$

$$10(9 - T_W) + 5T_W = 65$$

$$90 - 10T_W + 5T_W = 65$$

$$25 = 5T_W$$

$$5 = T_W$$

$$T_R = 9 - 5 = 4$$

$$D_R = \mathbf{40 \text{ km}}; \quad D_W = \mathbf{25 \text{ km}}$$

3. $0.79(800) - E = 0.30(800 - E)$

$$632 - E = 240 - 0.3E$$

$$392 = 0.7E$$

$$\mathbf{560 \text{ liters}} = E$$

4. (a) $10N_F = 2N_S - 140$

(b) $\dfrac{1}{2}N_S = 3N_F + 10$

(b′) $N_S = 6N_F + 20$

Substitute (b′) into (a) and get:

(a′) $10N_F = 2(6N_F + 20) - 140$

$$10N_F = 12N_F + 40 - 140$$

$$100 = 2N_F$$

$$\mathbf{50} = N_F$$

(b′) $N_S = 6(50) + 20 = \mathbf{320}$

5. $\dfrac{740}{100} \times S = 592$

$$S = 592 \cdot \frac{100}{740}$$

$$S = \mathbf{80}$$

6. $-x - 5 \le -3$ and $x - 3 < 1$; $D = \{\text{Integers}\}$

$$-x \le 2 \quad \text{and} \quad x < 4$$

$$x \ge -2 \quad \text{and} \quad x < 4$$

7. $-4 < x - 4 \le 1$; $D = \{\text{Reals}\}$

$$0 < x \le 5$$

8. $-x \not\ge 2$ or $-x < 1$; $D = \{\text{Reals}\}$

$$-x < 2 \quad \text{or} \quad -x < 1$$

$$x > -2 \quad \text{or} \quad x > -1$$

9. $7x = 2(14)$

$$7x = 28$$

$$x = \mathbf{4}$$

10. $4(10) = 2(y + 2)$

$$40 = 2y + 4$$

$$36 = 2y$$

$$\mathbf{18} = y$$

11. (a) $BT_D + 5T_D = 57$

(b) $\dfrac{BT_D - 5T_D = 27}{}$

$$2BT_D = 84$$

(c) $BT_D = 42$

(a) $(42) + 5T_D = 57$

$$5T_D = 15$$

$$T_D = \mathbf{3}$$

(c) $B(3) = 42$

$$B = \mathbf{14}$$

12. (a) $x^2 + y^2 = 3$

(b) $x - y = 2$

$$x = y + 2$$

(b′) $x^2 = (y + 2)^2 = y^2 + 4y + 4$

Substitute (b′) into (a) and get:

(a′) $(y^2 + 4y + 4) + y^2 = 3$

$$2y^2 + 4y + 1 = 0$$

Solve this equation by using the quadratic formula.

$$y = \frac{-4 \pm \sqrt{4^2 - 4(2)(1)}}{2(2)} = -1 \pm \frac{\sqrt{2}}{2}$$

Substitute these values of y into (b) and solve for x.

(b) $x = \left(-1 + \dfrac{\sqrt{2}}{2}\right) + 2$

$$x = 1 + \frac{\sqrt{2}}{2}$$

(b) $x = \left(-1 - \dfrac{\sqrt{2}}{2}\right) + 2$

$$x = 1 - \frac{\sqrt{2}}{2}$$

$$\left(\mathbf{1} + \frac{\sqrt{2}}{2}, \mathbf{-1} + \frac{\sqrt{2}}{2}\right) \text{ and}$$

$$\left(\mathbf{1} - \frac{\sqrt{2}}{2}, \mathbf{-1} - \frac{\sqrt{2}}{2}\right)$$

13. $\dfrac{x^a y^{2b} (x^{a+2})^{1/2}}{y^{3b}} = x^{a+a/2+1} y^{2b-3b} = \mathbf{x^{3a/2+1} y^{-b}}$

14. $\dfrac{(y^{a+2})^a (y^a)^a}{y^{2+a}} = y^{a^2+2a+a^2-2-a}$

$= \mathbf{y^{2a^2+a-2}}$

15. $\dfrac{k}{m + \dfrac{m}{a + \dfrac{1}{m}}} = \dfrac{k}{m + \dfrac{m^2}{am + 1}}$

$= \dfrac{\mathbf{k(am + 1)}}{\mathbf{am^2 + m^2 + m}}$

16. $\dfrac{m}{a + \dfrac{x}{b + \dfrac{d}{c}}} = \dfrac{m}{a + \dfrac{cx}{bc + d}}$

$= \dfrac{\mathbf{m(bc + d)}}{\mathbf{abc + ad + cx}}$

17. $\dfrac{3 - 2i}{i - 4} \cdot \dfrac{i + 4}{i + 4} = \dfrac{3i + 2 + 12 - 8i}{-1 - 16}$

$= \dfrac{14 - 5i}{-17} = \mathbf{-\dfrac{14}{17} + \dfrac{5}{17}i}$

18. $\dfrac{2 - 3i}{4i - 1} \cdot \dfrac{4i + 1}{4i + 1} = \dfrac{8i + 12 + 2 - 3i}{-16 - 1}$

$= \dfrac{14 + 5i}{-17} = \mathbf{-\dfrac{14}{17} - \dfrac{5}{17}i}$

19. (a) $3x + 2y + z = 9$
(b) $x - 2y - 2z = -3$
(c) $2x + z = 0$
$\qquad z = -2x$

Substitute (c) into (a) and (b) and get:

(a′) $x + 2y = 9$
(b′) $\underline{5x - 2y = -3}$
$\qquad 6x \quad\;\; = \;\; 6$
$\qquad\qquad x = 1$

(a′) $(1) + 2y = 9$
$\qquad\qquad y = 4$

(c) $z = -2(1)$
$\quad\; z = -2$

(1, 4, −2)

20.
$\sqrt{s} = 3 + \sqrt{s - 21}$
$s - 6\sqrt{s} + 9 = s - 21$
$\qquad 30 = 6\sqrt{s}$
$\qquad\; 5 = \sqrt{s}$
$\qquad \mathbf{25 = s}$

Check: $\sqrt{25} = 3 + \sqrt{25 - 21}$
$\qquad\quad 5 = 3 + 2$
$\qquad\quad 5 = 5 \quad$ check

21.

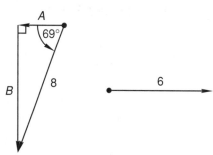

$A = 8 \cos 69° \approx 2.87$
$B = 8 \sin 69° \approx 7.47$

$\begin{array}{r} -2.87R - 7.47U \\ \underline{6.00R + 0.00U} \\ \mathbf{3.13R - 7.47U} \end{array}$

$\tan \theta = \dfrac{-7.47}{3.13}$
$\qquad \theta \approx -67.27°$

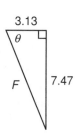

$F = \sqrt{(3.13)^2 + (-7.47)^2}$
$F \approx 8.10$

$\mathbf{8.10\,\underline{/\!-67.27°}}$

22. $\dfrac{4 + 2\sqrt{3}}{3\sqrt{3} - 2} \cdot \dfrac{3\sqrt{3} + 2}{3\sqrt{3} + 2}$

$= \dfrac{12\sqrt{3} + 18 + 8 + 4\sqrt{3}}{27 - 4} = \dfrac{\mathbf{26 + 16\sqrt{3}}}{\mathbf{23}}$

23. $\dfrac{x}{a + c} = \dfrac{mx}{y} + cm$

$xy = amx + cmx + acmy + c^2 my$
$xy - acmy - c^2 my = amx + cmx$
$\qquad\qquad y = \dfrac{\mathbf{amx + cmx}}{\mathbf{x - acm - c^2 m}}$

24. $\dfrac{x}{a + c} = \dfrac{mx + mcy}{y}$

$xy = amx + cmx + acmy + c^2 my$
$\qquad xy - cmx - c^2 my = amx + acmy$
$\qquad \dfrac{\mathbf{xy - cmx - c^2 my}}{\mathbf{mx + cmy}} = a$

25. $3i^3 + 5i - \sqrt{-2}\sqrt{2} = -3i + 5i - 2i = \mathbf{0}$

26. $3\sqrt{\dfrac{3}{8}} + 4\sqrt{\dfrac{8}{3}} - 2\sqrt{24}$

$= \dfrac{3\sqrt{3}}{2\sqrt{2}} \dfrac{\sqrt{2}}{\sqrt{2}} + \dfrac{8\sqrt{2}}{\sqrt{3}} \dfrac{\sqrt{3}}{\sqrt{3}} - 4\sqrt{6}$

$= \dfrac{3\sqrt{6}}{4} + \dfrac{8\sqrt{6}}{3} - 4\sqrt{6}$

$= \dfrac{9\sqrt{6}}{12} + \dfrac{32\sqrt{6}}{12} - \dfrac{48\sqrt{6}}{12} = \mathbf{-\dfrac{7\sqrt{6}}{12}}$

27. $-7x^2 = -x - 5$

$\qquad 0 = 7x^2 - x - 5$

$\qquad x = \dfrac{1 \pm \sqrt{1 - 4(7)(-5)}}{2(7)} = \dfrac{1 \pm \sqrt{141}}{14}$

28. $\dfrac{10 \text{ ft}^3}{\text{sec}} \times \dfrac{60 \text{ sec}}{1 \text{ min}} \times \dfrac{12 \text{ in.}}{1 \text{ ft}} \times \dfrac{12 \text{ in.}}{1 \text{ ft}} \times \dfrac{12 \text{ in.}}{1 \text{ ft}}$

$\qquad = 10(60)(12)(12)(12) \; \dfrac{\text{in.}^3}{\text{min}}$

29. $(6x + 20)° + (4x + 10)° = 180°$

$\qquad 6x + 20 + 4x + 10 = 180$

$\qquad\qquad\qquad\qquad 10x = 150$

$\qquad\qquad\qquad\qquad\quad x = \mathbf{15}$

$\qquad (4x + 10)° = A°$

$\qquad 4(15) + 10 = A$

$\qquad\qquad\quad 70 = A$

$\qquad (6x + 20)° = B°$

$\qquad 6(15) + 20 = B$

$\qquad\qquad\quad 110 = B$

30. $(x_1, y_1) = (-2, 3)$

$\qquad (x_2, y_2) = (8, 4)$

$\qquad D = \sqrt{(x_2 - x_1)^2 + (y_2 - y_1)^2}$

$\qquad D = \sqrt{(8 - (-2))^2 + (4 - 3)^2}$

$\qquad\quad = \sqrt{100 + 1} = \sqrt{101}$

PROBLEM SET 90

1. $R_R = R_P + 50;$

$R_P T_P = 300;$

$R_R T_R = 160;$

$T_P = 5T_R;\; R_P = R_R - 50$

$\qquad (R_R - 50)(5T_R) = 300$

$\qquad 5R_R T_R - 250T_R = 300$

$\qquad\quad 800 - 250T_R = 300$

$\qquad\qquad\quad 500 = 250T_R$

$\qquad\qquad\qquad 2 = T_R$

$R_R = \mathbf{80}$ **mph**; $R_P = \mathbf{30}$ **mph**;

$T_R = \mathbf{2}$ **hr**; $T_P = \mathbf{10}$ **hr**

2. $PV = nRT$

$\qquad V = \dfrac{nRT}{P}$

$\qquad V = \dfrac{(0.832)(0.0821)(400)}{3}$ liters $= \mathbf{9.11}$ **liters**

3. (a) $N_R = 3N_B - 5$

$\quad$ (b) $6N_B = 10N_R - 70$

-2(a) $-2N_R = -6N_B + 10$

$\quad$ (b) $\underline{\;10N_R = \quad 6N_B + 70\;}$

$\qquad\qquad 8N_R = \qquad\qquad\quad 80$

$\qquad\qquad\quad N_R = 10$

$\quad$ (a) $10 = 3N_B - 5$

$\qquad\quad 15 = 3N_B$

$\qquad\quad\; 5 = N_B$

$\quad N_B = \mathbf{5};\; N_R = \mathbf{10}$

4. (a) $0.1P_N + 0.4D_N = 104$

$\quad$ (b) $P_N + D_N = 800$

$\quad$ Substitute $D_N = 800 - P_N$ into (a) and get:

$\quad$ (a') $0.1P_N + 0.4(800 - P_N) = 104$

$\qquad\qquad\qquad\qquad -0.3P_N = -216$

$\qquad\qquad\qquad\qquad\quad\; P_N = 720$

$\quad$ (b) $(720) + D_N = 800$

$\qquad\qquad\qquad D_N = 80$

$\quad$ **720 liters 10%, 80 liters 40%**

5. $\dfrac{7}{16} \times SP = 672$

$\qquad SP = 672 \cdot \dfrac{16}{7} = \mathbf{1536}$

6. (a) $x + y - z = 3$

$\quad$ (b) $-x - 2y - 2z = 0$

$\quad$ (c) $x - 2y - 2z = 4$

$\quad$ (a) $\quad x + y - z = 3$

$\quad$ (b) $\underline{-x - 2y - 2z = 0}$

$\quad$ (d) $\qquad - y - 3z = 3$

$\quad$ (b) $-x - 2y - 2z = 0$

$\quad$ (c) $\underline{\; x - 2y - 2z = 4}$

$\quad$ (e) $\qquad - 4y - 4z = 4$

4(d) $-4y - 12z = 12$

-1(e) $\underline{\; 4y + \; 4z = -4}$

$\qquad\qquad - 8z = \; 8$

$\qquad\qquad\quad\; z = -1$

$\quad$ (d) $-y - 3(-1) = 3$

$\qquad\qquad\quad -y = 0$

$\qquad\qquad\quad\; y = 0$

$\quad$ (a) $x + (0) - (-1) = 3$

$\qquad\qquad x + 1 = 3$

$\qquad\qquad\quad\; x = 2$

$\quad$ **(2, 0, –1)**

7. (a) $2x - y + z = 2$
(b) $x + 2y + 2z = 3$
(c) $2x - 2y + z = 0$

2(a) $4x - 2y + 2z = 4$
(b) $\underline{\ x + 2y + 2z = 3\ }$
(d) $5x \qquad + 4z = 7$

(b) $x + 2y + 2z = 3$
(c) $\underline{2x - 2y + \ z = 0}$
(e) $3x \qquad + 3z = 3$

3(d) $\quad 15x + 12z = \quad 21$
-4(e) $\underline{-12x - 12z = -12}$
$\qquad 3x \qquad\quad = \qquad 9$
$\qquad\qquad x = 3$

(e) $3(3) + 3z = 3$
$\qquad 9 + 3z = 3$
$\qquad\quad 3z = -6$
$\qquad\quad\ z = -2$

(c) $2(3) - 2y + (-2) = 0$
$\qquad 6 - 2y - 2 = 0$
$\qquad\qquad 4 = 2y$
$\qquad\qquad 2 = y$

$(3, 2, -2)$

8. $-1 \le x - 1 < 4; \ D = \{\text{Integers}\}$
$\quad 0 \le x < 5$

9. $x - 2 \not\ge 0$ or $x - 2 > 2; \ D = \{\text{Reals}\}$
$\quad x - 2 < 0$ or $\qquad x > 4$
$\qquad x < 2$ or $\qquad x > 4$

10. (a) $BT_D + 3T_D = 28$
(b) $\underline{BT_D - 3T_D = 16}$
$\qquad 2BT_D \qquad = 44$
(c) $\qquad\quad BT_D = 22$

(a) $(22) + 3T_D = 28$
$\qquad\quad 3T_D = 6$
$\qquad\quad\ T_D = 2$

(c) $B(2) = 22$
$\quad\ B = 11$

11. (a) $x^2 + y^2 = 18$
(b) $y - x = 4$
$\qquad x = y - 4$
(b') $x^2 = (y - 4)^2 = y^2 - 8y + 16$
Substitute (b') into (a) and get:
(a') $\left(y^2 - 8y + 16 + y^2\right) = 18$
$\qquad 2y^2 - 8y - 2 = 0$
$\qquad\ y^2 - 4y - 1 = 0$
Solve this equation by using the quadratic formula.

$y = \dfrac{4 \pm \sqrt{16 + 4}}{2} = 2 \pm \sqrt{5}$

Substitute these values of y into (b) and solve for x.
(b) $x = (2 + \sqrt{5}) - 4$
$\qquad x = -2 + \sqrt{5}$
(b) $x = (2 - \sqrt{5}) - 4$
$\qquad x = -2 - \sqrt{5}$

$(-2 + \sqrt{5}, 2 + \sqrt{5})$ and $(-2 - \sqrt{5}, 2 - \sqrt{5})$

12. $\dfrac{a^2 \, a^{x/2} \, (a^2)^x}{(a^{-3})^{-x}} = a^{2 + x/2 + 2x - 3x} = \boldsymbol{a^{2 - x/2}}$

13. $\dfrac{y^c \, (m^{-b})^2}{m^{-b/2}} = y^c m^{-2b + b/2} = \boldsymbol{y^c m^{-3b/2}}$

14. $\dfrac{x}{xm - \dfrac{m}{m - \dfrac{1}{x}}} = \dfrac{x}{mx - \dfrac{mx}{mx - 1}}$

$= \dfrac{x(mx - 1)}{m^2 x^2 - 2mx} = \boldsymbol{\dfrac{mx - 1}{m^2 x - 2m}}$

15. $\dfrac{p}{x - \dfrac{xp}{1 - \dfrac{p}{x}}} = \dfrac{p}{x - \dfrac{px^2}{x - p}}$

$= \boldsymbol{\dfrac{p(x - p)}{x^2 - px - px^2}}$

16. $\dfrac{1 + 2i}{-5 - i} \cdot \dfrac{-5 + i}{-5 + i} = \dfrac{-5 - 10i + i - 2}{25 + 1}$

$= \boldsymbol{-\dfrac{7}{26} - \dfrac{9}{26}i}$

17. Since the $m\widehat{AC} = 120°$, the $m\widehat{ABC} = 240°$.

Length of $240°$ arc $= \dfrac{240}{360} \times 10\pi$ cm $= \boldsymbol{\dfrac{20}{3}\pi}$ **cm**

18. $\sqrt{s} - \sqrt{s - 15} = 3$
$\quad s - 6\sqrt{s} + 9 = s - 15$
$\qquad\qquad\quad 24 = 6\sqrt{s}$
$\qquad\qquad\qquad 4 = \sqrt{s}$
$\qquad\qquad\quad \boldsymbol{16 = s}$

Check: $\sqrt{16} - \sqrt{16 - 15} = 3$
$\qquad\qquad\qquad 4 - 1 = 3$
$\qquad\qquad\qquad\quad 3 = 3 \quad$ check

19.

$A = 4 \cos 40° \approx 3.06$
$B = 4 \sin 40° \approx 2.57$

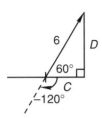

$C = 6 \cos 60° = 3.00$
$D = 6 \sin 60° \approx 5.20$

$\underline{\begin{array}{l} 3.06R + 2.57U \\ 3.00R + 5.20U \end{array}}$
6.06R + 7.77U

20. $\tan \theta = \dfrac{-10}{-4}$

$\theta \approx 68.20°$
Since θ is a third-quadrant angle:
$\theta = 68.20 + 180 = 248.2°$

$H = \sqrt{(-4)^2 + (-10)^2}$
$H = 2\sqrt{29}$
$2\sqrt{29}\,\underline{/248.2°}$

21. $\dfrac{-2 - \sqrt{3}}{2 - 2\sqrt{3}} \cdot \dfrac{2 + 2\sqrt{3}}{2 + 2\sqrt{3}}$

$= \dfrac{-4 - 2\sqrt{3} - 4\sqrt{3} - 6}{4 - 12} = \dfrac{-10 - 6\sqrt{3}}{-8}$

$= \dfrac{\mathbf{5 + 3\sqrt{3}}}{\mathbf{4}}$

22. $\dfrac{4 + \sqrt{3}}{3 - 2\sqrt{3}} \cdot \dfrac{3 + 2\sqrt{3}}{3 + 2\sqrt{3}}$

$= \dfrac{12 + 3\sqrt{3} + 8\sqrt{3} + 6}{9 - 12} = \dfrac{18 + 11\sqrt{3}}{-3}$

$= \dfrac{\mathbf{-18 - 11\sqrt{3}}}{\mathbf{3}}$

23. $-2i^2 + \sqrt{-4}\sqrt{4} - \sqrt{-3}\sqrt{-3} - 2i^5$
$= 2 + 4i + 3 - 2i = \mathbf{5 + 2i}$

24.
$$\dfrac{x}{y} - m = p\left(\dfrac{r}{c} - \dfrac{1}{b}\right)$$
$$\dfrac{x}{y} - m = \dfrac{pr}{c} - \dfrac{p}{b}$$
$$bcx - bcmy = bpry - cpy$$
$$bcx - bcmy + cpy = bpry$$
$$c = \dfrac{\boldsymbol{bpry}}{\boldsymbol{bx - bmy + py}}$$

25. $\sqrt[3]{9\sqrt[4]{3}} = (3^2 3^{1/4})^{1/3} = (3^{9/4})^{1/3} = \mathbf{3^{3/4}}$

26. (a) $3(12) = 4(4 + x)$
$36 = 16 + 4x$
$20 = 4x$
$\mathbf{5} = x$

(b) $6 \cdot 3 = 9 \cdot x$
$18 = 9x$
$\mathbf{2} = x$

27. $-6x^2 - x - 5 = 0$

$\left(x^2 + \dfrac{x}{6} + \quad\right) = -\dfrac{5}{6}$

$x^2 + \dfrac{x}{6} + \dfrac{1}{144} = -\dfrac{120}{144} + \dfrac{1}{144}$

$\left(x + \dfrac{1}{12}\right)^2 = -\dfrac{119}{144}$

$x + \dfrac{1}{12} = \pm\dfrac{\sqrt{119}}{12}i$

$x = -\dfrac{\mathbf{1}}{\mathbf{12}} \pm \dfrac{\sqrt{\mathbf{119}}}{\mathbf{12}}i$

28. $42 \text{ ft}^3 \times \dfrac{12 \text{ in.}}{1 \text{ ft}} \times \dfrac{12 \text{ in.}}{1 \text{ ft}} \times \dfrac{12 \text{ in.}}{1 \text{ ft}} \times \dfrac{2.54 \text{ cm}}{1 \text{ in.}}$

$\times \dfrac{2.54 \text{ cm}}{1 \text{ in.}} \times \dfrac{2.54 \text{ cm}}{1 \text{ in.}}$

$= \mathbf{42(12)(12)(12)(2.54)(2.54)(2.54) \ cm^3}$

29. $A = \sqrt{13^2 - 12^2} = \sqrt{25} = 5$

$13 \times \overrightarrow{SF} = 19$

$\overrightarrow{SF} = \dfrac{19}{13}$

$B = 5 \times \overrightarrow{SF}$

$B = 5 \cdot \dfrac{19}{13} = \dfrac{\mathbf{95}}{\mathbf{13}}$

30. Write the equation of the given line in slope-intercept form.
$x + 5y = 7$

$y = -\dfrac{1}{5}x + \dfrac{7}{5}$

Since the slopes of perpendicular lines are negative reciprocals of each other:
$m_\perp = 5$
$y = 5x + b$
$7 = 5(-2) + b$
$17 = b$
$\mathbf{y = 5x + 17}$

PROBLEM SET 91

1. $\dfrac{P_1}{T_1} = \dfrac{P_2}{T_2}$

$P_2 = \dfrac{P_1 T_2}{T_1}$

$P_2 = \dfrac{(740)(1600)}{400}$

$P_2 = \textbf{2960 mm Hg}$

2. (a) $0.6P_N + 0.3D_N = 126$
   (b) $P_N + D_N = 300$

Substitute $D_N = 300 - P_N$ into (a) and get:
(a′) $0.6P_N + 0.3(300 - P_N) = 126$
$\qquad\qquad\qquad 0.3P_N = 36$
$\qquad\qquad\qquad\quad P_N = 120$

(b) $(120) + D_N = 300$
$\qquad\qquad D_N = 180$

180 ml 30%, 120 ml 60%

3. Arsenic: $4 \times 75 = 300$
   Oxygen: $6 \times 16 = 96$
   Total: $\qquad\quad = 396$

Percent As $= \dfrac{\text{As}}{\text{Total}} \times 100\%$

$= \dfrac{300}{396} \cdot 100 \approx \textbf{75.8\%}$

4. $T_L = T_S + 1;$
$R_L T_L = 3000;$
$R_L = 3R_S;$
$R_S T_S = 800$

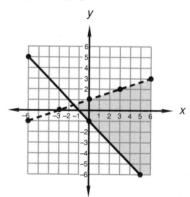

$3R_S(T_S + 1) = 3000$
$3R_S T_S + 3R_S = 3000$
$2400 + 3R_S = 3000$
$\qquad\quad 3R_S = 600$
$\qquad\qquad R_S = 200$

$R_L = \textbf{600 mph};\ R_S = \textbf{200 mph};$
$T_L = \textbf{5 hr};\ T_S = \textbf{4 hr}$

5. Variation method:

$G = \dfrac{k}{A}$

$35 = \dfrac{k}{70}$

$2450 = k$

$G = \dfrac{2450}{50}$

$G = \textbf{49}$

Equal ratio method:

$\dfrac{G_1}{G_2} = \dfrac{A_2}{A_1}$

$\dfrac{35}{G_2} = \dfrac{50}{70}$

$G_2 = \textbf{49}$

6. (a) $y \geq -x - 1$
   (b) $y < \dfrac{1}{3}x + 1$
   The first step is to graph each of these lines.

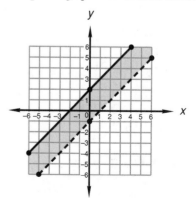

The region we wish to find is on or above the solid line and below the dashed line. This region is shaded in the previous figure.

7. (a) $y > x - 1$
   (b) $y \leq x + 2$
   The first step is to graph each of these lines.

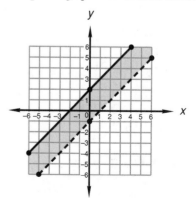

The region we wish to find is on or below the solid line and above the dashed line. This region is shaded in the previous figure.

8. (a) $x + 2y - z = 1$
(b) $2x - y + 2z = 9$
(c) $x - 2y - 3z = -9$

(a) $\quad x + 2y - z = 1$
2(b) $\underline{4x - 2y + 4z = 18}$
(d) $5x \qquad + 3z = 19$

(a) $\quad x + 2y - z = 1$
(c) $\underline{x - 2y - 3z = -9}$
(e) $2x \qquad - 4z = -8$

2(d) $\quad 10x + 6z = 38$
-5(e) $\underline{-10x + 20z = 40}$
$\qquad \quad 26z = 78$
$\qquad \quad\;\; z = 3$

(e) $2x - 4(3) = -8$
$\qquad \quad 2x = 4$
$\qquad \quad\;\; x = 2$

(a) $(2) + 2y - (3) = 1$
$\qquad \qquad 2y = 2$
$\qquad \qquad\;\; y = 1$

(2, 1, 3)

9. $4 < x + 4 < 6$; $D = \{\text{Integers}\}$
$0 < x < 2$

10. $x + 2 \not\leq 5$ or $x + 5 < 6$; $D = \{\text{Reals}\}$
$x + 2 > 5$ or $x + 5 < 6$
$\quad x > 3$ or $\qquad x < 1$

11. (a) $BT_D + 2T_D = 51$
(b) $\underline{BT_D - 2T_D = 39}$
$\quad 2BT_D \qquad = 90$
(c) $\qquad BT_D = 45$

(a) $(45) + 2T_D = 51$
$\qquad \quad 2T_D = 6$
$\qquad \quad\; T_D = 3$

(c) $B(3) = 45$
$\quad\; B = \mathbf{15}$

12. (a) $x^2 + y^2 = 12$
(b) $x + y = 4$
$\qquad x = 4 - y$
(b') $x^2 = (4 - y)^2 = 16 - 8y + y^2$
Substitute (b') into (a) and get:
(a') $\left(16 - 8y + y^2\right) + y^2 = 12$
$\qquad \quad 2y^2 - 8y + 4 = 0$
$\qquad \qquad y^2 - 4y + 2 = 0$
Solve this equation by using the quadratic formula.

$$y = \frac{4 \pm \sqrt{16 - 4(2)(1)}}{2(1)}$$

$$= 2 \pm \frac{\sqrt{8}}{2} = 2 \pm \sqrt{2}$$

Substitute these values of y into (b) and solve for x.
(b) $x = 4 - (2 + \sqrt{2})$
$\qquad x = 4 - 2 - \sqrt{2}$
$\qquad x = 2 - \sqrt{2}$
(b) $x = 4 - (2 - \sqrt{2})$
$\qquad x = 4 - 2 + \sqrt{2}$
$\qquad x = 2 + \sqrt{2}$

$\mathbf{(2 - \sqrt{2}, 2 + \sqrt{2})}$ and $\mathbf{(2 + \sqrt{2}, 2 - \sqrt{2})}$

13. $\dfrac{(x^b)^{2-a}\, x^{ab}}{x^{ab/2}} = x^{2b-ab+ab-ab/2} = x^{2b-ab/2}$

14. $\dfrac{a}{b + \dfrac{1}{cx + \dfrac{1}{x}}} = \dfrac{a}{b + \dfrac{x}{cx^2 + 1}}$

$$= \frac{a(cx^2 + 1)}{bcx^2 + b + x}$$

15. $\dfrac{m}{a + \dfrac{ma}{a + \dfrac{m}{a}}} = \dfrac{m}{a + \dfrac{a^2 m}{a^2 + m}}$

$$= \frac{m(a^2 + m)}{a^3 + am + a^2 m}$$

16. $\dfrac{2i + 7}{i + 2} \cdot \dfrac{i - 2}{i - 2} = \dfrac{-2 + 7i - 4i - 14}{-1 - 4}$

$$= \frac{-16 + 3i}{-5} = \frac{16}{5} - \frac{3}{5}i$$

17. $\dfrac{3i - 6}{-2i + 1} \cdot \dfrac{-2i - 1}{-2i - 1}$

$= \dfrac{6 + 12i - 3i + 6}{-4 - 1}$

$= \dfrac{12 + 9i}{-5} = -\dfrac{12}{5} - \dfrac{9}{5}i$

18. $\dfrac{1 - \sqrt{2}}{4 - 5\sqrt{2}} \cdot \dfrac{4 + 5\sqrt{2}}{4 + 5\sqrt{2}}$

$= \dfrac{4 - 4\sqrt{2} + 5\sqrt{2} - 10}{16 - 50} = \dfrac{-6 + \sqrt{2}}{-34}$

$= \dfrac{6 - \sqrt{2}}{34}$

19. $\dfrac{4 + 3\sqrt{2}}{-\sqrt{2}} \cdot \dfrac{\sqrt{2}}{\sqrt{2}} = \dfrac{4\sqrt{2} + 6}{-2} = -3 - 2\sqrt{2}$

20. $\dfrac{x}{my} = d\left(\dfrac{r}{a} + b\right)$

$\dfrac{x}{my} = \dfrac{dr}{a} + db$

$ax = dmry + abdmy$

$ax = y(dmr + abdm)$

$y = \dfrac{ax}{dmr + abdm}$

21. $\dfrac{x}{my} = d\left(\dfrac{r}{a} + b\right)$

$\dfrac{x}{my} = \dfrac{dr}{a} + db$

$ax = dmry + abdmy$

$ax - abdmy = dmry$

$a(x - bdmy) = dmry$

$a = \dfrac{dmry}{x - bdmy}$

22.

$A = 4 \cos 40° \approx 3.06$

$B = 4 \sin 40° \approx 2.57$

$\quad -3.06R + 2.57U$

$\quad\underline{6.00R + 0.00U}$

$\quad 2.94R + 2.57U$

$\tan \theta = \dfrac{2.57}{2.94}$

$\theta \approx 41.16°$

$F = \sqrt{(2.57)^2 + (2.94)^2} \approx 3.9$

3.9/41.16°

23. There is never an exact solution to these problems. One possible solution is given here.

$Op = mIp + b$

Use the graph to find the slope.

$m = \dfrac{100 - 30}{30 - 20} = \dfrac{70}{10} = 7$

$Op = 7Ip + b$

Use the point (20, 30) for Ip and Op.

$30 = 7(20) + b$

$-110 = b$

$Op = 7Ip - 110$

24. $\dfrac{-9^{3/2}}{27^{2/3}} = \dfrac{-(9^{1/2})^3}{(27^{1/3})^2} = \dfrac{-27}{9} = -3$

25. $-2i^5 + 3i^3 - \sqrt{-9} + \sqrt{-4} - \sqrt{-2}\sqrt{-2}$

$= -2i - 3i - 3i + 2i + 2 = 2 - 6i$

26. $2\sqrt{\dfrac{9}{5}} + 3\sqrt{\dfrac{5}{9}} + 3\sqrt{45} = \dfrac{6}{\sqrt{5}} + \sqrt{5} + 9\sqrt{5}$

$= \dfrac{6\sqrt{5}}{5} + \dfrac{5\sqrt{5}}{5} + \dfrac{45\sqrt{5}}{5} = \dfrac{56\sqrt{5}}{5}$

27. See Lesson 71.

28. $\dfrac{400 \text{ ml}}{\text{sec}} \times \dfrac{60 \text{ sec}}{1 \text{ min}} \times \dfrac{1 \text{ cm}^3}{1 \text{ ml}} \times \dfrac{1 \text{ in.}}{2.54 \text{ cm}}$

$\times \dfrac{1 \text{ in.}}{2.54 \text{ cm}} \times \dfrac{1 \text{ in.}}{2.54 \text{ cm}}$

$= \dfrac{400(60)}{(2.54)(2.54)(2.54)} \dfrac{\text{in.}^3}{\text{min}}$

29.

$$x + 2 \overline{) 3x^3 + 0x^2 - 2x + 2}\qquad 3x^2 - 6x + 10 - \dfrac{18}{x+2}$$

$\underline{3x^3 + 6x^2}$

$\quad\ -6x^2 - 2x$

$\quad\ \underline{-6x^2 - 12x}$

$\qquad\quad 10x + 2$

$\qquad\quad \underline{10x + 20}$

$\qquad\qquad\ \ - 18$

30. (a) $x \cdot x = 3(9 + 3)$

$x^2 = 36$

$x = 6$

(b) $2(2 + 4) = 1(1 + x)$

$12 = 1 + x$

$11 = x$

PROBLEM SET 92

1. (a) $(B + W)T_D = D_D$
$\qquad (10 + W)T_D = 70$

(b) $(B - W)T_U = D_U$
$\qquad (10 - W)T_U = 30$

$T_D = T_U$ so we use T in both equations.

(a') $\underline{10T + WT = 70}$
(b') $\underline{10T - WT = 30}$
$\qquad 20T \qquad = 100$
$\qquad\qquad T = \mathbf{5 \ hr}$

(a') $10(5) + W(5) = 70$
$\qquad\qquad 5W = 20$
$\qquad\qquad W = \mathbf{4 \ mph}$

2. (a) $(B + W)T_D = D_D$
$\qquad (B + W)4 = 60$
(a') $4B + 4W = 60$
(b) $(B - W)T_U = D_U$
$\qquad (B - W)5 = 55$
(b') $5B - 5W = 55$

5(a') $20B + 20W = 300$
4(b') $\underline{20B - 20W = 220}$
$\qquad 40B \qquad = 520$
$\qquad\qquad B = \mathbf{13 \ mph}$

(a') $4(13) + 4W = 60$
$\qquad\qquad 4W = 8$
$\qquad\qquad W = \mathbf{2 \ mph}$

3. (a) $(B + W)T_D = D_D$
$\qquad (B + 7)T_D = 35$
(b) $(B - W)T_U = D_U$
$\qquad (B - 7)T_U = 21$

$T_D = T_U$ so we use T in both equations.

(a') $BT + 7T = 35$
(b') $\underline{BT - 7T = 21}$
$\qquad 2BT \qquad = 56$
(c) $\qquad BT = 28$

(a') $(28) + 7T = 35$
$\qquad\qquad 7T = 7$
$\qquad\qquad T = 1 \ hr$

(c) $B(1) = 28$
$\qquad B = \mathbf{28 \ kph}$

4. Equal ratio method:
$$\frac{G_1}{G_2} = \frac{C_1}{C_2}$$
$$\frac{500}{1750} = \frac{20}{C_2}$$
$$C_2 = \mathbf{70}$$

Variation method:
$\qquad G = kC$
$\qquad 500 = k(20)$
$\qquad 25 = k$

$\qquad 1750 = 25(C)$
$\qquad \mathbf{70} = C$

5. $D_W = R_W T_W = 6T_W;$
$\quad D_H = D_W + 50;$
$\quad D_H = R_H T_H = 8T_H;$
$\quad T_W = T_H$ so use T

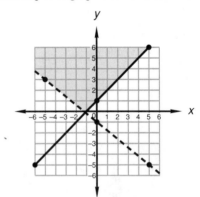

$\qquad 8T = 6T + 50$
$\qquad 2T = 50$
$\qquad\quad T = \mathbf{25 \ sec}$

6. (a) $y \geq x + 1$

(b) $y > -\dfrac{4}{5}x - 1$

The first step is to graph each of these lines.

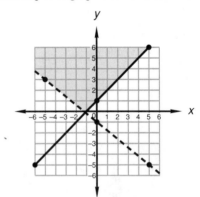

The region we wish to find is on or above the solid line and above the dashed line. This region is shaded in the previous figure.

7. (a) $y \leq \dfrac{1}{2}x + 2$

(b) $x < 2$

The first step is to graph each of these lines.

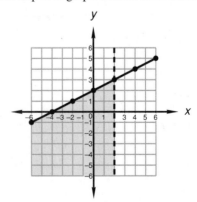

The region we wish to find is on or below the solid line and to the left of the dashed line. This region is shaded in the previous figure.

8. (a) $x + y + z = 6$
(b) $3x - y + z = 8$
(c) $x - 2y + z = 0$

(a) $x + y + z = 6$
(b) $3x - y + z = 8$
(d) $\overline{4x \qquad + 2z = 14}$

2(a) $2x + 2y + 2z = 12$
(c) $\underline{x - 2y + z = 0}$
(e) $3x \qquad + 3z = 12$

3(d) $12x + 6z = 42$
(−2)(e) $\underline{-6x - 6z = -24}$
$6x \qquad = 18$
$x = 3$
(e) $3(3) + 3z = 12$
$3z = 3$
$z = 1$
(a) $(3) + y + (1) = 6$
$y = 2$
(3, 2, 1)

9. (a) $x^2 + y^2 = 10$
(b) $x + y = 4$
$y = 4 - x$
(b′) $y^2 = (4 - x)^2 = 16 - 8x + x^2$
Substitute (b′) into (a) and get:
(a′) $x^2 + (16 - 8x + x^2) = 10$
$2x^2 - 8x + 6 = 0$
Solve this equation by using the quadratic formula.

$x = \dfrac{8 \pm \sqrt{64 - 4(12)}}{4} = 2 \pm 1 = 3, 1$

Substitute these values of x into (b) and solve for y.
(b) $y = 4 - (3) = 1$
(b) $y = 4 - (1) = 3$
(3, 1) and **(1, 3)**

10. $-4 \le x - 2 < 2;\ D = \{\text{Reals}\}$
$-2 \le x < 4$

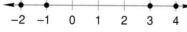

11. $x - 1 \not< 2$ or $x + 2 \not\ge 2;\ D = \{\text{Integers}\}$
$x - 1 \ge 2$ or $x + 2 < 2$
$x \ge 3$ or $\qquad x < 0$

12. $\dfrac{(y^{a+2})^2\, x^{2b/3}}{x^b\, y^{-a}} = y^{2a+4+a} x^{2b/3-b} = \mathbf{y^{3a+4} x^{-b/3}}$

13. $\dfrac{x}{y + \dfrac{y}{\dfrac{1}{x} + y}} = \dfrac{x}{y + \dfrac{xy}{1 + xy}}$

$= \dfrac{x(1 + xy)}{y + xy^2 + xy}$

14. $\dfrac{m}{a + \dfrac{a}{1 + \dfrac{a}{m}}} = \dfrac{m}{a + \dfrac{am}{m + a}}$

$= \dfrac{m(m + a)}{2am + a^2}$

15. $\dfrac{i - 5}{7 - i} \cdot \dfrac{7 + i}{7 + i} = \dfrac{7i - 1 - 35 - 5i}{49 + 1}$

$= \dfrac{-36 + 2i}{50} = -\dfrac{18}{25} + \dfrac{1}{25}i$

16. $\dfrac{3i + 2}{2i - 3} \cdot \dfrac{2i + 3}{2i + 3} = \dfrac{-6 + 9i + 4i + 6}{-4 - 9}$

$= \dfrac{13i}{-13} = -i$

17. $\sqrt{z} - 3 = \sqrt{z - 27}$
$z - 6\sqrt{z} + 9 = z - 27$
$36 = 6\sqrt{z}$
$6 = \sqrt{z}$
$\mathbf{36 = z}$

Check: $\sqrt{36} - 3 = \sqrt{36 - 27}$
$6 - 3 = \sqrt{9}$
$3 = 3 \quad$ check

18.

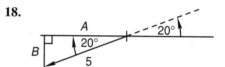

$A = 5 \cos 20° \approx 4.70$
$B = 5 \sin 20° \approx 1.71$

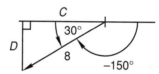

$C = 8 \cos 30° \approx 6.93$
$D = 8 \sin 30° = 4.00$

$-4.70R - 1.71U$
$\underline{-6.93R - 4.00U}$
$\mathbf{-11.63R - 5.71U}$

19. $\tan \theta = -\dfrac{20}{5}$
$\theta \approx -75.96°$
Since θ is a second-quadrant angle:
$\theta = -75.96 + 180 = 104.04°$

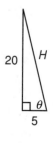

$H = \sqrt{(20)^2 + (-5)^2}$
$H = 5\sqrt{17}$

$\mathbf{5\sqrt{17}\, /\underline{104.04°}}$

20. $\dfrac{3 - 5\sqrt{2}}{2 - \sqrt{2}} \cdot \dfrac{2 + \sqrt{2}}{2 + \sqrt{2}}$

$= \dfrac{6 + 3\sqrt{2} - 10\sqrt{2} - 10}{4 - 2} = \dfrac{-4 - 7\sqrt{2}}{2}$

21. $\dfrac{8 - \sqrt{2}}{4 - \sqrt{8}} = \dfrac{8 - \sqrt{2}}{4 - 2\sqrt{2}} \cdot \dfrac{4 + 2\sqrt{2}}{4 + 2\sqrt{2}}$

$= \dfrac{32 + 16\sqrt{2} - 4\sqrt{2} - 4}{16 - 8} = \dfrac{28 + 12\sqrt{2}}{8}$

$= \dfrac{7 + 3\sqrt{2}}{2}$

22. $\sqrt[5]{x^4 y^3}\,\sqrt[3]{xy^2} = x^{4/5}y^{3/5}x^{1/3}y^{2/3} = x^{17/15}y^{19/15}$

23.
$$\frac{a}{by} = x\left(\frac{1}{R_1} + \frac{1}{R_2}\right)$$
$$\frac{a}{by} = \frac{x}{R_1} + \frac{x}{R_2}$$
$$aR_1R_2 = bxyR_2 + bxyR_1$$
$$bxyR_2 = aR_1R_2 - bxyR_1$$
$$bxyR_2 = R_1(aR_2 - bxy)$$
$$\frac{bxyR_2}{aR_2 - bxy} = R_1$$

24.
$$\frac{a}{by} = x\left(\frac{1}{R_1} + \frac{1}{R_2}\right)$$
$$\frac{a}{by} = \frac{x}{R_1} + \frac{x}{R_2}$$
$$aR_1R_2 = byxR_2 + byxR_1$$
$$aR_1R_2 = y(bxR_2 + bxR_1)$$
$$\frac{aR_1R_2}{bxR_2 + bxR_1} = y$$

25. $i^3 - 2i^4 + \sqrt{-9} - \sqrt{-2}\sqrt{-2}$
$= -i - 2 + 3i + 2 = 2i$

26. $\sqrt{\dfrac{5}{12}} - 3\sqrt{\dfrac{12}{5}} + 2\sqrt{60}$

$= \dfrac{\sqrt{5}}{2\sqrt{3}}\dfrac{\sqrt{3}}{\sqrt{3}} - \dfrac{6\sqrt{3}}{\sqrt{5}}\dfrac{\sqrt{5}}{\sqrt{5}} + 4\sqrt{15}$

$= \dfrac{\sqrt{15}}{6} - \dfrac{6\sqrt{15}}{5} + 4\sqrt{15}$

$= \dfrac{5\sqrt{15}}{30} - \dfrac{36\sqrt{15}}{30} + \dfrac{120\sqrt{15}}{30} = \dfrac{89\sqrt{15}}{30}$

27. $-3x^2 - x = 5$
$-3x^2 - x - 5 = 0$

$x = \dfrac{1 \pm \sqrt{1 - 4(15)}}{-6} = -\dfrac{1}{6} \pm \dfrac{\sqrt{59}}{6}i$

28. $\dfrac{15\text{ cm}}{\text{sec}} \times \dfrac{1\text{ in.}}{2.54\text{ cm}} \times \dfrac{1\text{ ft}}{12\text{ in.}} \times \dfrac{1\text{ yd}}{3\text{ ft}} \times \dfrac{60\text{ sec}}{1\text{ min}}$

$= \dfrac{15(60)}{(2.54)(12)(3)}\dfrac{\text{yd}}{\text{min}}$

29. $\dfrac{4}{x^2 - 9} - \dfrac{3x}{-3 + x}$

$= \dfrac{4}{(x + 3)(x - 3)} - \dfrac{3x(x + 3)}{(x - 3)(x + 3)}$

$= \dfrac{4 - 3x^2 - 9x}{(x + 3)(x - 3)} = \dfrac{-3x^2 - 9x + 4}{x^2 - 9}$

30. $x = \dfrac{100 + 20}{2} = 60$

$3 \cdot 7 = 4 \cdot y$
$21 = 4y$
$\dfrac{21}{4} = y$

PROBLEM SET 93

1. Downsteam: $(B + W)T_D = D_D$ (a)
Upstream: $(B - W)T_U = D_U$ (b)
Since $T_D = T_U$ we use T_D in both equations.

(a′) $BT_D + 5T_D = 45$
(b′) $\dfrac{BT_D - 5T_D = 15}{2BT_D \qquad = 60}$
(c) $\qquad BT_D = 30$

(a′) $(30) + 5T_D = 45$
$\qquad 5T_D = 15$
$\qquad T = 3$

(c) $B(3) = 30$
$\qquad B = \mathbf{10\ mph}$

2. Downstream: $(B + W)T_D = D_D$ (a)
Upstream: $(B - W)T_U = D_U$ (b)
(a′) $4B + 4W = 48$
(b′) $8B - 8W = 64$

2(a′) $8B + 8W = 96$
(b′) $\dfrac{8B - 8W = 64}{16B \qquad = 160}$
$\qquad B = \mathbf{10\ mph}$

(a′) $4(10) + 4W = 48$
$\qquad 4W = 8$
$\qquad W = \mathbf{2\ mph}$

3. Downstream: $(B + W)T_D = D_D$ (a)

Upstream: $(B - W)T_U = D_U$ (b)

Since $T_U = 2T_D$ we substitute and get:

(a') $40T_D + WT_D = 210$

(b') $80T_D - 2WT_D = 380$

2(a') $80T_D + 2WT_D = 420$

(b') $\underline{80T_D - 2WT_D = 380}$

$\,160T_D = 800$

$T_D = 5$

(a') $40(5) + W(5) = 210$

$5W = 10$

$W = \textbf{2 mph}$

4. $R_C T_C = 40$;

$T_R = T_C - 4$;

$R_R T_R = 48$; $R_R = 2R_C$

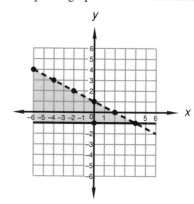

$2R_C(T_C - 4) = 48$

$2R_C T_C - 8R_C = 48$

$2(40) - 8R_C = 48$

$32 = 8R_C$

$4 = R_C$

$R_C = \textbf{4 mph}$; $R_R = \textbf{8 mph}$;

$T_C = \textbf{10 hr}$; $T_R = \textbf{6 hr}$

5. $\dfrac{P_1}{T_1} = \dfrac{P_2}{T_2}$

$T_2 = \dfrac{P_2 T_1}{P_1}$

$T_2 = \dfrac{(1)(4000)}{5} = \textbf{800 K}$

6. $x^2 = -5x + 1$

$x^2 + 5x - 1 = 0$

$b^2 - 4ac = (5)^2 - 4(1)(-1) = 25 + 4 = 29$

$b^2 - 4ac > 0$

Two real number solutions.

7. $-3x = -2x^2 - 5$

$2x^2 - 3x + 5 = 0$

$b^2 - 4ac = (-3)^2 - 4(2)(5) = 9 - 40 = -31$

$b^2 - 4ac < 0$

Two complex number solutions which are conjugates.

8. (a) $x + 2y < 2$

$2y < -x + 2$

$y < -\dfrac{1}{2}x + 1$

(b) $y \geq -1$

The first step is to graph each of these lines.

The region we wish to find is on or above the solid line and below the dashed line. This region is shaded in the previous figure.

9. $-3 < x - 4 < 2$; $D = \{\text{Integers}\}$

$\phantom{-3 <}1 < x < 6$

10. (a) $x + y - 2z = -3$

(b) $2x + y + z = 7$

(c) $3x - y - z = 13$

(−1)(a) $-x - y + 2z = 3$

(b) $\underline{2x + y + z = 7}$

(d) $x + 3z = 10$

(b) $2x + y + z = 7$

(c) $\underline{3x - y - z = 13}$

$5x = 20$

$x = 4$

(d) $(4) + 3z = 10$

$3z = 6$

$z = 2$

(a) $(4) + y - 2(2) = -3$

$y = -3$

$\textbf{(4, −3, 2)}$

11. (a) $x^2 + y^2 = 1$

(b) $x - 2y = 2$

$\quad x = 2y + 2$

(b') $x^2 = (2y + 2)^2 = 4y^2 + 8y + 4$

Substitute (b') into (a) and get:

(a') $(4y^2 + 8y + 4) + y^2 = 1$

$\quad\quad 5y^2 + 8y + 3 = 0$

Solve this equation by using the quadratic formula.

$$y = \frac{-8 \pm \sqrt{8^2 - 4(3)(5)}}{2(5)} = -\frac{4}{5} \pm \frac{1}{5}$$

$$y = -\frac{3}{5}, -1$$

Substitute these values of y into (b) and solve for x.

(b) $x = 2\left(-\dfrac{3}{5}\right) + \dfrac{10}{5} = \dfrac{4}{5}$

(b) $x = 2(-1) + 2 = 0$

$$\mathbf{(0, -1)} \text{ and } \left(\dfrac{\mathbf{4}}{\mathbf{5}}, -\dfrac{\mathbf{3}}{\mathbf{5}}\right)$$

12. $(y^{a+2})^2 y^{a/3} y^2 = y^{2a+4+a/3+2} = \mathbf{y^{7a/3+6}}$

13.
$$\dfrac{x}{1 + \dfrac{a}{b + \dfrac{1}{c}}} = \dfrac{x}{1 + \dfrac{ac}{bc + 1}}$$

$$= \dfrac{x(bc + 1)}{bc + 1 + ac}$$

14.
$$\dfrac{m}{2x + \dfrac{3}{3 + \dfrac{1}{m}}} = \dfrac{m}{2x + \dfrac{3m}{3m + 1}}$$

$$= \dfrac{m(3m + 1)}{6mx + 2x + 3m}$$

15. $\dfrac{3i - 5}{i - 7} \cdot \dfrac{i + 7}{i + 7} = \dfrac{-3 - 5i + 21i - 35}{-1 - 49}$

$$= \dfrac{-38 + 16i}{-50} = \dfrac{19}{25} - \dfrac{8}{25}i$$

16. $\dfrac{2i + 4}{3i + 2} \cdot \dfrac{3i - 2}{3i - 2} = \dfrac{-6 + 12i - 4i - 8}{-9 - 4}$

$$= \dfrac{-14 + 8i}{-13} = \dfrac{14}{13} - \dfrac{8}{13}i$$

17.
$$\sqrt{p} = 5 + \sqrt{p - 35}$$
$$\sqrt{p} - 5 = \sqrt{p - 35}$$
$$p - 10\sqrt{p} + 25 = p - 35$$
$$-10\sqrt{p} = -60$$
$$\sqrt{p} = 6$$
$$p = \mathbf{36}$$

Check: $\sqrt{36} = 5 + \sqrt{36 - 35}$

$\quad\quad\quad 6 = 5 + 1$

$\quad\quad\quad 6 = 6 \quad$ check

18.

$A = 12 \cos 32° \approx 10.18$

$B = 12 \sin 32° \approx 6.36$

$$\begin{array}{r} -10.18R - 6.36U \\ 15.00R + 0.00U \\ \hline \mathbf{4.82R - 6.36U} \end{array}$$

$\tan \theta = \dfrac{-6.36}{4.82}$

$\quad \theta = -52.84°$

$H = \sqrt{(-6.36)^2 + (4.82)^2} \approx 7.98$

$\mathbf{7.98\underline{/-52.84°}}$

19. $\dfrac{3 + 4\sqrt{2}}{3\sqrt{2} - 4} \cdot \dfrac{3\sqrt{2} + 4}{3\sqrt{2} + 4}$

$$= \dfrac{9\sqrt{2} + 24 + 12 + 16\sqrt{2}}{18 - 16}$$

$$= \dfrac{\mathbf{36 + 25\sqrt{2}}}{\mathbf{2}}$$

20. $\dfrac{5 + \sqrt{5}}{5 + 2\sqrt{5}} \cdot \dfrac{5 - 2\sqrt{5}}{5 - 2\sqrt{5}}$

$$= \dfrac{25 + 5\sqrt{5} - 10\sqrt{5} - 10}{25 - 20} = \dfrac{15 - 5\sqrt{5}}{5}$$

$$= \mathbf{3 - \sqrt{5}}$$

21. $\sqrt[3]{4\sqrt[5]{2}} = \left(2^2(2^{1/5})\right)^{1/3} = 2^{2/3}2^{1/15} = \mathbf{2^{11/15}}$

22. $\sqrt[4]{xy^6}\sqrt{x^3 y^5} = x^{1/4}y^{3/2}x^{3/2}y^{5/2} = \mathbf{x^{7/4}y^4}$

23. $\dfrac{a + c}{m} = m\left(\dfrac{1}{x} + \dfrac{b}{d}\right)$

$$\dfrac{a + c}{m} = \dfrac{m}{x} + \dfrac{bm}{d}$$

$$adx + cdx = dm^2 + bm^2x$$

$$adx = dm^2 + bm^2x - cdx$$

$$a = \dfrac{dm^2 + bm^2x - cdx}{dx}$$

24. $\dfrac{a + c}{m} = m\left(\dfrac{1}{x} + \dfrac{b}{d}\right)$

$$\dfrac{a + c}{m} = \dfrac{m}{x} + \dfrac{bm}{d}$$

$$adx + cdx = dm^2 + bm^2x$$

$$adx + cdx - bm^2x = dm^2$$

$$x = \dfrac{dm^2}{ad + cd - bm^2}$$

25. $i^3 - 2i^2 - \sqrt{-9} + \sqrt{-4}\sqrt{-4} + 2$
$= -i + 2 - 3i - 4 + 2 = \mathbf{-4i}$

26. $3\sqrt{\dfrac{2}{9}} + \sqrt{\dfrac{9}{2}} - 3\sqrt{18} = \sqrt{2} + \dfrac{3\sqrt{2}}{2} - 9\sqrt{2}$

$= \dfrac{2\sqrt{2}}{2} + \dfrac{3\sqrt{2}}{2} - \dfrac{18\sqrt{2}}{2} = \mathbf{-\dfrac{13\sqrt{2}}{2}}$

27.
$$3x^2 = -x - 2$$
$$(3x^2 + x + \quad) = -2$$
$$x^2 + \frac{1}{3}x + \frac{1}{36} = -\frac{24}{36} + \frac{1}{36}$$
$$\left(x + \frac{1}{6}\right)^2 = -\frac{23}{36}$$
$$x + \frac{1}{6} = \pm\frac{\sqrt{23}}{6}i$$
$$x = \mathbf{-\frac{1}{6} \pm \frac{\sqrt{23}}{6}i}$$

28. $\dfrac{10\ m}{sec} \times \dfrac{60\ sec}{1\ min} \times \dfrac{100\ cm}{1\ m} \times \dfrac{1\ in.}{2.54\ cm}$

$\times \dfrac{1\ ft}{12\ in.} = \dfrac{\mathbf{10(60)(100)}}{\mathbf{(2.54)(12)}}\ \dfrac{\mathbf{ft}}{\mathbf{min}}$

29.
$$
\require{enclose}
\begin{array}{r}
x^2 - 6x + 18 - \frac{56}{x+3} \\
x + 3 \enclose{longdiv}{x^3 - 3x^2 + 0x - 2} \\
\underline{x^3 + 3x^2} \\
-6x^2 + 0x \\
\underline{-6x^2 - 18x} \\
18x - 2 \\
\underline{18x + 54} \\
-56
\end{array}
$$

30. $\dfrac{90 - x}{2} = 25$
$$90 - x = 50$$
$$x = \mathbf{40}$$

$$5(5 + y) = 4(11)$$
$$25 + 5y = 44$$
$$5y = 19$$
$$y = \mathbf{\frac{19}{5}}$$

PROBLEM SET 94

1. $PV = nRT$

$n = \dfrac{PV}{RT}$

$n = \dfrac{(2)(10)}{(0.0821)(473)} = \mathbf{0.515\ mole}$

2. $0.2(40) + 0.6A = 0.44(40 + A)$
$$8 + 0.6A = 17.6 + 0.44A$$
$$0.16A = 9.6$$
$$A = \mathbf{60\ gallons}$$

3. Downstream: $(B + W)T_D = D_D$ (a)
Upstream: $(B - W)T_U = D_U$ (b)

Since $T_D = T_U$ we use T_D in both equations.

(a') $10T_D + WT_D = 48$
(b') $\dfrac{10T_D - WT_D = 32}{20T_D \qquad = 80}$
$$T_D = 4$$

(a') $10(4) + W(4) = 48$
$$4W = 8$$
$$W = \mathbf{2\ mph}$$

4. $P_1V_1 = P_2V_2$
$$(700)(1400) = (2800)V_2$$
$$V_2 = \dfrac{(700)(1400)}{2800} = \mathbf{350\ ml}$$

5. $P = \dfrac{k}{R^2}$

$4 = \dfrac{k}{10^2}$

$400 = k$

$P = \dfrac{400}{5^2} = \mathbf{16}$

6. **(a)**, **(b)**, and **(d)** are functions.
(c) is not a function because 6 has two images.

7. $p(x) = x^2 - 2x$
$$p(-2) = (-2)^2 - 2(-2)$$
$$p = 4 + 4$$
$$p = \mathbf{8}$$

8.
$$-x^2 = 4x + 4$$
$$0 = x^2 + 4x + 4$$
$$b^2 - 4ac = 4^2 - 4(1)(4) = 16 - 16 = 0$$

One real number solution.

9. (a) $2x + 3y > -6$

$$3y > -2x - 6$$

$$y > -\frac{2}{3}x - 2$$

(b) $x - 3y \geq -6$

$$-3y \geq -x - 6$$

$$y \leq \frac{1}{3}x + 2$$

The first step is to graph each of these lines.

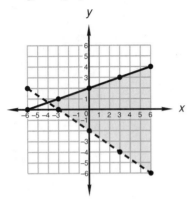

The region we wish to find is on or below the solid line and above the dashed line. This region is shaded in the previous figure.

10. $-x + 3 \not< -2$ or $-x + 3 < -5$; $D = \{\text{Integers}\}$

$$-x + 3 \geq -2 \quad \text{or} \quad -x + 3 < -5$$

$$-x \geq -5 \quad \text{or} \quad -x < -8$$

$$x \leq 5 \quad \text{or} \quad x > 8$$

```
  ←●──●──●──●────●──●→
    4  5  6  7  8  9  10
```

11. (a) $x + 2y - z = 2$

(b) $2x - y + z = 2$

(c) $3x - y + 2z = 4$

(a) $x + 2y - z = 2$

(b) $\underline{2x - y + z = 2}$

(d) $3x + y \phantom{{}+ z} = 4$

2(a) $2x + 4y - 2z = 4$

(c) $\underline{3x - y + 2z = 4}$

(e) $5x + 3y \phantom{{}+ 2z} = 8$

3(d) $9x + 3y = 12$

(−1)(e) $\underline{-5x - 3y = -8}$

$$4x \phantom{{}+ 3y} = 4$$

$$x = 1$$

(d) $3(1) + y = 4$

$$y = 1$$

(b) $2(1) - (1) + z = 2$

$$1 + z = 2$$

$$z = 1$$

(1, 1, 1)

12. (a) $x^2 + y^2 = 3$

(b) $x - y = 2$

$$x = y + 2$$

(b') $x^2 = (y + 2)^2 = y^2 + 4y + 4$

Substitute (b') into (a) and get:

(a') $(y^2 + 4y + 4) + y^2 = 3$

$$2y^2 + 4y + 1 = 0$$

Solve this equation by using the quadratic formula.

$$y = \frac{-4 \pm \sqrt{16 - 8}}{4} = -1 \pm \frac{\sqrt{2}}{2}$$

$$y = -1 + \frac{\sqrt{2}}{2}, -1 - \frac{\sqrt{2}}{2}$$

Substitute these values of y into (b) and solve for x.

(b) $x = \left(-1 + \dfrac{\sqrt{2}}{2}\right) + 2 = 1 + \dfrac{\sqrt{2}}{2}$

(b) $x = \left(-1 - \dfrac{\sqrt{2}}{2}\right) + 2 = 1 - \dfrac{\sqrt{2}}{2}$

$$\left(1 + \frac{\sqrt{2}}{2}, -1 + \frac{\sqrt{2}}{2}\right) \text{ and}$$

$$\left(1 - \frac{\sqrt{2}}{2}, -1 - \frac{\sqrt{2}}{2}\right)$$

13. $\dfrac{x^{a/3}y^2}{x^{3a/2}(y^2)^a} = x^{a/3 - 3a/2}y^{2 - 2a} = x^{-7a/6}y^{2 - 2a}$

14. $\dfrac{p}{m + \dfrac{m}{p - \dfrac{1}{mp}}} = \dfrac{p}{m + \dfrac{m^2p}{mp^2 - 1}}$

$$= \frac{p(mp^2 - 1)}{m^2p^2 - m + m^2p}$$

15. $\dfrac{x}{a + \dfrac{b}{ab - \dfrac{1}{b}}} = \dfrac{x}{a + \dfrac{b^2}{ab^2 - 1}}$

$$= \frac{x(ab^2 - 1)}{a^2b^2 - a + b^2}$$

16. $\dfrac{2i + 7}{-3i} \cdot \dfrac{3i}{3i} = \dfrac{-6 + 21i}{9} = -\dfrac{2}{3} + \dfrac{7}{3}i$

17. $\dfrac{-5i - 2}{-2i + 5} \cdot \dfrac{-2i - 5}{-2i - 5}$

$$= \frac{-10 + 4i + 25i + 10}{-4 - 25} = \frac{29i}{-29} = -i$$

18. $\sqrt{k} + \sqrt{k - 21} = 7$

$$\sqrt{k - 21} = 7 - \sqrt{k}$$

$$k - 21 = 49 - 14\sqrt{k} + k$$

$$14\sqrt{k} = 70$$

$$\sqrt{k} = 5$$

$$k = 25$$

Check: $\sqrt{25} + \sqrt{25 - 21} = 7$
$5 + 2 = 7$
$7 = 7$ check

19.

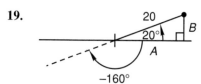

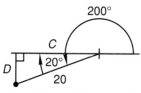

$A = 20 \cos 20° \approx 18.79$
$B = 20 \sin 20° \approx 6.84$

$C = 20 \cos 20° \approx 18.79$
$D = 20 \sin 20° \approx 6.84$

$\begin{array}{r} 18.79R + 6.84U \\ -18.79R - 6.84U \\ \hline 0R + \quad 0U \end{array}$

20. $\tan \theta = -\dfrac{8}{4}$

$\theta \approx -63.43°$

Since θ is a second-quadrant angle:

$\theta = -63.43 + 180 = 116.57°$

$H = \sqrt{8^2 + (-4)^2}$
$H = 4\sqrt{5}$

$\mathbf{4\sqrt{5}\underline{/116.57°}}$

21. $\dfrac{2 - 5\sqrt{2}}{3\sqrt{2} - 2} \cdot \dfrac{3\sqrt{2} + 2}{3\sqrt{2} + 2}$

$= \dfrac{6\sqrt{2} - 30 + 4 - 10\sqrt{2}}{18 - 4}$

$= \dfrac{-26 - 4\sqrt{2}}{14} = \dfrac{-13 - 2\sqrt{2}}{7}$

22. $4\sqrt{\dfrac{5}{12}} + 3\sqrt{\dfrac{12}{5}} - 2\sqrt{240}$

$= \dfrac{2\sqrt{5}}{\sqrt{3}} + \dfrac{6\sqrt{3}}{\sqrt{5}} - 8\sqrt{15}$

$= \dfrac{2\sqrt{15}}{3} + \dfrac{6\sqrt{15}}{5} - 8\sqrt{15}$

$= \dfrac{10\sqrt{15}}{15} + \dfrac{18\sqrt{15}}{15} - \dfrac{120\sqrt{15}}{15} = -\dfrac{\mathbf{92\sqrt{15}}}{\mathbf{15}}$

23.

$\dfrac{a + b}{x} = \left(\dfrac{1}{R_1} + \dfrac{c}{R_2} \right)$

$aR_1R_2 + bR_1R_2 = xR_2 + cxR_1$
$aR_1R_2 + bR_1R_2 - xR_2 = cxR_1$

$\dfrac{\mathbf{aR_1R_2 + bR_1R_2 - xR_2}}{\mathbf{xR_1}} = c$

24.

$\dfrac{a + b}{x} = \dfrac{1}{R_1} + \dfrac{c}{R_2}$

$aR_1R_2 + bR_1R_2 = xR_2 + cxR_1$
$aR_1R_2 + bR_1R_2 - cxR_1 = xR_2$

$R_1 = \dfrac{xR_2}{aR_2 + bR_2 - cx}$

25. $3i^2 + 2i^5 - 2i + \sqrt{-9} - \sqrt{-2}\sqrt{2}$
$= -3 + 2i - 2i + 3i - 2i = \mathbf{-3 + i}$

26. (a) Since this is a 30-60-90 triangle:
$\sqrt{3} \times \overrightarrow{SF} = 4$

$\overrightarrow{SF} = \dfrac{4\sqrt{3}}{3}$

$M = 1 \cdot \dfrac{4\sqrt{3}}{3} = \dfrac{\mathbf{4\sqrt{3}}}{\mathbf{3}}$

$N = 2 \cdot \dfrac{4\sqrt{3}}{8} = \dfrac{\mathbf{8\sqrt{3}}}{\mathbf{3}}$

(b) Since this is a 45-45-90 triangle:
$1 \times \overrightarrow{SF} = 4$
$\overrightarrow{SF} = 4$

$C = 1 \cdot 4 = \mathbf{4}$
$D = \sqrt{2} \cdot 4 = \mathbf{4\sqrt{2}}$

27. $-2x^2 = -7 + x$
$0 = 2x^2 + x - 7$

$x = \dfrac{-1 \pm \sqrt{1 + 56}}{4} = -\dfrac{\mathbf{1}}{\mathbf{4}} \pm \dfrac{\mathbf{\sqrt{57}}}{\mathbf{4}}$

28. $\dfrac{20 \text{ liters}}{\text{sec}} \times \dfrac{1000 \text{ cm}^3}{1 \text{ liter}} \times \dfrac{1 \text{ in.}}{2.54 \text{ cm}} \times \dfrac{1 \text{ in.}}{2.54 \text{ cm}}$

$\times \dfrac{1 \text{ in.}}{2.54 \text{ cm}} \times \dfrac{60 \text{ sec}}{1 \text{ min}}$

$= \dfrac{\mathbf{20(1000)(60)}}{\mathbf{(2.54)(2.54)(2.54)}} \dfrac{\text{in.}^3}{\text{min}}$

29. There is never an exact solution to these problems. One possible solution is given here.
$$N = mR + b$$

Use the graph to find the slope.
$$m = \frac{70 - 0}{120 - 180} \approx -1.2$$

$$N = (-1.2)R + b$$
Use the point (120, 70) for R and N.
$$70 = (-1.2)(120) + b$$
$$214 = b$$
$$\mathbf{N = -1.2R + 214}$$

30. $\dfrac{3}{a^4 x^3} + \dfrac{4}{x^2 - 9} - \dfrac{3x}{-3 + x}$

$$= \frac{3(x^2 - 9) + 4a^4 x^3 - 3a^4 x^4 (x + 3)}{a^4 x^3 (x^2 - 9)}$$

PROBLEM SET 95

1. $R_W T_W = D_W;\ D_W = D_R;$
$T_W = 10 - T_R;$
$R_R T_R = D_R;$
$R_W = 6;\ R_R = 24$

$$6(10 - T_R) = 24T_R$$
$$60 - 6T_R = 24T_R$$
$$60 = 30T_R$$
$$2 = T_R$$
$$D_R = 2(24) = \mathbf{48\ km}$$

2. $0.6A + 0.2(200) = 0.52(A + 200)$
$$0.6A + 40 = 0.52A + 104$$
$$0.08A = 64$$
$$A = \frac{64}{0.08}$$
$$A = \mathbf{800\ ml}$$

3. $\dfrac{V_1}{T_1} = \dfrac{V_2}{T_2}$

$$V_2 = \frac{V_1 T_2}{T_1}$$

$$V_2 = \frac{(400)(873)}{473} = \mathbf{738.27\ ml}$$

4. $R_R T_R = 360;\ R_J T_J = 480;$
$T_J = 4T_R;\ R_J = R_R - 60$

$$(R_R - 60)(4T_R) = 480$$
$$4(360) - 240T_R = 480$$
$$-240T_R = -960$$
$$T_R = 4$$
$T_R = \mathbf{4\ hr};\ T_J = \mathbf{16\ hr};$
$R_R = \mathbf{90\ mph};\ R_J = \mathbf{30\ mph}$

5. Downstream: $(B + W)T_D = D_D$ \quad (a)
Upstream: $(B - W)T_U = D_U$ \quad (b)
(a′) $4B + 4W = 60$
(b′) $8B - 8W = 72$

$$
\begin{array}{rl}
2(a′) & 8B + 8W = 120 \\
(b′) & 8B - 8W = 72 \\
\hline
& 16B \qquad = 192 \\
& B = \mathbf{12\ mph}
\end{array}
$$

(a′) $4(12) + 4W = 60$
$$4W = 12$$
$$W = \mathbf{3\ mph}$$

6. (a) $4x - y = 3$
(b) $xy = 6$

$$x = \frac{6}{y}$$

Substitute (b) into (a) and get:

(a′) $\quad 4\left(\dfrac{6}{y}\right) - y = 3$

$$\frac{24}{y} - y = 3$$

$$24 - y^2 = 3y$$
$$y^2 + 3y - 24 = 0$$

Solve this equation by using the quadratic formula.

$$y = \frac{-3 \pm \sqrt{9 - 4(-24)}}{2} = -\frac{3}{2} \pm \frac{\sqrt{105}}{2}$$

Substitute these values of y into (a) and solve for x.

(a) $4x = \left(-\dfrac{3}{2} + \dfrac{\sqrt{105}}{2}\right) + 3$

$$x = \frac{3}{8} + \frac{\sqrt{105}}{8}$$

(a) $4x = \left(-\dfrac{3}{2} - \dfrac{\sqrt{105}}{2}\right) + 3$

$$x = \frac{3}{8} - \frac{\sqrt{105}}{8}$$

$\left(\dfrac{3}{8} + \dfrac{\sqrt{105}}{8}, -\dfrac{3}{2} + \dfrac{\sqrt{105}}{2}\right)$ and
$\left(\dfrac{3}{8} - \dfrac{\sqrt{105}}{8}, -\dfrac{3}{2} - \dfrac{\sqrt{105}}{2}\right)$

7. (a) $x^2 + y^2 = 16$
   (b) $2x^2 - y^2 = -1$

   (a) $x^2 + y^2 = 16$
   (b) $\underline{2x^2 - y^2 = -1}$
   $\overline{3x^2 = 15}$
   $x^2 = 5$
   $x = \sqrt{5}, -\sqrt{5}$

   (a) $(\sqrt{5})^2 + y^2 = 16$
   $\phantom{(a)(\sqrt{5})^2}y^2 = 11$
   $\phantom{(a)(\sqrt{5})^2}y = \pm\sqrt{11}$

   (a) $(-\sqrt{5})^2 + y^2 = 16$
   $\phantom{(a)(-\sqrt{5})^2}y^2 = 11$
   $\phantom{(a)(-\sqrt{5})^2}y = \pm\sqrt{11}$

   $(\sqrt{5}, \pm\sqrt{11}), (-\sqrt{5}, \pm\sqrt{11})$

8. (a) $x + 2y - z = 4$
   (b) $2x - y + z = -3$
   (c) $x - y + z = -4$

   (a) $x + 2y - z = 4$
   (b) $\underline{2x - y + z = -3}$
   (d) $\overline{3x + y = 1}$

   (a) $x + 2y - z = 4$
   (c) $\underline{x - y + z = -4}$
   (e) $\overline{2x + y = 0}$
   $y = -2x$

   (d) $3x + (-2x) = 1$
   $x = 1$

   (e) $y = -2(1) = -2$

   (b) $2 + 2 + z = -3$
   $z = -7$

   $(1, -2, -7)$

9. (a) and (d) are functions. (b) is not a function because 5 has two images. (c) is not a function because m has no image.

10. $3x^2 - x + 5 = 0$
   $b^2 - 4ac = 1 - 60 = -59$
   Two complex number solutions which are conjugates.

11. (a) $x > -2$
   (b) $2x + 3y > -3$
   $3y > -2x - 3$
   $y > -\frac{2}{3}x - 1$

   The first step is to graph each of these lines.

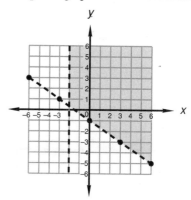

   The region we wish to find is to the right of the vertical dashed line and above the slanted dashed line. This region is shaded in the previous figure.

12. $-2 < -x - 2 < 0$; $D = \{\text{Integers}\}$
   $0 < -x < 2$
   $0 > x > -2$

13. $\dfrac{(a^{x+2})^{1/2}\, x^{2a}}{a^x\, x^a} = a^{x/2+1-x}x^{2a-a} = \boldsymbol{a^{-x/2+1}x^a}$

14. $\dfrac{x}{4 + \dfrac{4}{1 + \dfrac{x}{4}}} = \dfrac{x}{4 + \dfrac{16}{4 + x}}$

   $= \dfrac{x(4 + x)}{16 + 4x + 16} = \dfrac{\boldsymbol{x(4 + x)}}{\boldsymbol{32 + 4x}}$

15. $\dfrac{5}{2 + \dfrac{1}{2 + \dfrac{2}{x}}} = \dfrac{5}{2 + \dfrac{x}{2x + 2}}$

   $= \dfrac{5(2x + 2)}{4x + 4 + x} = \dfrac{\boldsymbol{5(2x + 2)}}{\boldsymbol{5x + 4}}$

16. $\dfrac{3i - 5}{3 - 5i} \cdot \dfrac{3 + 5i}{3 + 5i}$

   $= \dfrac{9i - 15 - 15 - 25i}{9 + 25} = \dfrac{-30 - 16i}{34}$

   $= -\dfrac{\boldsymbol{15}}{\boldsymbol{17}} - \dfrac{\boldsymbol{8}}{\boldsymbol{17}}\boldsymbol{i}$

17. $\dfrac{4 + i}{-i} \cdot \dfrac{i}{i} = \dfrac{4i - 1}{1} = \boldsymbol{-1 + 4i}$

18. $\sqrt{x^2 + 2x + 34} - x = 4$

$\qquad \sqrt{x^2 + 2x + 34} = x + 4$

$\qquad\qquad x^2 + 2x + 34 = x^2 + 8x + 16$

$\qquad\qquad\qquad\qquad 18 = 6x$

$\qquad\qquad\qquad\qquad \mathbf{3} = x$

Check: $\sqrt{3^2 + 2(3) + 34} - 3 = 4$

$\qquad\qquad\qquad \sqrt{49} - 3 = 4$

$\qquad\qquad\qquad\qquad 7 - 3 = 4$

$\qquad\qquad\qquad\qquad 4 = 4$ check

19.

$C = 12 \cos 35° \approx 9.83$

$D = 12 \sin 35° \approx 6.88$

$14.00R + 0.00U$

$\underline{9.83R - 6.88U}$

$\mathbf{23.83R - 6.88U}$

$\tan \theta = -\dfrac{6.88}{23.83}$

$\theta \approx -16.10°$

$H = \sqrt{(-6.88)^2 + (23.83)^2} \approx 24.80$

$\mathbf{24.80/{-16.10}°}$

20. $\dfrac{2\sqrt{2} - 5}{2\sqrt{2} + 3} \cdot \dfrac{2\sqrt{2} - 3}{2\sqrt{2} - 3}$

$= \dfrac{8 - 10\sqrt{2} - 6\sqrt{2} + 15}{8 - 9} = \dfrac{23 - 16\sqrt{2}}{-1}$

$= \mathbf{-23 + 16\sqrt{2}}$

21. $\dfrac{3 - \sqrt{2}}{\sqrt{2} + 4} \cdot \dfrac{\sqrt{2} - 4}{\sqrt{2} - 4}$

$= \dfrac{3\sqrt{2} - 2 - 12 + 4\sqrt{2}}{2 - 16}$

$= \dfrac{-14 + 7\sqrt{2}}{-14} = \dfrac{\mathbf{2} - \sqrt{\mathbf{2}}}{\mathbf{2}}$

22. $\dfrac{m(a + b)}{x} = \left(\dfrac{1}{p} + \dfrac{r}{q}\right)$

$mapq + mbpq = qx + prx$

$mapq + mbpq - qx = prx$

$\dfrac{\mathbf{ampq} + \mathbf{bmpq} - \mathbf{qx}}{\mathbf{px}} = r$

23. $\dfrac{m(a + b)}{x} = \left(\dfrac{1}{p} + \dfrac{r}{q}\right)$

$mapq + mbpq = qx + prx$

$mapq + mbpq - prx = qx$

$p = \dfrac{qx}{amq + bmq - rx}$

24. $4 - 3i^2 + \sqrt{-9} + \sqrt{-3}\sqrt{-3}$

$= 4 + 3 + 3i - 3 = \mathbf{4 + 3i}$

25. $2\sqrt{\dfrac{7}{3}} - 3\sqrt{\dfrac{3}{7}} - 2\sqrt{84}$

$= \dfrac{2\sqrt{21}}{3} - \dfrac{3\sqrt{21}}{7} - 4\sqrt{21}$

$= \dfrac{14\sqrt{21}}{21} - \dfrac{9\sqrt{21}}{21} - \dfrac{84\sqrt{21}}{21} = -\dfrac{\mathbf{79}\sqrt{\mathbf{21}}}{\mathbf{21}}$

26. $-x = x^2 - 3x - 4$

$\qquad 4 = x^2 - 2x$

$\qquad 4 + 1 = (x^2 - 2x + 1)$

$\qquad\qquad 5 = (x - 1)^2$

$\qquad\quad \pm\sqrt{5} = x - 1$

$\qquad \mathbf{1 \pm \sqrt{5}} = x$

27. $PV = nRT$

$V = \dfrac{nRT}{P}$

$V = \dfrac{(1.32)(0.0821)(600)}{5} = \mathbf{13 \ liters}$

28. Write the equation of the given line in slope-intercept form.

$3y - 2x = 5$

$\qquad 3y = 2x + 5$

$\qquad\quad y = \dfrac{2}{3}x + \dfrac{5}{3}$

Since the slopes of perpendicular lines are negative reciprocals of each other:

$m_\perp = -\dfrac{3}{2}$

$y = -\dfrac{3}{2}x + b$

$2 = -\dfrac{3}{2}(4) + b$

$8 = b$

$y = -\dfrac{3}{2}x + \mathbf{8}$

29.
$$4(4 + x) = 5(12)$$
$$16 + 4x = 60$$
$$4x = 44$$
$$x = \mathbf{11}$$

$$6(6) = 4(4 + y)$$
$$36 = 16 + 4y$$
$$20 = 4y$$
$$\mathbf{5} = y$$

30.
$$\frac{30 \text{ mi}}{\text{hr}} \times \frac{5280 \text{ ft}}{1 \text{ mi}} \times \frac{12 \text{ in.}}{1 \text{ ft}} \times \frac{2.54 \text{ cm}}{1 \text{ in.}}$$
$$\times \frac{1 \text{ hr}}{60 \text{ min}} \times \frac{1 \text{ min}}{60 \text{ sec}}$$
$$= \frac{30(5280)(12)(2.54)}{(60)(60)} \; \frac{\mathbf{cm}}{\mathbf{sec}}$$

PROBLEM SET 96

1. Variation method:
$$N_G = \frac{kN_T}{N_B}$$
$$65 = \frac{k(15)}{3}$$
$$k = 13$$

$$5 = \frac{13(100)}{N_B}$$
$$N_B = \mathbf{260}$$

Equal ratio method:
$$\frac{N_{G1}}{N_{G2}} = \frac{N_{T1}N_{B2}}{N_{T2}N_{B1}}$$
$$\frac{65}{5} = \frac{(15)N_{B2}}{100(3)}$$
$$N_{B2} = \mathbf{260}$$

2. Variation method:
$$S = kRM$$
$$100 = k(4)(5)$$
$$k = 5$$

$$20 = 5(2)R$$
$$R = \mathbf{2}$$

Equal ratio method:
$$\frac{S_1}{S_2} = \frac{R_1 M_1}{R_2 M_2}$$
$$\frac{100}{20} = \frac{4(5)}{R(2)}$$
$$R = \mathbf{2}$$

3. Downstream: $(B + W)T_D = D_D$ (a)
Upstream: $(B - W)T_U = D_U$ (b)

Since $T_U = 2T_D$ we substitute and get:
(a′) $BT_D + 6T_D = 80$
(b) $2BT_D - 12T_D = 40$

2(a′) $\quad 2BT_D + 12T_D = 160$
(b′) $\quad \underline{2BT_D - 12T_D = 40}$
(c) $\qquad 4BT_D \qquad\quad = 200$
$\qquad\qquad BT_D = 50$

(a′) $(50) + 6T_D = 80$
$\qquad\quad 6T_D = 30$
$\qquad\quad\; T_D = 5$

(c) $B(5) = 50$
$\qquad B = \mathbf{10 \text{ kph}}$

4. $R_G T_G = R_B T_B = 300;$
$T_G = T_B + 5; \; R_B = 2R_G$

$$2R_G(T_G - 5) = 300$$
$$2R_G T_G - 10R_G = 300$$
$$300 = 10R_G$$
$$30 = R_G$$

$R_G = \mathbf{30 \text{ mph}}; \; R_B = \mathbf{60 \text{ mph}};$
$T_G = \mathbf{10 \text{ hr}}; \; T_B = \mathbf{5 \text{ hr}}$

5. (a) $0.2P_N + 0.6D_N = 160$
(b) $P_N + D_N = 500$

Substitute $D_N = 500 - P_N$ into (a) and get:
(a′) $0.2P_N + 0.6(500 - P_N) = 160$
$\qquad\qquad -0.4P_N = -140$
$\qquad\qquad\quad P_N = 350$

(b) $(350) + D_N = 500$
$\qquad\qquad D_N = 150$

350 ml 20%, 150 ml 60%

6. (a) $x - 3y = 2$
    (b) $xy = 4$

    $$x = \frac{4}{y}$$

    Substitute (b) into (a) and get:

    (a') $\left(\dfrac{4}{y}\right) - 3y = 2$

    $$4 - 3y^2 = 2y$$
    $$0 = 3y^2 + 2y - 4$$

    Solve this equation by using the quadratic formula.

    $$y = \frac{-2 \pm \sqrt{4 + 48}}{6} = -\frac{1}{3} \pm \frac{\sqrt{13}}{3}$$

    Substitute these values of y into (a) and solve for x.

    (a) $x - 3\left(-\dfrac{1}{3} + \dfrac{\sqrt{13}}{3}\right) = 2$

    $$x + 1 - \sqrt{13} = 2$$
    $$x = 1 + \sqrt{13}$$

    (a) $x - 3\left(-\dfrac{1}{3} - \dfrac{\sqrt{13}}{3}\right) = 2$

    $$x + 1 + \sqrt{13} = 2$$
    $$x = 1 - \sqrt{13}$$

    $$\left(1 + \sqrt{13}, -\frac{1}{3} + \frac{\sqrt{13}}{3}\right) \text{ and }$$

    $$\left(1 - \sqrt{13}, -\frac{1}{3} - \frac{\sqrt{13}}{3}\right)$$

7. (a) $x^2 + y^2 = 4$
    (b) $\dfrac{4x^2 - y^2 = -4}{}$
    $$5x^2 \qquad = 0$$
    $$x^2 = 0$$
    $$x = 0$$

    (a) $(0)^2 + y^2 = 4$
    $$y^2 = 4$$
    $$y = \pm 2$$

    $(0, 2), (0, -2)$

8. (a) $x + 3y - z = 2$
    (b) $x + y + 2z = 6$
    (c) $2x + 2y - z = 2$

    2(a) $2x + 6y - 2z = 4$
    (b) $\dfrac{x + y + 2z = 6}{}$
    (d) $3x + 7y = 10$

    (b) $x + y + 2z = 6$
    2(c) $\dfrac{4x + 4y - 2z = 4}{}$
    (e) $5x + 5y = 10$

5(d) $15x + 35y = 50$
(−3)(e) $\dfrac{-15x - 15y = -30}{}$
$$20y = 20$$
$$y = 1$$

(e) $5x + 5(1) = 10$
$$5x = 5$$
$$x = 1$$

(b) $(1) + (1) + 2z = 6$
$$2z = 4$$
$$z = 2$$

$(1, 1, 2)$

9. **(a)** and **(b)** are functions. **(c)** is not a function because 4 has two images.

10. $g(x) = x^2 - 4;\ D = \{\text{Integers}\}$
    $g(-2) = (-2)^2 - 4$
    $g(-2) = 4 - 4$
    $g(-2) = \mathbf{0}$

11. (a) $x - y < -2$
    $$-y < -x - 2$$
    $$y > x + 2$$
    (b) $y \geq -2$
    The first step is to graph each of these lines.

    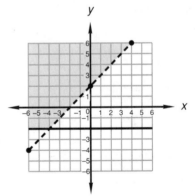

    The region we wish to find is on or above the solid line and above the dashed line. This region is shaded in the previous figure.

12. $x + 4 \not> 2$ or $x - 4 > -1;\ D = \{\text{Reals}\}$
    $x + 4 \leq 2$ or $x - 4 > -1$
    $\quad x \leq -2$ or $\quad x > 3$

    (number line from −3 to 4 with closed point at −2 shaded left, open point at 3 shaded right)

13. $\dfrac{(a^{x+4})^{1/2} b^{2x}}{a^{3/2} b^x} = a^{x/2 + 2 - 3/2} b^{2x - x} = a^{x/2 + 1/2} b^x$

14. $\dfrac{x}{a + \dfrac{b}{a^2 + \dfrac{1}{ab}}} = \dfrac{x}{a + \dfrac{ab^2}{a^3 b + 1}}$

$= \dfrac{x(a^3 b + 1)}{a^4 b + a + ab^2}$

15. $2\sqrt{8\sqrt[3]{2}} = 2\left(2^3(2^{1/3})\right)^{1/2} = 2(2^{10/3})^{1/2}$

$= 2(2^{5/3}) = \mathbf{2^{8/3}}$

16. $\dfrac{2i - 8}{4 - 6i} \cdot \dfrac{4 + 6i}{4 + 6i}$

$= \dfrac{8i - 32 - 12 - 48i}{16 + 36} = \dfrac{-44 - 40i}{52}$

$= -\dfrac{\mathbf{11}}{\mathbf{13}} - \dfrac{\mathbf{10}}{\mathbf{13}}i$

17. $\dfrac{i - i^2}{i} = \dfrac{i(1 - i)}{i} = \mathbf{1 - i}$

18. $\sqrt{s - 39} = 13 - \sqrt{s}$

$s - 39 = 169 - 26\sqrt{s} + s$

$26\sqrt{s} = 208$

$\sqrt{s} = 8$

$s = \mathbf{64}$

Check: $\sqrt{64 - 39} = 13 - \sqrt{64}$

$\sqrt{25} = 13 - 8$

$5 = 5 \quad$ check

19.

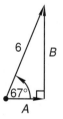

$A = 6 \cos 67° \approx 2.34$

$B = 6 \sin 67° \approx 5.52$

$\begin{array}{r} 2.34R + 5.52U \\ 0.00R - 2.00U \\ \hline \mathbf{2.34R + 3.52U} \end{array}$

$\tan \theta = \dfrac{3.52}{2.34}$

$\theta \approx 56.39°$

$H = \sqrt{(3.52)^2 + (2.34)^2}$

$H \approx 4.23$

$\mathbf{4.23/56.39°}$

20. $\dfrac{3\sqrt{5} - 1}{1 - \sqrt{5}} \cdot \dfrac{1 + \sqrt{5}}{1 + \sqrt{5}}$

$= \dfrac{3\sqrt{5} - 1 + 15 - \sqrt{5}}{1 - 5} = \dfrac{14 + 2\sqrt{5}}{-4}$

$= \dfrac{\mathbf{-7 - \sqrt{5}}}{\mathbf{2}}$

21. $\dfrac{2 - \sqrt{7}}{1 + \sqrt{7}} \cdot \dfrac{1 - \sqrt{7}}{1 - \sqrt{7}}$

$= \dfrac{2 - \sqrt{7} - 2\sqrt{7} + 7}{1 - 7} = \dfrac{9 - 3\sqrt{7}}{-6}$

$= \dfrac{\mathbf{-3 + \sqrt{7}}}{\mathbf{2}}$

22. $c = m\left(\dfrac{a + d}{p} - r\right)$

$c = \dfrac{am + dm}{p} - mr$

$cp = am + dm - mpr$

$cp + mpr = am + dm$

$p = \dfrac{\mathbf{am + dm}}{\mathbf{c + mr}}$

23. $c = m\left(\dfrac{a + d}{p} - r\right)$

$c = \dfrac{am + dm}{p} - mr$

$cp = am + dm - mpr$

$cp - dm + mpr = am$

$\dfrac{\mathbf{cp - dm + mpr}}{\mathbf{m}} = a$

24. $2i^4 - i^2 - \sqrt{-16} - \sqrt{-4}\sqrt{-4}$

$= 2 + 1 - 4i + 4 = \mathbf{7 - 4i}$

25. $PV = nRT$

$n = \dfrac{PV}{RT}$

$n = \dfrac{5(20)}{(0.0821)(673)} = \mathbf{1.81\ moles}$

26. $4x^2 + 6 = -x$

$4x^2 + x + 6 = 0$

$x = \dfrac{-1 \pm \sqrt{1 - 4(4)(6)}}{2(4)} = -\dfrac{1}{8} \pm \dfrac{\sqrt{95}}{8}i$

27. (a) $x + 2y = 6$

$2y = -x + 6$

$y = -\dfrac{1}{2}x + 3$

(b) $2x - 5y = -10$

$-5y = -2x - 10$

$y = \dfrac{2}{5}x + 2$

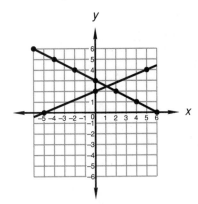

$\begin{array}{rl} 2(a) & 2x + 4y = 12 \\ (-1)(b) & -2x + 5y = 10 \\ \hline & 9y = 22 \\ & y = \dfrac{22}{9} \end{array}$

(a) $x + 2\left(\dfrac{22}{9}\right) = \dfrac{54}{9}$

$x = \dfrac{10}{9}$

$\left(\dfrac{\mathbf{10}}{\mathbf{9}}, \dfrac{\mathbf{22}}{\mathbf{9}}\right)$

28. $\dfrac{40\text{ in.}}{\text{hr}} \times \dfrac{2.54\text{ cm}}{1\text{ in.}} \times \dfrac{1\text{ m}}{100\text{ cm}} \times \dfrac{1\text{ km}}{1000\text{ m}}$

$\times \dfrac{1\text{ hr}}{60\text{ min}} = \dfrac{\mathbf{40(2.54)}}{\mathbf{(100)(1000)(60)}}\dfrac{\mathbf{km}}{\mathbf{min}}$

29. $A = \sqrt{8^2 - 6^2} = 2\sqrt{7}$

$8 \times \overrightarrow{SF} = 12$

$\overrightarrow{SF} = \dfrac{3}{2}$

$C = 2\sqrt{7} \cdot \dfrac{3}{2}$

$C = \mathbf{3\sqrt{7}}$

30. $A_{\text{Circle}} = \pi r^2 = 100\pi\text{ m}^2$

$r = 10\text{ m}$

Side of square $= 2r = 20\text{ m}$

$A_{\text{Square}} = (20\text{ m})^2 = \mathbf{400\text{ m}^2}$

---

1. (a) $2N_T = 3N_U + 12$

(b) $4N_U = 3N_T - 48$

$\begin{array}{rl} 3(a) & 6N_T = 9N_U + 36 \\ (-2)(b) & -6N_T = -8N_U - 96 \\ \hline & 0 = N_U - 60 \\ & \mathbf{60} = N_U \end{array}$

(a) $2N_T = 3(60) + 12$

$2N_T = 192$

$N_T = \mathbf{96}$

2. $0.3A + 0.2(300) = 0.24(300 + A)$

$0.3A + 60 = 72 + 0.24A$

$0.06A = 12$

$A = \mathbf{200\text{ ml}}$

3. $\dfrac{P_1}{T_1} = \dfrac{P_2}{T_2}$

$T_2 = \dfrac{P_2 T_1}{P_1}$

$T_2 = \dfrac{30(500)}{10} = \mathbf{1500\text{ K}}$

4. Variation method:

$B = \dfrac{kG}{W^2}$

$4 = \dfrac{k3}{2^2}$

$k = \dfrac{16}{3}$

$2 = \dfrac{\left(\dfrac{16}{3}\right)G}{4^2}$

$32 = \dfrac{16}{3}G$

$G = \mathbf{6}$

Equal ratio method:

$\dfrac{B_1}{B_2} = \dfrac{G_1 W_2^2}{G_2 W_1^2}$

$\dfrac{4}{2} = \dfrac{3(4^2)}{G_2(2^2)}$

$G_2 = \mathbf{6}$

5. Variation method:

$C = kFJ^2$

$1000 = k(100)(4)^2$

$k = 0.625$

$C = (0.625)(10)(20)^2 = \mathbf{2500}$

Equal ratio method:

$$\frac{C_1}{C_2} = \frac{F_1 J_1^2}{F_2 J_2^2}$$

$$\frac{1000}{C_2} = \frac{100\left(4^2\right)}{10\left(20^2\right)}$$

$$C_2 = \mathbf{2500}$$

6. (a) $2x + 3y = -7$

(b) $3x - 3y = 12$

$$y = x - 4$$

Substitute (b) into (a) and get:

(a') $2x + 3(x - 4) = -7$

$$2x + 3x - 12 = -7$$

$$5x = 5$$

$$x = 1$$

(b) $y = (1) - 4 = -3$

(1, –3)

7. (a) $4x - 2y = 8$

$$-2y = -4x + 8$$

$$y = 2x - 4$$

(b) $3x - 2y = -4$

Substitute (a) into (b) and get:

(b) $3x - 2(2x - 4) = -4$

$$3x - 4x + 8 = -4$$

$$-x = -12$$

$$x = 12$$

(a) $y = 2(12) - 4 = 20$

(12, 20)

8. (a) $y - 2x = 3$

$$y = 2x + 3$$

(b) $xy = 4$

Substitute (a) into (b) and get:

(b') $x(2x + 3) = 4$

$$2x^2 + 3x = 4$$

$$2x^2 + 3x - 4 = 0$$

Solve this equation by using the quadratic formula.

$$x = \frac{-3 \pm \sqrt{3^2 - 4(-4)(2)}}{2(2)} = -\frac{3}{4} \pm \frac{\sqrt{41}}{4}$$

Substitute these values of x into (a) and solve for y.

(a) $y = 2\left(-\dfrac{3}{4} + \dfrac{\sqrt{41}}{4}\right) + 3$

$$y = \frac{3}{2} + \frac{\sqrt{41}}{2}$$

(a) $y = 2\left(-\dfrac{3}{4} - \dfrac{\sqrt{41}}{4}\right) + 3$

$$y = \frac{3}{2} - \frac{\sqrt{41}}{2}$$

$\left(-\dfrac{3}{4} + \dfrac{\sqrt{41}}{4}, \dfrac{3}{2} + \dfrac{\sqrt{41}}{2}\right)$ and

$\left(-\dfrac{3}{4} - \dfrac{\sqrt{41}}{4}, \dfrac{3}{2} - \dfrac{\sqrt{41}}{2}\right)$

9. (a) $x^2 + y^2 = 11$

(b) $\dfrac{2x^2 - y^2 = -2}{}$

$$3x^2 \qquad = 9$$

$$x^2 = 3$$

$$x = \pm\sqrt{3}$$

(a) $(\sqrt{3})^2 + y^2 = 11$

$$3 + y^2 = 11$$

$$y^2 = 8$$

$$y = \pm 2\sqrt{2}$$

(a) $(-\sqrt{3})^2 + y^2 = 11$

$$3 + y^2 = 11$$

$$y^2 = 8$$

$$y = \pm 2\sqrt{2}$$

$(\sqrt{3}, \pm 2\sqrt{2})$ and $(-\sqrt{3}, \pm 2\sqrt{2})$

10. (a) $3x + y + z = 2$

(b) $2x - y - z = 3$

(c) $x + 2y - z = 8$

(a) $3x + y + z = 2$

(b) $\dfrac{2x - y - z = 3}{}$

$$5x \qquad = 5$$

$$x = 1$$

(a) $3x + y + z = 2$

(c) $x + 2y - z = 8$

(d) $\dfrac{}{4x + 3y \qquad = 10}$

(d) $4(1) + 3y = 10$

$$3y = 6$$

$$y = 2$$

(a) $3(1) + (2) + z = 2$

$$z = -3$$

(1, 2, –3)

11. $g(x) = x^2 - 2x + 2; \ D = \{\text{Reals}\}$

$$g(5) = 5^2 - 2(5) + 2$$

$$g(5) = \mathbf{17}$$

12. (a) $x - 2y < 2$

$-2y < -x + 2$

$y > \dfrac{1}{2}x - 1$

(b) $y \geq 0$

The first step is to graph each of these lines.

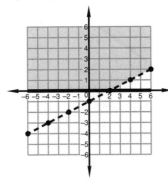

The region we wish to find is on or above the solid line and above the dashed line. This region is shaded in the previous figure.

13. $4 \ngtr x + 3 < 7;\ D = \{\text{Integers}\}$

$4 \leq x + 3 < 7$

$1 \leq x < 4$

14. $\dfrac{(x^{a+2})^{1/2}\,x^{3a/2}\,y^b}{y^{-b/2}} = x^{a/2+1+3a/2}y^{b+b/2}$

$= x^{2a+1}y^{3b/2}$

15. $\dfrac{\dfrac{p}{mx - \dfrac{m}{x + \dfrac{1}{mx}}}}{} = \dfrac{p}{mx - \dfrac{m^2 x}{mx^2 + 1}}$

$= \dfrac{p(mx^2 + 1)}{m^2 x^3 + mx - m^2 x}$

16. $\dfrac{5i - i^2}{2i^2 + i^3} = \dfrac{i(5 - i)}{i(2i + i^2)}$

$= \dfrac{5 - i}{2i - 1} \cdot \dfrac{2i + 1}{2i + 1} = \dfrac{10i + 2 + 5 - i}{-4 - 1}$

$= \dfrac{7 + 9i}{-5} = -\dfrac{7}{5} - \dfrac{9}{5}i$

17. $\sqrt[6]{8\sqrt{2}} = \left(2^3(2^{1/2})\right)^{1/6} = 2^{1/2}2^{1/12} = \mathbf{2^{7/12}}$

18. $\dfrac{5i - 2i^2}{-i} = \dfrac{i(5 - 2i)}{i(-1)}$

$= \dfrac{5 - 2i}{-1} = \mathbf{-5 + 2i}$

19. $\sqrt{z - 33} + \sqrt{z} = 11$

$\sqrt{z - 33} = 11 - \sqrt{z}$

$z - 33 = 121 - 22\sqrt{z} + z$

$22\sqrt{z} = 154$

$\sqrt{z} = 7$

$z = \mathbf{49}$

Check: $\sqrt{49 - 33} + \sqrt{49} = 11$

$\sqrt{16} + 7 = 11$

$4 + 7 = 11$

$11 = 11$ check

20.

$C = 4\cos 34° \approx 3.32$

$D = 4\sin 34° \approx 2.24$

$\begin{array}{r} -4.00R + 0.00U \\ 3.32R + 2.24U \\ \hline \mathbf{-0.68R + 2.24U} \end{array}$

$\tan\theta = \dfrac{2.24}{-0.68}$

$\theta \approx -73.11°$

Since θ is a second-quadrant angle:

$\theta = -73.11 + 180 = 106.89°$

$H = \sqrt{(-0.68)^2 + (2.24)^2}$

$H \approx 2.34$

$\mathbf{2.34/106.89°}$

21. $\dfrac{3 - 2\sqrt{2}}{4 + 2\sqrt{2}} \cdot \dfrac{4 - 2\sqrt{2}}{4 - 2\sqrt{2}}$

$= \dfrac{12 - 8\sqrt{2} - 6\sqrt{2} + 8}{16 - 8} = \dfrac{20 - 14\sqrt{2}}{8}$

$= \dfrac{\mathbf{10 - 7\sqrt{2}}}{\mathbf{4}}$

22. $\dfrac{3 - \sqrt{7}}{-\sqrt{7}} \cdot \dfrac{\sqrt{7}}{\sqrt{7}} = \dfrac{3\sqrt{7} - 7}{-7} = \dfrac{\mathbf{7 - 3\sqrt{7}}}{\mathbf{7}}$

23. $\sqrt{\dfrac{9}{3}} + 2\sqrt{\dfrac{3}{9}} - 5\sqrt{27} = \sqrt{3} + \dfrac{2\sqrt{3}}{3} - 15\sqrt{3}$

$= \dfrac{3\sqrt{3}}{3} + \dfrac{2\sqrt{3}}{3} - \dfrac{45\sqrt{3}}{3} = -\dfrac{\mathbf{40\sqrt{3}}}{\mathbf{3}}$

24. $-4i^2 - \sqrt{-9} - \sqrt{2}\sqrt{-2} - 5i^5$

$= 4 - 3i - 2i - 5i = \mathbf{4 - 10i}$

25.

$$\frac{a}{x + y} - c = \frac{1}{r^2}$$

$$ar^2 - cr^2x - cr^2y = x + y$$
$$ar^2 - x - cr^2x = y + cr^2y$$

$$\frac{ar^2 - x - cr^2x}{1 + cr^2} = y$$

26.

$$\frac{p}{x} = my\left(\frac{1}{a} + \frac{1}{c}\right)$$

$$\frac{p}{x} = \frac{my}{a} + \frac{my}{c}$$

$$pac = cmxy + amxy$$

$$pac - amxy = cmxy$$

$$a = \frac{cmxy}{cp - mxy}$$

27. $A_{\text{Circle}} = \pi r^2 = 49\pi \text{ cm}^2$

$$r = 7 \text{ cm}$$

Diagonal of square $= 2r = 14 \text{ cm}$

Side of square $= \dfrac{14}{\sqrt{2}} \text{ cm} = 7\sqrt{2} \text{ cm}$

$A_{\text{Square}} = (7\sqrt{2} \text{ cm})^2 = \textbf{98 cm}^2$

28. See Lesson 71.

29.

$$\frac{400 \text{ ml}}{\text{sec}} \times \frac{1 \text{ cm}^3}{1 \text{ ml}} \times \frac{1 \text{ in.}}{2.54 \text{ cm}} \times \frac{1 \text{ in.}}{2.54 \text{ cm}}$$

$$\times \frac{1 \text{ in.}}{2.54 \text{ cm}} \times \frac{60 \text{ sec}}{1 \text{ min}} \times \frac{60 \text{ min}}{1 \text{ hr}}$$

$$= \frac{(400)(60)(60)}{(2.54)(2.54)(2.54)} \frac{\text{in.}^3}{\text{hr}}$$

30.

$$\frac{x - 4}{2} - \frac{x - 6}{3} = 7$$

$$3x - 12 - 2x + 12 = 42$$

$$x = \textbf{42}$$

PROBLEM SET 98

1. $96 = R_W T_W + R_R T_R;$

$T_W = 12 - T_R;$

$R_R = 12; R_W = 6$

$$96 = 6(12 - T_R) + 12T_R$$
$$96 = 72 - 6T_R + 12T_R$$
$$24 = 6T_R$$
$$4 = T_R$$

$T_W = 8$

$D_R = 12(4) = \textbf{48 miles}$

$D_W = 6(8) = \textbf{48 miles}$

2. (a) $0.1P_N + 0.6D_N = 90$

(b) $P_N + D_N = 600$

Substitute $D_N = 600 - P_N$ into (a) and get:

(a') $0.1P_N + 0.6(600 - P_N) = 90$

$$-0.5P_N = -270$$
$$P_N = 540$$

(b) $(540) + D_N = 600$

$$D_N = 60$$

540 ml 10%, 60 ml 60%

3. $1200 = R_C T_C;$

$R_M = R_C + 10;$

$R_C = R_M - 10;$

$360 = R_M T_M; \ T_C = 4T_M$

$$(R_M - 10)(4T_M) = 1200$$
$$4R_M T_M - 40T_M = 1200$$
$$-40T_M = -240$$
$$T_M = 6$$

$T_C = \textbf{24 hr}; \ T_M = \textbf{6 hr};$

$R_C = \textbf{50 mph}; \ R_M = \textbf{60 mph}$

4. Downstream: $(B + W)T_D = D_D$ (a)

Upstream: $(B - W)T_U = D_U$ (b)

(a') $(B + W)4 = 28$

(b') $(B - W)8 = 40$

2(a') $8B + 8W = 56$

(b') $\underline{8B - 8W = 40}$

$16B \qquad = 96$

$B = \textbf{6 mph}$

(a') $4(6) + 4W = 28$

$$4W = 4$$
$$W = \textbf{1 mph}$$

5. Variation method:

$R = $ rabbits; $S = $ squirrels; $r = $ raccoons

$$R = \frac{kS}{r}$$

$$10 = \frac{k(40)}{2}$$

$$k = 0.5$$

$$5 = \frac{(0.5)(20)}{r}$$

$$r = \textbf{2}$$

Equal ratio method:

$$\frac{R_1}{R_2} = \frac{S_1 r_2}{S_2 r_1}$$

$$\frac{10}{5} = \frac{(40)r_2}{20(2)}$$

$$r_2 = \textbf{2}$$

6. $N = \dfrac{1}{4} + \dfrac{1}{2}\left(\dfrac{7}{12} - \dfrac{1}{4}\right)$

$= \dfrac{1}{4} + \dfrac{1}{2}\left(\dfrac{7}{12} - \dfrac{3}{12}\right)$

$= \dfrac{1}{4} + \dfrac{1}{2}\left(\dfrac{4}{12}\right)$

$= \dfrac{3}{12} + \dfrac{2}{12}$

$= \dfrac{5}{12}$

7. $N = 1\dfrac{1}{5} + \dfrac{2}{5}\left(2\dfrac{5}{6} - 1\dfrac{1}{5}\right)$

$= \dfrac{6}{5} + \dfrac{2}{5}\left(\dfrac{17}{6} - \dfrac{6}{5}\right)$

$= \dfrac{6}{5} + \dfrac{2}{5}\left(\dfrac{85}{30} - \dfrac{36}{30}\right)$

$= \dfrac{6}{5} + \dfrac{2}{5}\left(\dfrac{49}{30}\right)$

$= \dfrac{90}{75} + \dfrac{49}{75}$

$= \dfrac{139}{75}$

8. (a) $3x + 2y = 5$
(b) $5x + 6y = 7$

$6y = -5x + 7$

$y = \dfrac{-5x + 7}{6}$

Substitute (b) into (a) and get:

(a') $3x + 2\left(\dfrac{-5x + 7}{6}\right) = 5$

$9x + (-5x + 7) = 15$

$4x = 8$

$x = 2$

(b) $y = \dfrac{-5(2) + 7}{6} = -\dfrac{3}{6} = -\dfrac{1}{2}$

$\left(2, -\dfrac{1}{2}\right)$

9. (a) $y - 3x = 5$

$y = 3x + 5$

(b) $xy = 6$

Substitute (a) into (b) and get:

(b') $x(3x + 5) = 6$

$3x^2 + 5x - 6 = 0$

Solve this equation by using the quadratic formula.

$x = \dfrac{-5 \pm \sqrt{25 - 4(-18)}}{6} = -\dfrac{5}{6} \pm \dfrac{\sqrt{97}}{6}$

Substitute these values of x into (a) and solve for y.

(a) $y = 3\left(-\dfrac{5}{6} + \dfrac{\sqrt{97}}{6}\right) + 5$

$y = \dfrac{5}{2} + \dfrac{\sqrt{97}}{2}$

(a) $y = 3\left(-\dfrac{5}{6} - \dfrac{\sqrt{97}}{6}\right) + 5$

$y = \dfrac{5}{2} - \dfrac{\sqrt{97}}{2}$

$\left(-\dfrac{5}{6} + \dfrac{\sqrt{97}}{6}, \dfrac{5}{2} + \dfrac{\sqrt{97}}{2}\right)$ and

$\left(-\dfrac{5}{6} - \dfrac{\sqrt{97}}{6}, \dfrac{5}{2} - \dfrac{\sqrt{97}}{2}\right)$

10. (a) $x^2 + y^2 = 16$
(b) $2x^2 - y^2 = -4$

$3x^2 = 12$

$x^2 = 4$

$x = \pm 2$

(a) $(2)^2 + y^2 = 16$

$y^2 = 12$

$y = \pm 2\sqrt{3}$

(a) $(-2)^2 + y^2 = 16$

$y^2 = 12$

$y = \pm 2\sqrt{3}$

$(2, \pm 2\sqrt{3})$ and $(-2, \pm 2\sqrt{3})$

11. **(a)**, **(b)**, and **(c)** are all functions. (c) is a function because the last two ordered pairs are the same.

12. (a) $x + 2y > 4$

$y > -\dfrac{1}{2}x + 2$

(b) $y \geq 1$

The first step is to graph each of these lines.

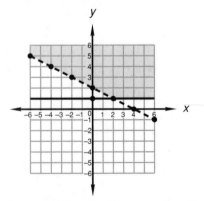

The region we wish to find is on or above the solid line and above the dashed line. This region is shaded in the previous figure.

13. $4 \not< -x + 2$ or $x + 3 < -1$; $D = \{\text{Integers}\}$
$4 \geq -x + 2$ or $x + 3 < -1$
$-2 \leq x$ or $x < -4$

$$\xrightarrow{\quad \bullet \quad \bullet \quad | \quad | \quad \bullet \quad \bullet \quad}$$
$$-6 \; -5 \; -4 \; -3 \; -2 \; -1$$

14. $\dfrac{(x^{2a})^{1/3} \, x^{2a}}{x^{a/2}} = x^{2a/3 + 2a - a/2} = x^{4a/6 + 12a/6 - 3a/6}$
$= x^{13a/6}$

15. $\dfrac{xy}{x + \dfrac{xy}{x + \dfrac{1}{y}}} = \dfrac{xy}{x + \dfrac{xy^2}{xy + 1}}$

$= \dfrac{xy(xy + 1)}{x^2 y + x + xy^2} = \dfrac{y(xy + 1)}{xy + 1 + y^2}$

16. $\sqrt[7]{4\sqrt[3]{2}} = \left(2^2(2^{1/3})\right)^{1/7} = 2^{2/7} 2^{1/21} = 2^{1/3}$

17. $\dfrac{2i - 3i^2}{-i} = -2 + 3i$

18. $\dfrac{2i^4 - i^3}{-2i} = \dfrac{2i^3 - i^2}{-2} = -\dfrac{1}{2} + i$

19. There is never an exact solution to these problems. One possible solution is given here.
$S = mP + b$
$m = \dfrac{25 - 175}{7 - 4} = \dfrac{-150}{3} = -50$
$S = -50P + b$
Use the point (4, 175) for P and S.
$175 = -50(4) + b$
$375 = b$
$S = -50P + 375$

20.

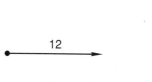

$C = 10 \cos 64° \approx 4.38$
$D = 10 \sin 64° \approx 8.99$

$12.00R + 0.00U$
$\underline{4.38R - 8.99U}$
$16.38R - 8.99U$

$\tan \theta = \dfrac{-8.99}{16.38}$
$\theta \approx -28.76°$
$H = \sqrt{(-8.99)^2 + (16.38)^2}$
$H \approx 18.68$

$\underline{18.68\,/\!-28.76°}$

21. $\dfrac{5 - \sqrt{2}}{2 - 5\sqrt{2}} \cdot \dfrac{2 + 5\sqrt{2}}{2 + 5\sqrt{2}}$

$= \dfrac{10 - 2\sqrt{2} + 25\sqrt{2} - 10}{4 - 50}$

$= \dfrac{23\sqrt{2}}{-46} = -\dfrac{\sqrt{2}}{2}$

22. $\dfrac{3 - \sqrt{3}}{3 - 2\sqrt{3}} \cdot \dfrac{3 + 2\sqrt{3}}{3 + 2\sqrt{3}}$

$= \dfrac{9 - 3\sqrt{3} + 6\sqrt{3} - 6}{9 - 12}$

$= \dfrac{3 + 3\sqrt{3}}{-3} = -1 - \sqrt{3}$

23. $\dfrac{a}{x + y} = z\left(\dfrac{1}{m} + \dfrac{1}{n}\right)$

$\dfrac{a}{x + y} = \dfrac{z}{m} + \dfrac{z}{n}$

$amn = nxz + nyz + mxz + myz$
$amn - mxz - myz = nxz + nyz$

$m = \dfrac{nxz + nyz}{an - xz - yz}$

24. $\dfrac{a}{x + y} = z\left(\dfrac{1}{m} + \dfrac{1}{n}\right)$

$\dfrac{a}{x + y} = \dfrac{z}{m} + \dfrac{z}{n}$

$nxz + nyz + mxz + myz = amn$
$amn - nxz - mxz = nyz + myz$

$\dfrac{amn - mxz - nxz}{mz + nz} = y$

25. $-\sqrt{-9} - \sqrt{-2}\sqrt{-2} + 3i^2 - 2i^3 + 4$
$= -3i + 2 - 3 + 2i + 4 = 3 - i$

26. $\sqrt{\dfrac{7}{3}} - 2\sqrt{\dfrac{3}{7}} + 5\sqrt{84} = \dfrac{\sqrt{21}}{3} - \dfrac{2\sqrt{21}}{7} + 10\sqrt{21}$

$= \dfrac{7\sqrt{21}}{21} - \dfrac{6\sqrt{21}}{21} + \dfrac{210\sqrt{21}}{21} = \dfrac{211\sqrt{21}}{21}$

27. $-5x^2 - x - 5 = 0$
$5x^2 + x = -5$
$\left(x^2 + \dfrac{1}{5}x + \quad\right) = -1$
$x^2 + \dfrac{1}{5}x + \dfrac{1}{100} = -\dfrac{100}{100} + \dfrac{1}{100}$
$\left(x + \dfrac{1}{10}\right)^2 = -\dfrac{99}{100}$
$x + \dfrac{1}{10} = -\dfrac{3\sqrt{11}}{10}i$
$x = -\dfrac{1}{10} \pm \dfrac{3\sqrt{11}}{10}i$

28. $\dfrac{60 \text{ in.}}{\text{sec}} \times \dfrac{2.54 \text{ cm}}{1 \text{ in.}} \times \dfrac{60 \text{ sec}}{1 \text{ min}}$

$= (60)(2.54)(60) \, \dfrac{\text{cm}}{\text{min}}$

29. $\dfrac{\dfrac{a}{xy} - \dfrac{x}{y^2}}{\dfrac{4}{x} - \dfrac{3}{xy^2}} = \dfrac{\dfrac{ay - x^2}{xy^2}}{\dfrac{4y^2 - 3}{xy^2}} = \dfrac{ay - x^2}{4y^2 - 3}$

30. (a) $\dfrac{47{,}162 \times 10^{-12}}{50{,}132 \times 10^5}$

Estimate: $\dfrac{5 \times 10^{-8}}{5 \times 10^9} = 1 \times 10^{-17}$

Calculator: $\mathbf{9.41 \times 10^{-18}}$

(b) $\sqrt[3.4]{311}$

Estimate: $310^{1/3} = 6.77$

Calculator: $\sqrt[3.4]{311} = \mathbf{5.41}$

PROBLEM SET 99

1. Silver (Ag): $2 \times 108 = 216$
Sulfur: $1 \times 32 = 32$
Total: $= 248$

$\% \text{ Ag} = \dfrac{\text{Ag}}{\text{Tot}} \times 100\% = \dfrac{216}{248} \times 100 \approx \mathbf{87.1\%}$

2. (a) $2N_P = 10N_M + 16$
(b) $13N_M = N_P + 8$
(b') $N_P = 13N_M - 8$
Substitute (b') into (a) and get:
(a') $2(13N_M - 8) = 10N_M + 16$
$26N_M - 16 = 10N_M + 16$
$16N_M = 32$
$N_M = \mathbf{2}$
(b') $N_P = \mathbf{18}$

3. Downstream: $(B + W)T_D = D_D$ (a)
Upstream: $(B - W)T_U = D_U$ (b)
Since $T_D = 2T_U$ we substitute and get:
(a') $(B + 8)2T_U = 280$
(b') $(B - 8)T_U = 60$

(a') $2BT_U + 16T_U = 280$
2(b') $\underline{2BT_U - 16T_U = 120}$
(c) $\quad 4BT_U \qquad\quad = 400$
(c) $\qquad\qquad BT_U = 100$

(a') $2(100) + 16T_U = 280$
$T_U = 5$
(c) $B(5) = 100$
$B = \mathbf{20 \text{ mph}}$

4. Equal ratio method:
$\dfrac{W_1}{W_2} = \dfrac{P_1 F_1}{P_2 F_2}$

$\dfrac{8{,}000}{16{,}000} = \dfrac{100(20)}{P_2(2)}$

$P_2 = \mathbf{2000}$

Variation method:
$W = kPF$
$8000 = k(100)(20)$
$4 = k$

$16{,}000 = (4)(P)(2)$
$\mathbf{2000} = P$

5. (a) $0.05P_N + 0.1D_N = 96$
(b) $P_N + D_N = 1200$
Substitute $D_N = 1200 - P_N$ into (a) and get:
(a') $0.05P_N + 0.1(1200 - P_N) = 96$
$-0.05P_N = -24$
$P_N = 480$
(b) $(480) + D_N = 1200$
$D_N = 720$

480 ml 5%, 720 ml 10%

6. $-|x| + 5 > 0; \; D = \{\text{Integers}\}$
$-|x| > -5$
$|x| < 5$
$x > -5 \text{ and } x < 5$

7. $-|x| + 1 < -3; \; D = \{\text{Reals}\}$
$-|x| < -4$
$|x| > 4$
$x > 4 \text{ or } x < -4$

8. (a) Since this is a 30-60-90 triangle:
$2 \times \overrightarrow{SF} = 124$
$\overrightarrow{SF} = 62$

$m = \sqrt{3}(62) = \mathbf{62\sqrt{3}}$
$n = 1(62) = \mathbf{62}$

(b) Since this is a 45-45-90 triangle:
$\sqrt{2} \times \overrightarrow{SF} = 5\sqrt{2}$
$\overrightarrow{SF} = 5$

$p = 1(5) = \mathbf{5}$
$q = 1(5) = \mathbf{5}$

9. $N = \dfrac{1}{2} + \dfrac{2}{7}\left(2\dfrac{1}{3} - \dfrac{1}{2}\right)$

$= \dfrac{1}{2} + \dfrac{2}{7}\left(\dfrac{11}{6}\right)$

$= \dfrac{21}{42} + \dfrac{22}{42} = \mathbf{\dfrac{43}{42}}$

10. (a) $3x + 3y = 9$

$y = -x + 3$

(b) $4x - 6y = -8$

Substitute (a) into (b) and get:

(b′) $4x - 6(-x + 3) = -8$

$4x + 6x - 18 = -8$

$10x = 10$

$x = 1$

(a) $y = -(1) + 3 = 2$

$\mathbf{(1, 2)}$

11. (a) $x - 3y = 2$

$y = \dfrac{1}{3}x - \dfrac{2}{3}$

(b) $xy = 8$

$y = \dfrac{8}{x}$

Substitute (b) into (a) and get:

(a′) $x - 3\left(\dfrac{8}{x}\right) = 2$

$x^2 - 24 = 2x$

$x^2 - 2x - 24 = 0$

$(x - 6)(x + 4) = 0$

$x = 6, -4$

Substitute these values of x into (a) and solve for y.

(a) $y = \dfrac{1}{3}(6) - \dfrac{2}{3}$

$y = \dfrac{4}{3}$

(a) $y = \dfrac{1}{3}(-4) - \dfrac{2}{3}$

$y = -2$

$\left(\mathbf{6, \dfrac{4}{3}}\right)$ and $\mathbf{(-4, -2)}$

12. (a) $x^2 + y^2 = 8$

(b) $\dfrac{2x^2 - y^2 = 7}{3x^2 \qquad = 15}$

$x^2 = 5$

$x = \sqrt{5}, -\sqrt{5}$

(a) $y^2 = 8 - (\sqrt{5})^2$

$y^2 = 3$

$y = \pm\sqrt{3}$

(a) $y^2 = 8 - (-\sqrt{5})^2$

$y^2 = 3$

$y = \pm\sqrt{3}$

$\mathbf{(\sqrt{5}, \pm\sqrt{3})}$ and $\mathbf{(-\sqrt{5}, \pm\sqrt{3})}$

13. (a) $3x + 2y - z = 1$

(b) $x + y - z = -1$

(c) $5x + 2y + 2z = 8$

 (a) $3x + 2y - z = 1$

-1(b) $\dfrac{-x - y + z = 1}{}$

(d) $2x + y \qquad = 2$

2(b) $2x + 2y - 2z = -2$

(c) $\dfrac{5x + 2y + 2z = 8}{}$

(e) $7x + 4y \qquad = 6$

-4(d) $-8x - 4y = -8$

(e) $\dfrac{7x + 4y = 6}{}$

$-x \qquad = -2$

$x = 2$

(d) $2(2) + y = 2$

$y = -2$

(b) $(2) + (-2) - z = -1$

$z = 1$

$\mathbf{(2, -2, 1)}$

14. $\varnothing$ **or { }** because 4 is not in the domain of $h(x)$.

15. (a) $x - y < -2$

$y > x + 2$

(b) $3x + 5y \le -5$

$y \le -\dfrac{3}{5}x - 1$

The first step is to graph each of these lines.

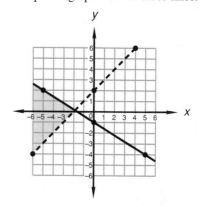

The region we wish to find is on or below the solid line and above the dashed line. This region is shaded in the previous figure.

16. $x + 4 \le 3$ or $x + 1 > 2$; $D = \{\text{Reals}\}$
   $x \le -1$ or $x > 1$

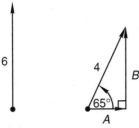

17. $(x^{2-a})^2 x^{a/4} = x^{4-2a} x^{a/4} = x^{4-7a/4}$

18. $\dfrac{ab}{a + \dfrac{b}{1 + \dfrac{a}{b^2}}} = \dfrac{ab}{a + \dfrac{b^3}{b^2 + a}}$

$= \dfrac{ab(b^2 + a)}{ab^2 + a^2 + b^3}$

19. $2\sqrt[5]{16\sqrt[3]{2}} = 2\sqrt[5]{2^4\sqrt[3]{2}} = 2(2^{4/5})(2^{1/15}) = 2^{28/15}$

20. $\dfrac{2 - 3i}{2i^3} = \dfrac{2 - 3i}{-2i} \cdot \dfrac{2i}{2i} = \dfrac{4i + 6}{4}$

$= \dfrac{3}{2} + i$

21. $\dfrac{2i - i^3}{-i} = \dfrac{2i + i}{-i} \cdot \dfrac{i}{i} = \dfrac{-2 - 1}{1} = -3$

22. $O = mI + b$
Use the graph to find the slope.

$m = \dfrac{200}{50} = 4$
$O = 4I + b$
Use the point $(45, 0)$ for I and O.
   $0 = 4(45) + b$
$-180 = b$
$O = 4I - 180$

23.

$A = 4 \cos 65° \approx 1.69$
$B = 4 \sin 65° \approx 3.63$

$\begin{array}{l} 0.00R + 6.00U \\ 1.69R + 3.63U \\ \hline 1.69R + 9.63U \end{array}$

$\tan \theta = \dfrac{9.63}{1.69}$
   $\theta \approx 80°$

$F = \sqrt{(1.69)^2 + (9.63)^2} \approx 9.78$

9.78 /80°

24. $\dfrac{3 - \sqrt{2}}{\sqrt{2} + 3} \cdot \dfrac{\sqrt{2} - 3}{\sqrt{2} - 3}$

$= \dfrac{3\sqrt{2} - 9 - 2 + 3\sqrt{2}}{2 - 9} = \dfrac{11 - 6\sqrt{2}}{7}$

25. $\dfrac{4 - 2\sqrt{3}}{-\sqrt{3}} \cdot \dfrac{\sqrt{3}}{\sqrt{3}} = \dfrac{4\sqrt{3} - 6}{-3}$

$= \dfrac{-4\sqrt{3} + 6}{3}$

26. $\dfrac{m}{c} = \dfrac{p}{a} + \dfrac{p}{b}$
   $abm = bcp + acp$

$\dfrac{abm}{bp + ap} = c$

27. $-\sqrt{-9} - 3i^3 + 2i - \sqrt{-4}\sqrt{4} + 3\sqrt{-4}$
$= -3i + 3i + 2i - 4i + 6i = 4i$

28. $-\sqrt{\dfrac{3}{10}} + 4\sqrt{\dfrac{10}{3}} - \sqrt{120}$

$= -\dfrac{\sqrt{3}}{\sqrt{10}} + \dfrac{4\sqrt{10}}{\sqrt{3}} - \sqrt{120}$

$= -\dfrac{\sqrt{30}}{10} + \dfrac{4\sqrt{30}}{3} - 2\sqrt{30}$

$= -\dfrac{3\sqrt{30}}{30} + \dfrac{40\sqrt{30}}{30} - \dfrac{60\sqrt{30}}{30} = -\dfrac{23\sqrt{30}}{30}$

29. $PV = nRT$

$V = \dfrac{nRT}{P}$

$V = \dfrac{(0.0163)(0.0821)(870)}{10} = 0.116 \text{ liter}$

30. $\dfrac{40 \text{ cm}}{\text{sec}} \times \dfrac{1 \text{ in.}}{2.54 \text{ cm}} \times \dfrac{1 \text{ ft}}{12 \text{ in.}} \times \dfrac{1 \text{ mi}}{5280 \text{ ft}}$

$\times \dfrac{60 \text{ sec}}{1 \text{ min}} \times \dfrac{60 \text{ min}}{1 \text{ hr}} = \dfrac{40(60)(60)}{(2.54)(12)(5280)} \dfrac{\text{mi}}{\text{hr}}$

PROBLEM SET 100

1. $R_B T_B = 320$;
   $R_S T_S = 240$;
   $R_B = 2R_S$;
   $T_B = T_S - 2$

D_B |————→| 320

D_S |————→| 240

   $2R_S(T_S - 2) = 320$
   $2(240) - 4R_S = 320$
           $4R_S = 160$
             $R_S = 40 \text{ mph}$

$R_B = 80 \text{ mph}$; $T_S = 6 \text{ hr}$; $T_B = 4 \text{ hr}$

2. $0.68(600) - 1(D_N) = 0.2(600 - D_N)$

$408 - D_N = 120 - 0.2D_N$

$0.8D_N = 288$

$D_N = \textbf{360 ml}$

3. (a) $10N_S = 6N_C + 40$

(b) $4N_C = 2N_S + 160$

(b') $N_S = 2N_C - 80$

Substitute (b') into (a) and get:

(a') $10(2N_C - 80) = 6N_C + 40$

$20N_C - 800 = 6N_C + 40$

$14N_C = 840$

$N_C = \textbf{60}$

(b') $N_S = 2(60) - 80$

$N_S = \textbf{40}$

4. $\dfrac{R_1}{R_2} = \dfrac{Y_1 G_2^2}{Y_2 G_1^2}$

$\dfrac{100}{R_2} = \dfrac{40(5)^2}{20(10)^2}$

$R_2 = \dfrac{100(20)(100)}{(40)(25)}$

$R_2 = \textbf{200}$

5. $N \qquad N + 1 \qquad N + 2$

$N(N + 2) = 8(N + 1) + 32$

$N^2 + 2N = 8N + 40$

$N^2 - 6N - 40 = 0$

$(N - 10)(N + 4) = 0$

$N = 10, -4$

The desired integers are $\textbf{10, 11, 12}$ and $\textbf{-4, -3, -2}$.

6. $y = x^2 - 6x + 1$

$y = (x^2 - 6x + 9) + 1 - 9$

$y = (x - 3)^2 - 8$

From this we see:

(a) Opens upward

(b) Axis of symmetry is $x = 3$

(c) y coordinate of vertex is -8

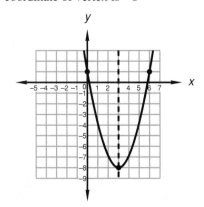

7. $y = -x^2 + 4x + 4$

$-y = (x^2 - 4x + 4) - 4 - 4$

$-y = (x - 2)^2 - 8$

$y = -(x - 2)^2 + 8$

From this we see:

(a) Opens downward

(b) Axis of symmetry is $x = 2$

(c) y coordinate of vertex is $+8$

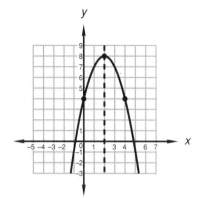

8. (a) Since this is a 30-60-90 triangle:

$\sqrt{3} \times \overrightarrow{SF} = 5$

$\overrightarrow{SF} = \dfrac{5\sqrt{3}}{3}$

$m = \dfrac{5\sqrt{3}}{3}(2) = \dfrac{\textbf{10}\sqrt{3}}{\textbf{3}}$

$n = \dfrac{5\sqrt{3}}{3 \cdot}(1) = \dfrac{\textbf{5}\sqrt{3}}{\textbf{3}}$

(b) Since this is a 45-45-90 triangle:

$1 \times \overrightarrow{SF} = 7$

$\overrightarrow{SF} = 7$

$p = 7(1) = \textbf{7}$

$q = 7(\sqrt{2}) = \textbf{7}\sqrt{\textbf{2}}$

9. $-|x| + 5 \nleq 3; \quad D = \{\text{Reals}\}$

$-|x| > -2$

$|x| < 2$

$x < 2$ and $x > -2$

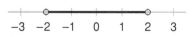

10. $N = 2\dfrac{1}{3} + \dfrac{5}{11}\left(3\dfrac{1}{6} - 2\dfrac{1}{3}\right)$

$= \dfrac{7}{3} + \dfrac{5}{11}\left(\dfrac{19}{6} - \dfrac{14}{6}\right)$

$= \dfrac{7}{3} + \dfrac{25}{66}$

$= \dfrac{154}{66} + \dfrac{25}{66} = \dfrac{\textbf{179}}{\textbf{66}}$

11. (a) $2x - 2y = -1$

$$x = y - \frac{1}{2}$$

(b) $4x + 3y = 5$

Substitute (a) into (b) and get:

(b') $4\left(y - \frac{1}{2}\right) + 3y = 5$

$$4y - 2 + 3y = 5$$
$$7y = 7$$
$$y = 1$$

(a) $x = (1) - \frac{1}{2} = \frac{1}{2}$

$$\left(\frac{1}{2}, 1\right)$$

12. (a) $x - y = 5$

(b) $xy = 2$

$$y = \frac{2}{x}$$

Substitute (b) into (a) and get:

(a') $x - \left(\dfrac{2}{x}\right) = 5$

$$x^2 - 2 = 5x$$
$$x^2 - 5x - 2 = 0$$

Solve this equation by using the quadratic formula.

$$x = \frac{-(-5) \pm \sqrt{(-5)^2 - 4(1)(-2)}}{2(1)}$$

$$x = \frac{5}{2} \pm \frac{\sqrt{33}}{2}$$

Substitute these values of x into (a) and solve for y.

(a) $y = \left(\dfrac{5}{2} + \dfrac{\sqrt{33}}{2}\right) - 5$

$$y = -\frac{5}{2} + \frac{\sqrt{33}}{2}$$

(a) $y = \left(\dfrac{5}{2} - \dfrac{\sqrt{33}}{2}\right) - 5$

$$y = -\frac{5}{2} - \frac{\sqrt{33}}{2}$$

$$\left(\frac{5}{2} + \frac{\sqrt{33}}{2}, -\frac{5}{2} + \frac{\sqrt{33}}{2}\right) \text{ and}$$

$$\left(\frac{5}{2} - \frac{\sqrt{33}}{2}, -\frac{5}{2} - \frac{\sqrt{33}}{2}\right)$$

13. (a) $x^2 + y^2 = 5$

(b) $\dfrac{2x^2 - y^2 = 4}{3x^2 \qquad = 9}$

$$x^2 = 3$$
$$x = \sqrt{3}, -\sqrt{3}$$

(a) $(\sqrt{3})^2 + y^2 = 5$

$$y^2 = 2$$
$$y = \pm\sqrt{2}$$

(a) $(-\sqrt{3})^2 + y^2 = 5$

$$y^2 = 2$$
$$y = \pm\sqrt{2}$$

$(\sqrt{3}, \pm\sqrt{2})$ and $(-\sqrt{3}, \pm\sqrt{2})$

14. (a) $2x + 2y - z = 0$

(b) $x + y - 2z = -12$

(c) $2x - y + z = 10$

(a) $\quad 2x + 2y - z = 0$

-2(b) $\dfrac{-2x - 2y + 4z = 24}{3z = 24}$

$$z = 8$$

(b) $\quad x + y - 2z = -12$

(c) $\dfrac{2x - y + z = 10}{}$

(d) $\overline{3x \qquad - z = -2}$

(d) $3x - (8) = -2$

$$3x = 6$$
$$x = 2$$

(a) $2(2) + 2y - (8) = 0$

$$2y = 4$$
$$y = 2$$

$(2, 2, 8)$

15. **None**

(a) is not a function because 2 has two images.

(b) is not a function because 2 has two images.

(c) is not a function because -3 has two images.

16. (a) $y \le -3$

(b) $4x + y < -2$

$$y < -4x - 2$$

The first step is to graph each of these lines.

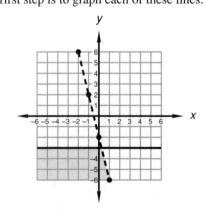

The region we wish to find is on or below the solid line and below the dashed line. This region is shaded in the previous figure.

17. $x + 2 \not\le 5$ or $x + 3 < 3$; $D = \{$Integers$\}$
  $x > 3$ or $x < 0$

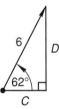

18. $\dfrac{a^{2x+4}a^{b/2}}{a^{1-b/2}} = a^{2x+4}a^{-1+b} = \mathbf{a^{2x+3+b}}$

19. $\dfrac{x}{x^2y - \dfrac{1}{1 + \dfrac{1}{xy}}} = \dfrac{x}{x^2y - \dfrac{xy}{xy + 1}}$

$= \dfrac{x(xy + 1)}{x^3y^2 + x^2y - xy} = \dfrac{\mathbf{xy + 1}}{\mathbf{x^2y^2 + xy - y}}$

20. $3\sqrt{9\sqrt{3}} = 3\sqrt{3^2\sqrt{3}} = 3 \cdot 3 \cdot 3^{1/4} = \mathbf{3^{9/4}}$

21. $\dfrac{2 - i^3}{-i} = \dfrac{2 + i}{-i} \cdot \dfrac{i}{i} = \dfrac{2i - 1}{-(-1)} = \mathbf{-1 + 2i}$

22. $\dfrac{3 - i}{i^2 - 3i} = \dfrac{3 - i}{-1 - 3i} \cdot \dfrac{-1 + 3i}{-1 + 3i}$

$= \dfrac{-3 + 9i + i + 3}{1 + 9} = \mathbf{i}$

23.

$A = 8\cos 28° \approx 7.06$
$B = 8\sin 28° \approx 3.76$

$C = 6\cos 62° \approx 2.82$
$D = 6\sin 62° \approx 5.30$

$\quad -7.06R + 3.76U$
$\quad \underline{2.82R + 5.30U}$
$\mathbf{-4.24R + 9.06U}$

$\tan\theta = \dfrac{9.06}{-4.24}$
$\quad \theta \approx -64.92°$
Since θ is a second-quadrant angle:
$\theta = (-64.92) + 180 = 115.08°$
$F = \sqrt{(-4.24)^2 + (9.06)^2} \approx 10$
$\mathbf{10/115.08°}$

24. $\dfrac{5 - 2\sqrt{2}}{\sqrt{2}} \cdot \dfrac{\sqrt{2}}{\sqrt{2}} = \dfrac{\mathbf{-4 + 5\sqrt{2}}}{\mathbf{2}}$

25. $\dfrac{3 - 4\sqrt{5}}{\sqrt{5} - 1} \cdot \dfrac{\sqrt{5} + 1}{\sqrt{5} + 1}$

$= \dfrac{3\sqrt{5} + 3 - 20 - 4\sqrt{5}}{5 - 1} = \dfrac{\mathbf{-17 - \sqrt{5}}}{\mathbf{4}}$

26. $\sqrt{\dfrac{2}{13}} + 3\sqrt{\dfrac{13}{2}} - 2\sqrt{104}$

$= \dfrac{\sqrt{2}}{\sqrt{13}}\dfrac{\sqrt{13}}{\sqrt{13}} + \dfrac{3\sqrt{13}}{\sqrt{2}}\dfrac{\sqrt{2}}{\sqrt{2}} - 4\sqrt{26}$

$= \dfrac{2\sqrt{26}}{26} + \dfrac{39\sqrt{26}}{26} - \dfrac{104\sqrt{26}}{26} = -\dfrac{\mathbf{63\sqrt{26}}}{\mathbf{26}}$

27. $-\sqrt{-7}\sqrt{-7} + 2\sqrt{-16} - 3i^5 + 2i^2$
$= 7 + 8i - 3i - 2 = \mathbf{5 + 5i}$

28. $\dfrac{a}{k + c} = \dfrac{mx}{y} + md$
$\quad ay = kmx + cmx + mdky + mdcy$
$ay - mdky - mdcy = kmx + cmx$
$\quad\quad y = \dfrac{\mathbf{kmx + cmx}}{\mathbf{a - mdk - mdc}}$

29. $\dfrac{400 \text{ ft}^3}{\text{min}} \times \dfrac{12 \text{ in.}}{1 \text{ ft}} \times \dfrac{12 \text{ in.}}{1 \text{ ft}} \times \dfrac{12 \text{ in.}}{1 \text{ ft}} \times \dfrac{1 \text{ min}}{60 \text{ sec}}$

$= \dfrac{\mathbf{400(12)(12)(12)}}{\mathbf{60}} \dfrac{\text{in.}^3}{\text{sec}}$

30. (a) Estimate: $\dfrac{2 \times 10^{-6}}{-5 \times 10^6} = -4 \times 10^{-13}$
  Calculator: $\mathbf{-3.92 \times 10^{-13}}$
(b) Estimate: $(4)^{-5}$ and $(5)^{-5}$; 10×10^{-4} and
  3×10^{-4}; Calculator: $\mathbf{7.46 \times 10^{-4}}$

PROBLEM SET 101

1. Selling price = purchase price + markup
$\quad\quad 78 = P_P + 0.3P_P$
$\quad\quad 78 = 1.3P_P$
$\quad\quad \mathbf{\$60} = P_P$
$M = 0.3(\$60) = \mathbf{\$18}$

2. Selling price = purchase price + markup
$\quad\quad S_P = 2100 + 0.6S_P$
$\quad 0.4S_P = 2100$
$\quad\quad S_P = \mathbf{\$5250}$

3. Selling price = purchase price + markup
$\quad 16{,}535 = P_P + 0.25P_P$
$\quad 16{,}535 = 1.25P_P$
$\quad \mathbf{\$13{,}228} = P_P$

4. $0.6(400) + 0.8(D_N) = 0.72(400 + D_N)$
$\quad 240 + 0.8D_N = 288 + 0.72D_N$
$\quad\quad 0.08D_N = 48$
$\quad\quad\quad D_N = \mathbf{600 \text{ liters}}$

5. Downstream: $(B + W)T_D = D_D$ (a)

Upstream: $(B - W)T_U = D_U$ (b)

Since $T_D = T_U$ we use T in both equations.

(a') $(20 + W)T = 104$

(b') $(20 - W)T = 56$

(a') $20T + WT = 104$

(b') $\underline{20T - WT = 56}$

$\;40T = 160$

$T = 4$

(a') $20(4) + 4W = 104$

$4W = 24$

$W = \textbf{6 mph}$

6. $y = -x^2 - 4x - 5$

$-y = (x^2 + 4x + 4) + 5 - 4$

$-y = (x + 2)^2 + 1$

$y = -(x + 2)^2 - 1$

From this we see:

(a) Opens downward

(b) Axis of symmetry is $x = -2$

(c) y coordinate of vertex is -1

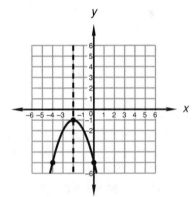

7. $y = x^2 + 2x + 2$

$y = (x^2 + 2x + 1) + 2 - 1$

$y = (x + 1)^2 + 1$

From this we see:

(a) Opens upward

(b) Axis of symmetry is $x = -1$

(c) y coordinate of vertex is $+1$

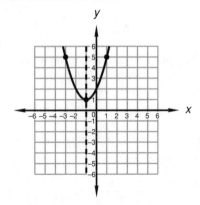

8. $V_{\text{Cone}} = \dfrac{1}{3}\pi r^2 h = 60\pi \text{ cm}^3$

$$h = \frac{60\pi(3) \text{ cm}^3}{(3 \text{ cm})^2\, \pi}$$

$$h = \frac{180\pi \text{ cm}^3}{9\pi \text{ cm}^2} = \textbf{20 cm}$$

9. $-|x| + 3 \not< 2; \; D = \{\text{Integers}\}$

$-|x| \geq -1$

$|x| \leq 1$

$x \leq 1 \text{ and } x \geq -1$

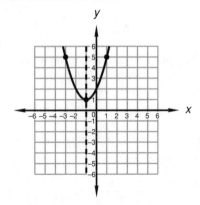

10. $N = 3\dfrac{1}{3} + \dfrac{2}{5}\left(4\dfrac{5}{6} - 3\dfrac{1}{3}\right)$

$ = \dfrac{10}{3} + \dfrac{2}{5}\left(\dfrac{29}{6} - \dfrac{20}{6}\right)$

$ = \dfrac{10}{3} + \dfrac{3}{5}$

$ = \dfrac{50}{15} + \dfrac{9}{15} = \dfrac{\textbf{59}}{\textbf{15}}$

11. (a) $6x + 5y = 8$

(b) $4x + 2y = 3$

$$y = -2x + \frac{3}{2}$$

Substitute (b) into (a) and get:

(a') $6x + 5\left(-2x + \dfrac{3}{2}\right) = 8$

$6x - 10x + \dfrac{15}{2} = 8$

$-8x + 15 = 16$

$-8x = 1$

$x = -\dfrac{1}{8}$

(b) $y = -2\left(-\dfrac{1}{8}\right) + \dfrac{3}{2}$

$y = \dfrac{1}{4} + \dfrac{6}{4} = \dfrac{7}{4}$

$$\left(-\frac{1}{8}, \frac{7}{4}\right)$$

12. (a) $3x - y = 4$

(b) $xy = 5$

$$y = \frac{5}{x}$$

Substitute (b) into (a) and get:

(a') $3x - \left(\dfrac{5}{x}\right) = 4$

$3x^2 - 5 = 4x$

$3x^2 - 4x - 5 = 0$

Solve this equation by using the quadratic formula.

$$x = \frac{-(-4) \pm \sqrt{(-4)^2 - 4(3)(-5)}}{2(3)}$$

$$x = \frac{2}{3} \pm \frac{\sqrt{19}}{3}$$

Substitute these values of x into (a) and solve for y.

(a) $y = 3\left(\dfrac{2}{3} + \dfrac{\sqrt{19}}{3}\right) - 4$

$\quad y = -2 + \sqrt{19}$

(a) $y = 3\left(\dfrac{2}{3} - \dfrac{\sqrt{19}}{3}\right) - 4$

$\quad y = -2 - \sqrt{19}$

$\left(\dfrac{\mathbf{2}}{\mathbf{3}} + \dfrac{\sqrt{\mathbf{19}}}{\mathbf{3}}, -\mathbf{2} + \sqrt{\mathbf{19}}\right)$ and

$\left(\dfrac{\mathbf{2}}{\mathbf{3}} - \dfrac{\sqrt{\mathbf{19}}}{\mathbf{3}}, -\mathbf{2} - \sqrt{\mathbf{19}}\right)$

13. (a) $x^2 + y^2 = 16$

(b) $\dfrac{2x^2 - y^2 = 2}{}$

$\quad \dfrac{3x^2 \qquad = 18}{}$

$\qquad x^2 = 6$

$\qquad x = \pm\sqrt{6}$

(a) $y^2 = 16 - (\sqrt{6})^2$

$\quad y^2 = 10$

$\quad y = \pm\sqrt{10}$

(a) $y^2 = 16 - (-\sqrt{6})^2$

$\quad y^2 = 10$

$\quad y = \pm\sqrt{10}$

$(\sqrt{\mathbf{6}}, \pm\sqrt{\mathbf{10}})$ and $(-\sqrt{\mathbf{6}}, \pm\sqrt{\mathbf{10}})$

14. (a) $3x + y + z = 7$

(b) $x - 2y - z = 2$

(c) $-x + y - z = -5$

(a) $3x + \quad y + z = 7$

(b) $\dfrac{x - 2y - z = 2}{}$

(d) $\dfrac{4x - \quad y \qquad = 9}{}$

(a) $3x + \quad y + z = \quad 7$

(c) $\dfrac{-x + \quad y - z = -5}{}$

(e) $\dfrac{2x + 2y \qquad = \quad 2}{}$

2(d) $8x - 2y = 18$

(e) $\dfrac{2x + 2y = \quad 2}{}$

$\quad \dfrac{10x \qquad = 20}{}$

$\qquad x = 2$

(e) $2(2) + 2y = 2$

$\qquad y = -1$

(c) $-(2) + (-1) - z = -5$

$\qquad z = 2$

$(\mathbf{2}, -\mathbf{1}, \mathbf{2})$

15. $p(x) = x^2 - 4; \; D = \{\text{Reals}\}$

$p\left(\dfrac{1}{2}\right) = \left(\dfrac{1}{2}\right)^2 - 4$

$\qquad = \dfrac{1}{4} - \dfrac{16}{4} = -\dfrac{\mathbf{15}}{\mathbf{4}}$

16. (a) $3y - 2x > -3$

$\qquad y > \dfrac{2}{3}x - 1$

(b) $x \geq -2$

The first step is to graph each of these lines.

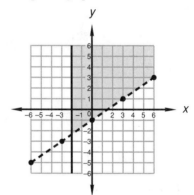

The region we wish to find is on or to the right of the solid line and above the dashed line. This region is shaded in the previous figure.

17. $-10 < x + 2 < -4; \; D = \{\text{Reals}\}$

$\quad -12 < x < -6$

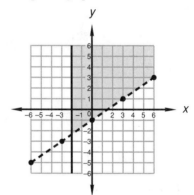

18. $\dfrac{(a^{x+2})^{1/2}\, y}{(y^2)^a} = \dfrac{a^{x/2+1}\, y}{y^{2a}} = a^{x/2+1}y^{1-2a}$

19. $\dfrac{m}{my - \dfrac{1}{1 - \dfrac{1}{my}}} = \dfrac{m}{my - \dfrac{my}{my - 1}}$

$\qquad = \dfrac{m(my - 1)}{m^2 y^2 - my - my} = \dfrac{\mathbf{my - 1}}{\mathbf{my^2 - 2y}}$

20. $7\sqrt{49\sqrt[3]{7}} = 7\sqrt{7^2\sqrt[3]{7}} = 7(7^2)^{1/2}(7^{1/6}) = \mathbf{7^{13/6}}$

21. $\dfrac{-2 - 3i}{3 + 2i} \cdot \dfrac{3 - 2i}{3 - 2i}$

$\quad = \dfrac{-6 + 4i - 9i - 6}{9 + 4} = -\dfrac{\mathbf{12}}{\mathbf{13}} - \dfrac{\mathbf{5}}{\mathbf{13}}i$

22. $\dfrac{i}{2i - 3} \cdot \dfrac{2i + 3}{2i + 3} = \dfrac{-2 + 3i}{-4 - 9}$

$= \dfrac{2}{13} - \dfrac{3}{13}i$

23.

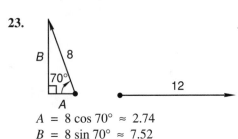

$A = 8 \cos 70° \approx 2.74$
$B = 8 \sin 70° \approx 7.52$

$\begin{array}{r} -2.74R + 7.52U \\ 12.00R + 0.00U \\ \hline \mathbf{9.26R + 7.52U} \end{array}$

$\tan \theta = \dfrac{7.52}{9.26}$
$\theta \approx 39.08°$
$F = \sqrt{(9.26)^2 + (7.52)^2} \approx 11.93$

11.93/39.08°

24. $\dfrac{2\sqrt{2} - \sqrt{3}}{\sqrt{6}} \cdot \dfrac{\sqrt{6}}{\sqrt{6}} = \dfrac{4\sqrt{3} - 3\sqrt{2}}{6}$

25. $\dfrac{3 - \sqrt{2}}{\sqrt{2} + 2} \cdot \dfrac{\sqrt{2} - 2}{\sqrt{2} - 2}$

$= \dfrac{3\sqrt{2} - 6 - 2 + 2\sqrt{2}}{2 - 4} = \dfrac{8 - 5\sqrt{2}}{2}$

26. $2\sqrt{\dfrac{2}{5}} - 4\sqrt{\dfrac{5}{2}} + 2\sqrt{40}$

$= \dfrac{2\sqrt{2}}{\sqrt{5}}\dfrac{\sqrt{5}}{\sqrt{5}} - \dfrac{4\sqrt{5}}{\sqrt{2}}\dfrac{\sqrt{2}}{\sqrt{2}} + 4\sqrt{10}$

$= \dfrac{4\sqrt{10}}{10} - \dfrac{20\sqrt{10}}{10} + \dfrac{40\sqrt{10}}{10} = \dfrac{12\sqrt{10}}{5}$

27. $-3i^2 - 2i^3 + 4 - i^5 + \sqrt{-9}$
$= 3 + 2i + 4 - i + 3i = \mathbf{7 + 4i}$

28. $\dfrac{m}{x} - c = \dfrac{1}{R_1} + \dfrac{1}{R_2}$

$mR_1R_2 - cR_1R_2x = R_2x + R_1x$

$mR_1R_2 - cR_1R_2x - R_1x = R_2x$

$R_1 = \dfrac{R_2x}{mR_2 - cR_2x - x}$

29. $-2x^2 = x + 4$
$2x^2 + x + 4 = 0$

$x = \dfrac{-1 \pm \sqrt{(1)^2 - 4(2)(4)}}{2(2)} = -\dfrac{1}{4} \pm \dfrac{\sqrt{31}}{4}i$

30. $\dfrac{2x^0 y^{-2} p^{-4} yp^4 m}{x^2 p^{-4} x^{-2}} - \dfrac{6x^0 y^{-2} p^{-4} xy^0}{x^2 p^{-4}}$

$= \dfrac{2y^{-1}m}{p^{-4}} - \dfrac{6xy^{-2}p^{-4}}{x^2 p^{-4}} = 2mp^4 y^{-1} - 6x^{-1}y^{-2}$

PROBLEM SET 102

1. $R_S T_S + 40 = R_{JR}T_{JR}$;
$R_S = 46$; $R_{JR} = 50$;
$T_S = T_{JR}$

$46(T_S) + 40 = 50(T_S)$
$4T_S = 40$
$T_S = 10$ hr
$D_S = R_S T_S = 46(10) = \mathbf{460\ miles}$

2. Downstream: $(B + W)T_D = D_D$ (a)
Upstream: $(B - W)T_U = D_U$ (b)
(a') $(10 + W)T = 78$
(b') $(10 - W)T = 42$

(a') $10T + WT = 78$
(b') $\underline{10T - WT = 42}$
$20T = 120$
$T = 6$

(a') $10(6) + 6W = 78$
$6W = 18$
$W = \mathbf{3\ mph}$

3. $\dfrac{H_1}{H_2} = \dfrac{G_1(P_2)^2}{G_2(P_1)^2}$

$\dfrac{5}{10} = \dfrac{2(6)^2}{G_2(4)^2}$

$G_2 = \dfrac{2(36)(10)}{5(16)} = \mathbf{9}$

4. Selling price = purchase price + markup
$1666 = P_P + 0.7P_P$
$1666 = 1.7P_P$
$\mathbf{\$980} = P_P$

5. Selling price = purchase price + markup
$1680 = P_P + 0.4P_P$
$1680 = 1.4P_P$
$\mathbf{\$1200} = P_P$

6. $\dfrac{V_1}{T_1} = \dfrac{V_2}{T_2}$

$\dfrac{450}{776} = \dfrac{V_2}{1016}$
$V_2 = \mathbf{589.2\ ml}$

7. $h(x) = x + 1$; $D = \{\text{Reals}\}$
$g(x) = x^2 - 6$; $D = \{\text{Negative integers}\}$

$h(-5) = -5 + 1 = -4$
$g(-5) = (-5)^2 - 6 = 19$
$hg(-5) = -4(19) = \mathbf{-76}$

8. $f(x) = x + 4$; $D = \{\text{Reals}\}$
$g(x) = x - 1$; $D = \{\text{Positive integers}\}$

$f(x)g(x) = (x + 4)(x - 1)$
$\quad fg(x) = x^2 + 3x - 4$

We cannot use this function to find $fg(-3)$ because
-3 is not a member of the domain of $g(x)$, so it is not
a member of the domain of $fg(x)$. Therefore, the
answer is either **∅ or { }**.

9. $y = x^2 + 4x + 2$
$y = (x^2 + 4x + 4) + 2 - 4$
$y = (x + 2)^2 - 2$
From this we see:
(a) Opens upward
(b) Axis of symmetry is $x = -2$
(c) y coordinate of vertex is -2

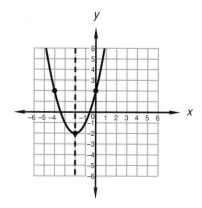

10. $y = -x^2 - 4x - 2$
$-y = (x^2 + 4x + 4) + 2 - 4$
$y = -(x + 2)^2 + 2$
From this we see:
(a) Opens downward
(b) Axis of symmetry is $x = -2$
(c) y coordinate of vertex is $+2$

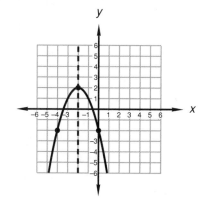

11. $-|x| - 2 \le -5$; $D = \{\text{Integers}\}$
$\quad -|x| \le -3$
$\quad |x| \ge 3$
$x \ge 3$ or $x \le -3$

12. (a) $2x + 3y = -3$
(b) $4x - 2y = 18$
$\quad\quad y = 2x - 9$

Substitute (b) into (a) and get:
(a') $2x + 3(2x - 9) = -3$
$\quad\quad 2x + 6x - 27 = -3$
$\quad\quad\quad\quad\quad 8x = 24$
$\quad\quad\quad\quad\quad\; x = 3$

(b) $y = 2(3) - 9 = -3$

(3, –3)

13. (a) $2x - y = 6$
(b) $xy = 4$
$\quad y = \dfrac{4}{x}$

Substitute (b) into (a) and get:

(a') $\quad 2x - \left(\dfrac{4}{x}\right) = 6$
$\quad\quad\quad 2x^2 - 4 = 6x$
$\quad\quad x^2 - 3x - 2 = 0$

Solve this equation by using the quadratic formula.

$$x = \frac{-(-3) \pm \sqrt{(-3)^2 - 4(1)(-2)}}{2(1)}$$

$$x = \frac{3}{2} \pm \frac{\sqrt{17}}{2}$$

Substitute these values of x into (a) and solve for y.

(a) $y = 2\left(\dfrac{3}{2} + \dfrac{\sqrt{17}}{2}\right) - 6$
$\quad\quad y = -3 + \sqrt{17}$

(a) $y = 2\left(\dfrac{3}{2} - \dfrac{\sqrt{17}}{2}\right) - 6$
$\quad\quad y = -3 - \sqrt{17}$

$\left(\dfrac{3}{2} + \dfrac{\sqrt{17}}{2}, -3 + \sqrt{17}\right)$ and

$\left(\dfrac{3}{2} - \dfrac{\sqrt{17}}{2}, -3 - \sqrt{17}\right)$

14. (a) $x^2 + y^2 = 12$

(b) $3x^2 - y^2 = 4$

(a) $x^2 + y^2 = 12$

$\underline{\text{(b) } 3x^2 - y^2 = 4}$

$4x^2 \qquad = 16$

$x^2 = 4$

$x = \pm 2$

(a) $y^2 = 12 - (2)^2$

$y^2 = 8$

$y = \pm 2\sqrt{2}$

(a) $y^2 = 12 - (-2)^2$

$y^2 = 8$

$y = \pm 2\sqrt{2}$

$(2, \pm 2\sqrt{2})$ and $(-2, \pm 2\sqrt{2})$

15. (a) $x + y + z = 8$

(b) $x + y - z = 0$

(c) $2x - y + z = 3$

(a) $x + y + z = 8$

$\underline{\text{(b) } x + y - z = 0}$

$2x + 2y \qquad = 8$

(d) $x + y \qquad = 4$

(b) $x + y - z = 0$

$\underline{\text{(c) } 2x - y + z = 3}$

$3x \qquad = 3$

$x = 1$

(d) $(1) + y = 4$

$y = 3$

(a) $(1) + (3) + z = 8$

$z = 4$

(1, 3, 4)

16. (a) $3x - 5y < 10$

$y > \dfrac{3}{5}x - 2$

(b) $y \geq -2$

The first step is to graph each of these lines.

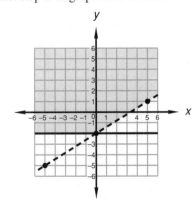

The region we wish to find is on or above the solid line and above the dashed line. This region is shaded in the previous figure.

17. $x + 2 < 0$ or $x + 3 \nleq 3$; $D = \{$Integers$\}$

$x < -2$ or $x > 0$

```
<---•---•---|---|---|---•---•--->
   -4  -3  -2  -1   0   1   2
```

18. $N = \dfrac{1}{5} + \dfrac{1}{8}\left(2\dfrac{1}{3} - \dfrac{1}{5}\right)$

$= \dfrac{1}{5} + \dfrac{1}{8}\left(\dfrac{35}{15} - \dfrac{3}{15}\right)$

$= \dfrac{1}{5} + \dfrac{32}{120}$

$= \dfrac{24}{120} + \dfrac{32}{120} = \dfrac{7}{15}$

19. $\dfrac{x^{a/2}\, y^{2a}}{x^{3a}\, y^{-2a/3}} = x^{-5a/2} y^{8a/3}$

20. $\dfrac{kx}{x - \dfrac{kx}{k - \dfrac{1}{x}}} = \dfrac{kx}{x - \dfrac{kx^2}{kx - 1}}$

$= \dfrac{kx(kx - 1)}{kx^2 - x - kx^2} = -k(kx - 1) = \mathbf{k - k^2 x}$

21. $\sqrt[3]{x^5 y}\,\sqrt[4]{xy^2} = x^{5/3}y^{1/3}x^{1/4}y^{1/2} = \mathbf{x^{23/12}y^{5/6}}$

22. $\dfrac{3i + 4}{-i^2 - i^5} = \dfrac{3i + 4}{1 - i} \cdot \dfrac{1 + i}{1 + i}$

$= \dfrac{3i + 3i^2 + 4 + 4i}{1 - i^2}$

$= \dfrac{1 + 7i}{2} = \dfrac{\mathbf{1}}{\mathbf{2}} + \dfrac{\mathbf{7}}{\mathbf{2}}\mathbf{i}$

23. $\dfrac{4i - 3}{i^3 - 2i^2} = \dfrac{4i - 3}{-i + 2} \cdot \dfrac{-i - 2}{-i - 2}$

$= \dfrac{4 - 8i + 3i + 6}{i^2 - 4}$

$= \dfrac{10 - 5i}{-5} = \mathbf{-2 + i}$

24. $H = \sqrt{10^2 + 4^2}$

$H = 2\sqrt{29}$

$\tan \theta = \dfrac{10}{4}$

$\theta \approx 68.20°$

$\mathbf{2\sqrt{29}\underline{/68.20°}}$

25. $\dfrac{a}{x} + \dfrac{b}{m + c} = ax$

$am + ac + bx = amx^2 + acx^2$

$ac - acx^2 = amx^2 - am - bx$

$c = \dfrac{amx^2 - am - bx}{a - ax^2}$

26. $\dfrac{2 - 2\sqrt{2}}{3\sqrt{2} - 2} \cdot \dfrac{3\sqrt{2} + 2}{3\sqrt{2} + 2}$

$= \dfrac{6\sqrt{2} + 4 - 12 - 4\sqrt{2}}{18 - 4} = \dfrac{-8 + 2\sqrt{2}}{14}$

$= \dfrac{-4 + \sqrt{2}}{7}$

27. $3i + 2 + 3i - 3i = \mathbf{2 + 3i}$

28. $4\sqrt{\dfrac{5}{12}} + 3\sqrt{\dfrac{12}{5}} - 3\sqrt{60}$

$= \dfrac{4\sqrt{5}}{\sqrt{12}} + \dfrac{3\sqrt{12}}{\sqrt{5}} - 3\sqrt{60}$

$= \dfrac{4\sqrt{60}}{12} + \dfrac{3\sqrt{60}}{5} - 3\sqrt{60}$

$= \dfrac{20\sqrt{60}}{60} + \dfrac{36\sqrt{60}}{60} - \dfrac{180\sqrt{60}}{60}$

$= -\dfrac{124\sqrt{60}}{60} = -\dfrac{\mathbf{62\sqrt{15}}}{\mathbf{15}}$

29. See Lesson 71.

30. **(c)** $\angle AED$ and $\angle CEA$ are supplementary angles and need not be equal.

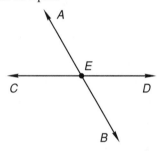

PROBLEM SET 103

1. Selling price = purchase price + markup
$1424 = P_P + 0.6P_P$
$1424 = 1.6P_P$
$\mathbf{\$890} = P_P$

2. Downstream: $(B + W)T_D = D_D$ (a)
Upstream: $(B - W)T_U = D_U$ (b)
(a′) $(18 + W)T = 132$
(b′) $(18 - W)T = 84$

(a′) $18T + WT = 132$
(b′) $\underline{18T - WT = 84}$
$36T = 216$
$T = 6$

(a′) $(18 + W)6 = 132$
$6W = 24$
$W = \mathbf{4\ mph}$

3. $R_D T_D = 720;$
$R_M T_M = 200;$
$T_D = 2T_M;$
$R_D = R_M + 40$

$(R_M + 40)2T_M = 720$
$2R_M T_M + 80T_M = 720$
$2(200) + 80T_M = 720$
$80T_M = 320$
$T_M = \mathbf{4\ hr}$

$T_D = \mathbf{8\ hr};\ R_M = \mathbf{50\ mph};\ R_D = \mathbf{90\ mph}$

4. $\dfrac{P_1}{T_1} = \dfrac{P_2}{T_2}$

$\dfrac{400}{300} = \dfrac{600}{T_2}$

$T_2 = \mathbf{450\ K}$

5. $11N \qquad 11(N + 1) \qquad 11(N + 2)$
$4(11N + 11N + 22) = 10(11N + 11) - 66$
$88N + 88 = 110N + 44$
$44 = 22N$
$2 = N$
The desired integers are **22**, **33** and **44**.

6.

$$3x + 2y \;\overline{\big)\; 27x^3 + 8y^3}$$

with quotient $9x^2 - 6xy + 4y^2$

$\underline{27x^3 + 18x^2y}$
$-18x^2y$
$\underline{-18x^2y - 12xy^2}$
$12xy^2 + 8y^3$
$\underline{12xy^2 + 8y^3}$

7. Since the lengths of tangent segments drawn to a circle from a point outside the circle are equal:
$y + 3 = 15$
$y = \mathbf{12}$
and
$y + 8 = x + 15$
$12 + 8 = x + 15$
$\mathbf{5} = x$

Perimeter of triangle $= x + 15 + y + 8 + x + x$
$ + y + 3 + 15$
$ = 3x + 2y + 41$
$ = 3(5) + 2(12) + 41 = \mathbf{80}$

8. Since 2 is not a member of the domain of $b(x)$, it is not a member of the domain of $ab(x)$. Therefore, the answer is either $\varnothing$ or $\{\ \}$.

9. $y = x^2 + 4x + 6$

$y = (x^2 + 4x + 4) + 6 - 4$

$y = (x + 2)^2 + 2$

From this we see:

(a) Opens upward

(b) Axis of symmetry is $x = -2$

(c) y coordinate of vertex is $+2$

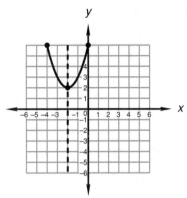

10. $y = -x^2 + 4x - 6$

$-y = (x^2 - 4x + 4) + 6 - 4$

$-y = (x - 2)^2 + 2$

$y = -(x - 2)^2 - 2$

From this we see:

(a) Opens downward

(b) Axis of symmetry is $x = 2$

(c) y coordinate of vertex is -2

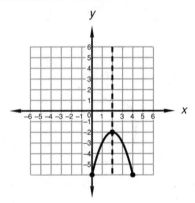

11. $x + 3 \geq 5$; $D = \{$Reals$\}$

$x \geq 2$

12. $N = \dfrac{1}{4} + \dfrac{2}{3}\left(2\dfrac{1}{2} - \dfrac{1}{4}\right)$

$= \dfrac{1}{4} + \dfrac{2}{3}\left(\dfrac{10}{4} - \dfrac{1}{4}\right)$

$= \dfrac{1}{4} + \dfrac{3}{2}$

$= \dfrac{1}{4} + \dfrac{6}{4} = \dfrac{7}{4}$

13. (a) $4x + 3y = 17$

(b) $2x - 3y = -5$

$y = \dfrac{2}{3}x + \dfrac{5}{3}$

Substitute (b) into (a) and get:

(a') $4x + 3\left(\dfrac{2}{3}x + \dfrac{5}{3}\right) = 17$

$4x + 2x + 5 = 17$

$6x = 12$

$x = 2$

(b) $y = \dfrac{2}{3}(2) + \dfrac{5}{3}$

$y = \dfrac{4}{3} + \dfrac{5}{3} = 3$

(2, 3)

14. (a) $x^2 + y^2 = 6$

(b) $x - y = 2$

$y = x - 2$

Substitute (b) into (a) and get:

(a') $x^2 + (x - 2)^2 = 6$

$x^2 + x^2 - 4x + 4 = 6$

$2x^2 - 4x - 2 = 0$

$x^2 - 2x - 1 = 0$

Solve this equation by using the quadratic formula.

$x = \dfrac{-(-2) \pm \sqrt{(-2)^2 - 4(1)(-1)}}{2(1)} = 1 \pm \sqrt{2}$

Substitute these values of x into (b) and solve for y.

(b) $y = (1 + \sqrt{2}) - 2$

$y = -1 + \sqrt{2}$

(b) $y = (1 - \sqrt{2}) - 2$

$y = -1 - \sqrt{2}$

$(1 + \sqrt{2}, -1 + \sqrt{2})$ and $(1 - \sqrt{2}, -1 - \sqrt{2})$

15. (a) $x^2 + y^2 = 10$

$y^2 = 10 - x^2$

(b) $2x^2 - 2y^2 = 5$

Substitute (a) into (b) and get:

(b') $2x^2 - 2(10 - x^2) = 5$

$2x^2 - 20 + 2x^2 = 5$

$4x^2 = 25$

$x^2 = \dfrac{25}{4}$

$x = \pm\dfrac{5}{2}$

Substitute these values of x into (a) and solve for y.

(a) $y^2 = 10 - \left(\dfrac{5}{2}\right)^2$

$y = \pm\dfrac{\sqrt{15}}{2}$

(a) $y^2 = 10 - \left(-\dfrac{5}{2}\right)^2$

$y = \pm\dfrac{\sqrt{15}}{2}$

$\left(\dfrac{5}{2}, \pm\dfrac{\sqrt{15}}{2}\right)$ and $\left(-\dfrac{5}{2}, \pm\dfrac{\sqrt{15}}{2}\right)$

16. (a) $x + 2y + z = -1$
(b) $3x - y + z = 6$
(c) $2x - 3y - z = 8$

$\dfrac{\begin{array}{l}\text{(b) } 3x - y + z = 6 \\ \text{(c) } 2x - 3y - z = 8\end{array}}{\text{(d) } 5x - 4y \qquad = 14}$

$\dfrac{\begin{array}{l}\text{(a) } x + 2y + z = -1 \\ \text{(c) } 2x - 3y - z = 8\end{array}}{\text{(e) } 3x - y \qquad = 7}$
(e) $y = 3x - 7$
Substitute (e) into (d) and get:
(d′) $5x - 4(3x - 7) = 14$
$\qquad 5x - 12x + 28 = 14$
$\qquad\qquad\qquad 7x = 14$
$\qquad\qquad\qquad\ x = 2$
(e) $y = 3(2) - 7 = -1$

(a) $(2) + 2(-1) + z = -1$
$\qquad\qquad\qquad\ z = -1$

(2, −1, −1)

17. (a) $x - 4y \le -4$

$y \ge \dfrac{1}{4}x + 1$

(b) $x < 3$
The first step is to graph each of these lines.

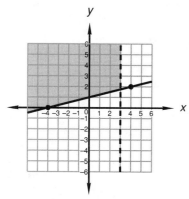

The region we wish to find is on or above the solid line and to the left of the dashed line. This region is shaded in the previous figure.

18. $-3 \le x - 3 \not> 4;\ D = \{\text{Integers}\}$
$0 \le x \le 7$

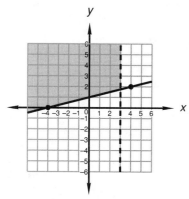

19. $\dfrac{x^{2ab-2b}}{x^{b/2}} = x^{2ab-5b/2}$

20. $\dfrac{m}{m^2 + \dfrac{m}{m^2 + \dfrac{1}{m}}} = \dfrac{m}{m^2 + \dfrac{m^2}{m^3 + 1}}$

$= \dfrac{m(m^3 + 1)}{m^5 + m^2 + m^2} = \dfrac{m^3 + 1}{m^4 + 2m}$

21. $\sqrt[5]{x^2 y^3}\ \sqrt[4]{xy} = x^{2/5}y^{3/5}x^{1/4}y^{1/4} = x^{13/20}y^{17/20}$

22. $\dfrac{-2 - i}{-i + 2} \cdot \dfrac{-i - 2}{-i - 2} = \dfrac{2i + 4 - 1 + 2i}{-1 - 4}$

$= \dfrac{3 + 4i}{-5} = -\dfrac{3}{5} - \dfrac{4}{5}i$

23. $\dfrac{2i - 5}{-5 - 2i} \cdot \dfrac{-5 + 2i}{-5 + 2i}$

$= \dfrac{-10i - 4 + 25 - 10i}{25 + 4} = \dfrac{21}{29} - \dfrac{20}{29}i$

24. $\dfrac{3 + 2\sqrt{5}}{5 - 2\sqrt{5}} \cdot \dfrac{5 + 2\sqrt{5}}{5 + 2\sqrt{5}}$

$= \dfrac{15 + 6\sqrt{5} + 10\sqrt{5} + 20}{25 - 20}$

$= \dfrac{35 + 16\sqrt{5}}{5}$

25.

$A = 6\cos 29° \approx 5.25$
$B = 6\sin 29° \approx 2.91$

$\dfrac{\begin{array}{l}-5.25R - 2.91U \\ 0.00R + 4.00U\end{array}}{-5.25R + \mathbf{1.09}U}$

$\tan\theta = -\dfrac{1.09}{5.25}$
$\quad \theta \approx -11.73°$
Since θ is a second-quadrant angle:
$\theta = (-11.73) + 180 = 168.27°$
$F = \sqrt{(-5.25)^2 + (1.09)^2} \approx 5.36$

5.36/168.27°

26. $a\left(\dfrac{b}{c} - \dfrac{1}{x}\right) = \dfrac{m}{p}$

$\dfrac{ab}{c} - \dfrac{a}{x} = \dfrac{m}{p}$

$abpx - acp = cmx$

$abpx - cmx = acp$

$x = \dfrac{acp}{abp - cm}$

27. $\sqrt{z} + \sqrt{z + 33} = 11$

$\sqrt{z + 33} = 11 - \sqrt{z}$

$z + 33 = 121 - 22\sqrt{z} + z$

$22\sqrt{z} = 88$

$\sqrt{z} = 4$

$z = \mathbf{16}$

Check: $\sqrt{16} + \sqrt{16 + 33} = 11$

$4 + 7 = 11$ check

28. $3\sqrt{\dfrac{4}{3}} - 2\sqrt{\dfrac{3}{4}} + 5\sqrt{48} = \dfrac{6}{\sqrt{3}} - \sqrt{3} + 5\sqrt{48}$

$= \dfrac{6\sqrt{3}}{3} - \sqrt{3} + 20\sqrt{3} = \mathbf{21\sqrt{3}}$

29. $4i - 2i(-3) - i = 4i + 6i - i = \mathbf{9i}$

30. Since there are $180°$ in a triangle:

$\angle ABD + \angle BDA + \angle BAD = 180°$

$\angle ABD + 90° + 40° = 180°$

$\angle ABD = 50°$

Since $\triangle ABC$ is isosceles:

$\angle DBC + \angle ABD = 70°$

$\angle DBC + 50° = 70°$

$\angle DBC = \mathbf{20°}$

PROBLEM SET 104

1. (a) $2N_D = 24N_G - 6$

(a') $N_D = 12N_G - 3$

(b) $10N_G = N_D - 5$

Substitute (a') into (b) and get:

(b') $10N_G = (12N_G - 3) - 5$

$2N_G = 8$

$N_G = \mathbf{4}$

(a') $N_D = 12(4) - 3 = \mathbf{45}$

2. $R_U T_U = R_B T_B;$

$R_U = 240; \; R_B = 360;$

$T_B = T_U - 4$

$240T_U = 360(T_U - 4)$

$120T_U = 1440$

$T_U = 12$

$D_U = R_U T_U = 240(12) = \mathbf{2880 \; miles}$

3. (a) $0.1P_N + 0.2D_N = 44$

(b) $P_N + D_N = 400$

Substitute $D_N = 400 - P_N$ into (a) and get:

(a') $0.1P_N + 0.2(400 - P_N) = 44$

$-0.1P_N = -36$

$P_N = 360$

(b) $(360) + D_N = 400$

$D_N = 40$

360 ml 10%, 40 ml 20%

4. Downstream: $(B + W)T_D = D_D$ (a)

Upstream: $(B - W)T_U = D_U$ (b)

(a') $12B + 12W = 168$

(b') $9B - 9W = 54$

$3(a') \; 36B + 36W = 504$

$4(b') \; \underline{36B - 36W = 216}$

$72B \qquad = 720$

$B = \mathbf{10 \; mph}$

(a') $12(10) + 12W = 168$

$12W = 48$

$W = \mathbf{4 \; mph}$

5. Selling price = purchase price + markup

$2400 = 400 + M_U$

$\$2000 = M_U$

% of cost $= \dfrac{2000}{400} \times 100\% = \mathbf{500\% \; of \; cost}$

% of $S_P = \dfrac{2000}{2400} \times 100\% = \mathbf{83.33\% \; of \; } S_P$

6. $0.00000512 \times \dfrac{10^8}{10^8} = \dfrac{\mathbf{512}}{\mathbf{100,000,000}}$

7. $N = 0.01432|32|32\ldots$

$100N = 1.432 \; 32 \; 32 \; 32\ldots$

$\underline{N = 0.014 \; 32 \; 32 \; 32\ldots}$

$99N = 1.418$

$N = \dfrac{\mathbf{1418}}{\mathbf{99,000}}$

8. $A_{\text{Triangle}} = \dfrac{1}{2}BH$

$H = \sqrt{\left(4\sqrt{3}\right)^2 - \left(2\sqrt{3}\right)^2}$

$H = \sqrt{48 - 12} = \sqrt{36} = 6 \; \text{cm}$

$A_{\text{Triangle}} = \dfrac{1}{2}(4\sqrt{3} \; \text{cm})(6 \; \text{cm}) = \mathbf{12\sqrt{3} \; cm^2}$

9. (a) Since this is a 45-45-90 triangle:

$$1 \times \overrightarrow{SF} = 3$$
$$\overrightarrow{SF} = 3$$

$$m = \sqrt{2}(3) = \mathbf{3\sqrt{2}}$$
$$n = 1(3) = \mathbf{3}$$

(b) Since this is a 30-60-90 triangle:

$$1 \times \overrightarrow{SF} = 4$$
$$\overrightarrow{SF} = 4$$

$$x = \sqrt{3}(4) = \mathbf{4\sqrt{3}}$$
$$y = 2(4) = \mathbf{8}$$

10.
$$\begin{array}{r} m^2 + mp + p^2 \\ m - p \overline{)\, m^3 \qquad\qquad\quad - p^3} \\ \underline{m^3 - m^2p} \\ m^2p \\ \underline{m^2p - mp^2} \\ mp^2 - p^3 \\ \underline{mp^2 - p^3} \end{array}$$

11. **(a)** and **(c)** are functions. **(b)** is not a function because -2 has two images.

12.
$$y = -x^2 + 4x - 2$$
$$-y = (x^2 - 4x + 4) + 2 - 4$$
$$-y = (x - 2)^2 - 2$$
$$y = -(x - 2)^2 + 2$$
From this we see:
(a) Opens downward
(b) Axis of symmetry is $x = 2$
(c) y coordinate of vertex is 2

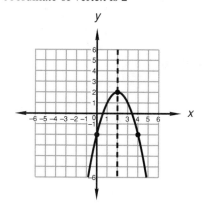

13. $-|x| + 2 > -2$; $D = \{\text{Integers}\}$
$$-|x| > -4$$
$$|x| < 4$$
$$x < 4 \text{ and } x > -4$$

14.
$$N = \frac{1}{8} + \frac{3}{7}\left(3\frac{1}{4} - \frac{1}{8}\right)$$
$$= \frac{1}{8} + \frac{3}{7}\left(\frac{26}{8} - \frac{1}{8}\right)$$
$$= \frac{1}{8} + \frac{3}{7}\left(\frac{25}{8}\right)$$
$$= \frac{7}{56} + \frac{75}{56} = \frac{\mathbf{41}}{\mathbf{28}}$$

15. (a) $\dfrac{1}{3}x - \dfrac{2}{5}y = -5$
(b) $0.005x - 0.04y = -0.755$

(a') $5x - 6y = -75$
(b') $5x - 40y = -755$

$$\begin{array}{r} \text{(a')} \quad 5x - 6y = -75 \\ -1\text{(b')} \quad -5x + 40y = 755 \\ \hline 34y = 680 \\ y = 20 \end{array}$$

(a') $5x - 6(20) = -75$
$$5x = 45$$
$$x = 9$$

(9, 20)

16. (a) $4x - y = 2$
(b) $xy = 3$
$$y = \frac{3}{x}$$
Substitute (b) into (a) and get:

(a') $4x - \left(\dfrac{3}{x}\right) = 2$
$$4x^2 - 3 = 2x$$
$$4x^2 - 2x - 3 = 0$$
Solve this equation by using the quadratic formula.

$$x = \frac{-(-2) \pm \sqrt{(-2)^2 - 4(4)(-3)}}{2(4)}$$

$$x = \frac{1}{4} \pm \frac{\sqrt{13}}{4}$$

Substitute these values of x into (a) and solve for y.

(a) $y = 4\left(\dfrac{1}{4} + \dfrac{\sqrt{13}}{4}\right) - 2$
$$y = -1 + \sqrt{13}$$

(a) $y = 4\left(\dfrac{1}{4} - \dfrac{\sqrt{13}}{4}\right) - 2$
$$y = -1 - \sqrt{13}$$

$$\left(\frac{1}{4} + \frac{\sqrt{13}}{4}, -1 + \sqrt{13}\right) \text{ and}$$

$$\left(\frac{1}{4} - \frac{\sqrt{13}}{4}, -1 - \sqrt{13}\right)$$

17. (a) $x - y - 2z = -14$
(b) $2x + y - z = 2$
(c) $-x + y - z = -4$

(a) $x - y - 2z = -14$
(b) $\underline{2x + y - z = \quad 2}$
$3x \qquad - 3z = -12$
(d) $x \qquad - z = -4$

(a) $x - y - 2z = -14$
(c) $\underline{-x + y - z = \quad -4}$
$\qquad - 3z = -18$
$\qquad z = 6$

(d) $x - (6) = -4$
$x = 2$

(a) $(2) - y - 2(6) = -14$
$y = 4$

(2, 4, 6)

18. (a) $x - 3y \le -6$
$y \ge \dfrac{1}{3}x + 2$

(b) $x \ge -2$
The first step is to graph each of these lines.

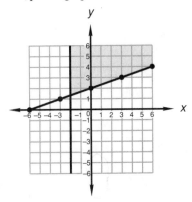

The region we wish to find is on or above the solid slanted line and on or to the right of the solid vertical line. This region is shaded in the previous figure.

19. $0 < x + 2 \not> 4; \ D = \{\text{Reals}\}$
$-2 < x \le 2$

$\begin{array}{ccccccc} \hline & \circ & & & & \bullet & \\ -3 & -2 & -1 & 0 & 1 & 2 & 3 \end{array}$

20. $\dfrac{(x^{-2})^{a+2} y^{-3a}}{y^{-a/2}} = \dfrac{x^{-2a-4} y^{-3a}}{y^{-a/2}} = x^{-2a-4} y^{-5a/2}$

21. $\dfrac{k}{k^2 x - \dfrac{1}{x - \dfrac{1}{k^2}}} = \dfrac{k}{k^2 x - \dfrac{k^2}{k^2 x - 1}}$

$= \dfrac{k(k^2 x - 1)}{k^4 x^2 - k^2 x - k^2} = \dfrac{k^2 x - 1}{k^3 x^2 - kx - k}$

22. $\sqrt[5]{4\sqrt[3]{2}} = (2^2)^{1/5} 2^{1/15} = 2^{2/5} 2^{1/15} = \mathbf{2^{7/15}}$

23. $\dfrac{6i - i^2}{-i^3 + 3} = \dfrac{6i + 1}{i + 3} \cdot \dfrac{i - 3}{i - 3}$

$= \dfrac{-6 - 18i + i - 3}{-1 - 9} = \dfrac{9}{10} + \dfrac{17}{10}i$

24. $\dfrac{2 + 3\sqrt{2}}{4 - \sqrt{18}} = \dfrac{2 + 3\sqrt{2}}{4 - 3\sqrt{2}} \cdot \dfrac{4 + 3\sqrt{2}}{4 + 3\sqrt{2}}$

$= \dfrac{8 + 6\sqrt{2} + 12\sqrt{2} + 18}{16 - 18} = \mathbf{-13 - 9\sqrt{2}}$

25. $-i^5 + \sqrt{-4}\sqrt{4} - 3\sqrt{-9} + 2i^4$
$= -i + 4i - 9i + 2 = \mathbf{2 - 6i}$

26. $3\sqrt{\dfrac{5}{2}} + 2\sqrt{\dfrac{2}{5}} - 4\sqrt{40} = \dfrac{3\sqrt{5}}{\sqrt{2}} + \dfrac{2\sqrt{2}}{\sqrt{5}} - 4\sqrt{40}$

$= \dfrac{3\sqrt{10}}{2} + \dfrac{2\sqrt{10}}{5} - 8\sqrt{10}$

$= \dfrac{15\sqrt{10}}{10} + \dfrac{4\sqrt{10}}{10} - \dfrac{80\sqrt{10}}{10} = \mathbf{-\dfrac{61\sqrt{10}}{10}}$

27.

$A = 30 \cos 45° \approx 21.21$
$B = 30 \sin 45° \approx 21.21$

$C = 20 \cos 20° \approx 18.79$
$D = 20 \sin 20° \approx 6.84$

$\begin{array}{l} 21.21R - 21.21U \\ \underline{18.79R - \ \ 6.84U} \\ \mathbf{40.00R - 28.05U} \end{array}$

$\tan \theta = -\dfrac{28.05}{40.00}$
$\theta = -35.04°$
$F = \sqrt{(40)^2 + (-28.05)^2} \approx 48.85$

$\mathbf{48.85\underline{/-35.04°}}$

28. $\dfrac{m}{x} + \dfrac{m}{y} = \dfrac{1}{c} + \dfrac{a}{b}$

$bcmy + bcmx = bxy + acxy$
$bcmy + bcmx - axcy = bxy$

$c = \dfrac{bxy}{bmy + bmx - axy}$

29.
$$-4x^2 = x - 5$$
$$4x^2 + x - 5 = 0$$
$$x^2 + \frac{1}{4}x - \frac{5}{4} = 0$$
$$\left(x^2 + \frac{1}{4}x + \frac{1}{64}\right) = \frac{5}{4} + \frac{1}{64}$$
$$\left(x + \frac{1}{8}\right)^2 = \frac{81}{64}$$
$$x + \frac{1}{8} = \pm\frac{9}{8}$$
$$x = -\frac{1}{8} \pm \frac{9}{8}$$
$$x = 1, -\frac{5}{4}$$

30.
$$\frac{4000\ ml}{sec} \times \frac{1\ cm^3}{1\ ml} \times \frac{1\ in.}{2.54\ cm} \times \frac{1\ in.}{2.54\ cm}$$
$$\times \frac{1\ in.}{2.54\ cm} \times \frac{1\ ft}{12\ in.} \times \frac{1\ ft}{12\ in.} \times \frac{1\ ft}{12\ in.}$$
$$\times \frac{60\ sec}{1\ min}$$
$$= \frac{4000(60)}{(2.54)(2.54)(2.54)(12)(12)(12)}\ \frac{ft^3}{min}$$

Problem Set 105

1. Carbon: $\ \ 3 \times 12 = 36$
Hydrogen: $\ \ 3 \times 1 = 3$
Chlorine: $\ \ 5 \times 35 = 175$
Total: $\qquad\quad = 214$
$$\frac{175}{214} = \frac{1050}{T}$$
$$T = \textbf{1284 grams}$$

2. $R_T T_T + R_W T_W = 64;$
$R_T = 10;\ R_W = 6;$
$T_T + T_W = 8$

$$10(8 - T_W) + 6T_W = 64$$
$$4T_W = 16$$
$$T_W = 4$$
$$T_T = 8 - (4) = 4$$
$$D_W = R_W T_W = 6(4) = \textbf{24 miles}$$
$$D_T = R_T T_T = 10(4) = \textbf{40 miles}$$

3. $0.2(140) + (D_N) = 0.44(140 + D_N)$
$$28 + D_N = 61.6 + 0.44D_N$$
$$0.56D_N = 33.6$$
$$D_N = \textbf{60 ml}$$

4.
$$\frac{P_1}{P_2} = \frac{M_1(F_1)^2}{M_2(F_2)^2}$$
$$\frac{750}{P_2} = \frac{5(5)^2}{10(10)^2}$$
$$P_2 = \textbf{6000}$$

5. Selling price = purchase price + markup
$$S_P = 2400 + 800$$
$$S_P = \$3200$$

$$\% \text{ of } S_P = \frac{800}{3200} \times 100\% = \textbf{25\% of } S_P$$
$$\% \text{ of cost} = \frac{800}{2400} \times 100\% = \textbf{33.33\% of cost}$$

6. $(g + h)(x) = g(x) + h(x)$
$$= x^2 + 1 + x - 5$$
$$= x^2 + x - 4$$

$$(g + h)(2) = (2)^2 + (2) - 4 = \textbf{2}$$

7. $hg(x) = h(x) \cdot g(x)$
$$= (x - 5)(x^2 + 1)$$
$$= x^3 + x - 5x^2 - 5$$
$$= \textbf{x}^3 - \textbf{5x}^2 + \textbf{x} - \textbf{5}$$

8. $N = 0.0001234|234|234\ldots$

$$1000N = 0.1234\ 234\ 234\ldots$$
$$\underline{N = 0.0001\ 234\ 234\ldots}$$
$$999N = 0.1233$$
$$N = \frac{\textbf{1233}}{\textbf{9,990,000}}$$

9. $N = 0.01651|651|651\ldots$

$$1000N = 16.51651\ 651\ 651\ldots$$
$$\underline{N = 0.01651\ 651\ 651\ldots}$$
$$999N = 16.5$$
$$N = \frac{\textbf{165}}{\textbf{9990}}$$

10.

$$
\begin{array}{r}
m^2 - mp + p^2 \\
m + p \overline{)\ m^3 \qquad\qquad\quad\ \ + p^3} \\
\underline{m^3 + m^2p} \\
-m^2p \\
\underline{-m^2p - mp^2} \\
mp^2 + p^3 \\
\underline{mp^2 + p^3}
\end{array}
$$

11. Divide $x^3 + y^3$ by $x + y$:

$$
\begin{array}{r}
x^2 - xy + y^2 \\
x + y \overline{) x^3 \qquad\qquad + y^3} \\
\underline{x^3 + x^2y} \\
- x^2y \\
\underline{- x^2y - xy^2} \\
xy^2 + y^3 \\
\underline{xy^2 + y^3}
\end{array}
$$

12. $y = x^2 + 6x + 8$

$y = (x^2 + 6x + 9) + 8 - 9$

$y = (x + 3)^2 - 1$

From this we see:

(a) Opens upward

(b) Axis of symmetry is $x = -3$

(c) y coordinate of vertex is -1

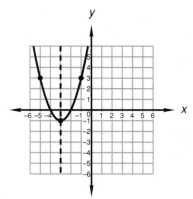

13. $-|x| + 2 > 1$; $D = \{$Integers$\}$

$\quad -|x| > -1$

$\quad\; |x| < 1$

$x < 1$ or $x > -1$

$$
\begin{array}{c}
\underset{-2 \quad -1 \quad\; 0 \quad\;\; 1 \quad\;\; 2}{\rule{3cm}{0.4pt}\;\bullet\;\rule{2cm}{0.4pt}}
\end{array}
$$

14. $N = 2\dfrac{1}{4} + \dfrac{3}{5}\left(4\dfrac{1}{2} - 2\dfrac{1}{4}\right)$

$\quad = \dfrac{9}{4} + \dfrac{3}{5}\left(\dfrac{18}{4} - \dfrac{9}{4}\right)$

$\quad = \dfrac{9}{4} + \dfrac{3}{5}\left(\dfrac{9}{4}\right) = \dfrac{45}{20} + \dfrac{27}{20} = \mathbf{\dfrac{18}{5}}$

15. (a) $\dfrac{2}{7}x - \dfrac{1}{4}y = -6$

(b) $0.07x + 0.14y = 6.58$

(a') $8x - 7y = -168$

(b') $7x + 14y = 658$

2(a') $16x - 14y = -336$

(b') $\underline{7x + 14y = 658}$

$\qquad 23x \qquad\quad = 322$

$\qquad\qquad\qquad x = 14$

(a') $8(14) - 7y = -168$

$\qquad\quad -7y = -280$

$\qquad\qquad y = 40$

(14, 40)

16. (a) $5x - y = 2$

(b) $xy = 6$

$\quad y = \dfrac{6}{x}$

Substitute (b) into (a) and get:

(a') $\quad 5x - \left(\dfrac{6}{x}\right) = 2$

$\qquad\quad 5x^2 - 6 = 2x$

$\qquad 5x^2 - 2x - 6 = 0$

Solve this equation by using the quadratic formula.

$x = \dfrac{-(-2) \pm \sqrt{(-2)^2 - 4(5)(-6)}}{2(5)}$

$x = \dfrac{1}{5} \pm \dfrac{\sqrt{31}}{5}$

Substitute these values of x into (a) and solve for y.

(a) $y = 5\left(\dfrac{1}{5} + \dfrac{\sqrt{31}}{5}\right) - 2$

$\quad\; y = -1 + \sqrt{31}$

(a) $y = 5\left(\dfrac{1}{5} - \dfrac{\sqrt{31}}{5}\right) - 2$

$\quad\; y = -1 - \sqrt{31}$

$\left(\dfrac{1}{5} + \dfrac{\sqrt{31}}{5}, -1 + \sqrt{31}\right)$ and

$\left(\dfrac{1}{5} - \dfrac{\sqrt{31}}{5}, -1 - \sqrt{31}\right)$

17. (a) $x^2 + y^2 = 8$

(b) $x - y = 2$

$\quad y = x - 2$

Substitute (b) into (a) and get:

(a') $\quad x^2 + (x - 2)^2 = 8$

$\qquad x^2 + x^2 - 4x + 4 = 8$

$\qquad\quad x^2 - 2x - 2 = 0$

Solve this equation by using the quadratic formula.

$x = \dfrac{-(-2) \pm \sqrt{(-2)^2 - 4(1)(-2)}}{2(1)} = 1 \pm \sqrt{3}$

Substitute these values of x into (b) and solve for y.

(b) $y = (1 + \sqrt{3}) - 2$

$\quad\; y = -1 + \sqrt{3}$

(b) $y = (1 - \sqrt{3}) - 2$

$\quad\; y = -1 - \sqrt{3}$

$\mathbf{(1 + \sqrt{3}, -1 + \sqrt{3})}$ and $\mathbf{(1 - \sqrt{3}, -1 - \sqrt{3})}$

18. (a) $2x - y - 2z = 2$
(b) $x + y - z = 7$
(c) $2x - y - z = 0$

(a) $2x - y - 2z = 2$
(b) $\underline{x + y - z = 7}$
$3x - 3z = 9$
(d) $x - z = 3$

(a) $2x - y - 2z = 2$
-1(c) $\underline{-2x + y + z = 0}$
$- z = 2$
$z = -2$

(d) $x - (-2) = 3$
$x = 1$

(b) $(1) + y - (-2) = 7$
$y = 4$

$\mathbf{(1, 4, -2)}$

19. (a) $3x + 5y = 4$
(b) $10x - 15y = -50$

$y = \dfrac{2}{3}x + \dfrac{10}{3}$

Substitute (b) into (a) and get:

(a') $3x + 5\left(\dfrac{2}{3}x + \dfrac{10}{3}\right) = 4$

$3x + \dfrac{10}{3}x + \dfrac{50}{3} = 4$

$\dfrac{19}{3}x = -\dfrac{38}{3}$

$\phantom{(a') 3x + \dfrac{19}{3}}x = -\dfrac{38}{3}\left(\dfrac{3}{19}\right) = -2$

(b) $y = \dfrac{2}{3}(-2) + \dfrac{10}{3}$

$y = -\dfrac{4}{3} + \dfrac{10}{3} = 2$

$\mathbf{(-2, 2)}$

20. (a) $x - 3y \geq 6$

$y \leq \dfrac{1}{3}x - 2$

(b) $x \geq -3$

The first step is to graph each of these lines.

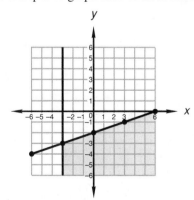

The region we wish to find is on or below the solid slanted line and on or to the right of the solid vertical line. This region is shaded in the previous figure.

21. $3x^3 - 5x^2 - 2x = 0$

$x(3x^2 - 5x - 2) = 0$

Now, we multiply the coefficient of x^2 by the constant term.

$3(-2) = -6$

Find the factors of -6 whose sum is -5.

$1(-6) = -6$
$1 + (-6) = -5$

Replace $-5x$ with $-6x + x$ and get:

$x(3x^2 - 6x + x - 2) = 0$
$x(3x(x - 2) + 1(x - 2)) = 0$
$x(3x + 1)(x - 2) = 0$

Complete the solution by using the zero factor theorem.

$x = 0$

$3x + 1 = 0$

$x = -\dfrac{1}{3}$

$x - 2 = 0$
$x = 2$

$\mathbf{0, -\dfrac{1}{3}, 2}$

22. $6x^2 - 10x - 4 = 0$

$2(3x^2 - 5x - 2) = 0$

Now, we multiply the coefficient of x^2 by the constant term.

$3(-2) = -6$

Find the factors of -6 whose sum is -5.

$1(-6) = -6$
$1 + (-6) = -5$

Replace $-5x$ with $-6x + x$ and get:

$3x^2 - 6x + x - 2 = 0$
$3x(x - 2) + 1(x - 2) = 0$
$(3x + 1)(x - 2) = 0$

Complete the solution by using the zero factor theorem.

$3x + 1 = 0$

$x = -\dfrac{1}{3}$

$x - 2 = 0$
$x = 2$

$\mathbf{-\dfrac{1}{3}, 2}$

23. $3x^2 + 8x + 4 = 0$

Now, we multiply the coefficient of x^2 by the constant term.

$3(4) = 12$

Find the factors of 12 whose sum is 8.

$6(2) = 12$

$6 + 2 = 8$

Replace $8x$ with $6x + 2x$ and get:

$3x^2 + 6x + 2x + 4 = 0$

$3x(x + 2) + 2(x + 2) = 0$

$(3x + 2)(x + 2) = 0$

Complete the solution by using the zero factor theorem.

$3x + 2 = 0$

$$x = -\frac{2}{3}$$

$x + 2 = 0$

$x = -2$

$$-\frac{2}{3}, -2$$

24. $9x^3 + 24x^2 + 12x = 0$

$3x(3x^2 + 8x + 4) = 0$

Now, we multiply the coefficient of x^2 by the constant term.

$3(4) = 12$

Find the factors of 12 whose sum is 8.

$6(2) = 12$

$6 + 2 = 8$

Replace $8x$ with $6x + 2x$ and get:

$3x(3x^2 + 6x + 2x + 4) = 0$

$3x(3x(x + 2) + 2(x + 2)) = 0$

$3x(3x + 2)(x + 2) = 0$

Complete the solution by using the zero factor theorem.

$3x = 0$

$x = 0$

$3x + 2 = 0$

$$x = -\frac{2}{3}$$

$x + 2 = 0$

$x = -2$

$$0, -\frac{2}{3}, -2$$

25. $2p^2 - 3p - 5 = 0$

Now, we multiply the coefficient of p^2 by the constant term.

$2(-5) = -10$

Find the factors of -10 whose sum is -3.

$2(-5) = -10$

$2 + (-5) = -3$

Replace $-3p$ with $2p - 5p$ and get:

$2p^2 + 2p - 5p - 5 = 0$

$2p(p + 1) - 5(p + 1) = 0$

$(2p - 5)(p + 1) = 0$

Complete the solution by using the zero factor theorem.

$2p - 5 = 0$

$$p = \frac{5}{2}$$

$p + 1 = 0$

$p = -1$

$$\frac{5}{2}, -1$$

26. $4x^2 + 18x + 8 = 0$

$2(2x^2 + 9x + 4) = 0$

Now, we multiply the coefficient of x^2 by the constant term.

$2(4) = 8$

Find the factors of 8 whose sum is 9.

$8(1) = 8$

$8 + 1 = 9$

Replace $9x$ with $8x + x$ and get:

$2x^2 + 8x + x + 4 = 0$

$2x(x + 4) + 1(x + 4) = 0$

$(2x + 1)(x + 4) = 0$

Complete the solution by using the zero factor theorem.

$2x + 1 = 0$

$$x = -\frac{1}{2}$$

$x + 4 = 0$

$x = -4$

$$-\frac{1}{2}, -4$$

27. $\dfrac{x^a (x^{a/2+4})^2 y^b}{y^{b/3} x^{a/6}} = \dfrac{x^a x^{a+8} y^b}{y^{b/3} x^{a/6}} = \dfrac{x^{2a+8} y^b}{y^{b/3} x^{a/6}}$

$= x^{11a/6+8} y^{2b/3}$

28. $\dfrac{2 - 3i^3}{i + 2i^2 + 3i^3} = \dfrac{2 + 3i}{-2i - 2} \cdot \dfrac{-2i + 2}{-2i + 2}$

$= \dfrac{-4i + 4 + 6 + 6i}{-4 - 4} = -\dfrac{5}{4} - \dfrac{1}{4}i$

29. $\dfrac{4 + 2\sqrt{5}}{5 - 3\sqrt{5}} \cdot \dfrac{5 + 3\sqrt{5}}{5 + 3\sqrt{5}}$

$= \dfrac{20 + 12\sqrt{5} + 10\sqrt{5} + 30}{25 - 45}$

$= \dfrac{-25 - 11\sqrt{5}}{10}$

30. Area of circles $= 2(\pi(1)^2) + \pi(2)^2$

$\qquad\qquad\qquad = 2\pi + 4\pi = 6\pi \text{ units}^2$

Area of circles $=$ Area of triangle

$6\pi = \dfrac{3(H)}{2}$

$4\pi \text{ units} = H$

Problem Set 106

1. Selling price $=$ purchase price $+$ markup

$715 = P_P + 0.3P_P$

$715 = 1.3\,P_P$

$\$550 = P_P$

2. $R_P T_P = 1200$; $R_C T_C = 250$;

$R_P = 6R_C$; $T_P = T_C - 1$

$6R_C(T_C - 1) = 1200$

$6(250) - 6R_C = 1200$

$6R_C = 300$

$R_C = 50 \text{ mph}$; $R_P = 300 \text{ mph}$;

$T_C = 5 \text{ hr}$; $T_P = 4 \text{ hr}$

$\dfrac{D_P}{} \rightarrow 1200$

$\dfrac{D_C}{} \rightarrow 250$

3. Downstream: $(B + W)T_D = D_D$ (a)

Upstream: $(B - W)T_U = D_U$ (b)

(a′) $(B + 2)T_D = 56$

(b′) $(B - 2)2T_D = 80$

2(a′) $2BT_D + 4T_D = 112$

(b′) $\underline{2BT_D - 4T_D = 80}$

(c) $4BT_D = 192$

$BT_D = 48$

(a′) $(48) + 2T_D = 56$

$T_D = 4$

(c) $B(4) = 48$

$B = 12 \text{ mph}$

4. $P_1V_1 = P_2V_2$

$700(500) = P_2(1000)$

$P_2 = \textbf{350 torr}$

5. (a) $3x - 3y = 9$

(b) $4x + z = 5$

(c) $4y + 2z = -10$

$-2(b) -8x - 2z = -10$

(c) $ 4y + 2z = -10$

$\overline{-8x + 4y = -20}$

(d) $-2x + y = -5$

$3(d) -6x + 3y = -15$

(a) $\underline{3x - 3y = 9}$

$-3x = -6$

$x = 2$

(d) $-2(2) + y = -5$

$y = -1$

(c) $4(-1) + 2z = -10$

$z = -3$

$(\mathbf{2, -1, -3})$

6. (a) $2x - 2y - z = 9$

(b) $3x + 3y - z = 6$

(c) $x + y + z = -2$

(a) $2x - 2y - z = 9$

(c) $\underline{x + y + z = -2}$

(d) $3x - y = 7$

(b) $3x + 3y - z = 6$

(c) $\underline{x + y + z = -2}$

$4x + 4y = 4$

(e) $x + y = 1$

(d) $3x - y = 7$

(e) $\underline{x + y = 1}$

$4x = 8$

$x = 2$

(e) $(2) + y = 1$

$y = -1$

(c) $(2) + (-1) + z = -2$

$z = -3$

$(\mathbf{2, -1, -3})$

7. $A_{\text{Triangle}} = \dfrac{bH}{2}$

$H = \sqrt{(7\pi)^2 - (4\pi)^2} = \sqrt{49\pi^2 - 16\pi^2}$

$ = \sqrt{33}\pi$

$A_{\text{Triangle}} = \dfrac{8\pi(\sqrt{33}\pi)}{2} = 4\sqrt{33}\pi^2 \text{ cm}^2$

8. (a) Since this is a 45-45-90 triangle:
$$\sqrt{2} \times \overrightarrow{SF} = 3$$
$$\overrightarrow{SF} = \frac{3\sqrt{2}}{2}$$

$$m = 1\left(\frac{3\sqrt{2}}{2}\right) = \frac{3\sqrt{2}}{2}$$
$$n = 1\left(\frac{3\sqrt{2}}{2}\right) = \frac{3\sqrt{2}}{2}$$

(b) Since this is a 30-60-90 triangle:
$$1 \times \overrightarrow{SF} = 7$$
$$\overrightarrow{SF} = 7$$

$$p = \sqrt{3}(7) = 7\sqrt{3}$$
$$q = 2(7) = 14$$

9.
$$1000N = 0.7013\ 013\ 013 \ldots$$
$$\underline{N = 0.0007\ 013\ 013 \ldots}$$
$$999N = 0.7006$$
$$N = \frac{7006}{9{,}990{,}000}$$

10.
$$100N = 410.26\ 26\ 26\ 26 \ldots$$
$$\underline{N = \quad 4.10\ 26\ 26\ 26 \ldots}$$
$$99N = 406.16$$
$$N = \frac{40{,}616}{9900}$$

11.
$$-y = x^2 + 2x + 3$$
$$-y = (x^2 + 2x + 1) + 3 - 1$$
$$-y = (x + 1)^2 + 2$$
$$y = -(x + 1)^2 - 2$$
From this we see:
(a) Opens downward
(b) Axis of symmetry is $x = -1$
(c) y coordinate of vertex is -2

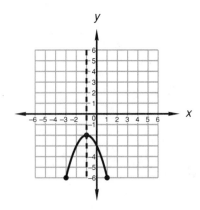

12. $-|x| - 4 \leq 0;\ D = \{\text{Integers}\}$
$$-|x| \leq 4$$
$$|x| \geq -4$$
All integers, because any integer (even zero) has an absolute value greater than -4.

13.
$$N = 3\frac{1}{3} + \frac{1}{10}\left(6\frac{1}{2} - 3\frac{1}{3}\right)$$
$$= \frac{10}{3} + \frac{1}{10}\left(\frac{13}{2} - \frac{10}{3}\right)$$
$$= \frac{10}{3} + \frac{1}{10}\left(\frac{19}{6}\right)$$
$$= \frac{200}{60} + \frac{19}{60} = \frac{73}{20}$$

14. (a) $\frac{3}{5}x - \frac{1}{4}y = 5$
(b) $0.012x + 0.07y = 2.20$

(a′) $12x - 5y = 100$
(b′) $12x + 70y = 2200$

$$\begin{array}{r} -1(a') \quad -12x + 5y = -100 \\ (b') \quad \underline{12x + 70y = 2200} \\ 75y = 2100 \\ y = 28 \end{array}$$

(a′) $12x - 5(28) = 100$
$$x = 20$$

(20, 28)

15. (a) $5x - y = 3$
(b) $xy = 4$
$$y = \frac{4}{x}$$
Substitute (b) into (a) and get:
(a′)
$$5x - \left(\frac{4}{x}\right) = 3$$
$$5x^2 - 4 = 3x$$
$$5x^2 - 3x - 4 = 0$$
Solve this equation by using the quadratic formula.
$$x = \frac{-(-3) \pm \sqrt{(-3)^2 - 4(5)(-4)}}{2(5)}$$
$$x = \frac{3}{10} \pm \frac{\sqrt{89}}{10}$$
Substitute these values of x into (a) and solve for y.
(a) $y = 5\left(\frac{3}{10} + \frac{\sqrt{89}}{10}\right) - 3$
$$y = -\frac{3}{2} + \frac{\sqrt{89}}{2}$$
(a) $y = 5\left(\frac{3}{10} - \frac{\sqrt{89}}{10}\right) - 3$
$$y = -\frac{3}{2} - \frac{\sqrt{89}}{2}$$
$$\left(\frac{3}{10} + \frac{\sqrt{89}}{10}, -\frac{3}{2} + \frac{\sqrt{89}}{2}\right) \text{ and}$$
$$\left(\frac{3}{10} - \frac{\sqrt{89}}{10}, -\frac{3}{2} - \frac{\sqrt{89}}{2}\right)$$

16. (a) $x^2 + y^2 = 7$
(b) $2x - y = 2$
$$y = 2x - 2$$

Substitute (b) into (a) and get:
(a') $\quad x^2 + (2x - 2)^2 = 7$
$$x^2 + 4x^2 - 8x + 4 = 7$$
$$5x^2 - 8x - 3 = 0$$

Solve this equation by using the quadratic formula.

$$x = \frac{-(-8) \pm \sqrt{(-8)^2 - 4(5)(-3)}}{2(5)}$$

$$x = \frac{4}{5} \pm \frac{\sqrt{31}}{5}$$

Substitute these values of x into (b) and solve for y.

(b) $y = 2\left(\dfrac{4}{5} + \dfrac{\sqrt{31}}{5}\right) - 2$

$$y = -\frac{2}{5} + \frac{2\sqrt{31}}{5}$$

(b) $y = 2\left(\dfrac{4}{5} - \dfrac{\sqrt{31}}{5}\right) - 2$

$$y = -\frac{2}{5} - \frac{2\sqrt{31}}{5}$$

$$\left(\frac{4}{5} + \frac{\sqrt{31}}{5}, -\frac{2}{5} + \frac{2\sqrt{31}}{5}\right) \text{ and }$$

$$\left(\frac{4}{5} - \frac{\sqrt{31}}{5}, -\frac{2}{5} - \frac{2\sqrt{31}}{5}\right)$$

17. (a) $5x - 3y = 27$
$$y = \frac{5}{3}x - 9$$
(b) $2x - 5y = 26$

Substitute (a) into (b) and get:

(b') $2x - 5\left(\dfrac{5}{3}x - 9\right) = 26$

$$\frac{6}{3}x - \frac{25}{3}x + 45 = 26$$

$$\frac{19}{3}x = 19$$

$$x = 19\left(\frac{3}{19}\right) = 3$$

(a) $y = \dfrac{5}{3}(3) - 9 = -4$

(3, -4)

18. (a) $3x - 4y \geq 8$
$$y \leq \frac{3}{4}x - 2$$
(b) $y > -2$
The first step is to graph each of these lines.

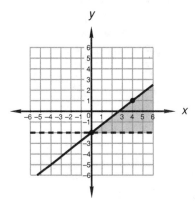

The region we wish to find is on or below the solid line and above the dashed line. This region is shaded in the previous figure.

19. $\dfrac{3 - 2i^2 - i}{3i^3 + 3i + 2} = \dfrac{5 - i}{2} = \dfrac{5}{2} - \dfrac{1}{2}i$

20. $$x^2 - \frac{1}{3}x = \frac{7}{3}$$
$$\left(x^2 - \frac{1}{3}x + \frac{1}{36}\right) = \frac{84}{36} + \frac{1}{36}$$
$$\left(x - \frac{1}{6}\right)^2 = \frac{85}{36}$$
$$x - \frac{1}{6} = \pm\frac{\sqrt{85}}{6}$$
$$x = \frac{1}{6} \pm \frac{\sqrt{85}}{6}$$

21.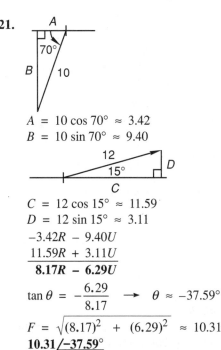

$A = 10 \cos 70° \approx 3.42$
$B = 10 \sin 70° \approx 9.40$

$C = 12 \cos 15° \approx 11.59$
$D = 12 \sin 15° \approx 3.11$
$-3.42R - 9.40U$
$\underline{11.59R + 3.11U}$
$\mathbf{8.17R - 6.29U}$

$\tan \theta = -\dfrac{6.29}{8.17} \quad \rightarrow \quad \theta \approx -37.59°$

$F = \sqrt{(8.17)^2 + (6.29)^2} \approx 10.31$
$\mathbf{10.31 \underline{/-37.59°}}$

22. **Yes**, because no two pairs have the same first element and different second elements.

23. Since $\frac{1}{2}$ is not a member of the domain of $\theta(x)$, it is not a member of the domain of $\psi\theta(x)$. Therefore, the answer is either $\varnothing$ or { }.

24.

$$\begin{array}{r} m^2 + mp + p^2 \\ m - p \overline{\smash{\big)}\ m^3 \qquad\qquad\quad - p^3} \\ \underline{m^3 - m^2p} \\ m^2p \\ \underline{m^2p - mp^2} \\ mp^2 - p^3 \\ \underline{mp^2 - p^3} \end{array}$$

25.
$$3x^2 + 7x + 2 = 0$$
$$3x^2 + 6x + x + 2 = 0$$
$$3x(x + 2) + 1(x + 2) = 0$$
$$(3x + 1)(x + 2) = 0$$

$$3x + 1 = 0 \qquad x + 2 = 0$$
$$x = -\frac{1}{3} \qquad x = -2$$
$$-\frac{1}{3}, -2$$

26.
$$3x^2 + x - 2 = 0$$
$$3x^2 + 3x - 2x - 2 = 0$$
$$3x(x + 1) - 2(x + 1) = 0$$
$$(3x - 2)(x + 1) = 0$$

$$3x - 2 = 0 \qquad x + 1 = 0$$
$$x = \frac{2}{3} \qquad x = -1$$
$$\frac{2}{3}, -1$$

27.
$$2z^2 + 13z + 15 = 0$$
$$2z^2 + 10z + 3z + 15 = 0$$
$$2z(z + 5) + 3(z + 5) = 0$$
$$(2z + 3)(z + 5) = 0$$

$$2z + 3 = 0 \qquad z + 5 = 0$$
$$z = -\frac{3}{2} \qquad z = -5$$
$$-\frac{3}{2}, -5$$

28.
$$6p^3 + 33p^2 + 45p = 0$$
$$3p(2p^2 + 11p + 15) = 0$$
$$3p(2p^2 + 6p + 5p + 15) = 0$$
$$3p(2p(p + 3) + 5(p + 3)) = 0$$
$$3p(2p + 5)(p + 3) = 0$$

$$3p = 0 \qquad 2p + 5 = 0 \qquad p + 3 = 0$$
$$p = 0 \qquad p = -\frac{5}{2} \qquad p = -3$$
$$0, -\frac{5}{2}, -3$$

29.
$$3p^2 - 13p - 10 = 0$$
$$3p^2 - 15p + 2p - 10 = 0$$
$$3p(p - 5) + 2(p - 5) = 0$$
$$(3p + 2)(p - 5) = 0$$

$$3p + 2 = 0 \qquad p - 5 = 0$$
$$p = -\frac{2}{3} \qquad p = 5$$
$$-\frac{2}{3}, 5$$

30.
$$2a^2 - 11a + 15 = 0$$
$$2a^2 - 6a - 5a + 15 = 0$$
$$2a(a - 3) - 5(a - 3) = 0$$
$$(2a - 5)(a - 3) = 0$$

$$2a - 5 = 0 \qquad a - 3 = 0$$
$$a = \frac{5}{2} \qquad a = 3$$
$$\frac{5}{2}, 3$$

PROBLEM SET 107

1. (a) $0.3P_N + 0.6D_N = 144$
(b) $P_N + D_N = 400$
Substitute $D_N = 400 - P_N$ into (a) and get:
(a') $0.3P_N + 0.6(400 - P_N) = 144$
$$-0.3P_N = -96$$
$$P_N = 320$$
(b) $(320) + D_N = 400$
$$D_N = 80$$

320 ml 30%, 80 ml 60%

2. Carbon: $\quad 1 \times 12 = 12$
Hydrogen: $\quad 3 \times 1 = 3$
Bromine: $\quad 1 \times 80 = 80$
Total: $\qquad\qquad\ = 95$

$$\%Br = \frac{80}{95} \times 100\% \approx \mathbf{84.21\%}$$

3. T = tens' digit
U = units' digit
$10T + U$ = original number
$10U + T$ = reversed number

(a) $T + U = 15$
$\qquad\ T = 15 - U$
(b) $10U + T = 10T + U + 9$

Substitute (a) into (b) and get:
(b′) $10U + (15 - U) = 10(15 - U) + U + 9$
$$9U + 15 = -9U + 159$$
$$18U = 144$$
$$U = 8$$
(a) $T = 15 - (8) = 7$
Original number = **78**

4. T = tens' digit
U = units' digit
$10T + U$ = original number
$10U + T$ = reversed number

(a) $T + U = 13$
$\quad\quad T = 13 - U$
(b) $10U + T = 10T + U - 9$
Substitute (a) into (b) and get:
(b′) $10U + (13 - U) = 10(13 - U) + U - 9$
$$9U + 13 = -9U + 121$$
$$18U = 108$$
$$U = 6$$
(a) $T = 13 - (6) = 7$
Original number = **76**

5. $\quad S_P = P_P + M_U$
$\quad 1400 = P_P + 0.7(1400)$
$\quad \$420 = P_P$

$\quad M_U = 0.7(1400) = \$980$

6. (a) $2x - y = -6$
(b) $3y + 2z = 12$
(c) $x - 3z = -11$

$\quad\quad$ (a) $\quad 2x - y \quad\quad = -6$
-2(c) $\underline{-2x \quad\quad + 6z = 22}$
$\quad\quad$ (d) $\quad\quad - y + 6z = 16$

$\quad\quad$ (b) $\quad 3y + 2z = 12$
3(d) $\underline{-3y + 18z = 48}$
$\quad\quad\quad\quad\quad\quad 20z = 60$
$\quad\quad\quad\quad\quad\quad\quad z = 3$

(d) $-y + 6(3) = 16$
$\quad\quad\quad\quad y = 2$

(a) $2x - (2) = -6$
$\quad\quad\quad x = -2$

(−2, 2, 3)

7. (a) $5x - y - z = 2$
(b) $x - 5y + z = -2$
(c) $-x + y - z = -2$

(b) $\quad x - 5y + z = -2$
(c) $\underline{-x + \quad y - z = -2}$
$\quad\quad\quad -4y \quad\quad = -4$
$\quad\quad\quad\quad\quad y = 1$

(a) $5x - y - z = 2$
(b) $\underline{x - 5y + z = -2}$
(d) $\overline{6x - 6y \quad\quad = 0}$

(d) $6x - 6(1) = 0$
$\quad\quad\quad x = 1$

(c) $-(1) + (1) - z = -2$
$\quad\quad\quad\quad\quad\quad z = 2$

(1, 1, 2)

8. $\quad 1000N = 1.213\ 213\ 213\ldots$
$\quad\quad\quad N = 0.001\ 213\ 213\ldots$
$\quad\overline{\ 999N = 1.212}$
$$N = \frac{\mathbf{1212}}{\mathbf{999{,}000}}$$

9.
$$x + y\ \overline{)\ \begin{array}{c} x^2 - xy + y^2 \\ x^3 \quad\quad\quad\quad\quad + y^3 \end{array}}$$
$$\underline{x^3 + x^2y}$$
$$- x^2y$$
$$\underline{- x^2y - xy^2}$$
$$xy^2 + y^3$$
$$\underline{xy^2 + y^3}$$

10. $-y = x^2 - 2x - 1$
$-y = (x^2 - 2x + 1) - 1 - 1$
$-y = (x - 1)^2 - 2$
$\ y = -(x - 1)^2 + 2$
From this we see:
(a) Opens downward
(b) Axis of symmetry is $x = 1$
(c) y coordinate of vertex is 2

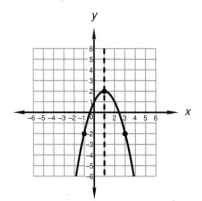

11. $x - 5 \geq -4;\ D = \{\text{Reals}\}$
$\quad\quad x \geq 1$

12. $N = 2 + \dfrac{3}{4}\left(6\dfrac{2}{3} - 2\right)$

$\qquad = 2 + \dfrac{3}{4}\left(\dfrac{14}{3}\right)$

$\qquad = \dfrac{4}{2} + \dfrac{7}{2} = \dfrac{11}{2}$

13. (a) $2x - y = 15$
$\qquad\qquad y = 2x - 15$
$\quad$ (b) $-6x - 40y = -432$
$\quad$ Substitute (a) into (b) and get:
$\quad$ (b') $-6x - 40(2x - 15) = -432$
$\qquad\qquad\qquad\qquad 86x = 1032$
$\qquad\qquad\qquad\qquad\; x = 12$

$\quad$ (a) $y = 2(12) - 15 = 9$

(12, 9)

14. (a) $4y - x = 2$
$\quad$ (b) $xy = 5$
$\qquad\qquad x = \dfrac{5}{y}$
$\quad$ Substitute (b) into (a) and get:

$\quad$ (a') $\quad 4y - \left(\dfrac{5}{y}\right) = 2$

$\qquad\qquad 4y^2 - 5 = 2y$

$\qquad\; 4y^2 - 2y - 5 = 0$

$\quad$ Solve this equation by using the quadratic formula.

$\quad y = \dfrac{-(-2) \pm \sqrt{(-2)^2 - 4(4)(-5)}}{2(4)}$

$\quad y = \dfrac{1}{4} \pm \dfrac{\sqrt{21}}{4}$

$\quad$ Substitute these values of y into (a) and solve for x.

$\quad$ (a) $x = 4\left(\dfrac{1}{4} + \dfrac{\sqrt{21}}{4}\right) - 2$

$\qquad\quad x = -1 + \sqrt{21}$

$\quad$ (a) $x = 4\left(\dfrac{1}{4} - \dfrac{\sqrt{21}}{4}\right) - 2$

$\qquad\quad x = -1 - \sqrt{21}$

$\left(-1 + \sqrt{21},\; \dfrac{1}{4} + \dfrac{\sqrt{21}}{4}\right)$ and

$\left(-1 - \sqrt{21},\; \dfrac{1}{4} - \dfrac{\sqrt{21}}{4}\right)$

15. (a) $x^2 + y^2 = 4$
$\quad$ (b) $x - 2y = 1$
$\qquad\qquad x = 2y + 1$
$\quad$ Substitute (b) into (a) and get:
$\quad$ (a') $\quad (2y + 1)^2 + y^2 = 4$
$\qquad\quad 4y^2 + 4y + 1 + y^2 = 4$
$\qquad\qquad\quad 5y^2 + 4y - 3 = 0$

Solve this equation by using the quadratic formula.

$y = \dfrac{-(4) \pm \sqrt{(4)^2 - 4(5)(-3)}}{2(5)}$

$y = -\dfrac{2}{5} \pm \dfrac{\sqrt{19}}{5}$

Substitute these values of y into (b) and solve for x.

$\quad$ (b) $x = 2\left(-\dfrac{2}{5} + \dfrac{\sqrt{19}}{5}\right) + 1$

$\qquad\quad x = \dfrac{1}{5} + \dfrac{2\sqrt{19}}{5}$

$\quad$ (b) $x = 2\left(-\dfrac{2}{5} - \dfrac{\sqrt{19}}{5}\right) + 1$

$\qquad\quad x = \dfrac{1}{5} - \dfrac{2\sqrt{19}}{5}$

$\left(\dfrac{1}{5} + \dfrac{2\sqrt{19}}{5},\; -\dfrac{2}{5} + \dfrac{\sqrt{19}}{5}\right)$ and

$\left(\dfrac{1}{5} - \dfrac{2\sqrt{19}}{5},\; -\dfrac{2}{5} - \dfrac{\sqrt{19}}{5}\right)$

16. (a) $5x - 3y = 32$
$\quad$ (b) $2x - 2y = 16$
$\qquad\qquad y = x - 8$
$\quad$ Substitute (b) into (a) and get:
$\quad$ (a') $5x - 3(x - 8) = 32$
$\qquad\qquad\qquad 2x = 8$
$\qquad\qquad\qquad\; x = 4$

$\quad$ (b) $y = (4) - 8 = -4$

(4, −4)

17. (a) $3x - 8y > -x$
$\qquad\qquad -8y > -4x$

$\qquad\qquad\quad y < \dfrac{1}{2}x$

$\quad$ (b) $x \le y$
$\qquad\quad y \ge x$
$\quad$ The first step is to graph each of these lines.

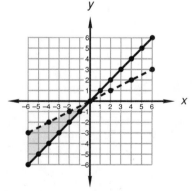

The region we wish to find is on or above the solid line and below the dashed line. This region is shaded in the previous figure.

18. $\dfrac{2i^3 - i + 2}{3 + 4i} = \dfrac{-3i + 2}{3 + 4i} \cdot \dfrac{3 - 4i}{3 - 4i}$

$= \dfrac{-9i - 12 + 6 - 8i}{9 + 16} = -\dfrac{6}{25} - \dfrac{17}{25}i$

19. $\sqrt[3]{x^5 y^2}\,\sqrt[4]{xy^3} = x^{5/3}y^{2/3}x^{1/4}y^{3/4} = x^{23/12}y^{17/12}$

20. $\sqrt{s - 48} = 8 - \sqrt{s}$

$s - 48 = 64 - 16\sqrt{s} + s$

$16\sqrt{s} = 112$

$\sqrt{s} = 7$

$s = 49$

Check: $\sqrt{49 - 48} = 8 - \sqrt{49}$

$1 = 1$ check

21. $H = \sqrt{(8)^2 + (4)^2} = 4\sqrt{5}$

$\tan \theta = \dfrac{8}{4}$

$\theta \approx 63.43°$

Since θ is a second-quadrant angle:

$\theta = -(63.43) + 180 = 116.57°$

$4\sqrt{5}\underline{/116.57°}$

22. Write the equation of the given line in slope-intercept form.

$5x - 3y = 4$

$y = \dfrac{5}{3}x - \dfrac{4}{3}$

Since the slopes of perpendicular lines are negative reciprocals of each other:

$m_\perp = -\dfrac{3}{5}$

$y = -\dfrac{3}{5}x + b$

$3 = -\dfrac{3}{5}(-2) + b$

$\dfrac{9}{5} = b$

$y = -\dfrac{3}{5}x + \dfrac{9}{5}$

23. $mx = \dfrac{1}{mr} + \dfrac{1}{mp}$

$m^2rpx = p + r$

$m^2rpx - r = p$

$r = \dfrac{p}{m^2 px - 1}$

24. $\dfrac{800 \text{ liters}}{\text{min}} \times \dfrac{1000 \text{ ml}}{1 \text{ liter}} \times \dfrac{1 \text{ cm}^3}{1 \text{ ml}}$

$\times \dfrac{1 \text{ in.}}{2.54 \text{ cm}} \times \dfrac{1 \text{ in.}}{2.54 \text{ cm}} \times \dfrac{1 \text{ in.}}{2.54 \text{ cm}}$

$\times \dfrac{1 \text{ ft}}{12 \text{ in.}} \times \dfrac{1 \text{ ft}}{12 \text{ in.}} \times \dfrac{1 \text{ ft}}{12 \text{ in.}} \times \dfrac{1 \text{ min}}{60 \text{ sec}}$

$= \dfrac{800(1000)}{(2.54)(2.54)(2.54)(12)(12)(12)(60)} \dfrac{\text{ft}^3}{\text{sec}}$

25. Area of rectangle = Area of right triangle

$10(5) = \dfrac{10(H)}{2}$

$10 \text{ m} = H$

$FG = \sqrt{(10)^2 + (10)^2} = 10\sqrt{2} \text{ m}$

26. Area of circle = Area of square

$\pi\left(2\sqrt{\pi}\right)^2 = s^2$

$4\pi^2 = s^2$

$2\pi \text{ cm} = s$

27. $4x^3 + 2x^2 - 30x = 0$

$2x(2x^2 + x - 15) = 0$

$2x(2x^2 + 6x - 5x - 15) = 0$

$2x(2x(x + 3) - 5(x + 3)) = 0$

$2x(2x - 5)(x + 3) = 0$

$2x = 0 \qquad 2x - 5 = 0 \qquad x + 3 = 0$

$x = 0 \qquad x = \dfrac{5}{2} \qquad x = -3$

$0, \dfrac{5}{2}, -3$

28. $8x^2 + 12x + 4 = 0$

$4(2x^2 + 3x + 1) = 0$

$2x^2 + 2x + x + 1 = 0$

$2x(x + 1) + 1(x + 1) = 0$

$(2x + 1)(x + 1) = 0$

$2x + 1 = 0 \qquad x + 1 = 0$

$x = -\dfrac{1}{2} \qquad x = -1$

$-\dfrac{1}{2}, -1$

29. $1000N = 74213.213\ 213\ 213\ldots$

$\underline{N = 74.213\ 213\ 213\ldots}$

$999N = 74,139$

$N = \dfrac{74,139}{999}$

30. $(h + g)(x) = (x + 2) + (x^3 + 2)$

$(h + g)(x) = x^3 + x + 4$

$(h + g)(-3) = (-3)^3 + (-3) + 4$

$(h + g)(-3) = -26$

PROBLEM SET 108

1.
$$S_P = P_P + M_U$$
$$6400 = P_P + 0.4(6400)$$
$$\mathbf{\$3840} = P_P$$

2. T = tens' digit
U = units' digit
$10T + U$ = original number
$10U + T$ = reversed number

(a) $T + U = 6$
$\quad T = 6 - U$
(b) $10U + T = 10T + U - 18$
Substitute (a) into (b) and get:
(b') $10U + (6 - U) = 10(6 - U) + U - 18$
$\quad\quad\quad 9U + 6 = -9U + 42$
$\quad\quad\quad\quad 18U = 36$
$\quad\quad\quad\quad\quad U = 2$
(a) $T = 6 - (2) = 4$
Original number = **42**

3. T = tens' digit
U = units' digit
$10T + U$ = original number
$10U + T$ = reversed number

(a) $T + U = 15$
$\quad T = 15 - U$
(b) $10U + T = 10T + U - 27$
Substitute (a) into (b) and get:
(b') $10U + (15 - U) = 10(15 - U) + U - 27$
$\quad\quad\quad 9U + 15 = -9U + 123$
$\quad\quad\quad\quad 18U = 108$
$\quad\quad\quad\quad\quad U = 6$
(a) $T = 15 - (6) = 9$
Original number = **96**

4. Equal ratio method:
$$\frac{Y_1}{Y_2} = \frac{(G_1)^2 B_2}{(G_2)^2 B_1}$$
$$\frac{100}{10} = \frac{(1)^2 B_2}{(10)^2 5}$$
$$B_2 = \mathbf{5000}$$

Variation method:
$$Y = k\frac{G^2}{B}$$
$$k = \frac{100(5)}{(1)^2} = 500$$

$$B = \frac{500(10)^2}{10}$$
$$B = \mathbf{5000}$$

5. Downstream: $(B + W)T_D = D_D$ (a)
Upstream: $(B - W)T_U = D_U$ (b)
(a') $(B + 3)T = 92$
(b') $(B - 3)T = 68$

$\quad$(a') $BT + 3T = 92$
$\quad$(b') $\underline{BT - 3T = 68}$
$\quad\quad\quad 2BT \quad\quad = 160$
$\quad$(c) $\quad\quad BT = 80$

(a') $(80) + 3T = 92$
$\quad\quad\quad\quad T = 4$ hr
(c) $B(4) = 80$
$\quad\quad B = \mathbf{20}$ **mph**
$T_D = T_U = \mathbf{4}$ **hr**

6. $27a^6p^{12} + y^3 = (3a^2p^4)^3 + (y)^3$
$\quad = \mathbf{(3a^2p^4 + y)(9a^4p^8 - 3a^2p^4y + y^2)}$

7. $8x^{12}z^6 - m^3y^9 = (2x^4z^2)^3 - (my^3)^3$
$\quad = \mathbf{(2x^4z^2 - my^3)(4x^8z^4 + 2mx^4y^3z^2 + m^2y^6)}$

8. $m^6y^9 - z^6 = (m^2y^3)^3 - (z^2)^3$
$\quad = \mathbf{(m^2y^3 - z^2)(m^4y^6 + m^2y^3z^2 + z^4)}$

9. $\sqrt{13^2 - 12^2} = \sqrt[n]{125}$
$\quad\quad\quad \sqrt{25} = \sqrt[n]{125}$
$\quad\quad\quad\quad 5 = \sqrt[n]{125}$
$\quad\quad\quad 5^n = 125$
$\quad\quad\quad\quad n = \mathbf{3}$

10. $10N = 41.23\ 3\ 3\ldots$
$\quad\ \underline{N = \ \ 4.12\ 3\ 3\ldots}$
$\quad 9N = 37.11$
$\quad\ N = \dfrac{\mathbf{3711}}{\mathbf{900}}$

11. $y = x^2 - 2x - 1$
$y = (x^2 - 2x + 1) - 1 - 1$
$y = (x - 1)^2 - 2$
From this we see:
(a) Opens upward
(b) Axis of symmetry is $x = 1$
(c) y coordinate of vertex is -2

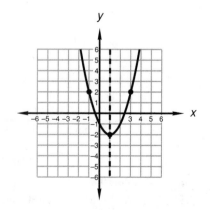

12. $-|x| - 3 < -5; \quad D = \{\text{Reals}\}$
$$-|x| < -2$$
$$|x| > 2$$
$$x > 2 \quad \text{or} \quad x < -2$$

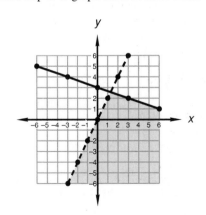

Number line: marks at $-3, -2, -1, 0, 1, 2, 3$ with open circles at -2 and 2, shaded to the left of -2 and to the right of 2.

13. (a) $\dfrac{2}{5}x - \dfrac{1}{4}y = 2$

(b) $-0.008x - 0.2y = -1.68$

(a′) $8x - \quad 5y = \quad 40$
(b′) $-8x - 200y = -1680$
$$\overline{\qquad\quad - 205y = -1640}$$
$$y = 8$$

(a′) $8x - 5(8) = 40$
$$8x = 80$$
$$x = 10$$
(10, 8)

14. (a) $x^2 + y^2 = 5$

(b) $y - 2x = 2$
$$y = 2x + 2$$
Substitute (b) into (a) and get:

(a′) $x^2 + (2x + 2)^2 = 5$
$$x^2 + 4x^2 + 8x + 4 = 5$$
$$5x^2 + 8x - 1 = 0$$
Solve this equation by using the quadratic formula.

$$x = \frac{-(8) \pm \sqrt{(8)^2 - 4(5)(-1)}}{2(5)}$$

$$x = -\frac{4}{5} \pm \frac{\sqrt{21}}{5}$$

Substitute these values of x into (b) and solve for y.

(b) $y = 2\left(-\dfrac{4}{5} + \dfrac{\sqrt{21}}{5}\right) + 2$

$$y = \frac{2}{5} + \frac{2\sqrt{21}}{5}$$

(b) $y = 2\left(-\dfrac{4}{5} - \dfrac{\sqrt{21}}{5}\right) + 2$

$$y = \frac{2}{5} - \frac{2\sqrt{21}}{5}$$

$$\left(-\frac{4}{5} + \frac{\sqrt{21}}{5}, \ \frac{2}{5} + \frac{2\sqrt{21}}{5}\right) \text{ and }$$

$$\left(-\frac{4}{5} - \frac{\sqrt{21}}{5}, \ \frac{2}{5} - \frac{2\sqrt{21}}{5}\right)$$

15. (a) $4x + 2y = 8$

(b) $3x - 3z = -9$

(c) $-3x + z = 1$

(b) $3x - 3z = -9$
(c) $\underline{-3x + \quad z = \quad 1}$
$$- 2z = -8$$
$$z = 4$$

(c) $-3x + (4) = 1$
$$x = 1$$

(a) $4(1) + 2y = 8$
$$y = 2$$
(1, 2, 4)

16. (a) $x - y - 3z = -2$

(b) $3x + y + z = 12$

(c) $2x - y + z = 5$

(a) $x - y - 3z = -2$
(b) $\underline{3x + y + \quad z = 12}$
(d) $4x \qquad\quad - 2z = 10$

(b) $3x + y + \quad z = 12$
(c) $\underline{2x - y + \quad z = \quad 5}$
(e) $5x \qquad\quad + 2z = 17$

(d) $4x - 2z = 10$
(e) $\underline{5x + 2z = 17}$
$$9x \qquad = 27$$
$$x = 3$$

(d) $4(3) - 2z = 10$
$$z = 1$$

(a) $(3) - y - 3(1) = -2$
$$y = 2$$
(3, 2, 1)

17. (a) $-x - 3y \geq -9$

$$y \leq -\frac{1}{3}x + 3$$

(b) $y < 2x$

The first step is to graph each of these lines.

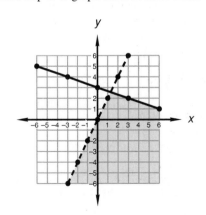

The region we wish to find is on or below the solid line and below the dashed line. This region is shaded in the previous figure.

18. $\dfrac{-i^3 - \sqrt{-2}\sqrt{-2}}{i^2 - 2i} = \dfrac{i + 2}{-1 - 2i} \cdot \dfrac{-1 + 2i}{-1 + 2i}$

$= \dfrac{-i - 2 - 2 + 4i}{1 + 4} = -\dfrac{4}{5} + \dfrac{3}{5}i$

19. $\dfrac{4 + 3\sqrt{5}}{2 + \sqrt{5}} \cdot \dfrac{2 - \sqrt{5}}{2 - \sqrt{5}}$

$= \dfrac{8 - 4\sqrt{5} + 6\sqrt{5} - 15}{4 - 5}$

$= 7 - 2\sqrt{5}$

20. $\sqrt[6]{9\sqrt[3]{3}} = (3^2)^{1/6}3^{1/18} = 3^{1/3}3^{1/18} = 3^{7/18}$

21. $\dfrac{5\sqrt{3}}{\sqrt{5}} + \dfrac{2\sqrt{5}}{\sqrt{3}} - \sqrt{60} = \dfrac{5\sqrt{15}}{5} + \dfrac{2\sqrt{15}}{3} - 2\sqrt{15}$

$= \dfrac{15\sqrt{15}}{15} + \dfrac{10\sqrt{15}}{15} - \dfrac{30\sqrt{15}}{15} = -\dfrac{\sqrt{15}}{3}$

22. $H = \sqrt{(4)^2 + (14)^2} = 2\sqrt{53}$

$\tan\theta = \dfrac{14}{4}$

$\theta \approx -74.05°$

$2\sqrt{53}/\!-\!74.05°$

23. $\dfrac{40\ \text{ft}}{\text{sec}} \times \dfrac{1\ \text{mi}}{5280\ \text{ft}} \times \dfrac{60\ \text{sec}}{1\ \text{min}} \times \dfrac{60\ \text{min}}{1\ \text{hr}}$

$= \dfrac{40(60)(60)}{5280} \dfrac{\text{mi}}{\text{hr}}$

24. $(x_1, y_1) = (-3, 7)$

$(x_2, y_2) = (5, 3)$

$D = \sqrt{(-3 - 5)^2 + (7 - 3)^2} = \sqrt{80}$

$= 4\sqrt{5}$

25. There is never an exact solution to these problems. One possible solution is given here.

$W = mE + b$

Use the graph to find the slope.

$m = \dfrac{150}{10} = 15$

$W = 15E + b$

Use the point (40, 100) for E and W.

$100 = 15(40) + b$

$-500 = b$

$W = 15E - 500$

26. $\dfrac{ab}{a^2 + \dfrac{ab^2}{a^3 + 1}} = \dfrac{ab(a^3 + 1)}{a^5 + a^2 + ab^2}$

$= \dfrac{a^3 b + b}{a^4 + a + b^2}$

27. See Lesson 71.

28.
$2x^2 + 9x + 9 = 0$
$2x^2 + 6x + 3x + 9 = 0$
$2x(x + 3) + 3(x + 3) = 0$
$(2x + 3)(x + 3) = 0$

$2x + 3 = 0 \qquad x + 3 = 0$

$x = -\dfrac{3}{2} \qquad\quad x = -3$

$-\dfrac{3}{2}, -3$

29.
$2b(3b^2 + 10b + 3) = 0$
$2b(3b^2 + 9b + b + 3) = 0$
$2b(3b(b + 3) + 1(b + 3)) = 0$
$2b(3b + 1)(b + 3) = 0$

$2b = 0 \qquad 3b + 1 = 0 \qquad\quad b + 3 = 0$

$b = 0 \qquad\quad b = -\dfrac{1}{3} \qquad\qquad b = -3$

$0, -\dfrac{1}{3}, -3$

30. $(h + p)(x) = x^2 + x^3$

$(h + p)(-3) = (-3)^2 + (-3)^3 = -18$

PROBLEM SET 109

1. T = tens' digit
U = units' digit
$10T + U$ = original number

(a) $10T + U = 8(T + U)$

(b) $6U = T + 5$
$T = 6U - 5$
Substitute (b) into (a) and get:
(a′) $10(6U - 5) + U = 8(6U - 5) + 8U$
$61U - 50 = 56U - 40$
$5U = 10$
$U = 2$

(b) $T = 6(2) - 5 = 7$
Original number = **72**

2. T = tens' digit
U = units' digit
$10T + U$ = original number
$10U + T$ = reversed number

(a) $T + U = 11$
$T = 11 - U$

(b) $10U + T = 10T + U - 27$
Substitute (a) into (b) and get:
(b′) $10U + (11 - U) = 10(11 - U) + U - 27$
$9U + 11 = -9U + 83$
$18U = 72$
$U = 4$

(a) $T = 11 - (4) = 7$
Original number = **74**

3. $R_L T_L = 63$; $R_C T_C = 60$;
$T_L = T_C + 11$; $R_C = 2R_L$

$$D_L \atop 63$$

$2R_L(T_L - 11) = 60$
$2(63) - 22R_L = 60$
$\quad -22R_L = -66$
$\qquad R_L = 3$

$$D_C \atop 60$$

$R_L = $ **3 mph**; $R_C = $ **6 mph**; $T_L = $ **21 hr**;
$T_C = $ **10 hr**

4. $\dfrac{P_1}{T_1} = \dfrac{P_2}{T_2}$

$\dfrac{800}{400} = \dfrac{P_2}{1200}$

$P_2 = $ **2400 torr**

5. $0.46(600) - 1(D_N) = 0.4(600 - D_N)$
$\qquad 276 - D_N = 240 - 0.4D_N$
$\qquad\quad 0.6D_N = 36$
$\qquad\qquad D_N = $ **60 ml**

6. Apply the power rule and get:
$27x^{3/4}y^{3/2}m^3$

7. $x^{1/4} + y^{1/4}$
$\dfrac{x^{1/4} + y^{1/4}}{}$
$x^{1/2} + \quad x^{1/4}y^{1/4}$
$\dfrac{\qquad\quad x^{1/4}y^{1/4} + y^{1/2}}{x^{1/2} + 2x^{1/4}y^{1/4} + y^{1/2}}$

8. $x^{1/4} + y^{-1/4}$
$\dfrac{x^{1/4} + y^{-1/4}}{}$
$x^{1/2} + \quad x^{1/4}y^{-1/4}$
$\dfrac{\qquad\quad x^{1/4}y^{-1/4} + y^{-1/2}}{x^{1/2} + 2x^{1/4}y^{-1/4} + y^{-1/2}}$

9. $x^3y^6 - 27m^3 = (xy^2)^3 - (3m)^3$
$\quad = (xy^2 - 3m)(x^2y^4 + 3mxy^2 + 9m^2)$

10. $64x^9y^6 + p^{12}z^3 = (4x^3y^2)^3 + (p^4z)^3$
$\quad = (4x^3y^2 + p^4z)(16x^6y^4 - 4p^4x^3y^2z + p^8z^2)$

11. Show that $1.023\overline{42}$ is a rational number.

$\quad 100N = 102.342\ 42\ 42\ldots$
$\quad \dfrac{N = \quad\ \ 1.023\ 42\ 42\ldots}{99N = 101.319}$

$\qquad N = \dfrac{\mathbf{101,319}}{\mathbf{99,000}}$

12. $-y = x^2 - 4x + 1$
$\quad -y = (x^2 - 4x + 4) + 1 - 4$
$\quad -y = (x - 2)^2 - 3$
$\qquad y = -(x - 2)^2 + 3$
From this we see:
(a) Opens downward
(b) Axis of symmetry is $x = 2$
(c) y coordinate of vertex is 3

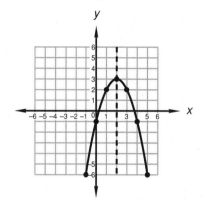

13. $-2 \geq x + 2 > -4$; $D = \{\text{Reals}\}$
$\quad -4 \geq x > -6$

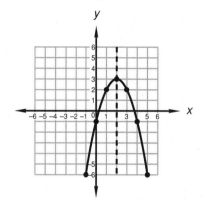

14. (a) $\dfrac{2}{5}x - \dfrac{1}{3}y = 1$
(b) $0.3x - 0.05y = 2.55$

(a′) $6x - 5y = 15$
(b′) $30x - 5y = 255$

$-1(a′)\ -6x + 5y = -15$
$\ (b′)\ \dfrac{30x - 5y = 255}{}$
$\qquad\quad 24x\qquad\ = 240$
$\qquad\qquad\quad x = 10$

(a′) $6(10) - 5y = 15$
$\qquad\qquad y = 9$

(10, 9)

15. (a) $x - 2y = 5$

(b) $xy = 3$

$$x = \frac{3}{y}$$

Substitute (b) into (a) and get:

(a′) $\left(\dfrac{3}{y}\right) - 2y = 5$

$$3 - 2y^2 = 5y$$

$$2y^2 + 5y - 3 = 0$$

Solve this equation by using the quadratic formula.

$$y = \frac{-(5) \pm \sqrt{(5)^2 - 4(2)(-3)}}{2(2)}$$

$$y = -\frac{5}{4} \pm \frac{7}{4} = \frac{1}{2}, -3$$

Substitute these values of y into (a) and solve for x.

(a) $x = 2\left(\dfrac{1}{2}\right) + 5$

$\quad x = 6$

(a) $x = 2(-3) + 5$

$\quad x = -1$

$\left(6, \dfrac{1}{2}\right)$ and $(-1, -3)$

16. (a) $x + y - z = 7$

(b) $4x + y + z = 4$

(c) $3x + y - z = 9$

(a) $\quad x + \quad y - z = \quad 7$

(b) $\underline{4x + \quad y + z = \quad 4}$

(d) $5x + 2y \qquad = 11$

(b) $4x + \quad y + z = \quad 4$

(c) $\underline{3x + \quad y - z = \quad 9}$

(e) $7x + 2y \qquad = 13$

-1(d) $-5x - 2y = -11$

(e) $\quad \underline{7x + 2y = \quad 13}$

$\qquad 2x \qquad = \quad 2$

$\qquad\quad x = 1$

(d) $5(1) + 2y = 11$

$\qquad\qquad y = 3$

(a) $(1) + (3) - z = 7$

$\qquad\qquad z = -3$

(1, 3, −3)

17. (a) $2x + 3y = 15$

(b) $x - 2z = -3$

(c) $3y - z = 6$

$\qquad z = 3y - 6$

Substitute (c) into (b) and get:

(b′) $x - 2(3y - 6) = -3$

$\qquad\quad x - 6y = -15$

2(a) $\quad 4x + 6y = \quad 30$

(b′) $\quad \underline{x - 6y = -15}$

$\qquad 5x \qquad = \quad 15$

$\qquad\quad x = 3$

(a) $2(3) + 3y = 15$

$\qquad\qquad y = 3$

(c) $z = 3(3) - 6 = 3$

(3, 3, 3)

18. (a) $2x - 5y \geq 15$

$$y \leq \frac{2}{5}x - 3$$

(b) $y \leq -x$

The first step is to graph each of these lines.

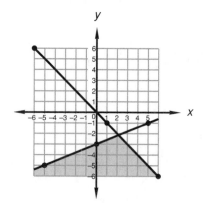

The region we wish to find is on or below the solid lines. This region is shaded in the previous figure.

19. $\dfrac{2i^2 - \sqrt{-9} + 2}{3 - \sqrt{-2}\sqrt{2}} = \dfrac{-3i}{3 - 2i} \cdot \dfrac{3 + 2i}{3 + 2i}$

$= \dfrac{-9i + 6}{9 + 4} = \dfrac{6}{13} - \dfrac{9}{13}i$

20. $\dfrac{3 + 2\sqrt{2}}{5\sqrt{2} - 2} \cdot \dfrac{5\sqrt{2} + 2}{5\sqrt{2} + 2}$

$= \dfrac{15\sqrt{2} + 6 + 20 + 4\sqrt{2}}{50 - 4} = \dfrac{26 + 19\sqrt{2}}{46}$

21. $\sqrt[3]{27}\sqrt[3]{3} = (3^3)^{1/3}3^{1/3} = 3(3^{1/3}) = 3^{4/3}$

22. $\sqrt{x^5 y}\sqrt{x^2 y} = x^{5/2}y^{1/2}xy^{1/2} = x^{7/2}y$

23. $\dfrac{x^{a/3 - 2/3}y^{b/2}}{x^{2a}y^{-b}} = x^{-5a/3 - 2/3}y^{3b/2}$

24. $\dfrac{\sqrt{2}}{\sqrt{3}} + \dfrac{4\sqrt{3}}{\sqrt{2}} - 6\sqrt{24} = \dfrac{\sqrt{6}}{3} + \dfrac{4\sqrt{6}}{2} - 12\sqrt{6}$

$\quad = \dfrac{2\sqrt{6}}{6} + \dfrac{12\sqrt{6}}{6} - \dfrac{72\sqrt{6}}{6} = -\dfrac{29\sqrt{6}}{3}$

25. $\dfrac{ka^2}{ka - \dfrac{a^3}{k^2a - 1}} = \dfrac{ka^2(k^2a - 1)}{k^3a^2 - ka - a^3}$

$\quad = \dfrac{ka(k^2a - 1)}{k^3a - k - a^2}$

26.

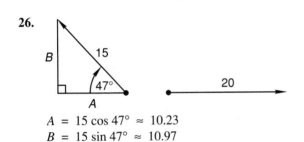

$A = 15\cos 47° \approx 10.23$
$B = 15\sin 47° \approx 10.97$

$\quad \begin{array}{r} -10.23R + 10.97U \\ 20.00R + 0.00U \\ \hline \mathbf{9.77R + 10.97U} \end{array}$

$\tan\theta = \dfrac{10.97}{9.77}$
$\quad \theta \approx 48.31°$

$F = \sqrt{(9.77)^2 + (10.97)^2} \approx 14.69$

14.69 /48.31°

27. $\dfrac{1000 \text{ liters}}{\text{min}} \times \dfrac{1000 \text{ ml}}{1 \text{ liter}} \times \dfrac{1 \text{ min}}{60 \text{ sec}}$

$\quad = \dfrac{\mathbf{1000(1000)}}{\mathbf{60}} \dfrac{\mathbf{ml}}{\mathbf{sec}}$

28. $\quad 6p^2 - 3p - 30 = 0$
$\quad\quad 3(2p^2 - p - 10) = 0$
$\quad\quad 2p^2 + 4p - 5p - 10 = 0$
$\quad\quad 2p(p + 2) - 5(p + 2) = 0$
$\quad\quad\quad (2p - 5)(p + 2) = 0$

$\quad 2p - 5 = 0 \quad\quad p + 2 = 0$
$\quad\quad p = \dfrac{5}{2} \quad\quad\quad p = -2$

$\dfrac{5}{2}, -2$

29. $\quad\quad 4x^3 - 14x^2 - 8x = 0$
$\quad\quad 2x(2x^2 - 7x - 4) = 0$
$\quad 2x(2x^2 - 8x + x - 4) = 0$
$\quad 2x(2x(x - 4) + 1(x - 4)) = 0$
$\quad\quad 2x(2x + 1)(x - 4) = 0$

$2x = 0 \quad\quad 2x + 1 = 0 \quad\quad x - 4 = 0$

$x = 0 \quad\quad\quad x = -\dfrac{1}{2} \quad\quad x = 4$

$\mathbf{0, -\dfrac{1}{2}, 4}$

30. $\quad\quad 3x^2 - 7x - 6 = 0$
$\quad\quad 3x^2 - 9x + 2x - 6 = 0$
$\quad 3x(x - 3) + 2(x - 3) = 0$
$\quad\quad (3x + 2)(x - 3) = 0$

$3x + 2 = 0 \quad\quad x - 3 = 0$
$\quad x = -\dfrac{2}{3} \quad\quad x = 3$

$-\dfrac{2}{3}, 3$

PROBLEM SET 110

1. $T =$ tens' digit
$U =$ units' digit
$10T + U =$ original number
$10U + T =$ reversed number

(a) $T + U = 9$
$\quad\quad T = 9 - U$
(b) $10U + T = 10T + U - 27$
Substitute (a) into (b) and get:
(b') $10U + (9 - U) = 10(9 - U) + U - 27$
$\quad\quad 9U + 9 = -9U + 63$
$\quad\quad\quad 18U = 54$
$\quad\quad\quad\quad U = 3$
(a) $T = 9 - (3) = 6$
Original number = **63**

2. $T =$ tens' digit
$U =$ units' digit
$10T + U =$ original number
$10U + T =$ reversed number

(a) $T + U = 7$
$\quad\quad T = 7 - U$
(b) $10U + T = 10T + U + 45$
Substitute (a) into (b) and get:
(b') $10U + (7 - U) = 10(7 - U) + U + 45$
$\quad\quad 9U + 7 = -9U + 115$
$\quad\quad\quad 18U = 108$
$\quad\quad\quad\quad U = 6$
(a) $T = 7 - (6) = 1$
Original number = **16**

3. Potassium (K): $1 \times 39 = 39$
Chromium (Cr): $2 \times 52 = 104$
Oxygen: $7 \times 16 = 112$
Total: $= 255$

$$\frac{112}{255} = \frac{336}{T}$$
$$T = \textbf{765 grams}$$

4. Downstream: $(B + W)T_D = D_D$ (a)
Upstream: $(B - W)T_U = D_U$ (b)
(a′) $(B + W)8 = 120$
(b′) $(B - W)9 = 63$

9(a′) $72B + 72W = 1080$
8(b′) $\underline{72B - 72W = 504}$
$144B = 1584$
$B = \textbf{11 mph}$

(a′) $8(11) + 8W = 120$
$ W = \textbf{4 mph}$

5. $S_P = P_P + M_U$
$S_P = 1400 + 700$
$S_P = \$2100$

$$\% \text{ of } S_P = \frac{700}{2100} \times 100\% = \textbf{33.33\% of } \textbf{\textit{S}}_{\textbf{\textit{P}}}$$

$$\% \text{ of cost} = \frac{700}{1400} \times 100\% = \textbf{50\% of cost}$$

6. $(x + 2)(x - 4) > 0$; $D = \{\text{Reals}\}$

(Pos)(Pos) > 0
$x + 2 > 0$ and $x - 4 > 0$
$ x > -2$ and $ x > 4$

(Neg)(Neg) > 0
$x + 2 < 0$ and $x - 4 < 0$
$ x < -2$ and $ x < 4$
Thus, the solution is $\textbf{\textit{x}} > \textbf{4}$ or $\textbf{\textit{x}} < \textbf{--2}$.

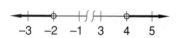

7. $x^2 - 4x - 5 \geq 0$; $D = \{\text{Integers}\}$
$(x - 5)(x + 1) \geq 0$

(Pos)(Pos) ≥ 0
$x - 5 \geq 0$ and $x + 1 \geq 0$
$ x \geq 5$ and $ x \geq -1$

(Neg)(Neg) ≥ 0
$x - 5 \leq 0$ and $x + 1 \leq 0$
$ x \leq 5$ and $ x \leq -1$
Thus, the solution is $\textbf{\textit{x}} \geq \textbf{5}$ or $\textbf{\textit{x}} \leq \textbf{--1}$.

8. Apply the power rule and get $\textbf{4\textit{x}}^\textbf{6}\textbf{\textit{y}}^\textbf{4}\textbf{\textit{z}}^\textbf{6}$

9. $x^{1/2} - y^{1/4}$

$$
\begin{array}{l}
x^{1/2} - y^{1/4} \\ \hline
x - x^{1/2}y^{1/4} \\
 - x^{1/2}y^{1/4} + y^{1/2} \\ \hline
x - 2x^{1/2}y^{1/4} + y^{1/2}
\end{array}
$$

10. $x^{1/2} + y^{1/2}$

$$
\begin{array}{l}
x^{1/2} - y^{1/2} \\ \hline
x + x^{1/2}y^{1/2} \\
 - x^{1/2}y^{1/2} - y \\ \hline
x - y = x - y
\end{array}
$$

11. $8x^9 - y^6p^3 = (2x^3)^3 - (y^2p)^3$
$ = (2x^3 - y^2p)(4x^6 + 2x^3y^2p + y^4p^2)$

12. $27x^{12}y^9 + p^6m^{15} = (3x^4y^3)^3 + (p^2m^5)^3$
$\phantom{27x^{12}y^9 + p^6m^{15}} = (3x^4y^3 + p^2m^5)(9x^8y^6 - 3x^4y^3p^2m^5 + p^4m^{10})$

13. $1000N = 13.62\ 362\ 362\ldots$
$\underline{N = 0.01\ 362\ 362\ldots}$
$999N = 13.61$
$$N = \frac{\textbf{1361}}{\textbf{99,900}}$$

14. $y = x^2 - 4x + 3$
$y = (x^2 - 4x + 4) + 3 - 4$
$y = (x - 2)^2 - 1$
From this we see:
(a) Opens upward
(b) Axis of symmetry is $x = 2$
(c) y coordinate of vertex is -1

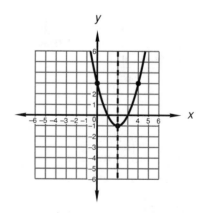

15. $3 > x + 2$ or $x + 5 \geq 8$; $D = \{\text{Reals}\}$
$1 > x$ or $x \geq 3$

16. (a) $\dfrac{2}{7}x - \dfrac{1}{5}y = 2$

(b) $0.03x + 0.07y = 1.12$

(a′) $10x - 7y = 70$

(b′) $\dfrac{3x + 7y = 112}{}$

$\overline{13x= 182}$

$x = 14$

(a′) $10(14) - 7y = 70$

$y = 10$

(14, 10)

17. (a) $x + y + z = 1$

(b) $4x - 2y - z = 6$

(c) $3x - y + z = -1$

(a) + (b): $5x - y = 7$ (d)

(b) + (c): $7x - 3y = 5$ (e)

-3(d) $-15x + 3y = -21$

(e) $\dfrac{7x - 3y = 5}{}$

$\overline{-8x = -16}$

$x = 2$

(d) $5(2) - y = 7$

$y = 3$

(a) $(2) + (3) + z = 1$

$z = -4$

(2, 3, –4)

18. (a) $x^2 + y^2 = 6$

(b) $x - y = 1$

$y = x - 1$

Substitute (b) into (a) and get:

(a′) $x^2 + (x - 1)^2 = 6$

$x^2 + x^2 - 2x + 1 = 6$

$2x^2 - 2x - 5 = 0$

Solve this equation by using the quadratic formula.

$$x = \dfrac{-(-2) \pm \sqrt{(-2)^2 - 4(2)(-5)}}{2(2)}$$

$$x = \dfrac{1}{2} \pm \dfrac{\sqrt{11}}{2}$$

Substitute these values of x into (b) and solve for y.

(b) $y = \left(\dfrac{1}{2} + \dfrac{\sqrt{11}}{2}\right) - 1$

$y = -\dfrac{1}{2} + \dfrac{\sqrt{11}}{2}$

(b) $y = \left(\dfrac{1}{2} - \dfrac{\sqrt{11}}{2}\right) - 1$

$y = -\dfrac{1}{2} - \dfrac{\sqrt{11}}{2}$

$\left(\dfrac{1}{2} + \dfrac{\sqrt{11}}{2}, \ -\dfrac{1}{2} + \dfrac{\sqrt{11}}{2}\right)$ and

$\left(\dfrac{1}{2} - \dfrac{\sqrt{11}}{2}, \ -\dfrac{1}{2} - \dfrac{\sqrt{11}}{2}\right)$

19. (a) $x - z = 3$

(b) $x + 2y = 5$

(c) $y + z = 0$

(a) + (c): $x + y = 3$ (d)

-2(d) $-2x - 2y = -6$

(b) $\dfrac{x + 2y = 5}{}$

$\overline{-x = -1}$

$x = 1$

(b) $(1) + 2y = 5$

$y = 2$

(a) $(1) - z = 3$

$z = -2$

(1, 2, –2)

20. (a) $2x \geq 6$

$x \geq 3$

(b) $x + y < 3$

$y < -x + 3$

The first step is to graph each of these lines.

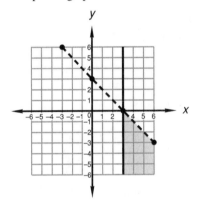

The region we wish to find is on and to the right of the solid line and below the dashed line. This region is shaded in the previous figure.

21. $\dfrac{2i^2 - 2i + 2}{\sqrt{-9} - \sqrt{-3}\sqrt{-3}} = \dfrac{-2i}{3i + 3} \cdot \dfrac{3i - 3}{3i - 3}$

$= \dfrac{6 + 6i}{-9 - 9} = -\dfrac{1}{3} - \dfrac{1}{3}i$

22. $\dfrac{3 + 4\sqrt{5}}{1 - \sqrt{5}} \cdot \dfrac{1 + \sqrt{5}}{1 + \sqrt{5}}$

$= \dfrac{3 + 3\sqrt{5} + 4\sqrt{5} + 20}{1 - 5} = \dfrac{-23 - 7\sqrt{5}}{4}$

23. $\dfrac{a^{4-2b}x^2}{x^{b/2}a^{b/2}} = a^{4-5b/2}x^{2-b/2}$

24. $\sqrt[4]{8\sqrt{2}} = (2^3)^{1/4}2^{1/8} = 2^{3/4}2^{1/8} = 2^{7/8}$

25. $\dfrac{\sqrt{2}}{\sqrt{3}} - \dfrac{5\sqrt{3}}{\sqrt{2}} + 3\sqrt{24} = \dfrac{\sqrt{6}}{3} - \dfrac{5\sqrt{6}}{2} + 6\sqrt{6}$

$\qquad = \dfrac{2\sqrt{6}}{6} - \dfrac{15\sqrt{6}}{6} + \dfrac{36\sqrt{6}}{6} = \dfrac{23\sqrt{6}}{6}$

26. $\qquad \sqrt{k} - 6 = \sqrt{k - 48}$

$\quad k - 12\sqrt{k} + 36 = k - 48$

$\qquad\qquad 12\sqrt{k} = 84$

$\qquad\qquad\quad \sqrt{k} = 7$

$\qquad\qquad\qquad k = \mathbf{49}$

Check: $\sqrt{49} - 6 = \sqrt{49 - 48}$

$\qquad\qquad\quad 1 = 1 \quad$ check

27. $H = \sqrt{(15)^2 + (4)^2} = \sqrt{241}$

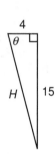

$\tan \theta = \dfrac{15}{4}$

$\quad \theta \approx -75.07°$

$\sqrt{241}\underline{/-75.07°}$

28. $\qquad 3s^2 + 11s + 10 = 0$

$\quad 3s^2 + 6s + 5s + 10 = 0$

$\quad 3s(s + 2) + 5(s + 2) = 0$

$\qquad\quad (3s + 5)(s + 2) = 0$

$3s + 5 = 0 \qquad s + 2 = 0$

$\quad s = -\dfrac{5}{3} \qquad\quad s = -2$

$-\dfrac{5}{3}, \mathbf{-2}$

29. $\qquad\quad 8x^3 + 6x^2 - 2x = 0$

$\qquad\quad 2x(4x^2 + 3x - 1) = 0$

$\qquad 2x(4x^2 + 4x - x - 1) = 0$

$\quad 2x(4x(x + 1) - 1(x + 1)) = 0$

$\qquad\quad 2x(4x - 1)(x + 1) = 0$

$2x = 0 \qquad 4x - 1 = 0 \qquad x + 1 = 0$

$x = 0 \qquad\quad x = \dfrac{1}{4} \qquad\quad x = -1$

$0, \dfrac{1}{4}, \mathbf{-1}$

30. Since -2 is not a member of the domain of $g(x)$, it is not a member of the domain of $(h + g)(x)$. Therefore, the answer is either $\varnothing$ or $\{ \}$.

PROBLEM SET 111

1. (a) $N_N + N_D + N_Q = 28$

(b) $5N_N + 10N_D + 25N_Q = 250$

(c) $N_N = 5N_Q$

Substitute (c) into (a) and (b) and get:

(a′) $N_D + 6N_Q = 28$

(b′) $10N_D + 50N_Q = 250$

$\begin{array}{rl} -10(\text{a}′) \quad -10N_D - 60N_Q = & -280 \\ (\text{b}′) \quad 10N_D + 50N_Q = & 250 \\ \hline -10N_Q = & -30 \\ N_Q = & \mathbf{3} \end{array}$

(c) $N_N = 5(3) = \mathbf{15}$

(a) $(15) + N_D + (3) = 28$

$\qquad\qquad N_D = \mathbf{10}$

2. (a) $N_B + N_G + N_Y = 10$

(b) $N_B + 4N_G + 5N_Y = 39$

(c) $N_Y = N_G + 2$

Substitute (c) into (a) and (b) and get:

(a′) $N_B + 2N_G = 8$

(b′) $N_B + 9N_G = 29$

$\begin{array}{rl} -1(\text{a}′) \quad -N_B - 2N_G = & -8 \\ (\text{b}′) \quad N_B + 9N_G = & 29 \\ \hline 7N_G = & 21 \\ N_G = & \mathbf{3} \end{array}$

(c) $N_Y = (3) + 2 = \mathbf{5}$

(a) $N_B + (3) + (5) = 10$

$\qquad\qquad N_B = \mathbf{2}$

3. $T = $ tens' digit

$U = $ units' digit

$10T + U = $ original number

(a) $10T + U = 4(T + U)$

(b) $U = T + 1$

Substitute (b) into (a) and get:

(a′) $10T + (T + 1) = 4T + 4(T + 1)$

$\qquad\qquad 11T + 1 = 8T + 4$

$\qquad\qquad\quad 3T = 3$

$\qquad\qquad\qquad T = 1$

(b) $U = (1) + 1 = 2$

Original number $= \mathbf{12}$

4. $N \qquad N + 1 \qquad N + 2$

$\qquad N(N + 2) = 5(N + 1) + 35$

$\qquad\quad N^2 + 2N = 5N + 40$

$N^2 - 3N - 40 = 0$

Solve this equation by factoring.

$(N - 8)(N + 5) = 0$

$\qquad\quad N = 8, -5$

The desired integers are $\mathbf{8, 9, 10}$ and $\mathbf{-5, -4, -3}$.

5. Equal ratio method:

$$\frac{S_1}{S_2} = \frac{T_1(A_1)^2}{T_2(A_2)^2}$$

$$\frac{1000}{S_2} = \frac{5(2)^2}{8(1)^2}$$

$$S_2 = \mathbf{400}$$

Variation method:

$$S = kTA^2$$

$$1000 = k(5)(2)^2$$

$$50 = k$$

$$S = (50)(8)(1)^2$$

$$S = \mathbf{400}$$

6. $(x + 4)(x - 2) > 0$; $D = \{\text{Integers}\}$

(Pos)(Pos) > 0
$x + 4 > 0$ and $x - 2 > 0$
$\quad x > -4$ and $\quad x > 2$

(Neg)(Neg) > 0
$x + 4 < 0$ and $x - 2 < 0$
$\quad x < -4$ and $\quad x < 2$

Thus, the solution is $x > 2$ or $x < -4$.

```
◄──●───●──┤/┤/┤──●───●──►
  -6  -5  -4  2   3   4
```

7. $x^2 - 5x + 6 > 0$; $D = \{\text{Integers}\}$
$(x - 3)(x - 2) > 0$

(Pos)(Pos) > 0
$x - 3 > 0$ and $x - 2 > 0$
$\quad x > 3$ and $\quad x > 2$

(Neg)(Neg) > 0
$x - 3 < 0$ and $x - 2 < 0$
$\quad x < 3$ and $\quad x < 2$

Thus, the solution is $x > 3$ or $x < 2$.

```
◄──●───●───┤───┤───●───●──►
   0   1   2   3   4   5
```

8. $\left(x^{1/2} + y^{1/4}\right)^2$

$$
\begin{array}{r}
x^{1/2} + y^{1/4} \\
x^{1/2} + y^{1/4} \\
\hline
x \quad + \quad x^{1/2}y^{1/4} \\
\quad\quad x^{1/2}y^{1/4} + y^{1/2} \\
\hline
x \quad + 2x^{1/2}y^{1/4} + y^{1/2}
\end{array}
$$

9. $\left(x^{1/2} - y^{-1/2}\right)^2$

$$
\begin{array}{r}
x^{1/2} - y^{-1/2} \\
x^{1/2} - y^{-1/2} \\
\hline
x \quad - \quad x^{1/2}y^{-1/2} \\
\quad\quad - \quad x^{1/2}y^{-1/2} + y^{-1} \\
\hline
x \quad - 2x^{1/2}y^{-1/2} + y^{-1}
\end{array}
$$

10. Apply the power rule and get xy^{-1}

11. $x^3 - m^6y^6 = (x)^3 - (m^2y^2)^3$
$\quad = (x - m^2y^2)(x^2 + m^2xy^2 + m^4y^4)$

12. $8x^6y^3 - 27m^3p^{12} = (2x^2y)^3 - (3mp^4)^3$
$\quad = (2x^2y - 3mp^4)(4x^4y^2 + 6x^2ymp^4 + 9m^2p^8)$

13.
$$
\begin{array}{r}
100N = 102.13\ 13\ 13\ldots \\
N = \quad 1.02\ 13\ 13\ldots \\
\hline
99N = 101.11 \\
\end{array}
$$
$$N = \frac{\mathbf{10{,}111}}{\mathbf{9900}}$$

14. $-y = x^2 + 4x + 1$
$-y = (x^2 + 4x + 4) + 1 - 4$
$-y = (x + 2)^2 - 3$
$\quad y = -(x + 2)^2 + 3$
From this we see:
(a) Opens downward
(b) Axis of symmetry is $x = -2$
(c) y coordinate of vertex is 3

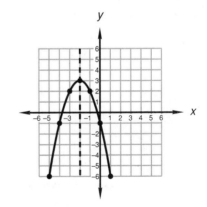

15. $-|x| - 3 \geq -7$; $D = \{\text{Reals}\}$,
$\quad\quad -|x| \geq -4$
$\quad\quad\quad |x| \leq 4$
$x \leq 4$ and $x \geq -4$

```
├────●──┤/┤/┤──●────┤
 -5  -4  -3  3  4   5
```

16. $-2 \leq x + 5 < 4$; $D = \{\text{Integers}\}$
$-7 \leq x < -1$

```
├──●──●──●──●──●──●──●──┤
 -8 -7 -6 -5 -4 -3 -2 -1
```

17. (a) $\dfrac{3}{5}x - \dfrac{2}{5}y = -10$

(b) $0.003x + 0.2y = 1.97$

(a') $3x - 2y = -50$

(b') $3x + 200y = 1970$

$$-1(a') \quad -3x + 2y = 50$$
$$(b') \quad \underline{3x + 200y = 1970}$$
$$202y = 2020$$
$$y = 10$$

(a') $3x - 2(10) = -50$

$$x = -10$$

(−10, 10)

18. (a) $x + 2y = 10$

(b) $x - 3z = -16$

(c) $y + 2z = 16$

$2(b) + 3(c): \quad 2x + 3y = 16$ (d)

$-2(a) + (d): \quad -y = -4$

$$y = 4$$

(a) $x + 2(4) = 10$

$$x = 2$$

(c) $(4) + 2z = 16$

$$z = 6$$

(2, 4, 6)

19. (a) $x^2 + y^2 = 4$

(b) $x - y = 1$

$$y = x - 1$$

Substitute (b) into (a) and get:

(a') $\quad x^2 + (x - 1)^2 = 4$

$$x^2 + x^2 - 2x + 1 = 4$$
$$2x^2 - 2x - 3 = 0$$

Solve this equation by using the quadratic formula.

$$x = \frac{-(-2) \pm \sqrt{(-2)^2 - 4(2)(-3)}}{2(2)}$$

$$x = \frac{1}{2} \pm \frac{\sqrt{7}}{2}$$

Substitute these values of x into (b) and solve for y.

(b) $y = \left(\dfrac{1}{2} + \dfrac{\sqrt{7}}{2}\right) - 1$

$$y = -\frac{1}{2} + \frac{\sqrt{7}}{2}$$

(b) $y = \left(\dfrac{1}{2} - \dfrac{\sqrt{7}}{2}\right) - 1$

$$y = -\frac{1}{2} - \frac{\sqrt{7}}{2}$$

$\left(\dfrac{1}{2} + \dfrac{\sqrt{7}}{2}, \; -\dfrac{1}{2} + \dfrac{\sqrt{7}}{2}\right)$ and

$\left(\dfrac{1}{2} - \dfrac{\sqrt{7}}{2}, \; -\dfrac{1}{2} - \dfrac{\sqrt{7}}{2}\right)$

20.
$$2x^2 - 3x - 5 = 0$$
$$x^2 - \frac{3}{2}x = \frac{5}{2}$$
$$\left(x^2 - \frac{3}{2}x + \frac{9}{16}\right) = \frac{40}{16} + \frac{9}{16}$$
$$\left(x - \frac{3}{4}\right)^2 = \frac{49}{16}$$
$$x - \frac{3}{4} = \pm\frac{7}{4}$$
$$x = \frac{3}{4} \pm \frac{7}{4}$$
$$x = \frac{5}{2}, -1$$

21. $\dfrac{40 \text{ in.}}{\text{sec}} \times \dfrac{2.54 \text{ cm}}{1 \text{ in.}} \times \dfrac{1 \text{ m}}{100 \text{ cm}} \times \dfrac{60 \text{ sec}}{1 \text{ min}}$

$\times \dfrac{60 \text{ min}}{1 \text{ hr}} = \dfrac{40(2.54)(60)(60)}{100} \dfrac{\text{m}}{\text{hr}}$

22. $\dfrac{2i^3 - \sqrt{-3}\sqrt{-3}}{4 - 3i^2} = \dfrac{-2i + 3}{7} = \dfrac{3}{7} - \dfrac{2}{7}i$

23. $\dfrac{2\sqrt{3} + 2}{3 - \sqrt{3}} \cdot \dfrac{3 + \sqrt{3}}{3 + \sqrt{3}}$

$= \dfrac{6\sqrt{3} + 6 + 6 + 2\sqrt{3}}{9 - 3} = \dfrac{6 + 4\sqrt{3}}{3}$

24. $\dfrac{a^{x/2}\,y^{1-x/2}}{a^{3x}\,y^{-2x}} = a^{-5x/2}y^{1+3x/2}$

25. $\sqrt{xy}\,\sqrt{x^2 y} = x^{1/2}y^{1/2}xy^{1/2} = x^{3/2}y$

26. $\dfrac{\sqrt{2}}{\sqrt{7}} - \dfrac{3\sqrt{7}}{\sqrt{2}} + 2\sqrt{126} = \dfrac{\sqrt{14}}{7} - \dfrac{3\sqrt{14}}{2} + 6\sqrt{14}$

$= \dfrac{2\sqrt{14}}{14} - \dfrac{21\sqrt{14}}{14} + \dfrac{84\sqrt{14}}{14} = \dfrac{65\sqrt{14}}{14}$

27. (a) $-y < 3$
  $y > -3$
(b) $3x + y \le 3$
  $y \le -3x + 3$
The first step is to graph each of these lines.

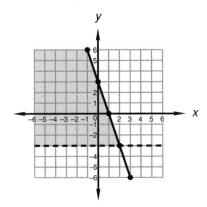

The region we wish to find is on and below the solid line and above the dashed line. This region is shaded in the previous figure.

28.

$A = 20 \cos 45° \approx 14.14$
$B = 20 \sin 45° \approx 14.14$

$-6.00R + 0.00U$
$\underline{14.14R + 14.14U}$
$\mathbf{8.14R + 14.14U}$

$\tan \theta = \dfrac{14.14}{8.14}$
  $\theta \approx 60.07°$

$F = \sqrt{(8.14)^2 + (14.14)^2} \approx 16.32$

16.32/60.07°

29. $2x^2 - x - 10 = 0$
  $2x^2 + 4x - 5x - 10 = 0$
  $2x(x + 2) - 5(x + 2) = 0$
  $(2x - 5)(x + 2) = 0$

$2x - 5 = 0 \qquad x + 2 = 0$
  $x = \dfrac{5}{2} \qquad\quad x = -2$

$\dfrac{5}{2}, -2$

30. $2x^3 - 7x^2 - 15x = 0$
  $x(2x^2 - 7x - 15) = 0$
  $x(2x^2 - 10x + 3x - 15) = 0$
  $x(2x(x - 5) + 3(x - 5)) = 0$
  $x(2x + 3)(x - 5) = 0$

$x = 0 \qquad 2x + 3 = 0 \qquad x - 5 = 0$
  $x = -\dfrac{3}{2} \qquad x = 5$

$\mathbf{0, -\dfrac{3}{2}, 5}$

PROBLEM SET 112

1. (a) $N_N + N_D + N_Q = 20$
(b) $5N_N + 10N_D + 25N_Q = 325$
(c) $N_Q = 2N_D$
Substitute (c) into (a) and (b) and get:
(a') $N_N + 3N_D = 20$
(b') $5N_N + 60N_D = 325$

$-5(a') \quad -5N_N - 15N_D = -100$
$(b') \quad \underline{5N_N + 60N_D = 325}$
$45N_D = 225$
$N_D = \mathbf{5}$

(c) $N_Q = 2(5) = \mathbf{10}$

(a) $N_N + (5) + (10) = 20$
  $N_N = \mathbf{5}$

2. $T = $ tens' digit
$U = $ units' digit
$10T + U = $ original number
$10U + T = $ reversed number

(a) $T + U = 7$
  $T = 7 - U$
(b) $10U + T = 10T + U - 9$
Substitute (a) into (b) and get:
(b') $10U + (7 - U) = 10(7 - U) + U - 9$
  $9U + 7 = -9U + 61$
  $18U = 54$
  $U = 3$
(a) $T = 7 - (3) = 4$
Original number = **43**

3. $S_P = P_P + M_U$
$1800 = P_P + 0.2(1800)$
$P_P = 1800 - 360$
$P_P = \mathbf{\$1440}$

4. Downstream: $(B + W)T_D = D_D$ (a)

Upstream: $(B - W)T_U = D_U$ (b)

$T_D = 2T_U$

(a') $(B + 3)2T_U = 230$

(b') $(B - 3)T_U = 85$

(a') $2BT_U + 6T_U = 230$

2(b') $\underline{2BT_U - 6T_U = 170}$

     $4BT_U \quad\quad = 400$

(c) $BT_U = 100$

(a') $2(100) + 6T_U = 230$

           $6T_U = 30$

          $T_U = \textbf{5 hr}; \; T_D = \textbf{10 hr}$

(c) $B(5) = 100$

    $B = \textbf{20 mph}$

5. $R_F T_F = 1800; \; R_H T_H = 1200;$

$R_H = R_F + 200; \; T_F = 3T_H$

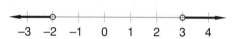

$(R_H - 200)3T_H = 1800$

$3(1200) - 600T_H = 1800$

      $-600T_H = -1800$

          $T_H = \textbf{3 hr}$

$T_F = \textbf{9 hr}; \; R_H = \textbf{400 mph}; \; R_F = \textbf{200 mph}$

6. $(x + 3)(x - 4) < 0; \; D = \{\text{Reals}\}$

(Neg)(Pos) < 0

$x + 3 < 0$ and $x - 4 > 0$

    $x < -3$ and $x > 4$

(Pos)(Neg) < 0

$x + 3 > 0$ and $x - 4 < 0$

    $x > -3$ and $x < 4$

There are no real numbers that satisfy the first conjunction, so the solution must be $\textbf{--3} < \textbf{\textit{x}} < \textbf{4.}$

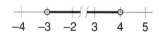

7. $(x - 6)(x + 1) < 0; \; D = \{\text{Integers}\}$

(Neg)(Pos) < 0

$x - 6 < 0$ and $x + 1 > 0$

    $x < 6$ and $x > -1$

(Pos)(Neg) < 0

$x - 6 > 0$ and $x + 1 < 0$

    $x > 6$ and $x < -1$

There are no integers that satisfy the second conjunction, so the solution must be $\textbf{--1} < \textbf{\textit{x}} < \textbf{6.}$

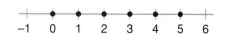

8. $(x + 2)(x - 3) > 0; \; D = \{\text{Reals}\}$

(Pos)(Pos) > 0

$x + 2 > 0$ and $x - 3 > 0$

    $x > -2$ and $x > 3$

(Neg)(Neg) > 0

$x + 2 < 0$ and $x - 3 < 0$

    $x < -2$ and $x < 3$

Thus, the solution is $\textbf{\textit{x}} > \textbf{3 or \textit{x}} < \textbf{--2.}$

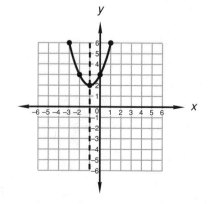

9. $\begin{array}{l} x^{1/2} + y^{1/2} \\ \underline{x^{1/2} - y^{-1/4}} \\ x \quad + x^{1/2}y^{1/2} \\ \quad\quad - x^{1/2}y^{-1/4} - y^{1/4} \\ \hline x \quad + x^{1/2}y^{1/2} - x^{1/2}y^{-1/4} - y^{1/4} \end{array}$

10. $p^6 x^6 - k^3 = (p^2 x^2)^3 - (k)^3$

     $= (p^2 x^2 - k)(p^4 x^4 + kp^2 x^2 + k^2)$

11. $100N = 401.43\ 43\ 43\ldots$

     $\underline{N = \quad\ 4.01\ 43\ 43\ldots}$

     $99N = 397.42$

        $N = \dfrac{\textbf{39,742}}{\textbf{9900}}$

12. $y = x^2 + 2x + 3$

     $y = (x^2 + 2x + 1) + 3 - 1$

     $y = (x + 1)^2 + 2$

From this we see:

(a) Opens upward

(b) Axis of symmetry is $x = -1$

(c) y coordinate of vertex is 2

13. $-|x| + 3 \le 0;\ D = \{\text{Reals}\}$

$-|x| \le -3$

$|x| \ge 3$

$x \ge 3$ or $x \le -3$

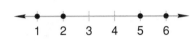

$$\begin{array}{c} -4 \quad -3 \quad -2 \quad 2 \quad 3 \quad 4 \end{array}$$

14. $x - 2 \le 0$ or $x + 4 > 8;\ D = \{\text{Integers}\}$

$x \le 2$ or $x > 4$

$$\begin{array}{c} 1 \quad 2 \quad 3 \quad 4 \quad 5 \quad 6 \end{array}$$

15. **(a)** and **(b)** are functions. (c) is not a function because -4 has two images.

16. $N = 2\dfrac{1}{8} + \dfrac{1}{3}\left(5 - 2\dfrac{1}{8}\right)$

$= \dfrac{17}{8} + \dfrac{1}{3}\left(\dfrac{40}{8} - \dfrac{17}{8}\right)$

$= \dfrac{51}{24} + \dfrac{23}{24}$

$= \dfrac{37}{12}$

17. $5x^2 + x + 4 = 0$

$x = \dfrac{-(1) \pm \sqrt{(1)^2 - 4(5)(4)}}{2(5)}$

$x = -\dfrac{1}{10} \pm \dfrac{\sqrt{79}}{10} i$

18. Write the equation of the given line in slope-intercept form.

$x + 3y - 4 = 0$

$$y = -\dfrac{1}{3}x + \dfrac{4}{3}$$

Since the slopes of perpendicular are negative reciprocals of each other:

$m_\perp = 3$

$y = 3x + b$

$-7 = 3(5) + b$

$-22 = b$

$y = 3x - 22$

19. (a) $2x + y = 28$

$y = 28 - 2x$

(b) $70x - 2y = 536$

Substitute (a) into (b) and get:

(b′) $70x - 2(28 - 2x) = 536$

$74x = 592$

$x = 8$

(a) $y = 28 - 2(8) = 12$

(8, 12)

20. (a) $x - 2y = 10$

(b) $3x - z = 11$

(c) $2y - 3z = -9$

(a) + (c): $x - 3z = 1$ (d)

$-3(d) + (b):\ 8z = 8$

$z = 1$

(d) $x - 3(1) = 1$

$x = 4$

(a) $(4) - 2y = 10$

$y = -3$

(4, −3, 1)

21. (a) $2x - y = 7$

(b) $xy = 4$

$y = \dfrac{4}{x}$

Substitute (b) into (a) and get:

(a′) $2x - \left(\dfrac{4}{x}\right) = 7$

$2x^2 - 4 = 7x$

$2x^2 - 7x - 4 = 0$

Solve this equation by using the quadratic formula.

$x = \dfrac{-(-7) \pm \sqrt{(-7)^2 - 4(2)(-4)}}{2(2)}$

$x = \dfrac{7}{4} \pm \dfrac{9}{4} = 4, -\dfrac{1}{2}$

Substitute these values of x into (a) and solve for y.

(a) $y = 2(4) - 7$

$y = 1$

(a) $y = 2\left(-\dfrac{1}{2}\right) - 7$

$y = -8$

(4, 1) and $\left(-\dfrac{1}{2}, -8\right)$

22. (a) $x + y + z = 1$

(b) $2x - y + z = -5$

(c) $3x + y + z = 5$

(a) + (b): $3x + 2z = -4$ (d)

(b) + (c): $5x + 2z = 0$ (e)

$-1(d) + (e):\ 2x = 4$

$x = 2$

(d) $3(2) + 2z = -4$

$z = -5$

(a) $(2) + y + (-5) = 1$

$y = 4$

(2, 4, −5)

23. $\dfrac{-3i^2 - 2i^3}{\sqrt{-3}\sqrt{-3} - \sqrt{-9}} = \dfrac{3 + 2i}{-3 - 3i} \cdot \dfrac{-3 + 3i}{-3 + 3i}$

$= \dfrac{-9 + 9i - 6i - 6}{9 + 9} = -\dfrac{5}{6} + \dfrac{1}{6}i$

24. $\dfrac{a^{x/2-1}y^2}{y^{x/2}a^{x/2}} = a^{-1}y^{2-x/2}$

25. $\sqrt[3]{x^5 y}\,\sqrt{x^5 y^2} = x^{5/3}y^{1/3}x^{5/2}y = x^{25/6}y^{4/3}$

26. $\dfrac{\sqrt{2}}{\sqrt{7}} + \dfrac{\sqrt{7}}{\sqrt{2}} - 3\sqrt{56} = \dfrac{\sqrt{14}}{7} + \dfrac{\sqrt{14}}{2} - 6\sqrt{14}$

$= \dfrac{2\sqrt{14}}{14} + \dfrac{7\sqrt{14}}{14} - \dfrac{84\sqrt{14}}{14} = -\dfrac{75\sqrt{14}}{14}$

27. $\dfrac{3\sqrt{2} - 2}{7\sqrt{2} - 3} \cdot \dfrac{7\sqrt{2} + 3}{7\sqrt{2} + 3}$

$= \dfrac{42 + 9\sqrt{2} - 14\sqrt{2} - 6}{98 - 9} = \dfrac{36 - 5\sqrt{2}}{89}$

28. $3x^3 + 5x^2 + 2x = 0$

$x(3x^2 + 5x + 2) = 0$

$x(3x^2 + 3x + 2x + 2) = 0$

$x(3x(x + 1) + 2(x + 1)) = 0$

$x(3x + 2)(x + 1) = 0$

$x = 0 \qquad 3x + 2 = 0 \qquad x + 1 = 0$

$x = -\dfrac{2}{3} \qquad x = -1$

$0, -\dfrac{2}{3}, -1$

29. $2x^2 - 3x - 2 = 0$

$2x^2 - 4x + x - 2 = 0$

$2x(x - 2) + 1(x - 2) = 0$

$(2x + 1)(x - 2) = 0$

$2x + 1 = 0 \qquad x - 2 = 0$

$x = -\dfrac{1}{2} \qquad x = 2$

$-\dfrac{1}{2}, 2$

30. $3x^2 + 8x + 4 = 0$

$3x^2 + 6x + 2x + 4 = 0$

$3x(x + 2) + 2(x + 2) = 0$

$(3x + 2)(x + 2) = 0$

$3x + 2 = 0 \qquad x + 2 = 0$

$x = -\dfrac{2}{3} \qquad x = -2$

$-\dfrac{2}{3}, -2$

PROBLEM SET 113

1. (a) $N_N + N_D + N_Q = 19$

(b) $5N_N + 10N_D + 25N_Q = 200$

(c) $N_N = 2N_D$

Substitute (c) into (a) and (b) and get:

(a′) $3N_D + N_Q = 19$

$N_Q = 19 - 3N_D$

(b′) $20N_D + 25N_Q = 200$

Substitute (a′) into (b′) and get:

(b″) $20N_D + 25(19 - 3N_D) = 200$

$-55N_D = -275$

$N_D = \mathbf{5}$

(a′) $N_Q = 19 - 3(5) = \mathbf{4}$

(a) $N_N + (5) + (4) = 19$

$N_N = \mathbf{10}$

2. $\dfrac{R_1}{R_2} = \dfrac{B_1(M_2)^2}{B_2(M_1)^2}$

$\dfrac{10}{3} = \dfrac{4(20)^2}{B_2(2)^2}$

$B_2 = \mathbf{120}$

3. (a) $0.7P_N + 0.6D_N = 126$

(b) $P_N + D_N = 200$

Substitute $D_N = 200 - P_N$ into (a) and get:

(a′) $0.7P_N + 0.6(200 - P_N) = 126$

$0.1P_N = 6$

$P_N = 60$

(b) $(60) + D_N = 200$

$D_N = 140$

60 ml 70%, 140 ml 60%

4. $R_C T_C = R_D T_D;$

$T_C = 8;\ T_D = 12;\ R_D = R_C - 20$

$R_C(8) = (R_C - 20)12$

$8R_C = 12R_C - 240$

$-4R_C = -240$

$R_C = \mathbf{60\ mph}$

$R_D = \mathbf{40\ mph};\ D_C = D_D = \mathbf{480\ miles}$

5. $\dfrac{V_1}{T_1} = \dfrac{V_2}{T_2}$

$\dfrac{1000}{1700} = \dfrac{2000}{T_2}$

$T_2 = \mathbf{3400\ K}$

6. (a) **−4.68**

(b) **4.14**

7. (a) e^{5163}

(b) **136.77**

8. $AE \times \overrightarrow{SF} = AB$

$6 \times \overrightarrow{SF} = 24$

$\overrightarrow{SF} = 4$

$AC = AD \times \overrightarrow{SF}$

$AC = 4(4) = $ **16 meters**

9. $(x + 2)(x - 3) < 0; \ D = \{\text{Reals}\}$

$(\text{Neg})(\text{Pos}) < 0$

$x + 2 < 0 \quad$ and $\quad x - 3 > 0$

$x < -2 \quad$ and $\qquad x > 3$

$(\text{Pos})(\text{Neg}) < 0$

$x + 2 > 0 \quad$ and $\quad x - 3 < 0$

$x > -2 \quad$ and $\qquad x < 3$

There are no real numbers that satisfy the first conjunction, so the solution must be **−2 < x < 3.**

10. $(x - 3)(x + 2) \geq 0: \ D = \{\text{Integers}\}$

$(\text{Pos})(\text{Pos}) \geq 0$

$x - 3 \geq 0 \quad$ and $\quad x + 2 \geq 0$

$x \geq 3 \quad$ and $\qquad x \geq -2$

$(\text{Neg})(\text{Neg}) \geq 0$

$x - 3 \leq 0 \quad$ and $\quad x + 2 \leq 0$

$x \leq 3 \quad$ and $\qquad x \leq -2$

Thus the solution is $x \geq 3$ **or** $x \leq -2$.

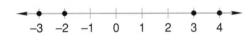

11. $(x - 4)(x + 2) \leq 0; \ D = \{\text{Integers}\}$

$(\text{Neg})(\text{Pos}) \leq 0$

$x - 4 \leq 0 \quad$ and $\quad x + 2 \geq 0$

$x \leq 4 \quad$ and $\qquad x \geq -2$

$(\text{Pos})(\text{Neg}) \leq 0$

$x - 4 \geq 0 \quad$ and $\quad x + 2 \leq 0$

$x \geq 4 \quad$ and $\qquad x \leq -2$

There are no integers that satisfy the second conjunction, so the solution must be **−2 ≤ x ≤ 4.**

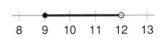

12. $|x| - 1 > 0; \ D = \{\text{Integers}\}$

$|x| > 1$

$x > 1$ or $x < -1$

13. $7 \leq x - 2 < 10: \ D = \{\text{Reals}\}$

$9 \leq x < 12$

14. $x^{1/3} + y^{2/3}$

$\underline{x^{2/3} + y^{1/3}}$

$x \quad + x^{2/3}y^{2/3}$

$\quad + x^{1/3}y^{1/3} + y$

$\underline{x \quad + x^{2/3}y^{2/3} + x^{1/3}y^{1/3} + y}$

15. $8p^6k^{15} - x^3m^6 = (2p^2k^5)^3 - (xm^2)^3$

$= (2p^2k^5 - xm^2)(4p^4k^{10} + 2p^2k^5xm^2 + x^2m^4)$

16. $100N = 0.316\ 16\ 16 \ldots$

$\underline{N = 0.003\ 16\ 16 \ldots}$

$99N = 0.313$

$N = \dfrac{313}{99,000}$

17. $-y = x^2 - 4x + 1$

$-y = (x^2 - 4x + 4) + 1 - 4$

$-y = (x - 2)^2 - 3$

$y = -(x - 2)^2 + 3$

From this we see:

(a) Opens downward

(b) Axis of symmetry is $x = 2$

(c) y coordinate of vertex is 3

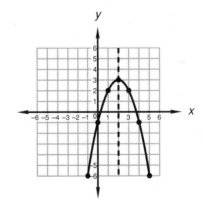

18. $N = 2 + \dfrac{2}{11}\left(4\dfrac{1}{6} - 2\right)$

$= 2 + \dfrac{2}{11}\left(\dfrac{25}{6} - \dfrac{12}{6}\right)$

$= \dfrac{66}{33} + \dfrac{13}{33} = \dfrac{79}{33}$

19.
$$3x^2 - x - 7 = 0$$
$$x^2 - \frac{1}{3}x = \frac{7}{3}$$
$$\left(x^2 - \frac{1}{3}x + \frac{1}{36}\right) = \frac{84}{36} + \frac{1}{36}$$
$$\left(x - \frac{1}{6}\right)^2 = \frac{85}{36}$$
$$x - \frac{1}{6} = \pm\frac{\sqrt{85}}{6}$$
$$x = \frac{1}{6} \pm \frac{\sqrt{85}}{6}$$

20. (a) $1\frac{1}{5}x + \frac{2}{3}y = 30$

(b) $-0.18x - 0.02y = -3.78$

(a′) $18x + 10y = 450$
(b′) $\underline{-18x - 2y = -378}$
$$8y = 72$$
$$y = 9$$

(a′) $18x + 10(9) = 450$
$$18x = 360$$
$$x = 20$$

(20, 9)

21. (a) $x^2 + y^2 = 4$

(b) $3x - y = 2$
$$y = 3x - 2$$

Substitute (b) into (a) and get:

(a′) $\quad x^2 + (3x - 2)^2 = 4$
$$x^2 + 9x^2 - 12x + 4 = 4$$
$$10x^2 - 12x = 0$$
$$2x(5x - 6) = 0$$
$$x = 0, \frac{6}{5}$$

Substitute these values of x into (b) and solve for y.

(b) $y = 3(0) - 2$
$$y = -2$$

(b) $y = 3\left(\frac{6}{5}\right) - 2$

$$y = \frac{8}{5}$$

$(0, -2)$ and $\left(\dfrac{6}{5}, \dfrac{8}{5}\right)$

22. (a) $x - 4y = -15$

(b) $3x + z = 20$

(c) $2y - z = 5$

(b) + (c): $3x + 2y = 25$ (d)

2(d) + (a): $7x = 35$
$$x = 5$$

(d) $3(5) + 2y = 25$
$$y = 5$$

(b) $3(5) + z = 20$
$$z = 5$$

(5, 5, 5)

23. (a) $x - 2y - z = -8$

(b) $3x - y - 2z = -5$

(c) $x + y + z = 9$

(a) + (c): $2x - y = 1$ (d)
(b) + 2(c): $\underline{5x + y = 13}$ (e)
$$7x = 14$$
$$x = 2$$

(e) $5(2) + y = 13$
$$y = 3$$

(c) $(2) + (3) + z = 9$
$$z = 4$$

(2, 3, 4)

24. $\dfrac{2i^3 - i}{-\sqrt{-3}\sqrt{-3} + 3} = -\dfrac{3i}{6} = -\dfrac{1}{2}i$

25. $\dfrac{\sqrt{2} - 5}{2\sqrt{2} - 4} \cdot \dfrac{2\sqrt{2} + 4}{2\sqrt{2} + 4}$

$$= \dfrac{4 + 4\sqrt{2} - 10\sqrt{2} - 20}{8 - 16}$$

$$= \dfrac{-16 - 6\sqrt{2}}{-8} = \dfrac{8 + 3\sqrt{2}}{4}$$

26. $\sqrt[6]{xy^3}\,\sqrt[3]{xy^2} = x^{1/6}y^{1/2}x^{1/3}y^{2/3} = x^{1/2}y^{7/6}$

27. $\dfrac{2}{\sqrt{3}} + \dfrac{2\sqrt{3}}{2} - 3\sqrt{48} = \dfrac{2\sqrt{3}}{3} + \sqrt{3} - 12\sqrt{3}$

$$= \dfrac{2\sqrt{3}}{3} + \dfrac{3\sqrt{3}}{3} - \dfrac{36\sqrt{3}}{3} = -\dfrac{31\sqrt{3}}{3}$$

28.
$$2x^3 + x^2 - 3x = 0$$
$$x(2x^2 + x - 3) = 0$$
$$x(2x^2 - 2x + 3x - 3) = 0$$
$$x(2x(x - 1) + 3(x - 1)) = 0$$
$$x(2x + 3)(x - 1) = 0$$

$x = 0 \qquad 2x + 3 = 0 \qquad x - 1 = 0$

$$x = -\frac{3}{2} \qquad x = 1$$

$0, -\dfrac{3}{2}, 1$

29.
$$3x^2 - x - 2 = 0$$
$$3x^2 - 3x + 2x - 2 = 0$$
$$3x(x - 1) + 2(x - 1) = 0$$
$$(3x + 2)(x - 1) = 0$$

$$3x + 2 = 0 \qquad x - 1 = 0$$
$$x = -\frac{2}{3} \qquad x = 1$$

$$-\frac{2}{3}, 1$$

30.
$$3x^3 + 7x^2 + 2x = 0$$
$$x(3x^2 + 7x + 2) = 0$$
$$x(3x^2 + 6x + x + 2) = 0$$
$$x(3x(x + 2) + 1(x + 2)) = 0$$
$$x(3x + 1)(x + 2) = 0$$

$$x = 0 \qquad 3x + 1 = 0 \qquad x + 2 = 0$$
$$x = -\frac{1}{3} \qquad x = -2$$

$$0, -\frac{1}{3}, -2$$

PROBLEM SET 114

1. $N \qquad N + 2 \qquad N + 4$
$$N(N + 4) = 8(N + 2) + 16$$
$$N^2 + 4N = 8N + 32$$
$$N^2 - 4N - 32 = 0$$
$$(N - 8)(N + 4) = 0$$
$$N = 8, -4$$
The desired integers are **8, 10, 12** and **−4, −2, 0.**

2. $0.2(240) + 1(D_N) = 0.52(240 + D_N)$
$$48 + D_N = 124.8 + 0.52D_N$$
$$0.48D_N = 76.8$$
$$D_N = \textbf{160 ml}$$

3. $R_F T_F = 4800; \quad R_S T_S = 2000;$
$T_F = T_S + 1; \quad R_F = 2R_S$

$\overset{D_F}{\underset{4800}{\longmapsto\!\!\longrightarrow}}$

$$2R_S(T_S + 1) = 4800$$
$$2(2000) + 2R_S = 4800$$

$\overset{D_S}{\underset{2000}{\longmapsto\!\!\longrightarrow}}$

$$2R_S = 800$$
$$R_S = \textbf{400 mph}$$
$$R_F = \textbf{800 mph}; \quad T_S = \textbf{5 hr}; \quad T_F = \textbf{6 hr}$$

4. Downstream: $(B + W)T_D = D_D$ (a)
Upstream: $(B - W)T_U = D_U$ (b)
$T_U = 3T_D$
(a′) $BT_D + 4T_D = 34$
(b′) $3BT_D - 12T_D = 54$

3(a′) + (b′): $6BT_D = 156$
$$BT_D = 26 \quad \text{(c)}$$

(a′) $(26) + 4T_D = 34$
$$T_D = 2$$

(c) $B(2) = 26$
$$B = \textbf{13 mph}$$

5. $T = $ tens' digit
$U = $ units' digit
$10T + U = $ original number
$10U + T = $ reversed number
(a) $T + U = 8$
$$T = 8 - U$$
(b) $10U + T = 10T + U + 54$
Substitute (a) into (b) and get:
(b′) $10U + (8 - U) = 10(8 - U) + U + 54$
$$9U + 8 = -9U + 134$$
$$18U = 126$$
$$U = 7$$
(a) $T = 8 - (7) = 1$
Original number $= \textbf{17}$

6. **Region A**, since this region includes all points on or above both the line and the parabola.

7. **Region A**, since this region includes all points on or above the parabola and all points below the line.

8. **Region B**, since this region includes all points on or inside the circle and all points on or above the line.

9. (a) **−4.77** (b) **4.47**

10. (a) e^{9185} (b) **1541.70**

11. $A_{\text{Square}} = s^2 = 9 \text{ cm}^2$
$$s = 3 \text{ cm}$$

$$AB = 3 \text{ cm} + \sqrt{7^2 - 3^2} \text{ cm} = (3 + 2\sqrt{10}) \text{ cm}$$

12. $7.15 \times 10^{-8} = 10^{-7.15}$
$\log 7.15 \times 10^{-8} = \textbf{−7.15}$

13. $S_P = P_P + M_U$
$$140 = P_P + 0.2(140)$$
$$\textbf{\$112} = P_P$$

14. $(x + 4)(x - 1) \geq 0; \quad D = \{\text{Integers}\}$

(Pos)(Pos) ≥ 0
$x + 4 \geq 0 \qquad$ and $\quad x - 1 \geq 0$
$x \geq -4 \qquad$ and $\qquad x \geq 1$

(Neg)(Neg) ≥ 0
$x + 4 \leq 0 \qquad$ and $\quad x - 1 \leq 0$
$x \leq -4 \qquad$ and $\qquad x \leq 1$
Thus, the solution is $x \geq 1$ or $x \leq -4.$

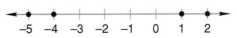

15. $(x - 4)(x + 1) \leq 0$; $D = \{\text{Integers}\}$

(Neg)(Pos) ≤ 0
$x - 4 \leq 0$ and $x + 1 \geq 0$
$x \leq 4$ and $x \geq -1$

(Pos)(Neg) ≤ 0
$x - 4 \geq 0$ and $x + 1 \leq 0$
$x \geq 4$ and $x \leq -1$

There are no integers that satisfy the second conjunction, so the solution must be **$-1 \leq x \leq 4$**

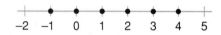

16. $-|x| - 3 > -7$; $D = \{\text{Integers}\}$
$-|x| > -4$
$|x| < 4$
$x < 4$ and $x > -4$

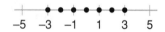

17. $6 \leq x - 4 < 8$; $D = \{\text{Integers}\}$
$10 \leq x < 12$

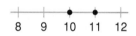

18. $100N = 0.1056\ 56\ 56\ldots$
$\underline{N = 0.0010\ 56\ 56\ldots}$
$99N = 0.1046$
$N = \dfrac{1046}{990,000}$

19. $y = x^2 - 4x + 7$
$y = (x^2 - 4x + 4) + 7 - 4$
$y = (x - 2)^2 + 3$
From this we see:
(a) Opens upward
(b) Axis of symmetry is $x = 2$
(c) y coordinate of vertex is 3

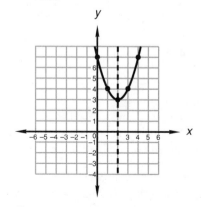

20. $N = 4\dfrac{1}{2} + \dfrac{3}{8}\left(6\dfrac{1}{4} - 4\dfrac{1}{2}\right)$

$= \dfrac{9}{2} + \dfrac{3}{8}\left(\dfrac{25}{4} - \dfrac{18}{4}\right)$

$= \dfrac{9}{2} + \dfrac{21}{32}$

$= \dfrac{144}{32} + \dfrac{21}{32} = \dfrac{\mathbf{165}}{\mathbf{32}}$

21. (a) $2\dfrac{1}{3}x + \dfrac{1}{5}y = 10$
(b) $0.03x - 0.03y = -0.36$

(a′) $35x + 3y = 150$
(b′) $\underline{3x - 3y = -36}$
$38x\ \ \ \ \ \ = 114$
$x = 3$
(b′) $3(3) - 3y = -36$
$y = 15$
(3, 15)

22. (a) $3x - z = 8$
(b) $2x - 2y = -4$
(c) $2y + 3z = 2$

(b) + (c): $2x + 3z = -2$ (d)

3(a) + (d): $11x = 22$
$x = 2$

(d) $2(2) + 3z = -2$
$z = -2$

(c) $2y + 3(-2) = 2$
$y = 4$
(2, 4, −2)

23. (a) $x + y = 6$
(b) $xy = -1$

$y = -\dfrac{1}{x}$

Substitute (b) into (a) and get:

(a′) $x + \left(-\dfrac{1}{x}\right) = 6$

$x^2 - 1 = 6x$
$x^2 - 6x - 1 = 0$
Solve this equation by using the quadratic formula.

$x = \dfrac{-(-6) \pm \sqrt{(-6)^2 - 4(1)(-1)}}{2(1)}$

$x = 3 \pm \sqrt{10}$
Substitute these values of x into (a) and solve for y.
(a) $y = 6 - (3 + \sqrt{10})$
$y = 3 - \sqrt{10}$
(a) $y = 6 - (3 - \sqrt{10})$
$y = 3 + \sqrt{10}$

$(3 + \sqrt{10}, 3 - \sqrt{10})$ and $(3 - \sqrt{10}, 3 + \sqrt{10})$

24. (a) $x - y + z = 3$
(b) $2x - y + 2z = 9$
(c) $-x + y + z = 1$

-1(a) + (b): $x + z = 6$ (d)

(b) + (c): $x + 3z = 10$ (e)

$\begin{aligned} -(d)\ -x\ -\ z &= -6 \\ (e)\ \underline{\ \ x + 3z} &= \underline{\ 10} \\ 2z &= 4 \\ z &= 2 \end{aligned}$

(d) $x + (2) = 6$
$x = 4$

(a) $(4) - y + (2) = 3$
$y = 3$

(4, 3, 2)

25. There is never an exact solution to these problems. One possible solution is given here.
$O = mI + b$
Use the graph to find the slope.

$m = \dfrac{225}{10} = 22.5$
$O = 22.5I + b$
Use the point (85, 400) for I and O.
$400 = 22.5(85) + b$
$-1512 = b$
$O = 22.5I - 1512$

26. $\dfrac{i^3 - i^2}{i^5 + 2} = \dfrac{-i + 1}{i + 2} \cdot \dfrac{i - 2}{i - 2}$

$= \dfrac{1 + 2i + i - 2}{-1 - 4} = \dfrac{1}{5} - \dfrac{3}{5}i$

27. $\dfrac{\sqrt{5}}{2\sqrt{2}} + \dfrac{4\sqrt{2}}{\sqrt{5}} - 3\sqrt{40}$

$= \dfrac{\sqrt{10}}{4} + \dfrac{4\sqrt{10}}{5} - 6\sqrt{10}$

$= \dfrac{5\sqrt{10}}{20} + \dfrac{16\sqrt{10}}{20} - \dfrac{120\sqrt{10}}{20}$

$= -\dfrac{99\sqrt{10}}{20}$

28. $\sqrt{x - 15} = 5 - \sqrt{x}$
$x - 15 = 25 - 10\sqrt{x} + x$
$-40 = -10\sqrt{x}$
$4 = \sqrt{x}$
$16 = x$

Check: $\sqrt{16 - 15} = 5 - \sqrt{16}$
$1 = 1$ check

29. $5x^3 + 7x^2 + 2x = 0$
$x(5x^2 + 7x + 2) = 0$
$x(5x^2 + 5x + 2x + 2) = 0$
$x(5x(x + 1) + 2(x + 1)) = 0$
$x(5x + 2)(x + 1) = 0$

$x = 0 \qquad 5x + 2 = 0 \qquad x + 1 = 0$
$x = -\dfrac{2}{5} \qquad x = -1$

$0, -\dfrac{2}{5}, -1$

30. $\begin{array}{r} x^{1/2} - y^{-1/2} \\ \underline{x^{1/2} - y^{-1/2}} \\ x \quad - x^{1/2}y^{-1/2} \\ \underline{\quad - x^{1/2}y^{-1/2} + y^{-1}} \\ x \quad - 2x^{1/2}y^{-1/2} + y^{-1} \end{array}$

PROBLEM SET 115

1. Potassium: $1 \times 39 = 39$
Manganese (Mn): $1 \times 55 = 55$
Oxygen: $4 \times 16 = 64$
Total: $= 158$

$\dfrac{Ox}{790} = \dfrac{64}{158}$
Ox = **320 grams**

2. $R_B T_B + R_S T_S = 280;$
$R_B = 20;\ R_S = 45;$
$T_B + T_S = 9$

$\begin{array}{c} D_B \qquad D_S \\ \vdash\!\!-\!\!-\!\!-\!\!-\!\!\dashv \\ 280 \end{array}$

$20T_B + 45(9 - T_B) = 280$
$-25T_B = -125$
$T_B = 5$
$T_S = 9 - (5) = 4$
$D_B = R_B T_B = 20(5) = $ **100 miles**
$D_S = R_S T_S = 45(4) = $ **180 miles**

3. $R_H T_H = 1200;\ R_B T_B = 480;$
$T_H = T_B + 1;\ R_H = 2R_B$

$\begin{array}{c} D_H \\ \vdash\!\!-\!\!-\!\!-\!\!\dashv \\ 1200 \end{array}$

$\begin{array}{c} D_B \\ \vdash\!\!-\!\!-\!\!\dashv \\ 480 \end{array}$

$2R_B(T_B + 1) = 1200$
$2(480) + 2R_B = 1200$
$2R_B = 240$
$R_B = 120$ mph
$R_H = $ **240 mph**; $T_B = $ **4 hr**; $T_H = $ **5 hr**

4. (a) $N_N + N_D + N_Q = 14$
(b) $5N_N + 10N_D + 25N_Q = 105$
(c) $N_D = 3N_Q$
Substitute (c) into (a) and (b) and get:
(a') $N_N + 4N_Q = 14$
(b') $5N_N + 55N_Q = 105$

-5(a') + (b'): $35N_Q = 35$
$N_Q = \mathbf{1}$

(a') $N_N + 4(1) = 14$
$N_N = \mathbf{10}$

(a) $(10) + N_D + (1) = 14$
$N_D = \mathbf{3}$

5. $T = $ tens' digit
$U = $ units' digit
$10T + U = $ original number
(a) $10T + U = 4(T + U)$
(b) $4U = T + 14$
$T = 4U - 14$
Substitute (b) into (a) and get:
(a') $10(4U - 14) + U = 4(4U - 14) + 4U$
$41U - 140 = 20U - 56$
$21U = 84$
$U = 4$
(b) $T = 4(4) - 14 = 2$
Original number $= \mathbf{24}$

6. $\mathbf{-4.88}$

7. $\mathbf{4.53}$

8. $10^{1.53} = 10^{x+3}$
$1.53 = x + 3$
$\mathbf{-1.47} = x$

9. $10^{3.412} = \mathbf{2582.26}$

10. $\dfrac{(e^{-7.89})(e^{6.18})}{(e^{-39.63})} = e^{-7.89+6.18+39.63}$

$= e^{37.92} = \mathbf{2.94 \times 10^{16}}$

11. $A_t = Pe^{rt}$

$A_9 = 100e^{0.09(9)}$

$A_9 = \mathbf{\$225}$

12. $A_t = A_0e^{kt}$

$60{,}000 = 16{,}000e^{k(3)}$

$3.75 = e^{3k}$

$1.32 = 3k$

$0.44 = k$

$A_{10} = 16{,}000e^{0.44(10)}$

$A_{10} = \mathbf{1{,}303{,}214}$

13. $H = \sqrt{\left(\dfrac{s}{2}\right)^2 + \left(\dfrac{s}{2}\right)^2} = \dfrac{s\sqrt{2}}{2}$ **units**

$\text{Area}_{HEFG} = H^2 = \left(\dfrac{s\sqrt{2}}{2}\right)^2 \text{ units}^2$

$= \dfrac{s^2}{2} \text{ units}^2$

14. **Region C**, since this region includes all points on or inside the circle and on or below the parabola.

15. $(x + 5)(x - 2) > 0$; $D = \{\text{Reals}\}$

(Pos)(Pos) > 0
$x + 5 > 0$ and $x - 2 > 0$
$x > -5$ and $x > 2$

(Neg)(Neg) > 0
$x + 5 < 0$ and $x - 2 < 0$
$x < -5$ and $x < 2$

Thus, the solution is $x > 2$ or $x < -5$.

$$\begin{array}{ccccccc} & -6 & -5 & -4 & 1 & 2 & 3 \end{array}$$

16. $(x + 5)(x + 2) < 0$; $D = \{\text{Reals}\}$

(Neg)(Pos) < 0
$x + 5 < 0$ and $x + 2 > 0$
$x < -5$ and $x > -2$

(Pos)(Neg) < 0
$x + 5 > 0$ and $x + 2 < 0$
$x > -5$ and $x < -2$

There are no real numbers that staisfy the first conjunction, so the solution must be $\mathbf{-5 < x < -2}$.

$$\begin{array}{cccccc} -6 & -5 & -4 & -3 & -2 & -1 \end{array}$$

17. $|x| + 2 \le 4$; $D = \{\text{Integers}\}$
$|x| \le 2$
$x \le 2$ and $x \ge -2$

$$\begin{array}{ccccccc} -3 & -2 & -1 & 0 & 1 & 2 & 3 \end{array}$$

18. $6 < x - 5 < 10$; $D = \{\text{Integers}\}$
$11 < x < 15$

$$\begin{array}{ccccc} 11 & 12 & 13 & 14 & 15 \end{array}$$

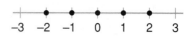

19. $\left(x^{1/4} - y^{-1/4}\right)^2$

$$
\begin{array}{r}
x^{1/4} - y^{-1/4} \\
x^{1/4} - y^{-1/4} \\
\hline
x^{1/2} - x^{1/4}y^{-1/4} \\
\quad - x^{1/4}y^{-1/4} + y^{-1/2} \\
\hline
x^{1/2} - 2x^{1/4}y^{-1/4} + y^{-1/2}
\end{array}
$$

20. $x^3m^{15} - 8p^3y^6 = (xm^5)^3 - (2py^2)^3$
$\quad = (xm^5 - 2py^2)(x^2m^{10} + 2xm^5py^2 + 4p^2y^4)$

21.
$$
\begin{array}{r}
100N = 104.747\ 47\ 47\ldots \\
N = \quad 1.047\ 47\ 47\ldots \\
\hline
99N = 103.7 \\
N = \dfrac{1037}{990}
\end{array}
$$

22. $y = x^2 + 2x + 1$
$y = (x^2 + 2x + 1) + 1 - 1$
$y = (x + 1)^2$
From this we see:
(a) Opens upward
(b) Axis of symmetry is $x = -1$
(c) y coordinate of vertex is 0

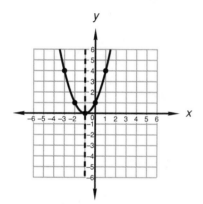

23. $N = 5\dfrac{2}{3} + \dfrac{1}{10}\left(6\dfrac{1}{2} - 5\dfrac{2}{3}\right)$

$\quad = \dfrac{17}{3} + \dfrac{1}{10}\left(\dfrac{39}{6} - \dfrac{34}{6}\right)$

$\quad = \dfrac{68}{12} + \dfrac{1}{12} = \dfrac{23}{4}$

24. (a) $1\dfrac{2}{3}x - 2\dfrac{1}{4}y = -7$
(b) $-0.2x + 0.05y = -1.8$

$\quad$ (a′) $\quad 20x - 27y = -84$
$\quad$ (b′) $\underline{-20x + 5y = -180}$
$\qquad\qquad\qquad -22y = -264$
$\qquad\qquad\qquad\quad\ y = 12$

$\quad$ (b′) $-20x + 5(12) = -180$
$\qquad\qquad\quad -20x = -240$
$\qquad\qquad\qquad\quad x = 12$

(12, 12)

25. (a) $4x - z = 12$
(b) $x - 3y = -10$
(c) $3y + z = 8$
(a) + (c): $4x + 3y = 20$ $\quad$ (d)

(b) + (d): $5x = 10$
$\qquad\qquad\ x = 2$

(d) $4(2) + 3y = 20$
$\qquad\qquad\quad y = 4$
(c) $3(4) + z = 8$
$\qquad\qquad\ z = -4$
(2, 4, −4)

26. (a) $x + y - z = 4$
(b) $2x + y + z = 10$
(c) $3x + y + z = 14$
(a) + (b): $3x + 2y = 14$ $\quad$ (d)
(a) + (c): $4x + 2y = 18$ $\quad$ (e)

-1(d) + (e): $x = 4$
(d) $3(4) + 2y = 14$
$\qquad\qquad\quad y = 1$

(b) $2(4) + (1) + z = 10$
$\qquad\qquad\qquad\qquad z = 1$
(4, 1, 1)

27. $-3 - 2i - (-2) + (-1) - 3i + 2i - 6$
$= \mathbf{-8 - 3i}$

28. $\dfrac{4\sqrt{2} + 1}{1 - 3\sqrt{2}} \cdot \dfrac{1 + 3\sqrt{2}}{1 + 3\sqrt{2}}$

$= \dfrac{4\sqrt{2} + 24 + 1 + 3\sqrt{2}}{1 - 18} = \dfrac{-25 - 7\sqrt{2}}{17}$

29.
$$3x^2 - x - 6 = 0$$
$$x^2 - \dfrac{1}{3}x = 2$$
$$\left(x^2 - \dfrac{1}{3}x + \dfrac{1}{36}\right) = \dfrac{72}{36} + \dfrac{1}{36}$$
$$\left(x - \dfrac{1}{6}\right)^2 = \dfrac{73}{36}$$
$$x - \dfrac{1}{6} = \pm\dfrac{\sqrt{73}}{6}$$
$$x = \dfrac{1}{6} \pm \dfrac{\sqrt{73}}{6}$$

30.
$$3x^2 + 2x - 8 = 0$$
$$3x^2 + 6x - 4x - 8 = 0$$
$$3x(x + 2) - 4(x + 2) = 0$$
$$(3x - 4)(x + 2) = 0$$
$$3x - 4 = 0 \qquad x + 2 = 0$$
$$x = \dfrac{4}{3} \qquad\qquad x = -2$$

$\dfrac{4}{3}, -2$

PROBLEM SET 116

1. $6N \qquad 6(N + 1) \qquad 6(N + 2)$

$6(6N + 6N + 12) = 10(6N + 6) - 84$

$72N + 72 = 60N - 24$

$12N = -96$

$N = -8$

The desired integers are $-48, -42$ and -36.

2. $\dfrac{V_1}{T_1} = \dfrac{V_2}{T_2}$

$\dfrac{400}{800} = \dfrac{200}{T_2}$

$T_2 = \mathbf{400\ K}$

3. Downstream: $(B + W)T_D = D_D$ (a)

Upstream: $(B - W)T_U = D_U$ (b)

(a') $(B + W)5 = 65$

(b') $(B - W)8 = 56$

$8(a') + 5(b')$: $80B = 800$

$B = \mathbf{10\ mph}$

(a') $(50) + 5W = 65$

$W = \mathbf{3\ mph}$

4. Equal ratio method:

$\dfrac{T_1}{T_2} = \dfrac{R_1(F_1)^2}{R_2(F_2)^2}$

$\dfrac{1000}{T_2} = \dfrac{2(1)^2}{1(2)^2}$

$T_2 = \mathbf{2000}$

Variation method:

$T = kRF^2$

$1000 = k(2)(1)^2$

$500 = k$

$T = 500(1)(2)^2$

$T = \mathbf{2000}$

5. T = tens' digit

U = units' digit

$10T + U$ = original number

(a) $10T + U = 2(T + U) + 7$

(b) $U = 3T + 3$

Substitute (b) into (a) and get:

(a') $10T + (3T + 3) = 2T + 2(3T + 3) + 7$

$13T + 3 = 8T + 13$

$5T = 10$

$T = 2$

(b) $U = 3(2) + 3 = 9$

Original number = $\mathbf{29}$

6. (a) $-\mathbf{4.95}$

(b) $\mathbf{5.86 \times 10^{-3}}$

(c) $10^{1.92} = 10^{x+4}$

$1.92 = x + 4$

$-\mathbf{2.08} = x$

7. $\dfrac{(e^{-7.40})(e^{6.36})}{(e^{-35.21})} = e^{-7.40 + 6.36 + 35.21}$

$= e^{34.17} = \mathbf{6.92 \times 10^{14}}$

8. (a) $\text{pH} = -\log H^+$

$\text{pH} = -\log 0.00204$

$\text{pH} = \mathbf{2.69}$

(b) $\text{pH} = -\log H^+$

$10^{-\text{pH}} = H^+$

$10^{-3.2} = H^+$

$H^+ = \mathbf{6.31 \times 10^{-4}} \dfrac{\mathbf{mole}}{\mathbf{liter}}$

9. $A_t = Pe^{rt}$

$A_7 = 1400e^{0.11(7)}$

$A_7 = \mathbf{\$3024}$

10. $A_t = A_0e^{kt}$

$1,200,000 = 400,000e^{k(3)}$

$3 = e^{3k}$

$1.10 = 3k$

$0.37 = k$

$A_8 = 400,000e^{0.37(8)}$

$A_8 = \mathbf{7,719,189}$

11.

9	8	7	6	5

$= \mathbf{15,120}$

12.

9	9	9	10	10	10	10

$= \mathbf{7,290,000}$

13. $m\angle BAC = 180 - 35 - 35 = 110°$

$m\angle 3 = 180 - 110 = 70°$

$m\angle 4 = 180 - 35 = 145°$

$m\angle 3 + m\angle 4 = 70 + 145 = \mathbf{215°}$

14. **Region A,** since this region includes all points on or above the line and on or above the parabola.

15. $x^2 + 2x - 3 \geq 0$; $D = \{\text{Integers}\}$

$(x + 3)(x - 1) \geq 0$

$(\text{Pos})(\text{Pos}) \geq 0$

$x + 3 \geq 0 \quad$ and $\quad x - 1 \geq 0$

$x \geq -3 \quad$ and $\qquad x \geq 1$

(Neg)(Neg) ≥ 0
$x + 3 \leq 0$ and $x - 1 \leq 0$
$x \leq -3$ and $x \leq 1$
Thus, the solution is $x \geq 1$ or $x \leq -3$.

16. $x^2 + 2x - 3 < 0$; $D = \{\text{Integers}\}$
$(x + 3)(x - 1) < 0$
(Neg)(Pos) < 0
$x + 3 < 0$ and $x - 1 > 0$
$x < -3$ and $x > 1$

(Pos)(Neg) < 0
$x + 3 > 0$ and $x - 1 < 0$
$x > -3$ and $x < 1$
There are no integers that satisfy the first
conjunction, so the solution must be $-3 < x < 1$.

17. $-|x| + 2 > -1$; $D = \{\text{Integers}\}$
$-|x| > -3$
$|x| < 3$
$x < 3$ and $x > -3$

18. $100N = 104.76\ 76\ 76\ldots$
$\underline{\quad N = \quad 1.04\ 76\ 76\ldots}$
$99N = 103.72$
$N = \dfrac{\mathbf{10{,}372}}{\mathbf{9900}}$

19. $-y = x^2 + 4x + 6$
$-y = (x^2 + 4x + 4) + 6 - 4$
$-y = (x + 2)^2 + 2$
$y = -(x + 2)^2 - 2$
From this we see:
(a) Opens downward
(b) Axis of symmetry is $x = -2$
(c) y coordinate of vertex is -2

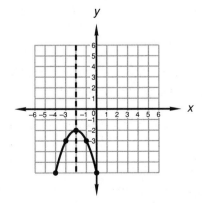

20. $\left(x^{1/2} - y^{3/4}\right)^2$

$\quad x^{1/2} - y^{3/4}$
$\quad \underline{x^{1/2} - y^{3/4}}$
$\quad x \quad - \quad x^{1/2}y^{3/4}$
$\qquad - \quad \underline{x^{1/2}y^{3/4} + y^{3/2}}$
$\quad x \quad - 2x^{1/2}y^{3/4} + y^{3/2}$

21. $27m^9p^3 - x^{12}y^3 = (3m^3p)^3 - (x^4y)^3$
$= (3m^3p - x^4y)(9m^6p^2 + 3m^3px^4y + x^8y^2)$

22. $N = \dfrac{1}{4} + \dfrac{2}{9}\left(3\dfrac{1}{2} - \dfrac{1}{4}\right)$
$= \dfrac{1}{4} + \dfrac{2}{9}\left(\dfrac{14}{4} - \dfrac{1}{4}\right)$
$= \dfrac{1}{4} + \dfrac{26}{36}$
$= \dfrac{9}{36} + \dfrac{26}{36} = \dfrac{35}{36}$

23.   (a) $\dfrac{3}{5}x - \dfrac{1}{7}y = 6$
(b) $-0.21x + 0.02y = -2.73$

(a') $21x - 5y =\ \ 210$
(b') $\underline{-21x + 2y = -273}$
$\qquad\quad -3y =\ \ -63$
$\qquad\qquad y = 21$

(a') $21x - 5(21) = 210$
$\qquad\quad 21x = 315$
$\qquad\qquad x = 15$
(15, 21)

24.   (a) $x - 2z = 7$
(b) $y + 2z = -9$
(c) $-x + 2y = -7$

(a) + (b): $x + y = -2$ (d)

(d) + (c): $3y = -9$
$\qquad\qquad y = -3$

(d) $x + (-3) = -2$
$\qquad\qquad x = 1$

(b) $(-3) + 2z = -9$
$\qquad\qquad z = -3$

(1, −3, −3)

25. (a) $2x - y + 2z = -9$
(b) $2x + 2y + z = -15$
(c) $x - 2y + z = 0$

(b) + (c): $3x + 2z = -15$ (d)

2(a) + (b): $6x + 5z = -33$ (e)

-2(d) + (e): $z = -3$

(d) $3x + 2(-3) = -15$
$x = -3$

(c) $(-3) - 2y + (-3) = 0$
$y = -3$

(−3, −3, −3)

26. $\dfrac{2i^3 - 3}{1 - \sqrt{-4}\sqrt{4}} = \dfrac{-2i - 3}{1 - 4i} \cdot \dfrac{1 + 4i}{1 + 4i}$

$= \dfrac{-2i + 8 - 3 - 12i}{1 + 16} = \dfrac{5}{17} - \dfrac{14}{17}i$

27. $\dfrac{3 - 2\sqrt{6}}{2 - \sqrt{6}} \cdot \dfrac{2 + \sqrt{6}}{2 + \sqrt{6}}$

$= \dfrac{6 + 3\sqrt{6} - 4\sqrt{6} - 12}{4 - 6} = \dfrac{6 + \sqrt{6}}{2}$

28. $\dfrac{a^{x-4}m^x}{m^{x/2}a^{x/2}} = a^{x/2-4}m^{x/2}$

29. $\sqrt{p - 45} = 9 - \sqrt{p}$

$p - 45 = 81 - 18\sqrt{p} + p$

$-126 = -18\sqrt{p}$

$7 = \sqrt{p}$

$\mathbf{49} = p$

Check: $\sqrt{49 - 45} = 9 - \sqrt{49}$
$2 = 2$ check

30. $3x^3 - x^2 - 2x = 0$
$x(3x^2 - x - 2) = 0$
$x(3x^2 - 3x + 2x - 2) = 0$
$x(3x(x - 1) + 2(x - 1)) = 0$
$x(3x + 2)(x - 1) = 0$

$x = 0$ $3x + 2 = 0$ $x - 1 = 0$
$x = -\dfrac{2}{3}$ $x = 1$

$\mathbf{0, -\dfrac{2}{3}, 1}$

PROBLEM SET 117

1. (a) $N_N + N_D + N_Q = 16$
(b) $5N_N + 10N_D + 25N_Q = 150$
(c) $N_D = 3N_Q$
Substitute (c) into (a) and (b) and get:
(a′) $N_N + 4N_Q = 16$
(b′) $5N_N + 55N_Q = 150$
$N_N + 11N_Q = 30$

-1(a′) + (b′): $7N_Q = 14$
$N_Q = \mathbf{2}$

(a′) $N_N + 4(2) = 16$
$N_N = \mathbf{8}$

(a) $(8) + N_D + (2) = 16$
$N_D = \mathbf{6}$

2. $S_P = P_P + M_U$
$5000 = 4000 + M_U$
$\$1000 = M_U$

% of $P_P = \dfrac{1000}{4000} \times 100\% = \mathbf{25\%}$ **of P_P**

% of $S_P = \dfrac{1000}{5000} \times 100\% = \mathbf{20\%}$ **of S_P**

3. $R_G T_G = 200$; $R_M T_M = 650$;
$T_M = 2T_G$; $R_M = R_G + 25$

$2T_G(R_G + 25) = 650$
$2(200) + 50T_G = 650$
$50T_G = 250$
$T_G = \mathbf{5\ hr}$
$T_M = \mathbf{10\ hr}$; $R_G = \mathbf{40\ mph}$; $R_M = \mathbf{65\ mph}$

(diagram: $D_G \to$, 200; $D_M \to$, 650)

4. (a) $0.3P_N + 0.6D_N = 234$
(b) $P_N + D_N = 600$
Substitute $D_N = 600 - P_N$ into (a) and get:
(a′) $0.3P_N + 0.6(600 - P_N) = 234$
$-0.3P_N = -126$
$P_N = 420$

(b) $(420) + D_N = 600$
$D_N = 180$

420 ml 30%, 180 ml 60%

5. $M_U = 0.8(4320) = \mathbf{\$3456}$

$S_P = P_P + M_U$
$4320 = P_P + 3456$
$\mathbf{\$864} = P_P$

6.

8	7	6	5	4

= 6720

7. $\dfrac{4}{36} \cdot \dfrac{26}{36} = \dfrac{13}{162}$

8. $x = 3(-5.50) = -16.50$

9. $4e^x = 24$
$e^x = 6$
$x = \ln 6$
$x = 1.79$

10. (a) $\text{pH} = -\log H^+$
$10^{-\text{pH}} = H^+$
$10^{-5.34} = H^+$
$H^+ = 4.57 \times 10^{-6} \dfrac{\text{mole}}{\text{liter}}$

(b) $\text{pH} = -\log H^+$
$10^{-\text{pH}} = H^+$
$10^{-0.00263} = H^+$
$H^+ = 0.994 \dfrac{\text{mole}}{\text{liter}}$

11. $A_t = Pe^{rt}$
$A_{11} = 1260e^{0.08(11)}$
$A_{11} = \$3038$

12. $A_t = A_0 e^{kt}$
$1700 = 85e^{k(12)}$
$20 = e^{12k}$
$3.00 = 12k$
$0.25 = k$

$A_{130} = 85e^{0.25(130)}$
$A_{130} = 1.11 \times 10^{16}$

13. **Region E,** since this region includes all points on or outside the circle and on or below the parabola.

14. $V = A_{\text{Base}} \times \text{height}$

$= \left[\dfrac{270}{360}\left(\pi(2 \text{ m})^2\right) + \dfrac{1}{2}(2 \text{ m})(2 \text{ m}) \right] \times 10\text{m}$

$= (3\pi + 2)(10) \text{ m}^3 \approx 114.2 \text{ m}^3$

$= \dfrac{114.2(100)(100)(100)}{(2.54)(2.54)(2.54)(12)(12)(12)} \text{ ft}^3$

$= 4032.93 \text{ ft}^3$

15. $6 \times \overrightarrow{SF} = 9$
$\overrightarrow{SF} = \dfrac{3}{2}$

$B \times \overrightarrow{SF} = 7$
$B \times \dfrac{3}{2} = 7$
$B = \dfrac{14}{3}$

$A = \sqrt{(6)^2 - \left(\dfrac{14}{3}\right)^2} = \dfrac{8\sqrt{2}}{3}$

$\dfrac{8\sqrt{2}}{3} \times \overrightarrow{SF} = A + C$

$\dfrac{8\sqrt{2}}{3} \cdot \dfrac{3}{2} = \dfrac{8\sqrt{2}}{3} + C$

$\dfrac{24\sqrt{2}}{6} = \dfrac{8\sqrt{2}}{3} + C$

$\dfrac{4\sqrt{2}}{3} = C$

16. $x + 1 > 1; \ D = \{\text{Integers}\}$
$x > 0$

17. $x^2 + 4x + 3 \geq 0; \ D = \{\text{Reals}\}$
$(x + 3)(x + 1) \geq 0$

(Pos)(Pos) ≥ 0
$x + 3 \geq 0$ and $x + 1 \geq 0$
$x \geq -3$ and $x \geq -1$

(Neg)(Neg) ≥ 0
$x + 3 \leq 0$ and $x + 1 \leq 0$
$x \leq -3$ and $x \leq -1$
Thus the solution is $x \geq -1$ or $x \leq -3$.

18. $x^2 + 4x + 3 < 0; \ D = \{\text{Reals}\}$
$(x + 3)(x + 1) < 0$

(Neg)(Pos) < 0
$x + 3 < 0$ and $x + 1 > 0$
$x < -3$ and $x > -1$

(Pos)(Neg) < 0
$x + 3 > 0$ and $x + 1 < 0$
$x > -3$ and $x < -1$
There are no real numbers that satisfy the first conjunction, so the solution must be $-3 < x < -1$.

19. (a) $y > x + 3$
(b) $y < -x$
The first step is to graph each of these lines.

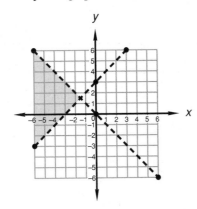

The region we wish to find is below the negatively sloped dashed line and above the positively sloped dashed line. This region is shaded in the previous figure.

20.
$$100N = 204.25\ 25\ 25\ \ldots$$
$$\underline{N = 2.04\ 25\ 25\ \ldots}$$
$$99N = 202.21$$
$$N = \frac{20{,}221}{9900}$$

21. $-y = x^2 - 4x + 2$
$-y = (x^2 - 4x + 4) + 2 - 4$
$-y = (x - 2)^2 - 2$
$y = -(x - 2)^2 + 2$
From this we see:
(a) Opens downward
(b) Axis of symmetry is $x = 2$
(c) y coordinate of vertex is 2

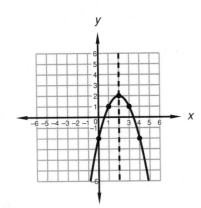

22. $x^{1/2} - y^{-1/2}$
$$\underline{x^{1/2} - y^{-1/2}}$$
$$x - x^{1/2}y^{-1/2}$$
$$\underline{ - x^{1/2}y^{-1/2} + y^{-1}}$$
$$x - 2x^{1/2}y^{-1/2} + y^{-1}$$

23. $m^3 - 8p^6k^9 = (m)^3 - (2p^2k^3)^3$
$ = (m - 2p^2k^3)(m^2 + 2mp^2k^3 + 4p^4k^6)$

24. $N = \dfrac{3}{5} + \dfrac{4}{9}\left(1\dfrac{2}{5} - \dfrac{3}{5}\right)$
$ = \dfrac{3}{5} + \dfrac{4}{9}\left(\dfrac{4}{5}\right) = \dfrac{27}{45} + \dfrac{16}{45} = \dfrac{43}{45}$

25. (a) $\dfrac{5}{8}x - \dfrac{2}{3}y = 4$
(b) $-0.15x + 0.2y = -0.6$

(a') $15x - 16y = 96$
(b') $\underline{-15x + 20y = -60}$
$4y = 36$
$y = 9$

(a') $15x - 16(9) = 96$
$x = 16$

(16, 9)

26. (a) $3x + y = 2$
(b) $2x - z = 0$
(c) $2y + z = -4$
(b) + (c): $2x + 2y = -4$ (d)

-2(a) + (d): $-4x = -8$
$x = 2$

(d) $2(2) + 2y = -4$
$y = -4$

(c) $2(-4) + z = -4$
$z = 4$

(2, −4, 4)

27. $\dfrac{\sqrt{-2}\sqrt{-2} - 3i^3}{i + 2} = \dfrac{-2 + 3i}{i + 2} \cdot \dfrac{i - 2}{i - 2}$
$ = \dfrac{-2i + 4 - 3 - 6i}{-1 - 4} = -\dfrac{1}{5} + \dfrac{8}{5}i$

28. $\dfrac{2\sqrt{2} - 4}{4 - \sqrt{2}} \cdot \dfrac{4 + \sqrt{2}}{4 + \sqrt{2}}$
$ = \dfrac{8\sqrt{2} + 4 - 16 - 4\sqrt{2}}{16 - 2} = \dfrac{-6 + 2\sqrt{2}}{7}$

29. $\sqrt[3]{x^5}\,\sqrt[4]{x^3y} = x^{5/3}x^{3/4}y^{1/4} = x^{29/12}y^{1/4}$

30. $2x^2 - 11x - 6 = 0$
$2x^2 - 12x + x - 6 = 0$
$2x(x - 6) + 1(x - 6) = 0$
$(2x + 1)(x - 6) = 0$

$2x + 1 = 0 \qquad x - 6 = 0$
$ x = -\dfrac{1}{2} \qquad x = 6$

$-\dfrac{1}{2}, 6$

PROBLEM SET 118

1. $\dfrac{9}{16} \times T = 6399$

$\qquad T = 11{,}376$

$S = 11{,}376 - 6399 = \textbf{4977}$

2. Carbon: $\quad 2 \times 12 = 24$
Hydrogen: $\quad 4 \times 1 = 4$
Oxygen: $\quad 1 \times 16 = 16$
Total: $\qquad\qquad = 44$

$\dfrac{4}{44} = \dfrac{H}{396}$
$\quad H = \textbf{36 grams}$

3. (a) $N_N + N_Q = 18$
(b) $5N_N + 25N_Q = 270$

$-5\text{(a)} + \text{(b): } 20N_Q = 180$
$\qquad\qquad\qquad N_Q = \textbf{9}$

(a) $N_N + (9) = 18$
$\qquad N_N = \textbf{9}$

4. Downstream: $(B + W)T_D = D_D$ (a)
Upstream: $(B - W)T_U = D_U$ (b)
$T_D = 2T_U$

(a′) $(B + 5)2T_U = 160$
(b′) $(B - 5)T_U = 40$

(a′) + 2(b′): $4BT_U = 240$
$\qquad\qquad\quad BT_U = 60$ (c)

(b′) $(60) - 5T_U = 40$
$\qquad\qquad T_U = \textbf{4 hr}; T_D = \textbf{8 hr}$

(c) $B(4) = 60$
$\qquad B = \textbf{15 mph}$

5. T = tens' digit
U = units' digit
$10T + U$ = original number

(a) $10T + U = 8(T + U) + 1$
(b) $3T = U + 11$
$\qquad U = 3T - 11$
Substitute (b) into (a) and get:
(a′) $10T + (3T - 11) = 8T + 8(3T - 11) + 1$
$\qquad\quad 13T - 11 = 32T - 87$
$\qquad\qquad\quad 76 = 19T$
$\qquad\qquad\quad\ 4 = T$
(b) $U = 3(4) - 11 = 1$
Original number = **41**

6. $A_t = Pe^{rt}$
$A_{10} = 15{,}000e^{0.095(10)}$
$A_{10} = \textbf{\$38,786}$

7.

11	10	9	8	7	6

$= \textbf{332,640}$

8. $\dfrac{3}{36} \cdot \dfrac{21}{36} = \dfrac{7}{\textbf{144}}$

9. $\ln x = 0.0144$
$\qquad x = \textbf{1.01}$

10. $\dfrac{(e^{-1.45})(e^{19.94})}{(e^{10.47})} = e^{-1.45+19.94-10.47}$

$= e^{\textbf{8.02}} = \textbf{3041.18}$

11. $A_t = A_0 e^{kt}$
$\quad 20 = 4e^{k5}$
$\qquad 5 = e^{5k}$
$\ 1.61 = 5k$
$0.322 = k$
$A_{40} = 4e^{0.322(40)}$
$A_{40} = \textbf{1,569,542}$

12. $pH = -\log H^+$
$pH = -\log 3.14 \times 10^{-3}$
$pH = \textbf{2.5}$

13. $H^+ = 10^{-pH}$
$H^+ = 10^{-5.042}$

$H^+ = \textbf{9.08} \times \textbf{10}^{-6} \dfrac{\textbf{mole}}{\textbf{liter}}$

14. Since $ABCD$ is a rhombus:
$m\angle ABC = m\angle ADC$
$m\angle ABD = m\angle CBD = m\angle BDA = m\angle BDC$
$2(m\angle ABD) + 34 = 180$
$\qquad 2(m\angle ABD) = 146$
$\qquad\quad m\angle ABD = 73°$
$m\angle BDA = m\angle ABD = \textbf{73°}$

15. $\log_7 (x + 5) + \log_7 3 = \log_7 60$
$\qquad \log_7 (x + 5)(3) = \log_7 60$
$\qquad\qquad\quad 3x + 15 = 60$
$\qquad\qquad\qquad\ 3x = 45$
$\qquad\qquad\qquad\ x = \textbf{15}$

16. $4\log_6 x = \log_6 64$
$\quad \log_6 x^4 = \log_6 64$
$\qquad\quad x^4 = 64$
$\qquad\qquad x = \sqrt[4]{64}$
$\qquad\qquad x = \textbf{2}\sqrt{\textbf{2}} \approx \textbf{2.83}$

17. Show that $0.00\overline{163}$ is a rational number:

$$100N = 0.163\ 63\ 63\ldots$$
$$N = 0.001\ 63\ 63\ldots$$
$$\overline{99N = 0.162}$$
$$N = \frac{162}{99{,}000}$$

18. $-y = x^2 + 6x + 10$
$-y = (x^2 + 6x + 9) + 10 - 9$
$-y = (x + 3)^2 + 1$
$y = -(x + 3)^2 - 1$

From this we see:
(a) Opens downward
(b) Axis of symmetry is $x = -3$
(c) y coordinate of vertex is -1

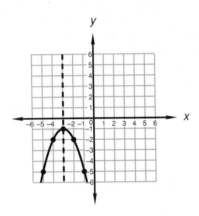

19. $|x| - 3 > -7;\ D = \{\text{Integers}\}$
$\qquad |x| > -4$

The solution is **all integers** because any integer (even zero) has an absolute value greater than -4.

20. (a) $x + y \geq -3$
$\qquad\qquad y \geq -x - 3$
(b) $x - 2y < -4$
$$y > \frac{1}{2}x + 2$$

The first step is to graph each of these lines.

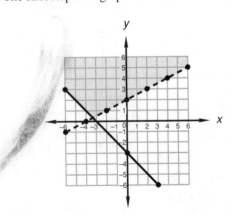

The region we wish to find is on and above the solid line and above the dashed line. This region is shaded in the previous figure.

21. $x^2 - 2x - 3 \geq 0;\ D = \{\text{Integers}\}$
$(x - 3)(x + 1) \geq 0$

(Pos)(Pos) ≥ 0
$x - 3 \geq 0$ and $x + 1 \geq 0$
$\quad x \geq 3$ and $\qquad x \geq -1$

(Neg)(Neg) ≥ 0
$x - 3 \leq 0$ and $x + 1 \leq 0$
$\quad x \leq 3$ and $\qquad x \leq -1$

Thus, the solution is $x \geq 3$ **or** $x \leq -1$.

22. $x^2 - 2x - 3 < 0;\ D = \{\text{Integers}\}$
$(x - 3)(x + 1) < 0$

(Neg)(Pos) < 0
$x - 3 < 0$ and $x + 1 > 0$
$\quad x < 3$ and $\qquad x > -1$

(Pos)(Neg) < 0
$x - 3 > 0$ and $x + 1 < 0$
$\quad x > 3$ and $\qquad x < -1$

There are no integers that satisfy the second conjunction, so the solution must be $-1 < x < 3$.

23. $x^2 + 6x + 10 = 0$
$(x^2 + 6x + 9) = -10 + 9$
$(x + 3)^2 = -1$
$x + 3 = \pm i$
$x = -3 \pm i$

24.
$$\begin{array}{r}
x^{1/4} - y^{1/4} \\
x^{1/4} - y^{1/4} \\
\hline
x^{1/2} - x^{1/4}y^{1/4} \\
- x^{1/4}y^{1/4} + y^{1/2} \\
\hline
x^{1/2} - 2x^{1/4}y^{1/4} + y^{1/2}
\end{array}$$

25. $x^6y^3 - 27p^6m^9 = (x^2y)^3 - (3p^2m^3)^3$
$= (x^2y - 3p^2m^3)(x^4y^2 + 3x^2yp^2m^3 + 9p^4m^6)$

26. $N = 4\dfrac{1}{5} + \dfrac{3}{8}\left(6\dfrac{1}{10} - 4\dfrac{1}{5}\right)$

$= \dfrac{21}{5} + \dfrac{3}{8}\left(\dfrac{61}{10} - \dfrac{42}{10}\right)$

$= \dfrac{21}{5} + \dfrac{57}{80}$

$= \dfrac{336}{80} + \dfrac{57}{80} = \dfrac{393}{80}$

27. (a) $3x - y - z = 9$
(b) $2x + y - z = 12$
(c) $2x - y + z = 0$

(a) + (b): $5x - 2z = 21$ (d)

(b) + (c): $4x = 12$
$x = 3$

(d) $5(3) - 2z = 21$
$z = -3$

(b) $2(3) + y - (-3) = 12$
$y = 3$

$(3, 3, -3)$

28. $\dfrac{2i + 2}{i + 1} \cdot \dfrac{i - 1}{i - 1} = \dfrac{-2 - 2i + 2i - 2}{-1 - 1}$
$= 2$

29. $\sqrt[5]{27\sqrt{3}} = 3^{3/5}3^{1/10} = 3^{7/10}$

30. $\dfrac{\sqrt{2} - 5}{3 - 2\sqrt{2}} \cdot \dfrac{3 + 2\sqrt{2}}{3 + 2\sqrt{2}}$

$= \dfrac{3\sqrt{2} + 4 - 15 - 10\sqrt{2}}{9 - 8} = -11 - 7\sqrt{2}$

PROBLEM SET 119

1. $\dfrac{V_1}{T_1} = \dfrac{V_2}{T_2}$

$\dfrac{500}{700} = \dfrac{V_2}{2100}$
$V_2 = \textbf{1500 ml}$

2. $R_G T_G = 250$;
$R_T T_T = 400$;
$T_T = 2T_G$;
$R_G = R_T + 10$

$2T_G(R_G - 10) = 400$
$2(250) - 20T_G = 400$
$-20T_G = -100$
$T_G = \textbf{5 hr}$
$T_T = \textbf{10 hr}$; $R_G = \textbf{50 mph}$; $R_T = \textbf{40 mph}$

3. Equal ratio method:
$\dfrac{P_1}{P_2} = \dfrac{(B_2)^2\, W_1}{(B_1)^2\, W_2}$

$\dfrac{2}{P_2} = \dfrac{(1)^2\, 4}{(10)^2\, 20}$
$P_2 = \textbf{1000}$

Variation method:
$P = \dfrac{kW}{B^2}$

$2 = \dfrac{k(4)}{(10)^2}$

$50 = k$

$P = \dfrac{50(20)}{(1)^2} = \textbf{1000}$

4. (a) $N_N + N_D + N_Q = 24$
(b) $5N_N + 10N_D + 25N_Q = 425$
(c) $N_N = N_D$
Substitute (c) into (a) and (b) and get:
(a') $2N_D + N_Q = 24$
$N_Q = 24 - 2N_D$
(b') $15N_D + 25N_Q = 425$

Substitute (a') into (b') and get:
(b'') $15N_D + 25(24 - 2N_D) = 425$
$-35N_D = -175$
$N_D = \textbf{5}$

(c) $N_N = N_D = \textbf{5}$

(a) $(5) + (5) + N_Q = 24$
$N_Q = \textbf{14}$

5. T = tens' digit
U = units' digit
$10T + U$ = original number
$10U + T$ = reversed number

(a) $T + U = 5$
$T = 5 - U$
(b) $10U + T = 10T + U - 27$
Substitute (a) into (b) and get:
(b') $10U + (5 - U) = 10(5 - U) + U - 27$
$9U + 5 = -9U + 23$
$18U = 18$
$U = 1$
(a) $T = 5 - (1) = 4$
Original number = **41**

6. $A_t = A_0 e^{kt}$
$640 = 40e^{k(3)}$
$16 = e^{3k}$
$2.77 = 3k$
$0.923 = k$

$A_{20} = 40e^{0.923(20)}$
$A_{20} = \textbf{4,160,409,959}$

7. $|x + 1| \leq 3$; $D = \{\text{Integers}\}$

$x + 1 \geq -3$ and $x + 1 \leq 3$

$x \geq -4$ and $x \leq 2$

```
     +   ●  ●  ●  ●  ●  ●   +
    -6  -4  -2   0   2   4
```

8. $|x - 3| > 5$; $D = \{\text{Reals}\}$

$x - 3 > 5$ or $x - 3 < -5$

$x > 8$ or $x < -2$

```
   ◄──────○──── | | | |──○──────►
        -3  -2  -1   7   8   9
```

9. $\ln(x + 3) + \ln 4 = \ln 40$

$\ln(x + 3)(4) = \ln 40$

$4x + 12 = 40$

$4x = 28$

$x = 7$

10. $3\log_{15} x = \log_{15} 27$

$\log_{15} x^3 = \log_{15} 27$

$x^3 = 27$

$x = 3$

11. $N = 6\dfrac{1}{10} + \dfrac{4}{5}\left(12\dfrac{3}{20} - 6\dfrac{1}{10}\right)$

$= \dfrac{61}{10} + \dfrac{4}{5}\left(\dfrac{243}{20} - \dfrac{122}{20}\right)$

$= \dfrac{61}{10} + \dfrac{484}{100}$

$= \dfrac{610}{100} + \dfrac{484}{100} = \dfrac{\mathbf{547}}{\mathbf{50}}$

12.
$$
\begin{array}{l}
x^{1/2} - y^{1/2} \\
\underline{x^{1/2} - y^{1/2}} \\
x \quad - x^{1/2}y^{1/2} \\
\quad - x^{1/2}y^{1/2} + y \\
\hline
x \quad - 2x^{1/2}y^{1/2} + y
\end{array}
$$

13. $x^3 y^9 - 64p^{12}m^9 = (xy^3)^3 - (4p^4m^3)^3$

$= (xy^3 - 4p^4m^3)(x^2y^6 + 4xy^3p^4m^3 + 16p^8m^6)$

14. (a) -5.45

(b) $e^{-4.13} = 1.61 \times 10^{-2}$

15. $\dfrac{(e^{-7.78})(e^{6.07})}{(e^{-42.28})} = e^{-7.78 + 6.07 + 42.28}$

$= e^{40.57} = 4.16 \times 10^{17}$

16. $m\angle CBD = 180 - 130 = 50°$

$m\angle CDB = 180 - 90 - 50 = 40°$

$m\angle x = 180 - 40 = \mathbf{140°}$

17. **Region B**, since this region includes all points on or above the line and on or below the parabola.

18. $\text{pH} = -\log \text{H}^+$

$\text{pH} = -\log 9.52 \times 10^{-12}$

$\text{pH} = \mathbf{11.02}$

19. $\text{H}^+ = 10^{-\text{pH}}$

$\text{H}^+ = 10^{-2.23}$

$\text{H}^+ = \mathbf{5.89 \times 10^{-3}} \dfrac{\textbf{mole}}{\textbf{liter}}$

20. $\dfrac{5000 \text{ lit}}{\text{min}} \times \dfrac{1000 \text{ ml}}{1 \text{ lit}} \times \dfrac{1 \text{ cm}^3}{1 \text{ ml}} \times \dfrac{1 \text{ in.}}{2.54 \text{ cm}}$

$\times \dfrac{1 \text{ in.}}{2.54 \text{ cm}} \times \dfrac{1 \text{ in.}}{2.54 \text{ cm}} \times \dfrac{1 \text{ min}}{60 \text{ sec}}$

$= \dfrac{\mathbf{5000(1000)}}{\mathbf{(2.54)(2.54)(2.54)(60)}} \dfrac{\text{in.}^3}{\text{sec}}$

21. $(-3 - \sqrt{2}, 4)$

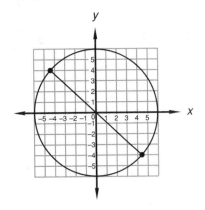

22. $A_{\text{Shaded}} = A_{\text{Sector } AOE} - A_{\text{Triangle } POE}$

$A_{\text{Sector } AOE} = \dfrac{30}{360}(\pi(3 \text{ cm})^2) = \dfrac{3\pi}{4}\text{cm}^2$

$\triangle DOE$ is equilateral because $\angle DOE = \angle OED = \angle EDO = 60°$. It follows that $PE = \frac{3}{2}$, since $\overline{OA}$ is the bisector of $\angle DOE$ and, therefore, the bisector of $\overline{DE}$.

Since triangles POE and POD are 30-60-90 triangles:

$2 \times \overrightarrow{SF} = 3$

$\overrightarrow{SF} = \dfrac{3}{2}$

$PO = \sqrt{3}\left(\dfrac{3}{2}\right) = \dfrac{3\sqrt{3}}{2}\text{cm}$

$A_{\text{Triangle } POE} = \dfrac{1}{2}\left(\dfrac{3}{2}\text{ cm}\right)\left(\dfrac{3\sqrt{3}}{2}\text{ cm}\right)$

$= \dfrac{9\sqrt{3}}{8}\text{ cm}^2$

$A_{\text{Shaded}} = \left(\dfrac{3\pi}{4} - \dfrac{9\sqrt{3}}{8}\right)\text{ cm}^2$

23. (a) $2x - y - z = 8$
(b) $3x + y - z = 9$
(c) $6x - y + z = 0$

(a) + (b): $5x - 2z = 17$ (d)

(b) + (c): $9x = 9$
$x = 1$

(d) $5(1) - 2z = 17$
$z = -6$

(b) $3(1) + y - (-6) = 9$
$y = 0$

(1, 0, –6)

24. $100N = 1.3\ 13\ 13 \ldots$
$\underline{N = 0.0\ 13\ 13 \ldots}$
$99N = 1.3$
$N = \dfrac{13}{990}$

25. $-y = x^2 + 2x - 1$
$-y = (x^2 + 2x + 1) - 1 - 1$
$-y = (x + 1)^2 - 2$
$y = -(x + 1)^2 + 2$
From this we see:
(a) Opens downward
(b) Axis of symmetry is $x = -1$
(c) y coordinate of vertex is 2

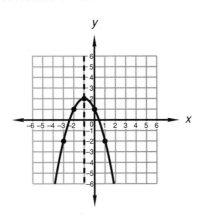

26. See Lesson 71.

27. $N = 3\dfrac{1}{3} + \dfrac{1}{5}\left(5\dfrac{2}{5} - 3\dfrac{1}{3}\right)$

$= \dfrac{10}{3} + \dfrac{1}{5}\left(\dfrac{81}{15} - \dfrac{50}{15}\right)$

$= \dfrac{10}{3} + \dfrac{31}{75}$

$= \dfrac{250}{75} + \dfrac{31}{75} = \dfrac{281}{75}$

28. $\dfrac{2i^3 - 2}{3 - \sqrt{-2}\sqrt{2}} = \dfrac{-2i - 2}{3 - 2i} \cdot \dfrac{3 + 2i}{3 + 2i}$

$= \dfrac{-6i + 4 - 6 - 4i}{9 + 4} = -\dfrac{2}{13} - \dfrac{10}{13}i$

29. $\dfrac{a^b\, x^b}{a^{b/2}\, x^{b/2}} = a^{b/2} x^{b/2}$

30. $3x^3 - 3x^2 - 6x = 0$
$3x(x^2 - x - 2) = 0$
$3x(x^2 - 2x + x - 2) = 0$
$3x(x(x - 2) + 1(x - 2)) = 0$
$3x(x + 1)(x - 2) = 0$

$3x = 0 \qquad x + 1 = 0 \qquad x - 2 = 0$
$x = 0 \qquad\quad x = -1 \qquad\quad x = 2$

0, –1, 2

PROBLEM SET 120

1. $0.76(300) - (D_N) = 0.2(300 - D_N)$
$228 - D_N = 60 - 0.2D_N$
$168 = 0.8D_N$
210 liters $= D_N$

2. $T = $ tens' digit
$U = $ units' digit
$10T + U = $ original number
$10U + T = $ reversed number

(a) $T + U = 11$
$T = 11 - U$
(b) $10U + T = 3(10T + U) + 5$
Substitute (a) into (b) and get:
(b′) $10U + (11 - U) = 30(11 - U) + 3U + 5$
$9U + 11 = -27U + 335$
$36U = 324$
$U = 9$
(a) $T = 11 - (9) = 2$
Original number $= $ **29**

3.

	Now	+ 15 Years
Man:	M_N	$M_N + 15$
Son:	S_N	$S_N + 15$

(a) $M_N = 18S_N$
(b) $M_N + 15 = 3(S_N + 15)$
Substitute (a) into (b) and get:
(b′) $(18S_N) + 15 = 3S_N + 45$
$15S_N = 30$
$S_N = $ **2**

(a) $M_N = 18(2) = $ **36**

4. Since the triangle is equilateral, each side must equal 8 meters.

$$H = \sqrt{(8 \text{ m})^2 - (4 \text{ m})^2} = 4\sqrt{3} \text{ m}$$

$$A = \frac{1}{2}(8 \text{ m})(4\sqrt{3} \text{ m}) = \textbf{16}\sqrt{\textbf{3}} \textbf{ m}^2$$

5.

	Now	–6 Years	–20 Years
Lucie:	L_N	$L_N - 6$	$L_N - 20$
Myrna:	M_N	$M_N - 6$	$M_N - 20$

(a) $L_N - 20 = 2(M_N - 20) + 2$

(b) $M_N - 6 = \frac{3}{4}(L_N - 6)$

(a′) $L_N - 2M_N = -18$

(b′) $3L_N - 4M_N = -6$

$$-2(\text{a}′) \quad -2L_N + 4M_N = 36$$
$$(\text{b}′) \quad \underline{3L_N - 4M_N = -6}$$
$$\phantom{-2(\text{b}′) 3} L_N = 30$$

(a′) $(30) - 2M_N = -18$
$$M_N = 24$$

$$L_N = \textbf{30}; \; M_N = \textbf{24}$$

6. $|x - 3| < 2; \; D = \{\text{Integers}\}$
$$x - 3 > -2 \quad \text{and} \quad x - 3 < 2$$
$$x > 1 \quad \text{and} \quad \phantom{x -3<} x < 5$$

```
     +---•---•---•---+
     1   2   3   4   5
```

7. $|x + 3| \le 2; \; D = \{\text{Reals}\}$
$$x + 3 \ge -2 \quad \text{and} \quad x + 3 \le 2$$
$$x \ge -5 \quad \text{and} \quad x \le -1$$

```
   +---•━━━━━━━━━━━•---+
  -6  -5  -4  -3  -2  -1   0
```

8. $A_t = Pe^{rt}$
$$A_8 = 2000e^{0.06(8)}$$
$$A_8 = \textbf{\$3232}$$

9.

26	25	24	10	9	8

$$= \textbf{11,232,000}$$

10. $\dfrac{e^{-15.91}}{e^{12.83}} = e^{-15.91-12.83} = e^{-28.74}$
$$= \textbf{3.30} \times \textbf{10}^{-13}$$

11. $(e^{7.11})^{1.5} = e^{10.665} = \textbf{4.3} \times \textbf{10}^4$

12. $4x = 3 \ln 0.0037$
$$x = \frac{3 \ln 0.0037}{4}$$
$$x = \textbf{-4.20}$$

13. $3 \ln x = -4.13$
$$\ln x = -\frac{4.13}{3}$$
$$x = \textbf{2.52} \times \textbf{10}^{-1}$$

14. $\text{pH} = -\log \text{H}^+$
$$\text{pH} = -\log 0.062$$
$$\text{pH} = \textbf{1.21}$$

15. $\text{H}^+ = 10^{-\text{pH}}$
$$\text{H}^+ = 10^{-3.13}$$
$$\text{H}^+ = \textbf{7.41} \times \textbf{10}^{-4} \, \frac{\textbf{mole}}{\textbf{liter}}$$

16. $\dfrac{40 \text{ ft}^3}{\text{min}} \times \dfrac{12 \text{ in.}}{1 \text{ ft}} \times \dfrac{12 \text{ in.}}{1 \text{ ft}} \times \dfrac{12 \text{ in.}}{1 \text{ ft}} \times \dfrac{1 \text{ min}}{60 \text{ sec}}$
$$= \frac{\textbf{40(12)(12)(12)}}{\textbf{60}} \; \frac{\textbf{in.}^3}{\textbf{sec}}$$

17. $1000N = 21.63 \, 163 \, 163 \ldots$
$$\underline{N = 0.02 \, 163 \, 163 \ldots}$$
$$999N = 21.61$$
$$N = \frac{\textbf{2161}}{\textbf{99,900}}$$

18. $(-3 + \sqrt{2}, 4 - \sqrt{2})$

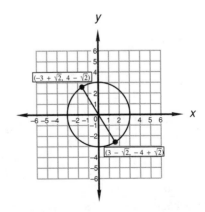

19. $-y = x^2 - 2x + 3$

$-y = (x^2 - 2x + 1) + 3 - 1$

$-y = (x - 1)^2 + 2$

$y = -(x - 1)^2 - 2$

From this we see:

(a) Opens downward

(b) Axis of symmetry is $x = 1$

(c) y coordinate of vertex is -2

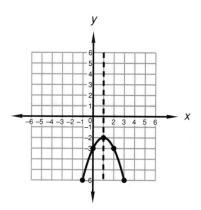

20. $5x^2 + 2x - 1 = 0$

$x^2 + \dfrac{2}{5}x = \dfrac{1}{5}$

$\left(x^2 + \dfrac{2}{5}x + \dfrac{1}{25}\right) = \dfrac{5}{25} + \dfrac{1}{25}$

$\left(x + \dfrac{1}{5}\right)^2 = \dfrac{6}{25}$

$x + \dfrac{1}{5} = \pm\dfrac{\sqrt{6}}{5}$

$\mathbf{x = -\dfrac{1}{5} \pm \dfrac{\sqrt{6}}{5}}$

21. (a) $\dfrac{3}{2}x + y = 13$

(b) $0.2x - 0.02y = 1.12$

(a') $3x + 2y = 26$
(b') $20x - 2y = 112$

$\overline{23x \quad\quad = 138}$

$x = 6$

(a') $3(6) + 2y = 26$

$y = 4$

(6, 4)

22. $H = \sqrt{(6)^2 + (4)^2} = 2\sqrt{13}$

$\tan \theta = \dfrac{6}{4}$

$\theta \approx -56.31°$

$\mathbf{2\sqrt{13}\underline{/-56.31°}}$

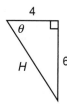

23.

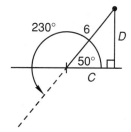

$A = 4 \cos 20° \approx 3.76$
$B = 4 \sin 20° \approx 1.37$

$C = 6 \cos 50° \approx 3.86$
$D = 6 \sin 50° \approx 4.60$

$3.76R + 1.37U$
$3.86R + 4.60U$
$\overline{7.62R + 5.97U}$

$F = \sqrt{(7.62)^2 + (5.97)^2} \approx 9.68$

$\tan \theta = \dfrac{5.97}{7.62}$

$\theta \approx 38.08°$

9.68$\underline{/38.08°}$

24. $5x^3 + 9x^2 - 2x = 0$

$x(5x^2 + 9x - 2) = 0$

$x(5x^2 + 10x - x - 2) = 0$

$x(5x(x + 2) - 1(x + 2)) = 0$

$x(5x - 1)(x + 2) = 0$

$x = 0 \quad 5x - 1 = 0 \quad x + 2 = 0$

$x = \dfrac{1}{5} \quad\quad x = -2$

$\mathbf{0, \dfrac{1}{5}, -2}$

25. $\log_4 (x - 3) - \log_4 7 = \log_4 31$

$\log_4 \dfrac{(x - 3)}{7} = \log_4 31$

$x - 3 = 217$

$x = \mathbf{220}$

26. $\ln (x + 5) + \ln 5 = \ln 65$

$\ln (5)(x + 5) = \ln 65$

$5x + 25 = 65$

$5x = 40$

$x = \mathbf{8}$

27. $\left(x^{3/2} - y^{3/2}\right)^2$

$$
\begin{array}{r}
x^{3/2} - y^{3/2} \\
x^{3/2} - y^{3/2} \\
\hline
x^3 - x^{3/2}y^{3/2} \\
- x^{3/2}y^{3/2} + y^3 \\
\hline
x^3 - 2x^{3/2}y^{3/2} + y^3
\end{array}
$$

28. $m^6p^{12} - 8y^3z^{15} = (m^2p^4)^3 - (2yz^5)^3$
$= (m^2p^4 - 2yz^5)(m^4p^8 + 2m^2p^4yz^5 + 4y^2z^{10})$

29. $OB = \dfrac{1}{2}BC = \mathbf{6}$

$AB = \sqrt{(12)^2 + (9)^2} = \mathbf{15}$
Since $\triangle BPO$ and $\triangle BCA$ are similar triangles:
$6 \times \overrightarrow{SF} = 15$

$$\overrightarrow{SF} = \frac{15}{6}$$

$$OP \times \frac{15}{6} = 9$$

$$OP = \frac{18}{5}$$

30. $m\angle B = \dfrac{1}{2}(180) = \mathbf{90°}$

$m\angle BAD = 180 - 90 - 54 = \mathbf{36°}$

$\dfrac{1}{2}m\angle BAO = \dfrac{1}{2}(36) = \mathbf{18°}$

$x = \dfrac{1}{2}m\angle BAO = \mathbf{18°}$

PROBLEM SET 121

1. $0.4(80) + (D_N) = 0.52(80 + D_N)$
$32 + D_N = 41.6 + 0.52D_N$
$0.48D_N = 9.6$
$D_N = \mathbf{20\ liters}$

2. $\dfrac{P_1}{T_1} = \dfrac{P_2}{T_2}$

$\dfrac{10}{600} = \dfrac{20}{T_2}$

$T_2 = \mathbf{1200\ K}$

3. Downstream: $(B + W)T_D = D_D$ (a)
Upstream: $(B - W)T_U = D_U$ (b)
(a') $6B + 6W = 84$
(b') $7B - 7W = 42$

$7(a') + 6(b')$: $84B = 840$
$B = \mathbf{10\ mph}$

(a') $6(10) + 6W = 84$
$W = \mathbf{4\ mph}$

4.

	Now	+ 10 Years
Garfunkel:	G_N	$G_N + 10$
Spot:	S_N	$S_N + 10$

(a) $G_N = 2S_N$
(b) $4(G_N + 10) = 3(S_N + 10) + 15$
Substitute (a) into (b) and get:
(b') $4(2S_N) + 40 = 3S_N + 45$
$5S_N = 5$
$S_N = \mathbf{1}$

(a) $G_N = 2(1) = \mathbf{2}$

5.

	Now	+ 10 Years
Rover Boy:	R_N	$R_N + 10$
Yolanda:	Y_N	$Y_N + 10$

(a) $R_N = Y_N + 5$
(b) $4(R_N + 10) = 2(Y_N + 10) + 50$
Substitute (a) into (b) and get:
(b') $4(Y_N + 5) + 40 = 2Y_N + 70$
$2Y_N = 10$
$Y_N = \mathbf{5}$

(a) $R_N = (5) + 5 = \mathbf{10}$

6.

9	8	7	6

$= \mathbf{3024}$

7. $\dfrac{3}{36} \times \dfrac{26}{36} = \dfrac{\mathbf{13}}{\mathbf{216}}$

8. $A_t = Pe^{rt}$
$A_7 = 1{,}000{,}000\,e^{0.13(7)}$
$A_7 = \mathbf{\$2{,}484{,}323}$

9. $A_t = A_0e^{kt}$
$140 = 40e^{k(5)}$
$3.5 = e^{5k}$
$1.25 = 5k$
$0.25 = k$

$A_{10} = 40e^{0.25(10)}$
$A_{10} = \mathbf{487}$

10. $|x + 2| \leq 3; \quad D = \{\text{Reals}\}$
$x + 2 \geq -3 \quad$ and $\quad x + 2 \leq 3$
$x \geq -5 \quad$ and $\quad x \leq 1$

11. Multiply both sides by $(m + 3)^2$

$$(m + 3)(m - 3) \leq 2(m + 3)^2$$
$$m^2 - 9 \leq 2m^2 + 12m + 18$$
$$m^2 + 12m + 27 \geq 0$$
$$(m + 9)(m + 3) \geq 0$$

$m = -3$ cannot be a solution since division by zero is not defined.

$$(\text{Pos})(\text{Pos}) \geq 0$$
$$m + 9 \geq 0 \quad \text{and} \quad m + 3 > 0$$
$$m \geq -9 \quad \text{and} \quad m > -3$$

$$(\text{Neg})(\text{Neg}) \geq 0$$
$$m + 9 \leq 0 \quad \text{and} \quad m + 3 < 0$$
$$m \leq -9 \quad \text{and} \quad m \leq -3$$

Thus, the solution is $m > -3$ or $m \leq -9$.

12. $\ln (x + 2) + \ln 6 = \ln 36$
$$\ln (6)(x + 2) = \ln 36$$
$$6x + 12 = 36$$
$$6x = 24$$
$$x = 4$$

13. $3 \log_{11} x = \log_{11} 27$
$$\log_{11} x^3 = \log_{11} 27$$
$$x^3 = 27$$
$$x = 3$$

14. $x = 6 \ln 0.003$
$$x = -34.85$$

15. $6 \ln x = -2.78$
$$\ln x = -\frac{2.78}{6}$$
$$x = 6.29 \times 10^{-1}$$

16. $\dfrac{e^{11.62}}{e^{-8.11}} = e^{11.62+8.11} = e^{19.73} = 3.70 \times 10^8$

17. $\angle PQR + \angle PRQ = 180 - 80$
$$\angle PQR + \angle PRQ = 100°$$
$$\frac{1}{2}(\angle PQR + \angle PRQ) = \frac{1}{2}(100°) = 50°$$
$$m\angle QSR = 180 - 50 = 130°$$

18. $\text{pH} = -\log \text{H}^+$
$$\text{pH} = -\log 0.053$$
$$\text{pH} = 1.28$$

19. $\text{H}^+ = 10^{-\text{pH}}$
$$\text{H}^+ = 10^{-7.24 \times 10^{-5}}$$
$$\text{H}^+ = 0.99 \, \frac{\text{mole}}{\text{liter}}$$

20. $P = 2 \times 62 = 124$

$x = 180 - 62 = 118$

$y = 180 - 28 - 118 = 34$

$m\widehat{BD} = 124 - 2(34) = 56°$
$m\angle A = \frac{1}{2}(56) = 28°$
$m\angle BDA = 180 - 62 - 28 = 90°$

21. $A_{\text{Triangle}} = \frac{1}{2}bh$
Since $\triangle OAB$ is equalateral:
$b = 8$ cm
$h = \sqrt{8^2 - 4^2} = 4\sqrt{3}$ cm

$A_{\text{Triangle}} = \frac{1}{2}(8 \text{ cm})(4\sqrt{3} \text{ cm}) = 16\sqrt{3} \text{ cm}^2$

$A_{\text{Hexagon}} = 6(A_{\text{Triangle}}) = 6(16\sqrt{3} \text{ cm}^2)$
$$= 96\sqrt{3} \text{ cm}^2$$

22. **Region B**, since this region includes all points above the line and below the parabola.

23. $-y = x^2 - 4x + 7$
$$-y = (x^2 - 4x + 4) + 7 - 4$$
$$-y = (x - 2)^2 + 3$$
$$y = -(x - 2)^2 - 3$$
From this we see:
(a) Opens downward
(b) Axis of symmetry is $x = 2$
(c) y coordinate of vertex is -3

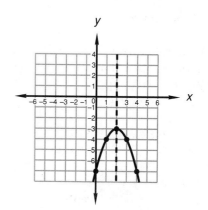

24. $A = 4 \cos 15° \approx 3.86$
$B = 4 \sin 15° \approx 1.03$

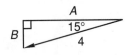

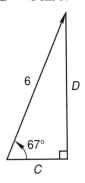

$C = 6 \cos 67° \approx 2.34$
$D = 6 \sin 67° \approx 5.52$

$-3.86R - 1.03U$
$\underline{2.34R + 5.52U}$
$\mathbf{-1.52R + 4.49U}$

$F = \sqrt{(-1.52)^2 + (4.49)^2} \approx 4.74$

$\tan \theta = -\dfrac{4.49}{1.52}$

$\theta \approx -71.30 + 180 = 108.70°$

$\mathbf{4.74 \underline{/108.70°}}$

25. (a) $3x - y + z = 1$
(b) $x - y - z = 1$
(c) $x - 2y - z = -2$

(a) + (b): $4x - 2y = 2$ (d)

(a) + (c): $4x - 3y = -1$ (e)

(d) + (−1)(e): $y = 3$

(d) $4x - 2(3) = 2$
$x = 2$

(b) $(2) - (3) - z = 1$
$z = -2$
$\mathbf{(2, 3, -2)}$

26. (a) $\dfrac{1}{2}x + \dfrac{1}{3}y = 5$
(b) $0.4x - 0.2y = -0.2$

(a′) $3x + 2y = 30$
(b′) $\underline{4x - 2y = -2}$
$7x \quad\quad = 28$
$x = 4$

(a′) $3(4) + 2y = 30$
$y = 9$
$\mathbf{(4, 9)}$

27. (a) $5x + y = 7$
(b) $2x - z = -1$
(c) $y + z = 5$

(b) + (c): $2x + y = 4$ (d)

-1(d) + (a): $3x = 3$
$x = 1$

(a) $5(1) + y = 7$
$y = 2$

(c) $(2) + z = 5$
$z = 3$

$\mathbf{(1, 2, 3)}$

28. $\dfrac{4 - \sqrt{5}}{\sqrt{5} + 2} \cdot \dfrac{\sqrt{5} - 2}{\sqrt{5} - 2}$

$= \dfrac{4\sqrt{5} - 8 - 5 + 2\sqrt{5}}{5 - 4} = \mathbf{-13 + 6\sqrt{5}}$

29. $\quad\quad 3x^3 - 4x^2 - 7x = 0$
$\quad\quad x(3x^2 - 4x - 7) = 0$
$\quad x(3x^2 + 3x - 7x - 7) = 0$
$x(3x(x + 1) - 7(x + 1)) = 0$
$\quad\quad x(3x - 7)(x + 1) = 0$

$x = 0 \quad\quad 3x - 7 = 0 \quad\quad x + 1 = 0$
$\quad\quad\quad\quad x = \dfrac{7}{3} \quad\quad\quad x = -1$

$\mathbf{0, \dfrac{7}{3}, -1}$

30. (a) $\mathbf{1.74 \times 10^4}$
(b) $\mathbf{4.84 \times 10^{-3}}$

PROBLEM SET 122

1. (a) $5N_N + 10N_D = 455$
(b) $N_N = N_D + 25$

Substitute (b) into (a) and get:
(a′) $5(N_D + 25) + 10N_D = 455$
$\quad\quad\quad\quad 15N_D = 330$
$\quad\quad\quad\quad N_D = \mathbf{22}$

(b) $N_N = (22) + 25 = \mathbf{47}$

2. Equal ratio method:

$\dfrac{M_1}{M_2} = \dfrac{A_1(V_1)^2}{A_2(V_2)^2}$

$\dfrac{400}{1600} = \dfrac{2(2)^2}{A_2\left(\dfrac{1}{2}\right)^2}$

$A_2 = \mathbf{128}$

Variation method:

$$M = kA(V)^2$$
$$400 = k(2)(2)^2$$
$$50 = k$$

$$1600 = (50)(A)\left(\frac{1}{2}\right)^2$$
$$A = \textbf{128}$$

3. T = tens' digit
U = units' digit
$10T + U$ = original number

(a) $10T + U = 2(T + U) + 8$
(b) $4U = T + 30$
$\quad\quad T = 4U - 30$
Substitute (b) into (a) and get:
(a') $10(4U - 30) + U = 2(4U - 30) + 2U + 8$
$\quad\quad 41U - 300 = 10U - 52$
$\quad\quad\quad\quad 31U = 248$
$\quad\quad\quad\quad\quad U = 8$
(b) $T = 4(8) - 30 = 2$
Original number = **28**

4.

	Now	+ 10 Years
Yehudi:	Y_N	$Y_N + 10$
Mohab:	M_N	$M_N + 10$

(a) $Y_N = M_N + 4$
(b) $2(Y_N + 10) = M_N + 10 + 24$
Substitute (a) into (b) and get:
(b') $2(M_N + 4) + 20 = M_N + 34$
$\quad\quad 2M_N + 28 = M_N + 34$
$\quad\quad\quad\quad M_N = 6$
(a) $Y_N = (6) + 4 = \textbf{10}$

5.

	Now	+ 10 Years
Petunia:	P_N	$P_N + 10$
Daisy:	D_N	$D_N + 10$

(a) $P_N = 2D_N$
(b) $2(P_N + 10) = D_N + 10 + 25$
Substitute (a) into (b) and get:
(b') $2(2D_N) + 20 = D_N + 35$
$\quad\quad\quad 3D_N = 15$
$\quad\quad\quad\quad D_N = \textbf{5}$
(a) $P_N = 2(5) = \textbf{10}$

6.

$$A_t = A_0 e^{kt}$$
$$1{,}500{,}000 = 1000\, e^{k(3)}$$
$$1500 = e^{3k}$$
$$2.44 = k$$

$$A_8 = 1000 e^{2.44(8)}$$
$$A_8 = \textbf{3.0} \times \textbf{10}^{\textbf{11}}$$

7.

7	6	5	4	= **840**

8.

3	3	3	3	3	3	= **729**

9. (a) $\log MN = \log M + \log N$

(b) $\log \dfrac{M}{N} = \log M - \log N$

(c) $\log M^N = N \log M$

10.
$$(x - 2)^2(2) \le -2(x - 2)$$
$$2x^2 - 8x + 8 \le -2x + 4$$
$$x^2 - 3x + 2 \le 0$$
$$(x - 2)(x - 1) \le 0$$

$x = 2$ cannot be a solution since division by zero is not defined.

(Neg)(Pos) ≤ 0
$x - 2 < 0 \quad$ and $\quad x - 1 \ge 0$
$\quad x < 2 \quad$ and $\quad\quad x \ge 1$

(Pos)(Neg) ≤ 0
$x - 2 > 0 \quad$ and $\quad x - 1 \le 0$
$\quad x > 2 \quad$ and $\quad\quad x \le 1$

There are no real numbers that satisfy the second conjunction, so the solution must be $\mathbf{1 \le x < 2.}$

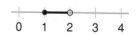

11. $(m - 3)(m + 3) \le 1(m - 3)^2$
$\quad\quad m^2 - 9 \le m^2 - 6m + 9$
$\quad\quad\quad\quad -18 \le -6m$
$\quad\quad\quad\quad\quad 3 \ge m$

$m = 3$ cannot be a solution since division by zero is not defined.
Thus, the solution is $\boldsymbol{m < 3}$.

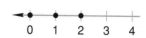

12. $|x - 2| < 1; \; D = \{\text{Reals}\}$
$x - 2 > -1 \quad$ and $\quad x - 2 < 1$
$\quad x > 1 \quad$ and $\quad\quad x < 3$

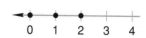

13. $|x + 3| \le 4; \; D = \{\text{Integers}\}$
$x + 3 \ge -4 \quad$ and $\quad x + 3 \le 4$
$\quad x \ge -7 \quad$ and $\quad\quad x \le 1$

14. $\dfrac{10^{9.71}}{10^{-7.15}} = 10^{9.71+7.15} = \mathbf{7.24 \times 10^{16}}$

15. $(10^{5.51})^{2/7} = 10^{1.57} = \mathbf{3.74 \times 10^1}$

16. $(10^{14.66})(10^{-19.52}) = 10^{14.66-19.52}$
$= 10^{-4.86} = \mathbf{1.39 \times 10^{-5}}$

17. $\text{pH} = -\log \text{H}^+$
$\text{pH} = -\log 3.26 \times 10^{-9}$
$\text{pH} = \mathbf{8.49}$

18. $\text{pH} = -\log \text{H}^+$
$\text{pH} = -\log 7.04 \times 10^{-5}$
$\text{pH} = \mathbf{4.15}$

19. $\text{pH} = -\log \text{H}^+$
$\text{pH} = -\log 0.0016$
$\text{pH} = \mathbf{2.79}$

20. $\text{H}^+ = 10^{-\text{pH}}$
$\text{H}^+ = 10^{-4.02}$
$\text{H}^+ = \mathbf{9.55 \times 10^{-5}\ \dfrac{mole}{liter}}$

21. $\text{H}^+ = 10^{-\text{pH}}$
$\text{H}^+ = 10^{-8.23}$
$\text{H}^+ = \mathbf{5.88 \times 10^{-9}\ \dfrac{mole}{liter}}$

22. $\text{H}^+ = 10^{-\text{pH}}$
$\text{H}^+ = 10^{-10.13}$
$\text{H}^+ = \mathbf{7.41 \times 10^{-11}\ \dfrac{mole}{liter}}$

23. $A_{\text{Shaded}} = A_{\text{Big Circle}} - 2A_{\text{Small Circle}}$
$= \pi(12\text{ yd})^2 - 2(\pi(6\text{ yd})^2)$
$= 72\pi\text{ yd}^2 \approx 226.08\text{ yd}^2$

$226.08\text{ yd}^2 = \mathbf{226.08(3)(3)(12)(12)\ in.^2}$

24. $\dfrac{1000\text{ in.}}{\text{sec}} \times \dfrac{2.54\text{ cm}}{1\text{ in.}} \times \dfrac{1\text{ m}}{100\text{ cm}} \times \dfrac{1\text{ km}}{1000\text{ m}}$
$\times \dfrac{60\text{ sec}}{1\text{ min}} \times \dfrac{60\text{ min}}{1\text{ hr}}$
$= \mathbf{\dfrac{1000(2.54)(60)(60)}{(100)(1000)}\ \dfrac{km}{hr}}$

25. $100N = 0.168\ 68\ 68\ldots$
$\underline{\quad N = 0.001\ 68\ 68\ldots}$
$99N = 0.167$
$N = \mathbf{\dfrac{167}{99,000}}$

26. $5x^2 + x + 4 = 0$
$x^2 + \dfrac{1}{5}x = -\dfrac{4}{5}$
$\left(x^2 + \dfrac{1}{5}x + \dfrac{1}{100}\right) = -\dfrac{80}{100} + \dfrac{1}{100}$
$\left(x + \dfrac{1}{10}\right)^2 = -\dfrac{79}{100}$
$x + \dfrac{1}{10} = \pm\dfrac{\sqrt{79}}{10}i$
$x = \mathbf{-\dfrac{1}{10} \pm \dfrac{\sqrt{79}}{10}i}$

27. $\dfrac{\sqrt{-3}\sqrt{-3} - i^3}{2 - \sqrt{-2}\sqrt{2}} = \dfrac{-3 + i}{2 - 2i} \cdot \dfrac{2 + 2i}{2 + 2i}$
$= \dfrac{-6 - 6i + 2i - 2}{4 + 4} = \mathbf{-1 - \dfrac{1}{2}i}$

28. $\dfrac{4\sqrt{2} - 5}{1 - \sqrt{2}} \cdot \dfrac{1 + \sqrt{2}}{1 + \sqrt{2}}$
$= \dfrac{4\sqrt{2} + 8 - 5 - 5\sqrt{2}}{1 - 2} = \mathbf{-3 + \sqrt{2}}$

29. $\sqrt[4]{xy^5}\,\sqrt[6]{x^3y} = x^{1/4}y^{5/4}x^{1/2}y^{1/6} = \mathbf{x^{3/4}y^{17/12}}$

30. (a) $P \cap Z$
The points that are members of both P and Z
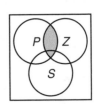

(b) $P \cap S$
The points that are members of both P and S
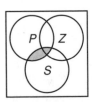

(c) $Z \cup P$
The points that are members of either Z or P or both
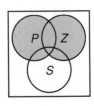

(d) $Z \cup S$
The points that are members of either Z or S or both
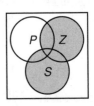

Problem Set 123

1. See Lesson 123.B.

2. See Lesson 123.B.

3. See Lesson 123.B.

4. See Lesson 123.B.

5. See Lesson 123.B.

6. See Lesson 123.B.

7. $N \quad N+2 \quad N+4$
$$N(N+4) = 10(N+2) - 13$$
$$N^2 + 4N = 10N + 7$$
$$N^2 - 6N - 7 = 0$$
$$(N-7)(N+1) = 0$$
$$N = 7, -1$$
The desired integers are **7, 9, 11** and **−1, 1, 3**.

8. $R_M T_M = 800$; $R_O T_O = 650$;
$T_M = 2T_O$; $R_M = R_O - 50$

$$(R_O - 50)2T_O = 800$$
$$2(650) - 100T_O = 800$$
$$-100T_O = -500$$
$$T_O = \textbf{5 hr}$$
$T_M = \textbf{10 hr}$; $R_O = \textbf{130 mph}$; $R_M = \textbf{80 mph}$

9. $R_H T_H = R_W T_W$;
$R_H = 16$; $R_W = 10$;
$T_H + T_W = 13$

$$16(T_H) = 10(13 - T_H)$$
$$26T_H = 130$$
$$T_H = 5$$
Distance $= R_H T_H = 16(5) = \textbf{80 miles}$

10. T = tens' digit
U = units' digit
$10T + U$ = original number

(a) $10T + U = 3(T + U) + 13$
(b) $U = T + 1$
Substitute (b) into (a) and get:
(a') $10T + (T + 1) = 3T + 3(T + 1) + 13$
$$11T + 1 = 6T + 16$$
$$5T = 15$$
$$T = 3$$
(b) $U = 4$
Original number = **34**

11.
	Now	+ 5 Years
Man:	M_N	$M_N + 5$
Son:	S_N	$S_N + 5$

(a) $M_N = 6S_N$
(b) $M_N + 5 = 3(S_N + 5) + 2$
Substitute (a) into (b) and get:
(b') $(6S_N) + 5 = 3S_N + 17$
$$3S_N = 12$$
$$S_N = \textbf{4}$$

(a) $M_N = \textbf{24}$

12. $P \cap K = \{$**5, 7, 13**$\}$

13. $A \cup B = \{$**1, 3, 5, 7, 8, 10**$\}$

14. $P \cup K = \{$**2, 3, 5, 7, 8, 9, 10, 11, 13, 15**$\}$

15. $(x-1)^2(1) \le -1(x-1)$
$$x^2 - 2x + 1 \le -x + 1$$
$$x^2 - x \le 0$$
$$x(x-1) \le 0$$

$x = 1$ cannot be a solution since division by zero is not defined.

(Neg)(Pos) ≤ 0
$x \le 0$ and $x > 1$

(Pos)(Neg) ≤ 0
$x \ge 0$ and $x < 1$

There are no real numbers that satisfy the first conjunction, so the solution must be $0 \le x < 1$.

16. $(m-1)(m+1) \le 1(m-1)^2$
$$m^2 - 1 \le m^2 - 2m + 1$$
$$-2 \le -2m$$
$$1 \ge m$$

$m = 1$ cannot be a solution since division by zero is not defined.
Thus, the solution is $1 > m$.

17. $|x+3| \le 3$; $D = \{$Reals$\}$
$x + 3 \ge -3$ and $x + 3 \le 3$
$x \ge -6$ and $x \le 0$

18. $\text{pH} = -\log \text{H}^+$

$\text{pH} = -\log 1.42 \times 10^{-11}$

$\text{pH} = \mathbf{10.85}$

19. $\text{H}^+ = 10^{-\text{pH}}$

$\text{H}^+ = 10^{-3.97}$

$\text{H}^+ = \mathbf{1.07 \times 10^{-4}} \dfrac{\textbf{mole}}{\textbf{liter}}$

20. $A_t = A_0 e^{kt}$

$480 = 240 e^{k(10)}$

$2 = e^{10k}$

$0.069 = k$

$A_{18} = 240 e^{0.069(18)}$

$A_{18} = \mathbf{831}$

21. **Region** C, since this region includes all points on or below the line and inside the circle.

22. $\log_7 (x + 2) + \log_7 3 = \log_7 15$

$\log_7 3(x + 2) = \log_7 15$

$3x + 6 = 15$

$3x = 9$

$x = \mathbf{3}$

23. $\ln (x - 9) + \ln 2 = \ln 45$

$\ln 2(x - 9) = \ln 45$

$2x - 18 = 45$

$2x = 63$

$x = \dfrac{\mathbf{63}}{\mathbf{2}}$

24.
$$
\begin{array}{r}
x^{4/5} - y^{4/5} \\
x^{4/5} - y^{4/5} \\
\hline
x^{8/5} - x^{4/5}y^{4/5} \\
- x^{4/5}y^{4/5} + y^{8/5} \\
\hline
x^{8/5} - 2x^{4/5}y^{4/5} + y^{8/5}
\end{array}
$$

25. $a^3 m^{27} - p^6 y^{36} = (am^9)^3 - (p^2 y^{12})^3$

$= (am^9 - p^2 y^{12})(a^2 m^{18} + am^9 p^2 y^{12} + p^4 y^{24})$

26. Show that $0.0012\overline{352}$ is a rational number:

$$
\begin{array}{r}
1000N = 1.2352\ 352\ 352\ldots \\
N = 0.0012\ 352\ 352\ldots \\
\hline
999N = 1.234 \\
\end{array}
$$

$N = \dfrac{\mathbf{1234}}{\mathbf{999,000}}$

27. $-y = x^2 - 4x + 1$

$-y = (x^2 - 4x + 4) + 1 - 4$

$-y = (x - 2)^2 - 3$

$y = -(x - 2)^2 + 3$

From this we see:

(a) Opens downward

(b) Axis of symmetry is $x = 2$

(c) y coordinate of vertex is 3

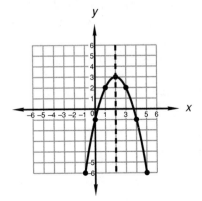

28. $3x^2 + 5x + 2 = 0$

$x = \dfrac{-(5) \pm \sqrt{(5)^2 - 4(3)(2)}}{2(3)}$

$x = \dfrac{-(5) \pm 1}{6}$

$x = -\dfrac{\mathbf{2}}{\mathbf{3}}, \mathbf{-1}$

29. $3 - 6i + 2i - 2i + 2i = \mathbf{3 - 4i}$

30. (a) Since this is a 30-60-90 triangle:

$2 \times \overrightarrow{SF} = 14$

$\overrightarrow{SF} = 7$

$m = 1(7) = \mathbf{7}$

$n = \sqrt{3}(7) = \mathbf{7\sqrt{3}}$

(b) Since this is a 45-45-90 triangle:

$1 \times \overrightarrow{SF} = 3$

$\overrightarrow{SF} = 3$

$c = \sqrt{2}(3) = \mathbf{3\sqrt{2}}$

$d = 1(3) = \mathbf{3}$

PROBLEM SET 124

1. $\triangle WAX \cong \triangle ZBY$ SAS
$\angle W \cong \angle Z$ CPCTC

2. $\triangle ABD \cong \triangle CBD$ SAS
$\overline{AD} \cong \overline{DC}$ CPCTC

3. $\triangle BAG \cong \triangle DEF$ SAS
$\overline{BG} \cong \overline{DF}$ CPCTC

4. $\triangle DCB \cong \triangle DAB$ SAS
$\angle CBD \cong \angle ABD$ CPCTC
$\overline{BD} \perp \overline{CA}$
Two equal adjacent angles whose sum is 180° are right angles.

5. $\triangle EAB \cong \triangle EDC$ SAS
$\overline{BE} \cong \overline{CE}$ CPCTC

6. See Lesson 123.B.

7. See Lesson 123.B.

8. See Lesson 123.B.

9. $(m - 3)(m + 3) \geq 1(m - 3)^2$
$$m^2 - 9 \geq m^2 - 6m + 9$$
$$-18 \geq -6m$$
$$3 \leq m$$

$m = 3$ cannot be a solution since division by zero is not defined.
Thus, the solution is **$3 < m$**.

10. $2(x - 1)^2 \leq -2(x - 1)$
$$2x^2 - 4x + 2 \leq -2x + 2$$
$$2x^2 - 2x \leq 0$$
$$2x(x - 1) \leq 0$$

$x = 1$ cannot be a solution since division by zero is not defined.

(NEG)(POS) ≤ 0
$x \leq 0$ and $x > 1$

(POS)(NEG) ≤ 0
$x \geq 0$ and $x < 1$

There are no integers that satisfy the first conjunction, so the solution must be **$0 \leq x < 1$.**

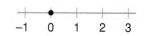

11. $|x + 2| < 3; \quad D = \{\text{Reals}\}$
$x + 2 > -3$ and $x + 2 < 3$
$x > -5$ and $x < 1$

12. $$38 = 10^{x+3}$$
$$10^{1.58} = 10^{x+3}$$
$$1.58 = x + 3$$
$$\mathbf{-1.42} = x$$

13. $\ln (x + 2) + \ln 3 = \ln 39$
$$\ln 3(x + 2) = \ln 39$$
$$3x + 6 = 39$$
$$x = \mathbf{11}$$

14. $4 \log_8 x = \log_8 48$
$$\log_8 x^4 = \log_8 48$$
$$x^4 = 48$$
$$x = \sqrt[4]{\mathbf{48}}$$

15. (a) $\log MN = \log M + \log N$

(b) $\log \dfrac{M}{N} = \log M - \log N$

(c) $\log M^N = N \log M$

16.

	Now	+ 12 YEARS
Man:	M_N	$M_N + 12$
Son:	S_N	$S_N + 12$

(a) $M_N = 11 S_N$
(b) $M_N + 12 = 3(S_N + 12)$
Substitute (a) into (b) and get:
(b') $(11 S_N) + 12 = 3 S_N + 36$
$$8 S_N = 24$$
$$S_N = \mathbf{3}$$

(a) $M_N = \mathbf{33}$

17.

5	5	5	5

$= 625$

18.

6	6	6

$= 216$

19. $N = \dfrac{3}{8} + \dfrac{2}{7}\left(2\dfrac{1}{5} - \dfrac{3}{8}\right)$

$ = \dfrac{3}{8} + \dfrac{2}{7}\left(\dfrac{88}{40} - \dfrac{15}{40}\right)$

$ = \dfrac{3}{8} + \dfrac{73}{140}$

$ = \dfrac{105}{280} + \dfrac{146}{280} = \dfrac{\mathbf{251}}{\mathbf{280}}$

20. $-y = x^2 - 2x + 4$

$-y = (x^2 - 2x + 1) + 4 - 1$

$-y = (x - 1)^2 + 3$

$y = -(x - 1)^2 - 3$

From this we see:

(a) Opens downward

(b) Axis of symmetry is $x = 1$

(c) y coordinate of vertex is -3

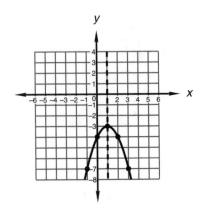

Problem Set 125

1.
1. Circle Q	1. Given
2. $\overline{PR} \perp \overline{ST}$	2. Given
3. $\overline{QS} \cong \overline{QT}$	3. Radii of same circle
4. $\overline{QR} \cong \overline{QR}$	4. Reflexive property
5. $\triangle QRS \cong \triangle QRT$	5. HL (3, 4)
6. $\overline{RS} \cong \overline{RT}$	6. CPCTC
7. $\angle PRS \cong \angle PRT$	7. CPCTC
8. $\overline{PR} \cong \overline{PR}$	8. Reflexive property
9. $\triangle PRS \cong \triangle PRT$	9. SAS (6, 7, 8)
10. $\angle S \cong \angle T$	10. CPCTC

2.
$\angle BED \cong \angle BDE$	Given
$\angle AEB \cong \angle CDB$	Given
$\angle AEB + \angle BED \cong \angle CDB + \angle BDE$	
$\angle AED \cong \angle CDE$	

3. $\angle ABE \cong \angle CBE$ and $\angle AEB \cong \angle CEB$

$\overline{BE} \cong \overline{BE}$ by reflexive property

$\triangle ABE \cong \triangle CBE$ by AAAS

$\overline{AE} \cong \overline{CE}$ by CPCTC

4. $\triangle OAM \cong \triangle OBM$ SSS

$\angle AMO \cong \angle BMO$ CPCTC

$\overline{OM} \perp \overline{AB}$

Two equal adjacent angles whose sum is $180°$ are right angles.

5. Begin by drawing $\overline{OA}, \overline{OB},$ and $\overline{OP}$.

$\triangle OAP \cong \triangle OBP$ HL

$\overline{AP} \cong \overline{BP}$ CPCTC

6. See Lesson 123.B.

7. See Lesson 123.B.

8. See Lesson 123.B.

9. See Lesson 123.B.

10. $A_t = Pe^{rt}$

$A_{11} = 1150e^{0.08(11)}$

$A_{11} = \mathbf{\$2773}$

11. $A_t = A_0 e^{kt}$

$1600 = 1000e^{k(3)}$

$1.6 = e^{3k}$

$0.157 = k$

$A_{10} = 1000e^{0.157(10)}$

$A_{10} = \mathbf{4791}$

12. $5x = 3 \ln 0.0035$

$x = \dfrac{3 \ln 0.0035}{5}$

$x = \mathbf{-3.39}$

13. $\ln x = -\dfrac{5.13}{3}$

$x = \mathbf{0.181}$

14. $|x - 5| < 3;\ D = \{\text{Integers}\}$

$x - 5 > -3$ and $x - 5 < 3$

$x > 2$ and $x < 8$

15. $x^2 - 6x + 9 > 0;\ D = \{\text{Reals}\}$

$(x - 3)(x - 3) > 0$

$(\text{Pos})(\text{Pos}) > 0$

$x > 3$

$(\text{Neg})(\text{Neg}) > 0$

$x < 3$

Thus, the solution is $\mathbf{3 < x < 3}$.

16.

	Now	−6 Years	−20 Years
Melvina:	M_N	$M_N - 6$	$M_N - 20$
Beula:	B_N	$B_N - 6$	$B_N - 20$

(a) $M_N - 20 = 2(B_N - 20) + 4$

(b) $B_N - 6 = \dfrac{3}{5}(M_N - 6)$

(a′) $M_N - 2B_N = -16$
(b′) $-3M_N + 5B_N = 12$

3(a′) + (b′): $B_N = 36$
$\qquad\qquad B_N = \mathbf{36}$
(a′) $M_N = \mathbf{56}$

17.

2	2	2	2	2	2	2	2	2	2	2	2

= **4096**

18.

4	3	2	1

= **24**

19. $\dfrac{(e^{-7.56})(e^{6.82})}{(e^{-27.89})} = e^{-7.56+6.82+27.89}$

$= e^{\mathbf{27.15}} = \mathbf{6.18 \times 10^{11}}$

20. $1000N = 31.54\ 154\ 154\ldots$
$N = 0.03\ 154\ 154\ldots$
$\overline{999N = 31.51}$
$\qquad N = \dfrac{\mathbf{3151}}{\mathbf{99,900}}$

PROBLEM SET 126

1. $\triangle BCE \cong \triangle DCE$ SAS
$\overline{BE} \cong \overline{ED}$ CPCTC

2. Draw radii OX and OW.
$\triangle OXY \cong \triangle OWY$ SSS
$\angle XYZ \cong \angle WYZ$ CPCTC
$\angle XZY \cong \angle WZY$ AA → AAA
$\overline{OY} \perp \overline{WX}$
Two equal adjacent angles whose sum is 180° are right angles.

3. Draw radii $AP, BP, CP,$ and DP.
$\triangle AXP \cong \triangle BXP \cong \triangle DYP \cong \triangle CYP$ HL
$\overline{AX} \cong \overline{BX} \cong \overline{CY} \cong \overline{DY}$ CPCTC
$\triangle APB \cong \triangle CPD$ SSS
$\overline{AB} \cong \overline{CD}$ CPCTC

4. $\triangle BED \cong \triangle BFD$ SAS
$\triangle BAE \cong \triangle BCF$ AAAS

5. $\triangle AFD \cong \triangle ABD$ SSS
$\angle 3 \cong \angle 4$ CPCTC
$\triangle DEF \cong \triangle DCB$ SAS
$\overline{EF} \cong \overline{BC}$ CPCTC

6. See Lesson 123.B.

7. See Lesson 123.B.

8. See Lesson 123.B.

9. See Lesson 123.B.

10.

9	8	7	6	5

= **15,120**

11.

3	3	3	3	3	3	3	3

= **6561**

12. (a) $\log MN = \log M + \log N$

(b) $\log \dfrac{M}{N} = \log M - \log N$

(c) $\log M^N = N \log M$

13. $x = \dfrac{3}{4} \ln 0.0069$
$x = \mathbf{-3.73}$

14. $\ln x = -\dfrac{4.98}{4}$
$\qquad x = \mathbf{0.288}$

15. $\text{pH} = -\log \text{H}^+$
$\text{pH} = -\log 0.0053$
$\text{pH} = \mathbf{2.28}$

16. $\text{H}^+ = 10^{-\text{pH}}$
$\text{H}^+ = 10^{-6.19}$

$\text{H}^+ = \mathbf{6.46 \times 10^{-7}\ \dfrac{mole}{liter}}$

17. $(p - 2)(p + 2) \le 2(p - 2)^2$
$\qquad\qquad p^2 - 4 \le 2p^2 - 8p + 8$
$\quad p^2 - 8p + 12 \ge 0$
$\quad (p - 2)(p - 6) \ge 0$

$p = 2$ cannot be a solution since division by zero is not defined.

(Pos)(Pos) ≥ 0
$p - 2 > 0$ and $p - 6 \ge 0$
$\quad p > 2$ and $\qquad p \ge 6$

(Neg)(Neg) ≥ 0
$p - 2 < 0$ and $p - 6 \le 0$
$\quad p < 2$ and $\qquad p \le 6$

Thus, the solution is $p \ge 6$ or $p < 2$.

18. $x^2 - 5x + 4 \le 0$; $D = \{$Reals$\}$
$(x - 4)(x - 1) \le 0$

(Neg)(Pos) ≤ 0
$x \le 4$ and $x \ge 1$

(Pos)(Neg) ≤ 0
$x \ge 4$ and $x \le 1$

There are no real numbers that satisfy the second conjunction, so the solution must be $\mathbf{1 \le x \le 4}$.

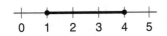

19. T = tens' digit
U = units' digit
$10T + U$ = original number
$10U + T$ = reversed number

(a) $T + U = 14$
$\qquad T = 14 - U$
(b) $10U + T = 2(10T + U) - 23$

Substitute (a) into (b) and get:
(b′) $10U + (14 - U) = 20(14 - U) + 2U - 23$
$\qquad\qquad 9U + 14 = -18U + 257$
$\qquad\qquad\qquad 27U = 243$
$\qquad\qquad\qquad\quad U = 9$

(a) $T = 5$
Original number = **59**

20. $y = x^2 - 6x + 3$
$y = (x^2 - 6x + 9) + 3 - 9$
$y = (x - 3)^2 - 6$
From this we see:
(a) Opens upward
(b) Axis of symmetry is $x = 3$
(c) y coordinate of vertex is -6

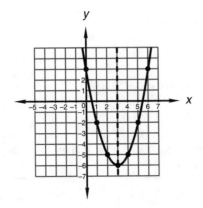

1. $\triangle APB \cong \triangle CPB$ Given
$\overline{AB} \cong \overline{CB}$ CPCTC

2. $\triangle BAP \cong \triangle CAP$ SAS
$\overline{AB} \cong \overline{CB}$ CPCTC
Thus, A is equidistant from B and C.

3. $\angle BYZ \cong \angle C$ Corresponding angles are equal.
$\angle C \cong \angle YAC$ Definition of isosceles triangle
$\angle YAC \cong \angle AYZ$ Alternate interior angles are equal.
$\angle BYZ \cong \angle AYZ$ Reflexive property
$\overline{YZ}$ bisects $\angle AYB$ Definition of angle bisector

4. $\overleftrightarrow{EP}$ is a perpendicular bisector of $\overline{XY}$.
Therefore, $\overline{EZ} \perp \overline{XY}$

5. $\triangle ABD \cong \triangle ACD$ SSS
$\triangle BDE \cong \triangle CDE$ AAAS
$\overline{BE} \cong \overline{CE}$ CPCTC
$\overline{AE}$ bisects $\overline{BC}$ Definition of bisector

6. See Lesson 123.B.

7. See Lesson 123.B.

8. See Lesson 123.B.

9. See Lesson 123.B.

10. $A_t = Pe^{rt}$
$A_9 = 1460e^{0.09(9)}$
$A_9 = \mathbf{\$3282}$

11. $A_t = A_0 e^{kt}$
$560 = 40e^{k(3)}$
$0.88 = k$

$A_9 = 40e^{0.88(9)}$
$A_9 = \mathbf{109{,}760}$

12. pH $= -\log H^+$
pH $= -\log 0.081$
pH $= \mathbf{1.09}$

13. $H^+ = 10^{-pH}$
$H^+ = 10^{-5.11}$

$H^+ = \mathbf{7.76 \times 10^{-6}} \; \dfrac{\textbf{mole}}{\textbf{liter}}$

14. $91 = 10^{x+2}$
$10^{1.96} = 10^{x+2}$
$1.96 = x + 2$
$\mathbf{-0.04} = x$

15. $\dfrac{e^{-20.49}}{e^{13.16}} = e^{-20.49 - 13.16}$

$= e^{-33.65} = \mathbf{2.43 \times 10^{-15}}$

16. (a) $|x + 1| \leq 3$; $D = \{\text{Reals}\}$

$$x + 1 \geq -3 \quad \text{and} \quad x + 1 \leq 3$$
$$x \geq -4 \quad \text{and} \qquad x \leq 2$$

(b) $(p + 3)(p - 3) \geq 2(p + 3)^2$

$$p^2 - 9 \geq 2p^2 + 12p + 18$$
$$p^2 + 12p + 27 \leq 0$$
$$(p + 9)(p + 3) \leq 0$$

$p = -3$ cannot be a solution since division by zero is not defined.

$(\text{Neg})(\text{Pos}) \leq 0$
$p \leq -9 \quad \text{and} \quad p > -3$

$(\text{Pos})(\text{Neg}) \leq 0$
$p \geq -9 \quad \text{and} \quad p < -3$
There are no real numbers that satisfy the first conjunction, so the solution must be
$-9 \leq p < -3$.

17. $-y = x^2 + 2x - 4$

$$-y = (x^2 + 2x + 1) - 4 - 1$$
$$-y = (x + 1)^2 - 5$$
$$y = -(x + 1)^2 + 5$$

From this we see:
(a) Opens downward
(b) Axis of symmetry is $x = -1$
(c) y coordinate of vertex is 5

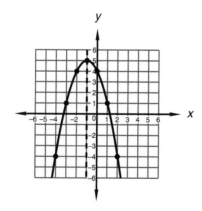

18. $N = \dfrac{5}{7} + \dfrac{3}{5}\left(3\dfrac{1}{4} - \dfrac{5}{7}\right)$

$$= \dfrac{5}{7} + \dfrac{3}{5}\left(\dfrac{91}{28} - \dfrac{20}{28}\right)$$
$$= \dfrac{5}{7} + \dfrac{213}{140}$$
$$= \dfrac{100}{140} + \dfrac{213}{140} = \dfrac{\mathbf{313}}{\mathbf{140}}$$

19.

7	6	5	4	= **840**

20. $A_{\text{Triangle}} = \dfrac{1}{2}bh$

$$h = \sqrt{(5)^2 - \left(\dfrac{5}{2}\right)^2} = \dfrac{5\sqrt{3}}{2} \text{ cm}$$

$$A_{\text{Triangle}} = \dfrac{1}{2}(5 \text{ cm})\left(\dfrac{5\sqrt{3}}{2} \text{ cm}\right) = \dfrac{\mathbf{25\sqrt{3}}}{\mathbf{4}} \text{ cm}^2$$

$$A_{\text{Shaded}} = A_{\text{Circle}} - A_{\text{Triangle}}$$
$$= \pi(5 \text{ cm})^2 - \dfrac{25\sqrt{3}}{4} \text{ cm}^2$$
$$= \left(\mathbf{25\pi} - \dfrac{\mathbf{25\sqrt{3}}}{\mathbf{4}}\right) \text{ cm}^2$$

Problem Set 128

1. $\triangle CBD \cong \triangle ABD$ SAS
$\angle ADB \cong \angle CDB$ CPCTC

2.
1. $\overline{AB} \cong \overline{BC}$	1. Given
2. $\overline{BD} \perp \overline{CA}$	2. Given
3. $\angle CBD \cong \angle ABD$	3. Perpendicular angles are equal (2).
4. $\overline{BD} \cong \overline{BD}$	4. Reflexive property
5. $\triangle CBD \cong \triangle ABD$	5. SAS (1, 3, 4)
6. $\angle ADB \cong \angle CDB$	6. CPCTC

3. $\triangle XDA \cong \triangle YCB$ SAS
$\angle AXD \cong \angle BYC$ CPCTC
Therefore, by definition, $\triangle OXY$ is isosceles.

4.
1. $ABCD$ is a rectangle	1. Given
2. $\overline{XD} \cong \overline{CY}$	2. Given
3. $\angle XDA \cong \angle YCB$	3. From 1
4. $\overline{AD} \cong \overline{BC}$	4. From 1
5. $\triangle XDA \cong \triangle YCB$	5. SAS (2, 3, 4)
6. $\triangle OXY$ is isosceles	6. From 5

5. $\triangle KNL \cong \triangle MNL$ SAS

6. $\triangle SOT \cong \triangle POT$ SAS
$\overline{ST} \cong \overline{PT}$ CPCTC
Therefore, by definition, $\triangle STP$ is isosceles

7.
1. $NPRS$ is a parallelogram	1. Given
2. $\overline{SO} \cong \overline{OP}$	2. Given
3. $\angle SOT \cong \angle POT$	3. Given
4. $\overline{TO} \cong \overline{TO}$	4. Reflexive property
5. $\triangle SOT \cong \triangle POT$	5. SAS (2, 3, 4)
6. $\overline{ST} \cong \overline{PT}$	6. CPCTC
7. $\triangle STP$ is isosceles	7. From 6

8. $\triangle BXP \cong \triangle AXP \cong \triangle CYP \cong \triangle DYP$ HL
   $\triangle BPA \cong \triangle CPD$ SAS
   $\overline{AB} \cong \overline{CD}$ CPCTC

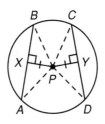

9. $\triangle AOP \cong \triangle BOP$ HL
   $\overline{AP} \cong \overline{BP}$ CPCTC

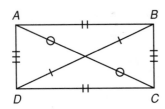

10. $\triangle ADB \cong \triangle BCA$ SSS
    $\angle ADB \cong \angle BCA$ CPCTC

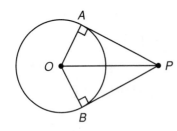

11. $PQ = RS$
    $2x + 6 = x + 8$
          $x = 2$
   $P_{PQRS} = 2(8) + (2x + 6) + (x + 8)$
   $P_{PQRS} = 16 + 3x + 14$
   $P_{PQRS} = 3x + 30$
   $P_{PQRS} = 3(2) + 30 = \mathbf{36}$

12. $AC = DB$
   $x - 12 = 0.2x$
     $-12 = -0.8x$
       $15 = x$

   $DB = 0.2(15) = \mathbf{3}$

13. $x + 60 = 90$
        $x = 30$
   $P_{DEFG} = 4(30 + 1) = \mathbf{124}$

14. $V = \pi r^2 h$
   $250\pi \text{ cm}^3 = \pi r^2 (1000 \text{ cm})$
   $\dfrac{250 \text{ cm}^3}{1000 \text{ cm}} = r^2$
      $\mathbf{0.5 \text{ cm}} = r$

15. $A = \pi r^2$
   $4\pi^2 \text{ cm}^2 = \pi r^2$
    $\mathbf{2\sqrt{\pi} \text{ cm}} = r$

16. Since this is a 45-45-90 triangle:
   $\sqrt{2} \times \overrightarrow{SF} = 10$
       $\overrightarrow{SF} = 5\sqrt{2}$

   $A = 1(5\sqrt{2}) = \mathbf{5\sqrt{2}}$
   $B = 1(5\sqrt{2}) = \mathbf{5\sqrt{2}}$

17. Since this is a 30-60-90 triangle:
   $1 \times \overrightarrow{SF} = 5$
      $\overrightarrow{SF} = 5$

   $C = 2(5) = \mathbf{10}$
   $D = \sqrt{3}(5) = \mathbf{5\sqrt{3}}$

18. $A = \dfrac{1}{2}bh$

   $h = \sqrt{(4\pi)^2 - (2\pi)^2} = 2\sqrt{3}\pi \text{ in.}$

   $A = \dfrac{1}{2}(4\pi \text{ in.})(2\sqrt{3}\pi \text{ in.}) = \mathbf{4\sqrt{3}\pi^2 \text{ in.}^2}$

19. $x = \dfrac{40 + 80}{2} = \mathbf{60}$

20. $y = \dfrac{120 - 40}{2} = \mathbf{40}$

21. $7x = 4(5)$
    $x = \dfrac{20}{7}$

22. $6(y + 6) = 7 \cdot 7$
   $6y + 36 = 49$
      $6y = 13$
       $y = \dfrac{13}{6}$

23. $4 \times \overrightarrow{SF} = 10$
     $\overrightarrow{SF} = \dfrac{5}{2}$

   $7\left(\dfrac{5}{2}\right) = x + 7$

   $\dfrac{21}{2} = x$

   $y = 9\left(\dfrac{5}{2}\right) = \dfrac{45}{2}$

24. $4 \times \overrightarrow{SF} = 6$

$\overrightarrow{SF} = \dfrac{3}{2}$

$M \times \dfrac{3}{2} = 7$

$M = \dfrac{\mathbf{14}}{\mathbf{3}}$

$N = 3 \cdot \dfrac{3}{2} = \dfrac{\mathbf{9}}{\mathbf{2}}$

25. $\widehat{ABC} = \dfrac{300}{360}(2\pi(6 \text{ cm}))$

$\widehat{ABC} = \mathbf{10\pi \text{ cm}} \approx \mathbf{31.4 \text{ cm}}$

PROBLEM SET 129

1. $A = \dfrac{1}{2}bh$

$h = \sqrt{\left(\sqrt{13}\right)^2 - \left(\dfrac{\sqrt{7}}{2}\right)^2} = \dfrac{3\sqrt{5}}{2} \text{ cm}$

$A = \dfrac{1}{2}(\sqrt{7} \text{ cm})\left(\dfrac{3\sqrt{5}}{2} \text{ cm}\right)$

$= \dfrac{\mathbf{3\sqrt{35}}}{\mathbf{4}} \text{ cm}^2 \approx \mathbf{4.43 \text{ cm}^2}$

2. $x + 2 = 3x - 6$

$8 = 2x$

$4 = x$

$p = (4 + 1) + (4 + 2) + (3(4) - 6) = \mathbf{17 \text{ cm}}$

3. $x = \dfrac{110 + 20}{2} = \mathbf{65}$

4. $4x = 5(3)$

$x = \dfrac{\mathbf{15}}{\mathbf{4}}$

5. $z = \dfrac{140 - 40}{2} = \mathbf{50}$

6. $4 \times \overrightarrow{SF} = 11$

$\overrightarrow{SF} = \dfrac{11}{4}$

$x \times \dfrac{11}{4} = 9$

$x = \dfrac{\mathbf{36}}{\mathbf{11}}$

$6 \cdot \dfrac{11}{4} = y + 6$

$y = \dfrac{\mathbf{21}}{\mathbf{2}}$

7. $2 \times \overrightarrow{SF} = 5$

$\overrightarrow{SF} = \dfrac{5}{2}$

$P \times \dfrac{5}{2} = 8$

$P = \dfrac{\mathbf{16}}{\mathbf{5}}$

$4 \cdot \dfrac{5}{2} = Q$

$\mathbf{10} = Q$

8. $A = \dfrac{40}{360}\pi(5 \text{ cm})^2 = \dfrac{\mathbf{25\pi}}{\mathbf{9}} \text{ cm}^2 \approx \mathbf{8.72 \text{ cm}^2}$

9. Since this is a 45-45-90 triangle:

$\sqrt{2} \times \overrightarrow{SF} = \sqrt{6}$

$\overrightarrow{SF} = \sqrt{3}$

$x = 1(\sqrt{3}) = \mathbf{\sqrt{3}}$

$y = 1(\sqrt{3}) = \mathbf{\sqrt{3}}$

10. Since this is a 30-60-90 triangle:

$2 \times \overrightarrow{SF} = \sqrt{7}$

$\overrightarrow{SF} = \dfrac{\sqrt{7}}{2}$

$P = 1 \cdot \dfrac{\sqrt{7}}{2} = \dfrac{\mathbf{\sqrt{7}}}{\mathbf{2}}$

$Q = \sqrt{3} \cdot \dfrac{\sqrt{7}}{2} = \dfrac{\mathbf{\sqrt{21}}}{\mathbf{2}}$

11. $A_{\text{Trapezoid}} = \dfrac{1}{2}(4)(3) + \dfrac{1}{2}(2)(3) = 9 \text{ units}^2$

$A_{\text{Circle}} = \pi r^2 = 9 \text{ units}^2$

$r = \dfrac{\mathbf{3\sqrt{\pi}}}{\mathbf{\pi}} \text{ units}$

12. See Lesson 123.B.

13. See Lesson 123.B.

14. See Lesson 123.B.

15. $\triangle ABX \cong \triangle CBX \quad$ SAS

$\angle AXB \cong \angle CXB \quad$ CPCTC

$\overline{BX} \perp \overline{AC}$

Two equal adjacent angles whose sum is 180° are right angles.

16. $LSA = 2\pi rh$

$6\pi^2 \text{ cm}^2 = 2\pi(3 \text{ cm})h$

$\pi \text{ cm} = h$

17. $\pi r^2 = 25\pi^2 \text{ cm}^2$

$r^2 = 25\pi \text{ cm}^2$

$r = 5\sqrt{\pi} \text{ cm}$

$d = 2r = 10\sqrt{\pi} \text{ cm}$

18. $PQ = RS$

$2x + 4 = 6x - 2$

$6 = 4x$

$\dfrac{3}{2} = x$

$P_{PQRS} = 2(7) + \left(2\left(\dfrac{3}{2}\right) + 4\right)$

$\qquad + \left(6\left(\dfrac{3}{2}\right) - 2\right)$

$P_{PQRS} = \mathbf{28}$

19. $2x + 40 = 90$

$x = 25$

$P_{ABCD} = 4(25 - 15) = \mathbf{40}$

20. $A_t = Pe^{rt}$

$A_5 = 1400e^{0.09(5)}$

$A_5 = \mathbf{\$2196}$

21. $A_t = A_0 e^{kt}$

$400 = 50e^{3k}$

$0.69 = k$

$A_{12} = 50e^{0.69(12)}$

$A_{12} = \mathbf{204{,}800}$

22. $75 = 10^{x+4}$

$10^{1.88} = 10^{x+4}$

$1.88 = x + 4$

$\mathbf{-2.12} = x$

23. $\dfrac{e^{-24.27}}{e^{13.34}} = e^{-24.27-13.34}$

$= e^{-37.61} = \mathbf{4.64 \times 10^{-17}}$

24. $x + 3 \geq 2;\; D = \{\text{Reals}\}$

$x \geq -1$

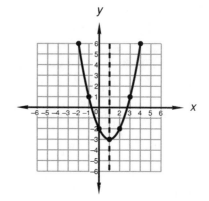

25. $y = x^2 - 2x - 2$

$y = (x^2 - 2x + 1) - 2 - 1$

$y = (x - 1)^2 - 3$

From this we see:

(a) Opens upward

(b) Axis of symmetry is $x = 1$

(c) y coordinate of vertex is -3